JLC FIELD GUIDE
TO RESIDENTIAL CONSTRUCTION

A Manual of Best Practice

Volume I

hanley▲wood

THE JOURNAL OF LIGHT CONSTRUCTION

A Journal of Light Construction Book

jlconline.com

First printing: October 2003
Second printing: December 2003

Edited by Clayton DeKorne
Illustrations by Tim Healey

International Standard Book Number: 1-928580-16-5
Library of Congress Control Number: 2003106388
Printed in the United States of America

A *Journal of Light Construction* Book.
The Journal of Light Construction
186 Allen Brook Lane
Williston, VT 05495
The Journal of Light Construction is a
trade name of Hanley Wood, LLC.

hanley wood

Production of this book was made possible in part by a grant from the CertainTeed Corporation, www.certainteed.com

CertainTeed
Quality made certain. Satisfaction guaranteed.

Disclaimer of Liability

Construction is inherently dangerous work and should be undertaken only by trained building professionals. This book is intended for expert building professionals who are competent to evaluate the information provided and who accept full responsibility for the application of this information. The techniques, practices, and tools described herein may or may not meet current safety requirements in your jurisdiction. The editors and publisher do not approve of the violation of any safety regulations and urge readers to follow all current codes and regulations as well as commonsense safety practices. An individual who uses the information contained in this book thereby accepts the risks inherent in such use and accepts the disclaimer of liability contained herein.

The editors and publisher hereby fully disclaim liability to any and all parties for any loss, and do not assume any liability whatsoever for any loss or alleged damages caused by the use or interpretation of the information found in this book, regardless of whether such information contains deficiencies, errors, or omissions, and regardless of whether such deficiencies, errors, or omissions result from negligence, accident, or any other cause that may be attributed to the editors or publisher.

Table of Contents

Acknowledgements . . . . iv.

Introduction . . . . v.

Navigating JLC's Field Guide . . . . vii.

Chapter 1: Foundations 1

Chapter 2: Framing 97

Chapter 3: Roofing 209

Chapter 4: Exteriors 289

Index 381

ACKNOWLEDGEMENTS

Several years ago, then *Journal of Light Construction* chief editor Steve Bliss assembled a group of editors to share thoughts on creating our own manual of best practice — the JLC Field Guide. We imagined this as a builder's trusty companion, ever present on the seat of the truck or in the toolbox, ready to answer the kinds of questions that come up on the job site every day. The book has taken shape slowly, but here at last is Volume 1, with chapters on Foundations, Framing, Roofing, and Exteriors. Volume 2, due out in 2005, will contain chapters on Plumbing, Electrical, HVAC, and Interior Finishes.

Thanks to Steve Bliss, who envisioned, mapped, and directed the project in its early stages; to Clayton DeKorne, who expertly executed the work that Steve had started; to Tim Healey for illustration; to JLC staff editors Ted Cushman and Charlie Wardell, who compiled large portions of the original manuscript; to Josie Masterson-Glen for editorial production and copyediting; to Jacinta Monniere for book design; to Barbara Nevins, Lyn Hoffelt, and Theresa Emerson for production work; to Ursula Jones for production support; and to Sal Alfano and Rick Strachan of Hanley Wood's Washington, D.C., office for executive management.

Finally, special thanks to all the authors and *JLC* editors over the years, too numerous to mention, whose work is the basis of this volume.

Don Jackson
JLC Editor

Introduction

Over the last 20 years, *The Journal of Light Construction* has amassed a wealth of first-hand, practical building knowledge from professionals who have dedicated themselves to custom residential projects. In this two-volume field guide, we have distilled this valuable knowledge into a single reference — selecting the critical data, fundamental principles, and rules of thumb that apply to strategic phases of residential building and remodeling. In Volume I, we have addressed the key aspects of building a sound structure and durable exterior shell. In Volume II, we will tackle house mechanicals and interior finishes.

Our intention is not to set building standards, but to provide the housebuilding trades with a compilation of practical details and proven methods that work for the many builders, remodelers, subcontractors, engineers, and architects who are committed to producing top quality, custom homes. The recommendations we have compiled in these volumes usually exceed the building code. Code compliance is essential to building a safe home — one that won't collapse or create unsafe living conditions for the occupants. However, we are striving to reach beyond this minimum standard by offering a record of best practice for residential construction: details and methods used not only to produce a safe building, but to create a long-lasting, fine-quality home.

While it is not our first focus, we have made every effort to uphold the building codes. The prescriptive recommendations in this book are generally consistent with the 2000 International Residential Code and the Wood Frame Construction Manual for One- and Two-Family Dwellings, published by the American Forest and Paper Association. Although these standards reflect the major U.S. model codes (CABO, BOCA, ICBO, and SBCCI), regional conditions have forced some municipalities to adopt more stringent requirements. Before taking the information in this volume as gospel, consult your local code authority.

As comprehensive as we have tried to make this resource, it will be imperfect. Certainly we have strived to limit any error. However, many variables, not just codes, affect local building and remodeling practices. Climate variability, material availability, land-use regulations, and native building traditions all impact how houses are built in each city, town, county, and region. To account for every variation would require a database of understanding far greater than the scope of this book. Instead, we focus here on some principles of physics, design, and craftsmanship that won't change by region or style. It is our hope that these principles, used alongside the building code, will guide professionals towards a greater understanding of best practice.

Clayton DeKorne
Editor

Navigating JLC's Field Guide

This volume is intended to be a reference book. It is organized in order of construction, and within each section we have provided several navigational tools to help you quickly locate the information you need, including margin headings, cross references, and listings for additional resources.

Table of Contents: A tabbed table of contents precedes each chapter. This provides an overview of the information in each chapter with appropriate page numbers.

Index: A detailed index of the entire volume can be found at the back of this book.

1 - FOUNDATIONS: Pier Foundations

PIER-AND-GRADE-BEAM FOUNDATIONS CAN BE AS SIMPLE AS A sonotube foundation for a porch or small addition, or as complex as an engineered foundation for expansive soils.

SIMPLE PIER FOUNDATIONS

Simple pier-and-grade-beam foundations can support decks, porches, pole barns, and small home additions, providing good insulation and ventilation (**Figure 1-92**).

Figure 1–92. Simple Pier Foundations

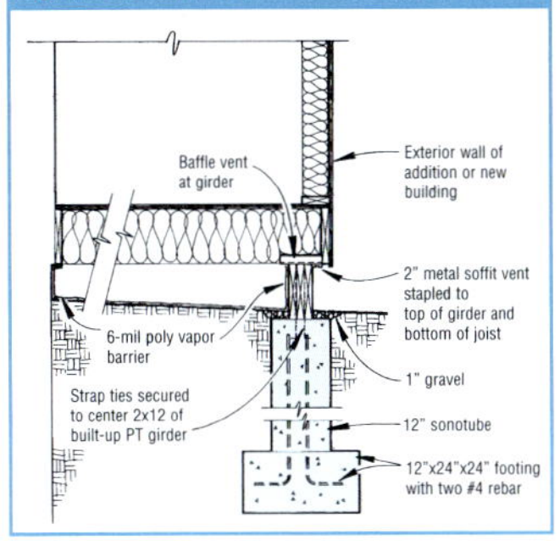

Sonotubes, placed on pier footings with a pressure-treated grade beam, can be installed easily without disturbing existing utility lines.

For more information: "Post and Pier Room Additions," JLC 7/94.

Sizing Piers

For sizing pier diameters for small foundations, the rule of thumb is "1 inch per foot of span." Thus, a deck that spans 8 ft. will stand comfortably on 8-in.-diameter piers, while a deck that spans 10 ft. requires 10-in.-diameter piers. For spans longer than 12 ft., add a second row of piers and a second girder at the center of the joist span.

For calculating concrete amounts, see "Estimating Concrete for Piers," page 1.

Pier Footings

Footings for shallow piers (less than 6 ft. deep) will help prevent the pier from settling. The pier footing should be as thick as the pier's diameter, with sides that measure twice that much. So an 8-in. pier, for example, should rest on a footing that's 8 in. thick and 16 in. square, while a 12-in. pier should rest on a footing that's 12 in. thick and 24 in. square.

Simple Pier Founc   tails

- Use strap ties to anchor piers to a triple pressure-treated 2-by girder. The ties wrap around the center stick in the built-up girder and lap over the top. (The girder looks better if the ties aren't exposed on its face.)
- Install a poly vapor barrier over the ground beneath the joists, and run it up and staple it to the inside of the girder. This helps prevent water vapor from building up in the shallow space beneath the joists.
- Additional moisture control can be added by cantilevering joists 2 in. beyond the rim of the girder, and installing a 2-in.-wide standard metal soffit vent on the underside of the overhanging joists.

82 JLC FIELD GUIDE

For more information listings: Printed in blue and located in the left-hand-margins of left-hand pages, these refer to in-depth articles from *The Journal of Light Construction*. The articles can be accessed for free from the Internet at www.jlconline.com/pub/jlcfieldguide. For information on subscribing to *JLC* or getting access to the full range of construction resources available to members of JLC-Online, visit www.jlconline.com.

3 - ROOFING: Flashing

Figure 3–81. Chimney Flashing

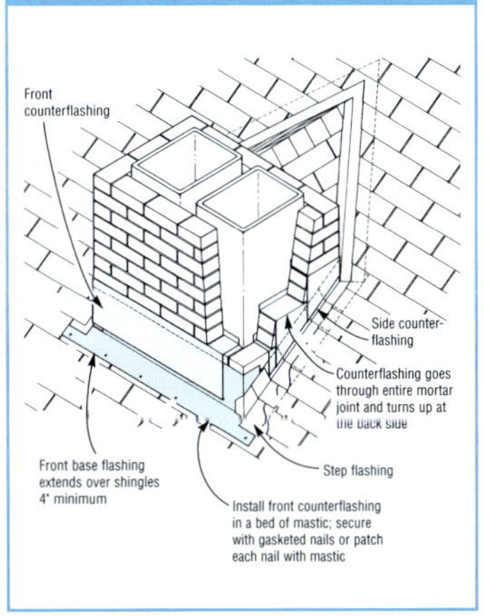

Never attach a flashing to both the chimney and the roof. The base flashing is attached to the roof only, the counterflashing to the chimney only. This allows the house to move independently of the chimney.

Figure 3–82. Skylight Opening

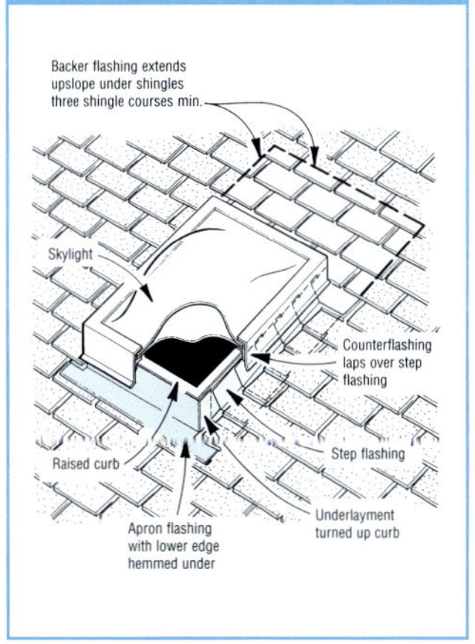

Unless it comes with a premanufactured flashing assembly, a skylight should rest on a curb or box that protrudes at least 6 in. above the roof surface.

Flashing Shed Dormers

Flashing Chimneys

Flashing Skylights

JLC FIELD GUIDE 281

Margin headings: Summary headings, printed in blue, are located in the right-hand margin of each right-hand page. Use these to target information by flipping through the pages of each chapter.

GOOD NATURAL SOILS CAN SUPPORT ALMOST ANY RESIDENTIAL foundation. However, problem soils can contribute to moisture problems, settlement, building movement, or foundation failure. Soil problems can turn up after construction has begun. When work starts, stay alert to soil conditions that may affect schedules or could cause damage to work in progress — especially subsidence, erosion, frost action, or drainage problems.

SOIL TYPES

Different soil types are classified mostly by the size of the soil particles (**Figure 1-4**).

Most soils are blends of different particle sizes, and so they have combined names such as "silty sand" or "clayey gravel."

Figure 1–4. Soil Types

Soil Type	Particle Size (fractions of in.)
Gravel	Larger than 1/5
Coarse Sand	1/5 to 8/100
Fine-Medium Sand	3/1,000 to 8/100
Silt	3/1,000 to 4/10,000
Clay	Smaller than 4/10,000

Different soil classification systems use different particle sizes to define soil types. Approximate values shown here are drawn from the Unified Soil Classification System. To determine particle sizes, samples are sifted through a series of screen with different mesh sizes.

Soil Characteristics

Characteristics of Soil Types

Different soils have different characteristics that make them more or less suitable as a support for a foundation or as a well-draining backfill. Refer to **Figure 1-5**.

Common Soil Terms

A few terms that may crop up in soil reports are defined here:

Cohesiveness. Soils that stick together when dry are called *cohesive*. Very fine-grained soils (clays and some silts) are cohesive soils. When dry, they solidify into a firm mass, but as water is added they soften and then become liquid.

Noncohesive or cohesionless soils. These are also called *granular* soils because of their large particle size and grainy shape. Granular soils include gravels and sands, as well as some silts. Granular soils are generally stronger and better-draining than cohesive soils.

Plasticity. As clays and some silts go from a dry to a wet state, they go through a state that is neither solid nor liquid — a soft and workable condition called the *plastic* state. Soils that are plastic through a high range of water contents are called *highly plastic* soils, or are said to have *high plasticity*. Highly plastic soils have low bearing strengths and exert a lot of pressure on foundation walls.

Gradation. The range or distribution of particle sizes in a soil is called *gradation* or *grading*. A gravel or sand with even amounts of different pebble or particle sizes will be called "well graded," while a soil with mostly one particle size or with one particle size missing will be called "poorly graded" or "gap graded."

Figure 1–5. Properties of Soils According to the Unified Soil Classification System

Soil Group	Unified Soil Classification System Symbol	Soil Description	Drainage Characteristics	Frost Heave Potential	Volume Change Potential (Expansion)
Group I	GW	Well-graded gravels, gravel-sand mixtures, little or no fines	Good	Low	Low
	GP	Poorly graded gravels or gravel-sand mixture, little or no fines	Good	Low	Low
	SW	Well-graded sands, gravelly sands, little or no fines	Good	Low	Low
	SP	Poorly graded sands or gravelly sands, little or no fines	Good	Low	Low
	GM	Silty gravels, gravel-sand mixtures	Good	Medium	Low
	SM	Silty sand, sand-silt mixtures	Good	Medium	Low
Group II	GC	Clayey gravels, gravel-sand-clay mixtures	Medium	Medium	Low
	SC	Clayey sands, sand-clay mixtures	Medium	Medium	Low
	ML	Inorganic silts and very fine sands, rock flour, silty or clayey fine sands or clayey silts with slight plasticity	Medium	High	Low
	CL	Inorganic clays of low to medium plasticity, gravelly clays, sandy clays, silty clays, lean clays	Medium	Medium	Medium to Low
Group III	CH	Inorganic clays of high plasticity, fat clays	Poor	Medium	High
	MH	Inorganic silts, micaceous or diatomaceous fine sandy or silty soils, elastic silts	Poor	High	High
Group IV	OL	Organic silts and organic silty clays of low plasticity	Poor	Medium	Medium
	OH	Organic clays of medium to high plasticity, organic silts	Unsatisfactory	Medium	High
	PT	Peat and other highly organic soils	Unsatisfactory	Medium	High

Note: The percolation rate for good drainage is over 4 in. per hour; medium drainage is 2- to 4-in. per hour; and poor drainage is less than 2 in. per hour.

The table above shows the two-letter designations for the soil types defined in the Unified Soil Classification System, which is widely used for construction engineering. A soil engineering report will usually use this scheme to identify soils on a site. In general, sands and gravels have good characteristics for bearing strength and drainage, while soils that contain more fine silt or clay are weaker and drain more slowly. Soils that contain large amounts of decaying plant material (the peats and organic clays) are no good for building and have to be removed.

BEARING CAPACITY OF SOILS

If the native soil type is known, codes allow builders to assume a bearing strength based on soil type (**Figure 1-6**). Footings must be placed on undisturbed original soil to use these assumed values. Soil testing may establish higher bearing strengths, in which case footing size sometimes can be reduced.

Bearing Capacity of Soils

To support foundation loads, the width of the footing depends on the loadbearing value of the soil. See "Footings," **page 38**.

Problem Soils

Organic soils. Soils containing decaying plant or animal matter, like topsoil or peat, are a special type called *organic soils*. Organic soils will not support a building and should be removed, or the building must be relocated.

Expansive clay. Certain clays that are highly attractive to water tend to expand with force when wet and shrink when dry; this is a special class of soil called *expansive clay.* Expansive soil swells when it gets wet and shrinks when it dries, leaving large cracks that channel water down into the hillside. The swelling action of expansive soil can be powerful enough to lift a house, and expansive soil used for perimeter fill can cause windows and doors to stick, and stucco, siding, drywall, and even the foundation to crack. A site with expansive soil may require specialized pier and grade beam foundations with extensive drainage, deeper piers and footings, and more heavily reinforced concrete slabs (see "Pier Foundations," **page 82**). In many cases, the expansive soil must be removed and replaced with non-expansive fill.

Natural fill vs. engineered fill. Soils that have been moved or brought in from outside, called *fill,* are not a soil type as such, but may be any type of soil that's been moved to a new location. Fill often contains junk, debris, or wood and tree stumps. Because it subsides, compresses, and washes out, fill is generally unsafe to build on. The only exception is *engineered fill,* or soil material that has been carefully placed and compacted under an engineer's supervision.

SOIL PRESSURE ON FOUNDATIONS

Pressure exerted by soil on the sides of foundation and basement walls is a major concern in the design and construction of foundation walls (**Figure 1-7**) and retaining walls. This is known as *lateral pressure,* and is defined in

Figure 1–6. Soil Bearing Capacities

Material	Loadbearing Value (pounds per sq. ft.)
Crystalline bedrock	12,000 psf
Sedimentary rock	4,000 psf
Sandy gravel or gravel	3,000 psf
Sand, silty sand, clayey sand, silty gravel, and clayey gravel	2,000 psf
Clay, sandy clay, silty clay, and clayey silt	1,500 psf

2000 International Residential Code

Loadbearing values indicate the amount of force that undisturbed, native soils can support. Refer to ***Figure 1-38 , page 38*** *for sizing concrete and masonry footings.*

terms of *equivalent fluid pressure.* Similar to the way water pressure increases at increased depth, lateral soil pressure increases with depth of soil (or height of fill).

Figure 1–7. Lateral Pressure on Foundations

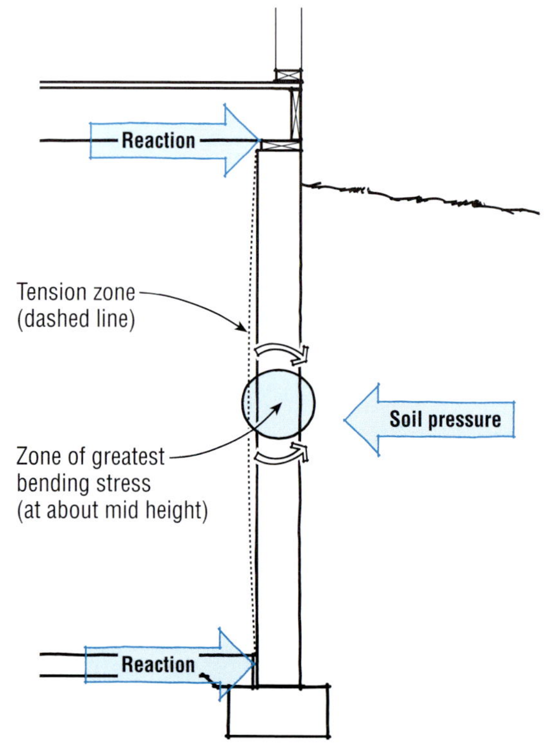

In all but the sandiest soils, the pressure of the earth can bow and crack a concrete foundation wall. Well-draining backfill reduces lateral pressure, while rebar strengthens the concrete and prevents cracks from appearing on the inside face.

For more information: "Avoiding Foundation Failures," JLC 7/96

Water in soil tends to increase lateral pressure:

- Silts and clays (especially clays) become more fluid-like as they become wet or saturated, exerting greater lateral pressure than they do when dry.
- Sands and gravels tend to be less influenced by water content.

Caution
In any soil, the risk of sudden trench collapse or foundation failure may increase as soils become saturated.

Equivalent Fluid Pressure

Soil pressures are usually the greatest loads on a foundation. Code rules of thumb can be used when building on sites with natural sand or gravel soils. Engineered design is recommended when natural soils are clay or silt.

As a rule of thumb, code assumes a lateral fluid pressure of 30 psf for block or concrete foundations, which is typical for a well-drained sand or gravel soil. But fine, cohesive soils can exert higher pressures. **Figure 1-8** shows typical fluid pressures associated with various soil types in different foundation cases.

Look for foundation soils that exert low pressures in the active and at-rest cases, and exert high resistance in the passive case. Sand and gravel meet these requirements. By contrast, soft clays — which exert high pressures in the active and at rest cases and are relatively weak in the passive case — are not good foundation soils. Foundation and retaining wall failure is a much greater threat when the structure is built on clay.

Brent Anderson Associates

Figure 1–8. Soil Fluid Pressures

Soil Classification	Friction Angle (degrees)	Density or Consistency	Unit Soil Weight (pcf)	Equivalent Fluid Pressure (psf/ft. of depth)		
				Active	At Rest	Passive
Course sand or sand and gravel	45	compact	140	24	41	820
	38	firm	120	29	46	510
	32	loose	90	28	42	290
Medium sand	40	compact	130	28	46	600
	34	firm	110	31	48	390
	30	loose	90	30	45	270
Fine sand	34	compact	130	37	57	460
	30	firm	100	33	50	300
	28	loose	85	31	41	280
Fine, silty sand or sandy silt	32	compact	130	40	61	420
	30	firm	100	33	50	300
	28	loose	85	31	45	280
Fine, uniform sand	30	compact	135	45	68	400
	28	firm	110	38	58	300
	26	loose	85	33	48	220
Clay silt	20	medium	120	59	79	245
	20	soft	90	44	59	183
Silty clay	15	medium	120	84	99	170
	15	soft	90	53	67	153
Clay	10-0	medium	120	84-120	99-120	170-120
	10-0	soft	90	63-90	74-90	153-90

Effect of Water on Soils

Equivalent Fluid Pressure

*This chart shows estimated pressures exerted by various types of soils under three conditions. The **active** case refers to situations where the structure can give somewhat to relieve the pressure, such as a foundation restrained by a wood frame floor. The **at rest** case applies if the structure cannot flex at all, as when a foundation wall is pinned in place by a concrete slab. The **passive** case applies when the structure wants to move but the soil is resisting that force.*

SOIL DRAINAGE

Coarse soils drain better than fine soils. The ability to drain is measured by a *permeability coefficient* (**Figure 1-9**). The higher the coefficient, the better a soil can drain.

EVALUATING SOILS ON SITE

It is hard to identify soils precisely in the field. If you are uncertain about the soil type, consult a soil engineer. A preliminary soil investigation may give enough information to go on, but sometimes an engineering soils report will be required.

Brent Anderson Associates

Figure 1–9. Permeability Rates in Soils

Soil	Permeability Coefficient
Fine to coarse, clean gravel	23 ft./min.
Uniform, fine gravel	11 ft./min.
Uniform, very coarse, clean sand	6.9 ft./min.
Uniform, coarse sand	1.0 ft./min.
Uniform, medium sand	14 ft./hr.
Clean, well-graded sand and gravel	1.4 ft./hr.
Uniform, fine sand	13 ft./day
Well-graded, silty sand and gravel	1.3 ft./day
Silty sand	9.8 ft./month
Sandy clay	5 ft./month
Silty clay	1.2 in./month
Clay	1.2 in./month

Soils with high permeability coefficients drain faster, and therefore exert less pressure against the foundation.

Soil Identification

Some rough information about soils can be learned from simple on-site tests:

- **Dirt-ball test:** To assess soil cohesiveness, take a moist double handful of soil and squeeze it into a ball, then drop it from a height of about 1 foot. If the soil will not form a ball or if the ball readily fragments when dropped, the soil is relatively non-cohesive and granular, with a low proportion of fine clay. However, if the soil forms a ball that holds together when dropped, it is more likely to contain a high percentage of cohesive clay.
- **Water suspension test:** Drop a scoop of soil into a large jar of water. Gravel and sand will settle to the bottom of the jar almost immediately. Finer silt particles will take fifteen minutes to an hour to settle. Clay particles will remain suspended in water for a day or longer. So if the water remains very cloudy for a long time, the soil probably contains a high percentage of clay.
- **Noodle test:** Roll a small quantity of soil into a thin noodle or string shape between your palms. If the soil

Caution

These casual tests are not a substitute for a soils laboratory report. Visual identification of soils is unreliable. For example, soils containing both clay and gravel may look like gravel but behave as clay. Soils containing 20% clay have the bearing strengths and drainage characteristics of a clay soil, and soils containing 30% clay are defined as clays even though large amounts of gravel may be present.

can be rolled as thin as 1 in. without breaking apart, it is probably a cohesive soil with a substantial percentage of clay.

Soil Borings

The building department may require soil borings taken by a qualified geotechnical engineering firm. Even if not required, an engineer's report is useful to establish the type of soil, the bearing strength of the soil, lateral soil pressures, drainage characteristics of the site, and the presence of rock below the surface.

- Soil borings, or "test pits," are recommended on low-lying sites near water, on sites that are suspected of having been filled in the past, or whenever soft, unstable, or expansive soils are encountered during excavation.
- Sample borings are especially important where piers are involved, because they help determine both the presence of water and the anticipated depth of the piers.
- A sample boring analysis can determine whether on-site soils will be useful as engineered fill.

Number of Borings

- On flat, well-drained sites, at least two test pits should be bored, at opposite corners of the proposed foundation. Intermediate borings may be required if the test holes are far apart. It is the site engineer's responsibility to determine if this is required.
- On hillsides or potentially wet sites, at least four or five borings should be taken, preferably on both the uphill and downhill sides of the proposed building.

Depth of Borings

- On flat, well-drained sites, borings should be made to a depth 5 to 8 ft. below the proposed footing depth.
- On hillsides, deep borings — 25 ft. deep or deeper — are essential. A deep boring can better define "bedrock" and help locate subterranean water moving horizontally through the hillside. A deep boring may also identify ancient landslides and waterways, which can affect the stability of the ground above.

Soil Drainage

Borings

COMPACTING SOIL

Compacting Soil

Any fill that is placed during or after construction, and any soil that is disturbed during excavation, must be compacted to prevent future settlement.

- Compact footing trenches and slab sub-bases before pouring footings or adding a gravel base (**Figure 1-10**).
- Compact backfill in utility trenches to prevent settlement.
- Compact backfill around foundation walls.
- It is best to use a gravel or sandy gravel for fill, because the compaction of silts and clays is too hard to control.

Compacting Different Soil Types

Some soils are easier to compact than others (**Figure 1-11**). In general, gravel sub-bases can be quickly compacted with a vibrating plate compactor. Gravel used for backfill will settle very little, even if not compacted. By contrast, clays typically require engineering and may not be practical to compact at all.

Compacting Equipment

For gravels and sands, use a vibrating plate compactor or vibrating smooth roller. For silts, use a vibrating compactor or an impact compactor (jumping jack). For clays, use an impact compactor.

Compact Soil in Lifts

Place and compact soil in lifts no higher than 6 in.

Effect of Moisture on Compaction
Different soils compact best at different moisture contents. Moisture in a granular soil lubricates the grains and helps them slide into a more tightly packed arrangement. In cohesive soils, moisture makes the soil somewhat moldable. However, too much moisture in a soil will hold particles apart and prevent full compaction. On site, properly moistened soils will hold together in a ball when squeezed, but will not release water. To accurately specify and monitor optimum soil moisture, however, engineering services are required.

Compaction Engineering
Jobs with soil engineering specifications usually call for a minimum soil compaction as measured by the *Proctor test*. This lab test determines the optimum moisture content of a specific type of compacted soil. An engineer can then measure field compaction with a density meter; 90% of the lab value is considered sufficient.

Figure 1–10. Soil Pressure Beneath Footings

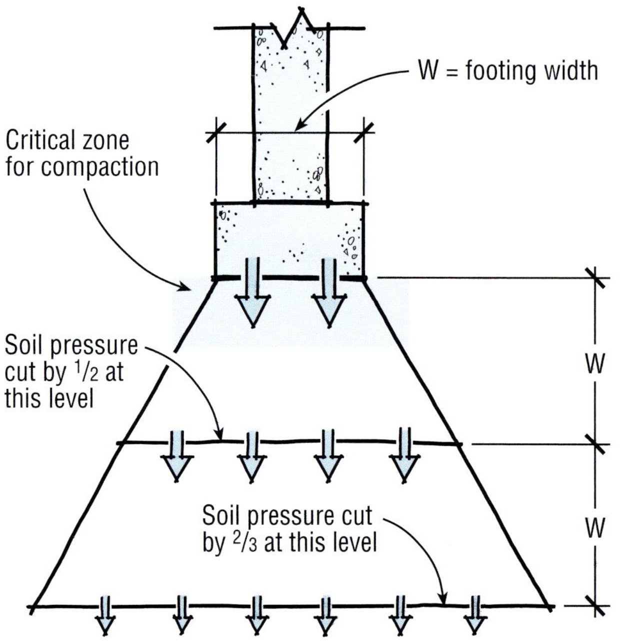

As the load under a footing spreads out, pressure on the soil diminishes. Soil directly under a footing takes the greatest load. To ensure against settlement, footing trenches should be thoroughly compacted.

For more information: "Compacting Problem Soils," JLC 4/97

Figure 1–11. Soil Compacting Characteristics

Type of Soil	**Compacting Characteristics and Equipment**
Gravel	Compacts readily with vibrating plate
Sand	Compacts well with vibrating plate
Silt	Can be compacted with difficulty; impact compactors work best
Clay	Difficult or impractical to compact (consult engineer)
Organic soil	Not compactible (do not use for building site)

Each type of soil responds best to a given type of compaction equipment. Any organic soil, such as topsoil or peat, cannot be compacted and should not be used.

SITE PREPARATION

Excavation must not begin until the locations of underground wires, cables, and pipes have been marked, or until you have verification from utility companies that the area is clear.

> **Caution**
> Look for existing utilities before any excavation begins. Any digging within 24 in. of utility locations must be done by hand.

Clearing the Site

Clear the site at least 20- to 25-ft. from the foundation. This involves more than cutting down the trees and pushing them into a pile with a bulldozer:

- Cut up trees and haul them way
- Chip the brush
- Remove the stumps and haul them away

Saving Topsoil

Retain as much topsoil as possible to preserve the environmental quality of a site. Topsoil helps maintain vegetation that will preserve the soil and limit erosion (see "Practical Erosion Control," **page 16**). When topsoil will be preserved, till the soil first, and then remove to a safe location with a front-end loader. On wooded sites, however, saving topsoil is often more difficult than it's worth because of the number of roots.

EXCAVATING SLAB FOUNDATIONS

Foundation excavations should be as level as possible, particularly for structural slabs. Set an elevation bench-

Excavation

Figure 1–12. Open Excavations

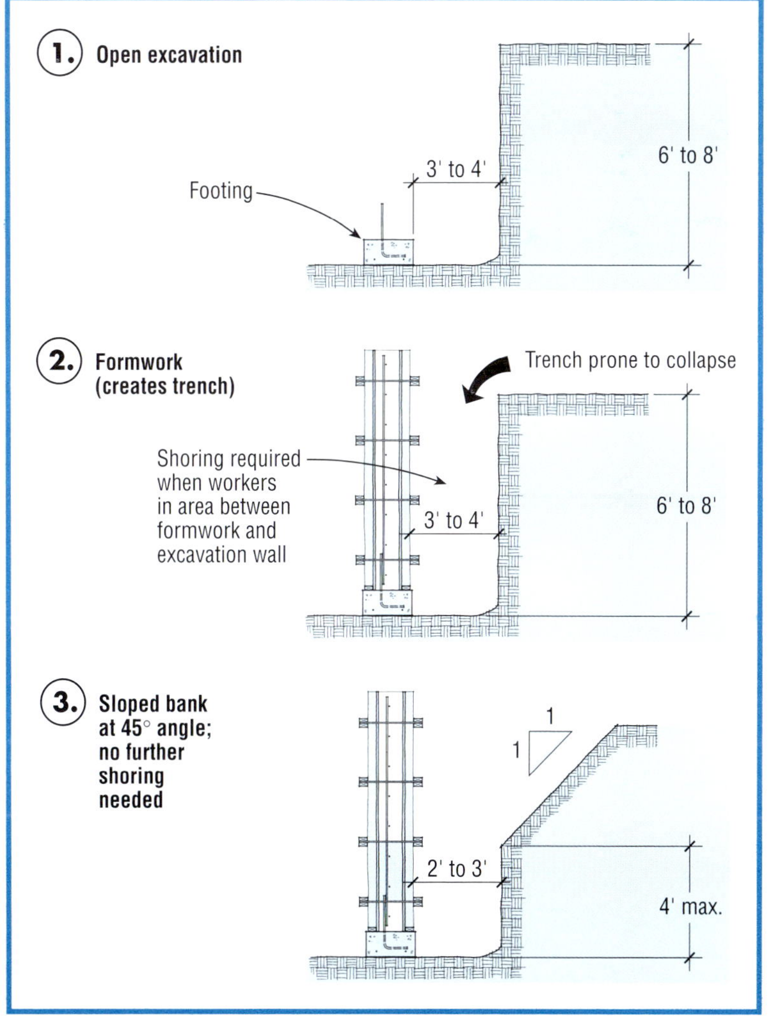

If the sides of an open excavation are higher than 7 ft., slope the sides back at an angle, or install shoring to protect workers who will be setting and stripping forms, applying dampproofing to the foundation, or laying foundation drainage.

mark prior to excavation. Place the benchmark somewhere convenient and make sure everybody on the site knows where it is during excavation.

Plan all utility layouts that will run beneath slabs prior to digging (see "Site Layout for Slab Foundations," **page 74**).

EXCAVATING FOUNDATION HOLES

Plan foundation drainage before excavating to identify the locations of cleanouts and daylight drains (see "Foundation Drainage," **page 68**).

Overdig foundation holes to give plenty of room for a well-draining backfill (see "Soil Drainage," **page 8**), and to provide room to work.

- If the depth of the excavation is 6 ft. or less, overdig by 3- to 4-ft.
- If the excavation is deeper than 7 ft., the sides must be sloped outwards at a 45-degree angle above 4 ft.

Figure 1-13. Hillside Cuts

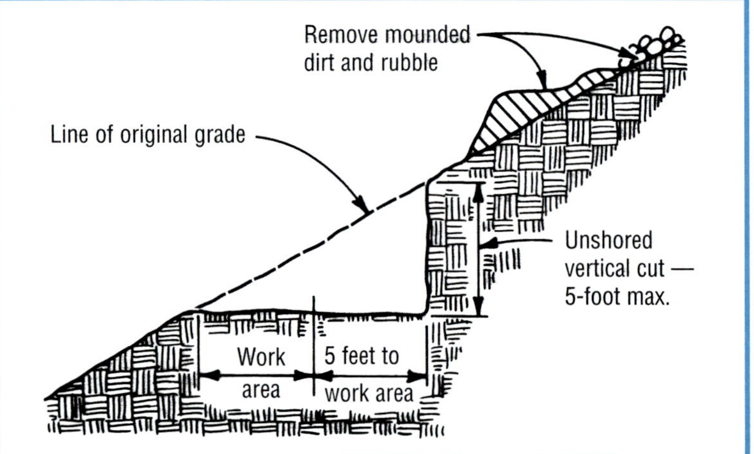

For more information: "Hillside Foundations, Part One," JLC 3/93

"Hillside Foundations, Part Two," JLC 4/93

(**Figure 1-12**, previous page), or install shoring (**Figure 1-16, page 15**).

EXCAVATING HILLSIDES

One of the problems with excavating steep sites is that as you dig into a hillside, you remove material that holds the hill together, creating an unstable and potentially dangerous situation.

Angle of Repose

Excavate the hill back to an angle of repose — the angle at which the slope is stable. Each type of soil has a natural angle of repose. For most soils it is usually safe to excavate back to a 2:1 slope. This is the easiest and least expensive way to stabilize a hillside, but it's often not practical on a single-family lot because there simply is not enough room.

Vertical Cuts vs. Shoring

An unshored vertical cut should not exceed 5 ft. on a sloped site (**Figure 1-13**). Be sure to reduce the danger of collapse by removing mounded dirt and rubble that can add a surcharge to the area above the cut. Install a series of steps or terraces — as required — to meet minimum height restrictions and reduce erosion (**Figure 1-14**).

Often, shoring is the only safe alternative and will eliminate worry about how delays, change orders, or bad weather will affect the hillside (see "Shoring," **page 14**).

Pressure Grouting

In porous soils, which are very much at risk of collapse, you may need to drill a pattern of small-diameter holes in the hillside and inject them with concrete. This kind of *pressure grouting* is expensive and is best left to an experienced specialty subcontractor.

EXCAVATION SAFETY

An excavation collapse poses an extreme danger to workers. Soil is heavy — typically weighing more than 100 pounds per cu. ft. This means that a 3x3x3-ft. hole will contain 27 cu. ft. of soil that can weigh almost 1 1/2 tons — as much as a car. If the soil is damp, wet, or filled with rock, it will weigh even more.

Excavation Failure

Keep an eye out for these signs of distress in and around trenches and open excavations:

- Cracks in the soil parallel to or in the face of an excavation
- Subsidence of the edge or bulging of the side of the excavation (this may be hard to see)
- Heaving or boiling of the bottom of the excavation, which is an indication of imminent failure
- Spalling or raveling of the face of the excavation (this probably indicates lack of proper sheeting and the lack of, or failure to, classify the soil)
- Water running into the excavation from the surface or face (do not allow workers in excavations with standing or running water)
- Bending, buckling, or groaning of any support member (if any movement of a support member can be seen or heard, an extremely dangerous situation exists)

Caution in Frozen Ground

Frozen soil is as capable of collapsing as thawed soil of the same type, and should not be considered a safe alternative to proper shoring.

Trench Safety Checklist

- The sides of trenches may need to be sloped back, depending on soil type and trench depth (**Figure 1-15**).
- Place excavated soil at least 2 ft. from the top edge of the trench, or position it behind a stable barrier.
- Trenches more than 4 ft. deep should contain an exit ladder which extends at least 3 ft. above ground level and is located within 10 ft. of a worker at all times.

Foundation Holes

Excavating Hillsides

Figure 1–14. Stepped and Terraced Slopes

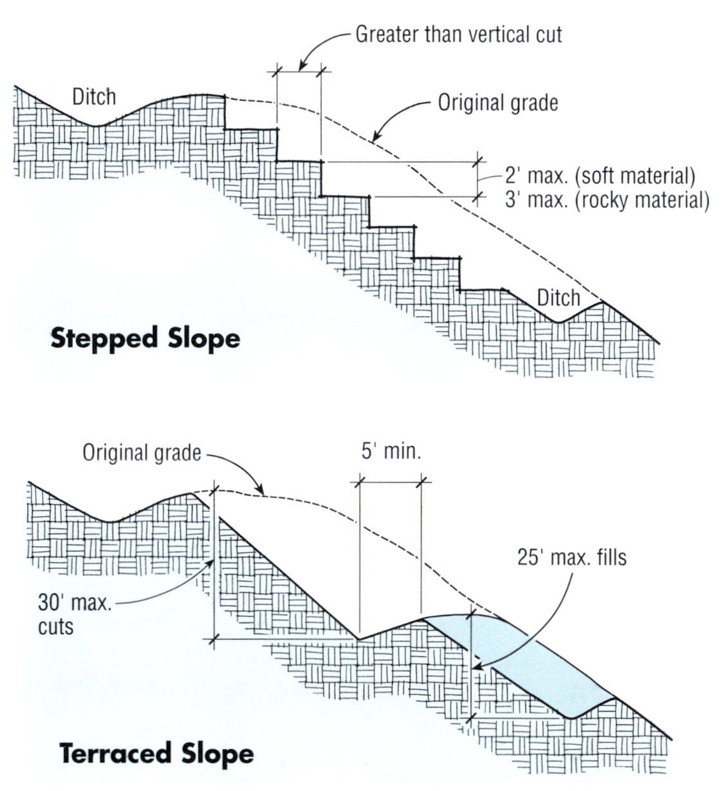

Figure 1–15. Trenching Cuts

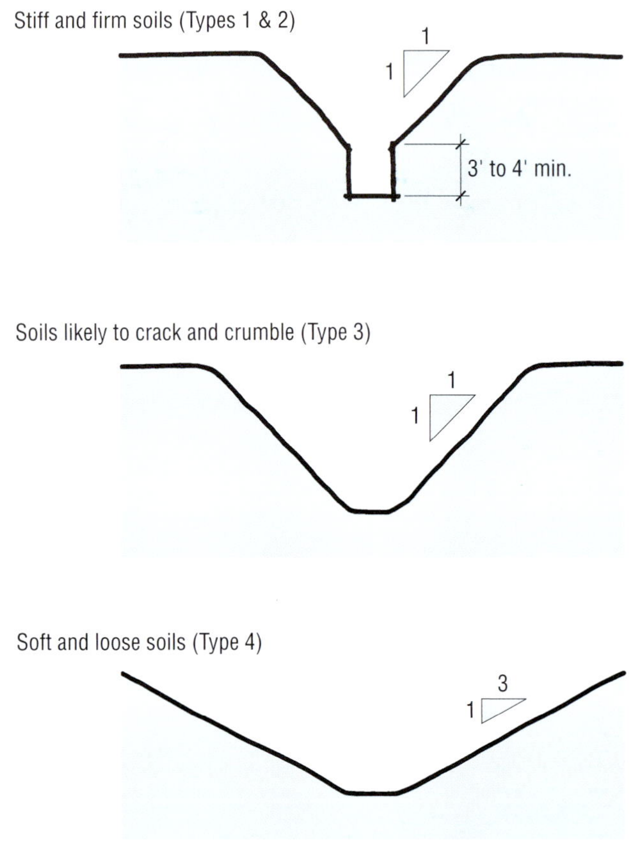

Stiff and firm soils: *For trenches deeper than 6 ft., slope sides back at a minimum 45-degree angle above 4 ft.* ***Soils likely to crack or crumble:*** *Slope entire height of trench side at a minimum 45-degree angle.* ***Soft and loose soils:*** *In sand, gravel, silt, organic soil, soft and wet clay, and loose fill, cut trench sides at a minimum 3:1 slope.*

- Cross-braces and trench jacks should be level and spaced vertically in order to prevent wall material from moving into the trench (**Figure 1-16**).
- Trench boxes or safety cages can provide alternate protection methods.

SHORING

In areas where vibrations or unstable conditions are likely, such as near a highway, railroad, or a backfilled trench, plan for very complete shoring.

Provide shoring in an open excavation (foundation hole) deeper than 7 ft. and in trenches deeper than 6 ft., unless the sides have been cut at a slope. On hillside sites, restrict unshored, vertical cuts to 5 ft. (see "Excavating Hillsides," **page 12**).

Installing Shoring

- When installing shoring, place the bucket of the excavation machine in the trench directly in front of the shoring being installed.
- When installing shoring struts or jacks, work from the top down, and remove struts working from the bottom up. Using this method, the worker is protected by the shoring already installed.
- Do not try to remove a jack under pressure or it may cause a sudden collapse. Backfill up to the bottom jack before it is removed; then up to the next jack and so forth.

Shoring Alternative

If shoring is not used, slope the walls of the excavations at a grade of 1:1 (45 degrees) or shallower (**Figure 1-12, page 11**).

Figure 1-16. Trench Shoring

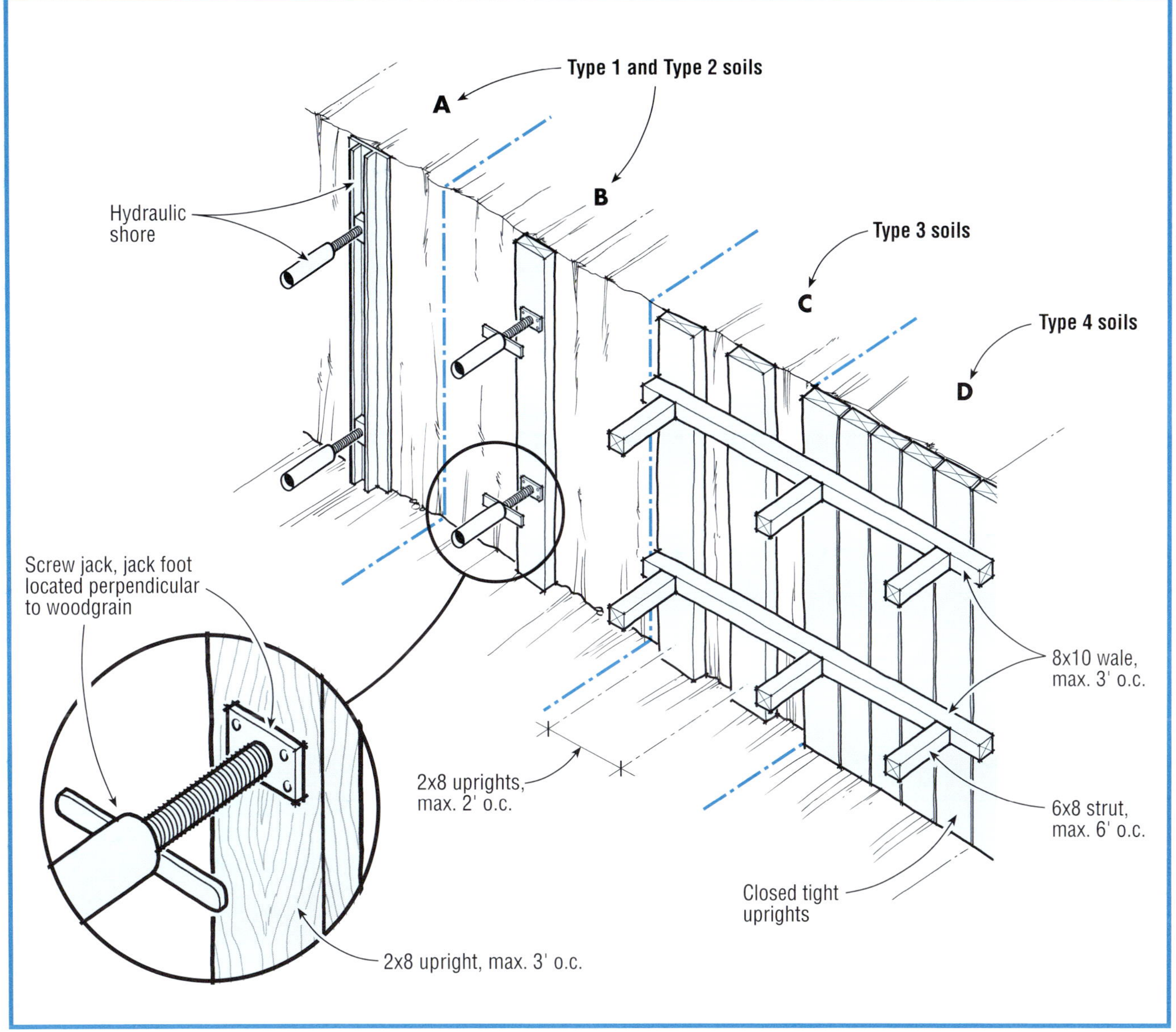

Adapted from Manitoba Labor & Immigration

Trenching

Shoring

For wood shoring components, use No. 2 grade or better lumber. Dimensions shown are suitable for trenches up to 60 in. wide.

Figure 1–17. Temporary Construction Entrance

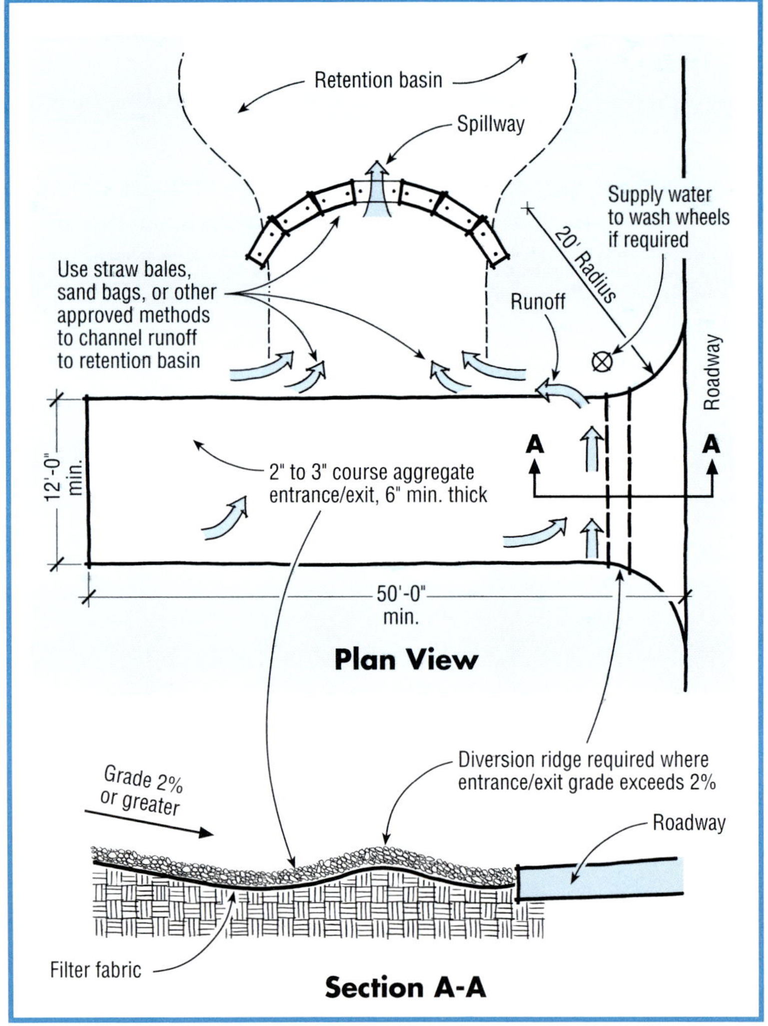

Drives should be installed before construction begins. If the schedule does not allow for this, use 2- to 3-in. of coarse gravel for a temporary roadbed, and use silt fences or straw-bales to trap silt and prevent erosion.

Foundation excavations should be sloped back anyway to provide room enough for a well-draining backfill.

PRACTICAL EROSION CONTROL

Cleared and excavated sites are prone to erosion from runoff that will carry silt and sand, which can pollute streams and wetlands or damage nearby properties. Fuel, pesticides, and other chemicals also may be carried into the environment by runoff. To avoid liability and reduce landscaping costs, take these practical steps to minimize erosion and pollution:

Preserve vegetation: Protect grass, trees, and other plants wherever possible. Maintain strips of grass or other growth across cleared slopes to catch water and soil runoff.

Schedule work in stages: Do not clear or excavate large areas too far in advance of construction; disturb vegetation only as needed, then reseed or replant as soon as is practical.

Control vehicle traffic: Limit equipment traffic in and out of the job site — do not drive over uncleared areas. Protect uncleared areas with snow fencing or other barriers. Place barriers around the drip-line of trees to prevent injury to root systems. If a permanent drive cannot be installed before construction begins, a temporary gravel driveway should be installed where vehicles enter the job site (**Figure 1-17**).

Channel runoff: Divert runoff around cleared areas by mounding earth berms and digging swales. Where concentrated runoff encounters a cleared slope, create a lined channel or install a drain pipe at the bottom of a swale and cover with gravel. Make drain channels as long and gently sloped as practical. Seed low-volume, shallow swales with grass. Use fabric or rock to protect higher-volume channels.

Install silt barriers: Erect silt fences or staked hay-bale barriers on the downhill edge of disturbed earth slopes (**Figure 1-18**). Straw bales placed on slope contours should be tightly butted and staked to prevent erosion or flow between and under bales (**Figure 1-19**).

Protect drains and watercourses: Set up silt fences or staked hay-bale barriers around any stream, drainage channel, or drain opening. Surround storm drains with straw-bale dikes to detain water so silt can settle out. If plans include new, permanent stormwater drainage, install the system as soon as possible, then protect inlets during construction.

Build sediment traps: Use sandbag and gravel structures or silt fencing to pond and detain runoff water, allowing sediment to settle out. Inspect sediment traps and excavate sediment as needed. Place excavated sediment in flat upland areas and seed immediately.

Erosion Control

Preserve topsoil: Topsoil is necessary for reseeded plants to thrive (see "Saving Topsoil," **page 11**). If the existing topsoil will not support vegetation, import good soil. Topsoil is not suitable for use as backfill or subgrade.

Temporary Construction Drives

Cover exposed soil: Reseed cleared areas as soon as possible. Seed soil (or excavated soil stockpiles) immediately if they will remain on the site for a period of time. If slopes or earth stockpiles cannot be reseeded immediately, cover with plastic sheeting (short-term only), mulch (heavy layer of wood chips or straw), or erosion-control fabric (mesh, netting, or geoblanket).

Figure 1–18. Silt Fence

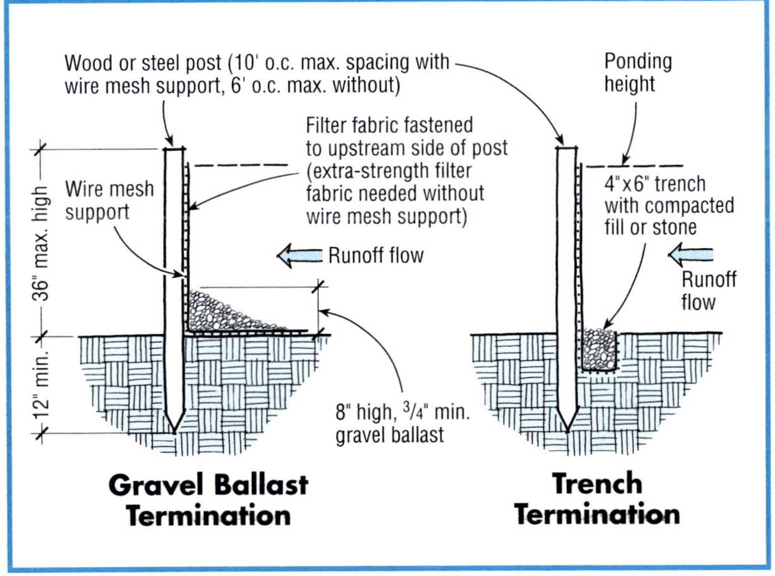

Figure 1–19. Straw-Bale Dike

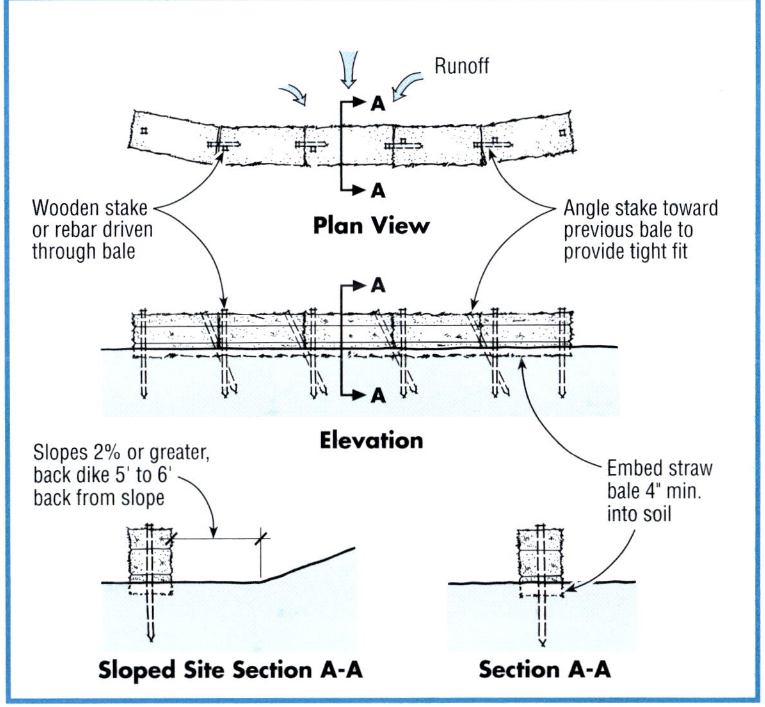

DRIVEWAY MATERIALS SERVE TO SPREAD OUT THE LOAD OF vehicle traffic over a wider area. The main choices for residential driveways are gravel, asphalt, and concrete. For all three types of driveways, performance depends, in large part, on the quality of the subgrade.

DRIVEWAY LAYOUT

Evaluate the drainage strategy for any drive at the layout stage (see "Subgrade Drainage," below).

Driveway Width

Plan for a 12- to 14-ft.-wide roadbed. With a paved roadbed, you could get away with 11 ft., but some drivers might feel a little restricted on such a narrow drive, especially if it's long.

If you have two or three houses at the end of the driveway, with two-way traffic likely, you'll need to go with a 22-ft.-wide road.

Turnarounds

On a long driveway, plan for parking and a turnaround so residents don't have to back out. Don't skimp on space in a turnaround; the minimum turning radius should be 24 ft. to allow room for larger cars or pickup trucks.

DRIVEWAY SUBGRADE

- All driveway beds should be cut in stable native soils.
- Cut out soil or place fill to create the proper drainage grade.
- Plan the cut so the finished driveway is at least 1 in. higher than the surrounding soil.
- Clear the roadbed of roots, twigs, bushes, grass, and topsoil that will decay and might cause the drive surface to sink or break up.

For more information: "Rural Driveways," JLC 6/90

Unsuitable Soils

Peat and expansive clays should be excavated and replaced with compacted granular fill. Unpaved drives built on soils with high clay content must be stabilized with lime (see "Subgrade Stabilization," **page 22**).

Existing Fill

On existing home sites, probe the soil with a steel rod. If resistance below the top 6 in. is less than resistance at the surface, deep fill may have been placed and compacted only at the surface, leaving an unreliable subgrade. Uncompacted fill should be excavated down to native soil and replaced with proper fill.

SUBGRADE DRAINAGE

A subsurface drainage system consists of trenches, drainpipe, stone or gravel aggregate, and geotextiles (**Figure 1-20**). A complete system is crucial to keeping unpaved drives from turning to mud, and for extending the life of more expensive paved toppings.

Drainage Patterns

Grade the subsurface so it slopes away from the building site, if possible, to allow water from rain or snow-melt to drain away from the house.

Drainage for Downhill Drives

If the building site is at the bottom of a slope, build a swale to collect water, or use a slot drain in front of the garage door and wherever else the water might flow

Figure 1–20. Roadbed Construction

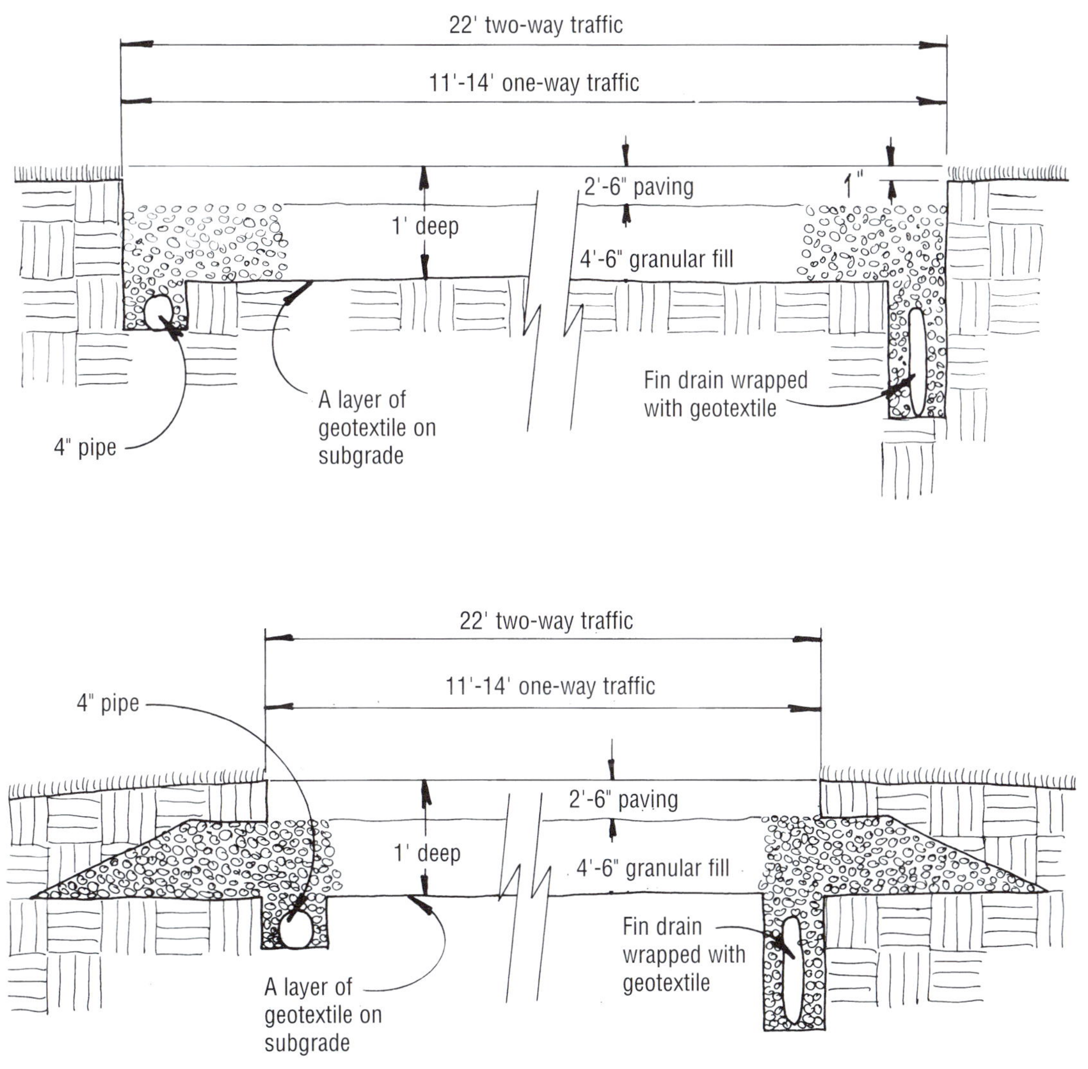

Driveway Construction

A roadbed must have a drainable aggregate base and good subsurface. Use either a 4-in. pipe drain or a fin drain to collect the water. To protect edges from becoming rutted, extend the base at least 18 in. from the surface edges (at bottom).

toward the building. Connect slot drains to a collection drain large enough to carry the water away from the building. If the drainage design might not completely handle all the water, plan on a step-up and ramp into the garage.

Drainage for Uphill Drives

On uphill driveways, route subsurface drainage to a storm sewer or ditch at the edge of the main road. On rural sites, the driveway usually passes over an entrance culvert where it intersects the main highway. Bring the drain pipe from your drain system to this collection point, so the ditch will carry away the run-off. Put a rodent screen over the drain pipe because field mice will damage polymeric materials.

Figure 1-21. Road Mix

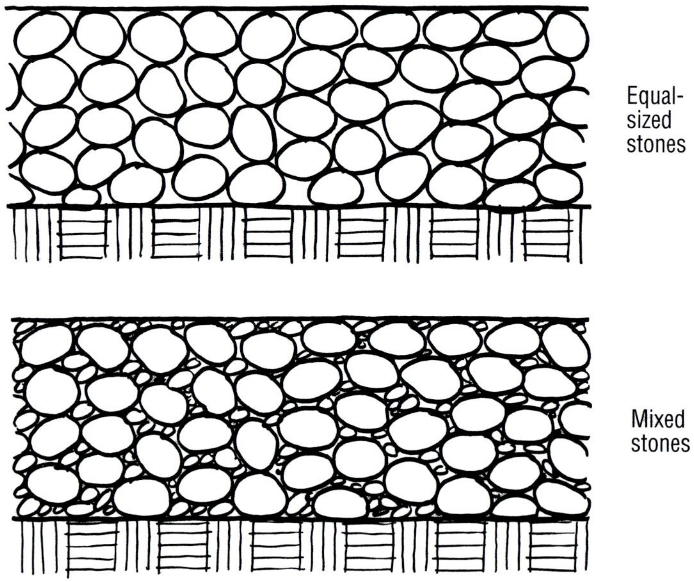

A roadbed made from stones of equal size (at top) will shift under the weight of vehicles. A dense mix of stones of different-sized aggregate (at bottom) will compact under load.

Drainage Trenching

After grading the roadbed, excavate drainage trenches on both sides of the road. A small, 6-in.-wide chain trencher works best. The trenches should be at least as deep as the drain pipe or fin drain, but deeper is better.

Geotextiles

After grading subgrade, lay down a good "road fabric" — a minimum 20-mil woven polypropylene. Geotextile fabric actually provides tension support across the bottom of the base materials, allowing the road to carry greater loads. Without road fabric, wheel action will eventually push the granular base materials into the mud, creating potholes in paved surfaces and corrugations in unpaved surfaces.

Drainage Pipe

Lay the geotextile across the subgrade and drainage trenches, then lay in drainage pipe — a standard 4-in. corrugated drain tubing or fin drains. (Fin drains can carry more water; see "Drainage," **page 68**). Wrap geotextile around the drain.

Driveway Shoulders

On many driveways, the owners will want to landscape right up to the edge of the paving. The best way to handle this is to extend the road base 18- to 24-in. beyond the road surface, then haul in topsoil, spreading it over the shoulders of the base and right to the edges of the finish surface (**Figure 1-20**, previous page). This method will help stabilize the edges of the surface lift.

Driveway Base Materials

Good road base requires a mix of aggregate sizes (**Figure 1-21**). Avoid using pea gravel and other natural gravels from river deposits. These aggregates are too round and smooth, and will not compact well, so they are likely to shift under traffic.

Placing Base Materials

Firm compaction of base fill is crucial for any areas that will carry vehicle traffic. Place up to 12 in. of fill in at least two, 6-in. lifts. Compact each lift with the wheels of a tractor or with construction trucks, or compact with a small crawler (roller) from a rental yard. If the schedule

Driveway Drainage

Base Materials

Figure 1-22. Driveway Construction for Wet Sites

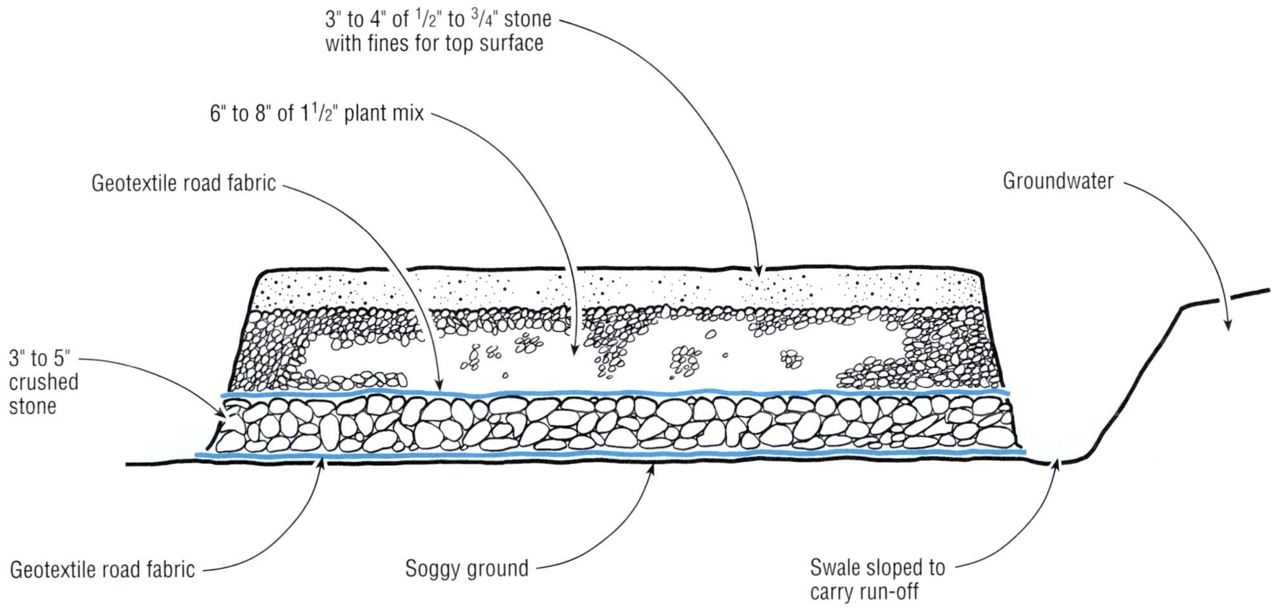

On wet sites, place road fabric directly on top of soggy ground and cover with a heavy stone base. Next, install a second layer of fabric over the stone before placing the gravel base and surface layers.

allows, let the driveway sit for several months to allow settlement before placing a final 3- to 4-in. lift of surface fill (gravel drives) or paving with asphalt or concrete.

Wet-Site Roadbeds

When building a road over wetlands, do not cut and fill. Not only is digging messy and expensive, but it may be difficult to get permits. Land planners frown on any roadbed with deep gravel fill that might create a dam and block wetlands from flowing.

Lay road fabric over the ground, and place a layer of 3- to 5-in. stone on top of the road fabric to help spread out traffic loads on the soggy land (**Figure 1-22**, previous page). Over this stone, spread another layer of geotextile, and place 6- to 8-in. of mixed gravel base material, topped by a 3- to 4-in. layer of gravel surface mix.

Base Materials for Unpaved Drives

The base material used for an unpaved drive is different than it is for paved roads, but good base and subgrade drainage material are important for both.

Road pack: A mix of gravel or crushed stone, road pack contains a high percentage of fine aggregate (called "fines"), and is the least expensive material used for unpaved drives. While road pack compacts well and is reasonably stable when dry, it has very poor internal drainage and gets soft and spongy when wet. Do not use in wet, low-lying areas. A roadbed made of this material should be graded several times a year to keep it from getting rough and to keep it free of potholes.

Open-graded aggregate: This is a gravel mix with particles ranging from about 1/4- to 1-in. in diameter (range of sizes in the mix varies according to local supplier), which will drain better than road pack, but is not stable enough to use as the surface course of a driveway.

Crushed stone: The best base material for a gravel drive is crushed stone, which is angular and compacts well, but doesn't clog as easily as road pack. The more angular the material, the less regrading will be required. Crushed stone mixes vary by region, and every region has its own numbering system for specifying aggregate.

GRAVEL DRIVES

Unpaved roads typically have the same surface and base materials (see "Base Materials for Unpaved Drives," at left). Proper subgrade preparation, good subsurface drainage, and the use of geotextiles are critical for maintaining a drivable surface over time.

Subgrade Stabilization

Subgrade soil stabilization will help make an unpaved driveway less muddy and will help prevent potholes and corrugation. Use hydrated lime, distributed from open bags off the tailgate of a pickup truck. Add 3% to 5% by weight of the soil. Mix in the lime to a depth of 4- to 6-in. with a garden tiller or a disk; then recompact the subgrade.

Surface Mix for Unpaved Roads

After the house is finished and the last load of topsoil has been hauled onto the job site, top the road with a surface mix. This can be the same material used for the base, but good practice calls for 2- to 4-in. of a "three-quarter-minus" mix — 3/4-in. stones, fine stones, and a small amount of clay or organic topsoil for a binder. Alternatively, use 4 in. of 1/2-in. stone "plant mix," which has a complete mix of aggregate to create a smooth surface.

Place the surface mix in thin, 2-in. lifts, and compact each lift with a crawler. The driveway surface should finish out to be at least 1 in. higher than the surrounding soil.

ASPHALT DRIVES

An asphalt, or "blacktop," drive is the best system to use when you have a long approach to a home. Blacktop is a bitumenized concrete — a mixture of asphalt and small stones, 1/2 in. or smaller — that should last for 12 to 14 years. Eventually, asphalt will begin to lose its ability to expand and contract with temperature changes, and cracks will form. To prolong the life of the driveway, the owner should be prepared to fill cracks and put on a sealer coat every two years.

For asphalt drives, prepare subgrade and install subsurface drainage, as described on **pages 18-22**.

Base Materials for Asphalt Drives

The aggregate mix used for a paved road is slightly different from that used for an unpaved road. Because blacktop will spread loads better than a gravel surface, the base does not need to be as tightly packed. Larger chunks of aggregate will drain better, since there is plenty of room for water to percolate through and travel to the drains at the sides of the road.

Use 3/4- to 1-in. gravel as a base for asphalt. Compact the base materials with a plate compactor or vibratory roller. This material is harder to spread and compact than dense-grade base mix, such as road pack. If it will be left for a few months before paving, add some smaller aggregate — a mix of 1/8- to 3/8-in. angular aggregate — at the surface.

Base Thickness for an Asphalt Drive

Place a minimum of 6- to 8-in. of aggregate base material over the roadbed, as well as the drain system. A base thickness of 8- to 12-in. of gravel will extend the life of the blacktop; it's cheaper to put in more base material and less asphalt.

Asphalt Mix

Asphalt hot mix should have an asphalt/cement ratio of 4.5 to 6% by weight for the first lift (the *tack course*) and 5% to 7% for the *wearing course.*

Do not use cold-mix asphalt. This is used only for patching.

Placing Asphalt Paving

- Before applying blacktop, roll the base to ensure even compaction. If soft spots are detected, add fill and compact. Crown the base course slightly to the center of the driveway width (1- to 2-in. over 12 ft.). This ensures that when the topping is applied, the surface will drain and there will be no low spots where puddles can form.
- Some asphalt mixes may require a primer.
- Place asphalt on a warm day. Asphalt that cools too fast won't compact well. Do not place on frozen base materials.
- On a residential drive, asphalt can be placed in a single lift. However, it is better to apply it in two thin lifts, rolling between the tack course and the wearing course.
- With a 10- to 12-in. base, the total asphalt thickness can be as thin as 1 1/2 in. For a 6- to 8-in. base, increase the topping thickness to about 2 1/2 in.
- Apply the top, wearing course at one time so there are no seams in it. Roll it out until all roll marks disappear.

Figure 1–23. Concrete Driveway Construction

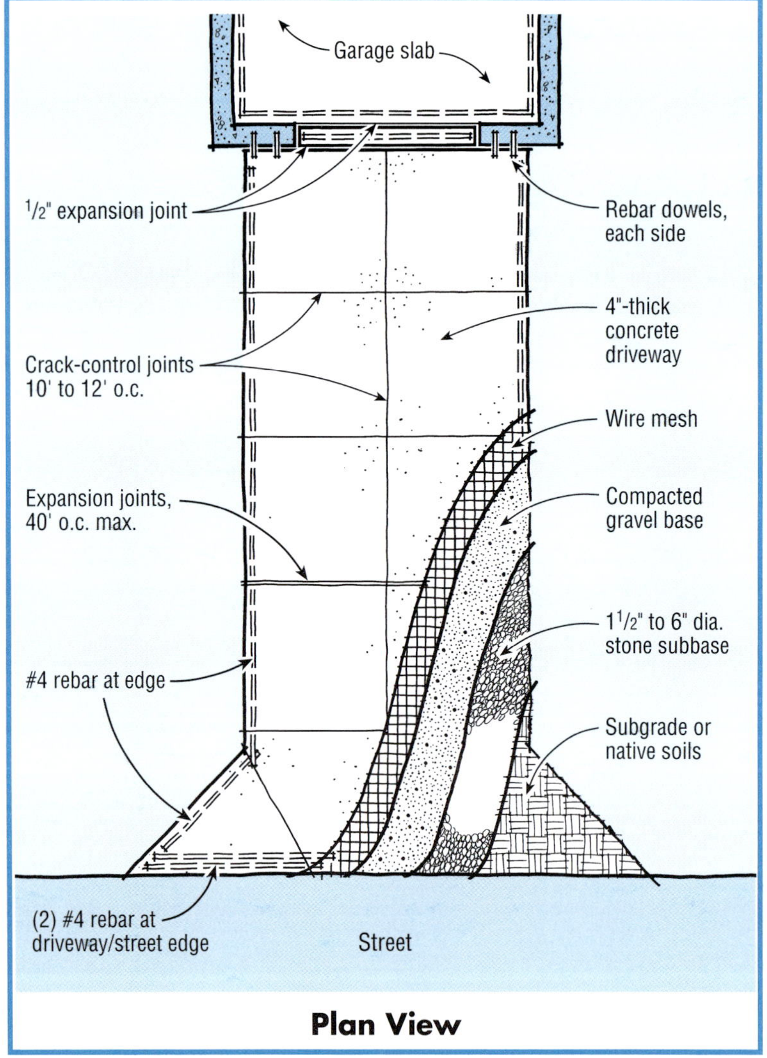

To prevent settling and cracking, a concrete driveway must be built on a well-compacted base and subbase. Use rebar to reinforce slab edges and control joints to limit surface cracking.

Asphalt Edges

Edge reinforcing is typically not required, but you must compact soil well along the edges, or else the outer few inches will chip off (see "Driveway Shoulders," **page 20**). Cover the gravel base at the edge with geotextile, then sod or topsoil.

CONCRETE DRIVEWAYS

Like other durable roadbeds, a concrete drive is built up in layers: A subgrade of native soil (if suitable) or compacted fill, a subbase of compacted stone, a base of compacted gravel, then a 4-in. concrete slab (**Figure 1-23**).

Subgrade and Subbase

Concrete is brittle and requires more support than a gravel or asphalt topping. If the native soils are free-draining sand or gravel, they can be compacted to serve as a subbase. Otherwise, 1 1/2-in. (maximum) stone should be brought in and compacted for the subbase.

Base Materials for Concrete Drives

Use a 3/4- to 1-in. gravel as a base for concrete, placed and compacted as described for other roadbeds (see "Placing Base Materials," **page 21**).

Preparing Slab Edges

Because concrete is so brittle, slab edges can chip or break if they are thin or unsupported. Before pouring concrete, scrape the outermost corners of the form to clear build-up of loose soil so that the slab is full thickness (or thicker) and supported on a well-compacted subgrade.

Where the driveway joins the street, thicken the concrete to 12 in. and reinforce with double rebar (**Figure 1-24**).

Concrete for Driveways

Driveway surfaces may not see heavy truck traffic, but they may take a lot of abuse from freeze-thaw cycles and road salt. Use a minimum 3,500-psi concrete mix with a maximum 4-in. slump (see "Concrete," **page 26**).

Isolation Joints

Concrete expands and contracts with temperature. To prevent damage to the driveway or other structures, place expansion joints where the driveway meets the street and the house, and within the driveway every 40 ft. or less (see "Control Joints," **page 55**).

Garage Transition

Where the driveway meets the garage, the driveway slab should be free to move up but not down. Even a piece of garage door trim that restrains the slab from heaving upward may crack the concrete. Expansion joints must also be placed on each side of a garage transition to isolate the driveway from the garage (**Figure 1-23**).

To support the slab edge and to keep it from sinking near the garage transition, drive rebar dowels into drilled holes in the foundation wall (#4 bar, 12-in.-long, driven minimum 2 in. into wall, 4 ft. o.c.).

Expansion joints in exterior slabs should be sealed with caulk to prevent water from infiltrating the joint (**Figure 1-91, page 81**).

Crack Control Joints

All concrete will crack from shrinkage as elements cool after placement and water evaporates. Control joints are provided to induce cracking to occur at the intended location. In driveways, place control joints in a roughly square grid with squares 10- to 12-ft. on a side. When cracks occur, seal the cracks with caulk. Advise homeowners that caulk must be renewed every few years to prevent water intrusion below the slab.

Crack Control Reinforcement

Support from the base, not from steel in the slab, gives the slab the strength to carry vehicle loads. Steel should be included to limit the size of shrinkage cracks. Use #3 or #4 rebar around the slab perimeter, 3 in. from the edge and 2 in. down from the top surface. In the field, use 6x6-in. welded wire mesh, 2 in. down from the top surface (**Figure 1-24**).

Figure 1–24. Concrete Driveway Subbase

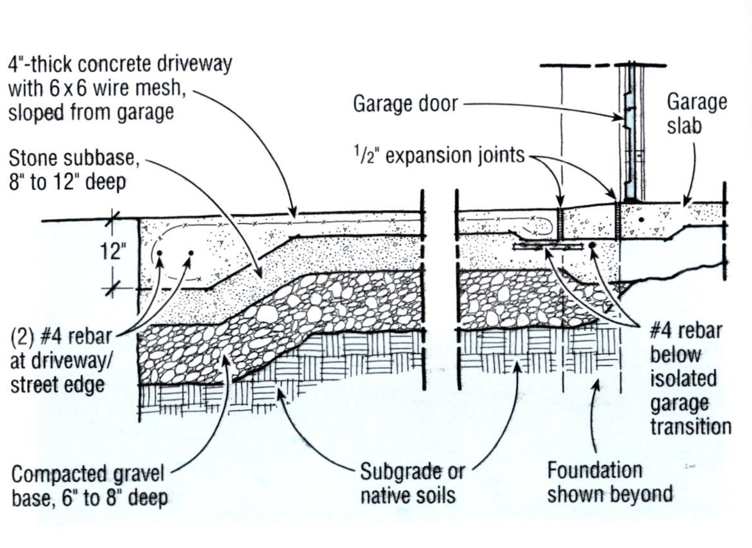

Concrete requires a subbase of compacted stone fill. The subbase adds support and serves as a capillary break for preventing frost heaves from cracking the brittle surface.

CONCRETE IS MADE UP OF PORTLAND CEMENT, WATER, AIR, AND aggregate (sand and gravel). Depending on the mix, the rough proportions by volume are about 10% cement, 20% water and air, 30% sand, and 40% gravel (**Figure 1-25**).

SPECIFYING READY-MIX

Depending on the needs of the project, ready-mix suppliers can provide hundreds of different concrete mixes. In general, it's a good idea to tell your supplier what the concrete will be used for, and follow the supplier's recommendations for the appropriate mix. However, a builder should understand the way mix adjustments affect the concrete's properties.

At a minimum, concrete specifications will usually call out the compressive strength and water/cement ratio (**Figure 1-26**), as well as the slump (**Figure 1-29, page 28**). Concrete mixes can vary in the type and quantity of cement, the ratio of water to cement, the percentage of entrained air, and the size and grading of aggregates. You may also want to order concrete with various admixtures for special circumstances (see "Admixtures," **page 30**).

Cement Types

Figure 1-27 shows the five standard types of cement in use today.

Figure 1–25. Components of Concrete

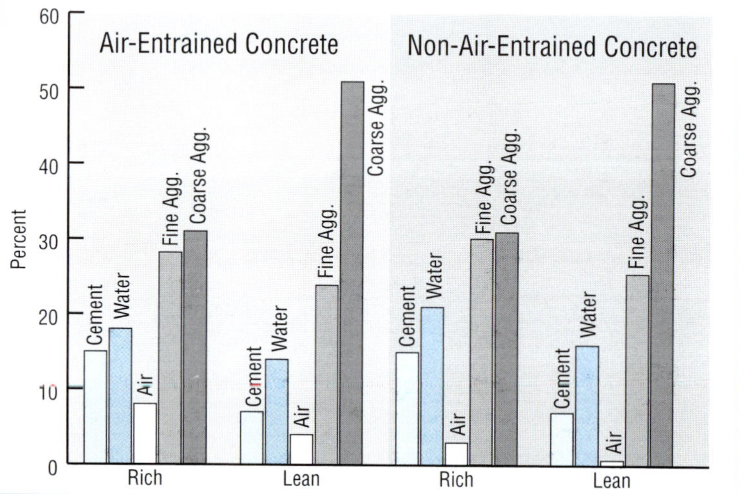

Percentages of cement, water, air, sand, and gravel in concrete mixes. *Rich mixes contain higher percentages of cement. Air-entrainment introduces tiny air bubbles that allow the concrete to flow easier with less water, which makes for a stronger mix. The trapped bubbles also absorb minor expansion and contraction, allowing the concrete to better resist freeze/thaw damage.*

For more information: "Concrete Basics," JLC 6/00

"Ordering Ready Mix," JLC 5/02

Figure 1–26. Selecting Residential Concrete

Structural Element	Minimum Compressive Strength Required	Practical Water/Cement Ratios
Foundations, basement walls, and slabs not exposed to weather	2,500–3,000 psi	.55
Foundations, basement walls, and slabs exposed to weather	3,500 psi	.45
Driveways, garage slabs, sidewalks, porches, patios, and steps exposed to weather	3,500–4,000 psi	.45

Aggregate

Sand and gravel are the strongest and cheapest ingredients in concrete. It is most economical to use aggregate that is large and well-graded (containing a good proportion of various sizes from large to small), because this reduces the required volume of cement paste. If reinforcing steel will be spaced close together, or if concrete must be pumped, maximum gravel sizes may have to be reduced.

Using more fine sand makes a concrete mix "creamier" and makes it easier to achieve a smooth finish; however, the mix will require more water, and therefore should have more cement added for adequate strength.

Specifying Ready-Mix

Water

Water used to mix concrete should be clean enough to drink. Adding water to a concrete mix can weaken concrete. Follow water/cement ratio guidelines below.

Water/Cement Ratio

The ratio of water to cement should be strictly controlled to ensure that the concrete reaches the specified strength (**Figure 1-28**).

Figure 1–27. Types of Cement

Type	Use
Type I, Type IA (air-entraining)	Good for most residential work
Type II, Type IIA (air-entraining)	Use where soils and ground water contain moderate sulfates that can attack concrete
Type III, Type IIIA (air-entraining)	Use when freezing is a risk or to speed up setting and curing
Type IV	Only needed for massive industrial placements (special order)
Type V	Use in extreme sulfate exposure conditions

Concrete Types I, II, and III will meet most residential needs. For freeze-thaw durability, order air-entraining cement or an air-entraining admixture.

Figure 1–28. Effect of Adding Water to Concrete

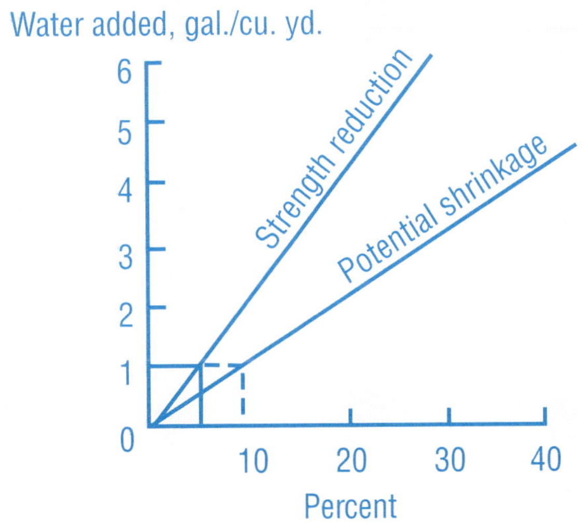

The more water added to a concrete mix on site, the weaker it will become, and the more it will shrink and crack. Adding 2 gallons per yard will cut compressive strength by around 10% and will increase shrinkage by close to 20%.

Figure 1–29. Concrete Slump

A slump test uses a standard cone — 12 in. high, 8 in. wide at the base, and 4 in. wide at the top. To perform the test, fill the cone in one-third lifts and "rod it" (churning by moving a piece of rebar up and down) 25 times between each lift. Remove the cone and measure the distance from the height of the cone to the height of the slumped concrete. Residential concrete should slump no more than 4 in.

On site, add the minimum amount of water needed to make the concrete workable. More water makes the concrete easier to handle, but also makes it much weaker and more prone to shrinkage and cracking (**Figure 1-28**).

Air Entrainment

Air entraining creates billions of microscopic air voids in hardened concrete, which serve to absorb the pressures caused by expanding ice or de-icing salts. Most ready-mix suppliers today add an air-entraining admixture to a standard cement mix.

Figure 1–30. Recommended Air Entrainment for Residential Concrete

Adapted from Portland Cement Assoc.

Nominal Maximum Aggregate Size (in.)	Air Content (%)*		
	Severe Exposure	**Moderate Exposure**	**Mild Exposure**
3/8	7 1/2	6	4 1/2
1/2	7	5 1/2	4
3/4	6	5	3 1/2
1	6	4 1/2	3
1 1/2	5 1/2	4 1/2	2 1/2
2	5	4	2
3	4 1/2	3 1/2	1 1/2

*Air content is specified as a percentage by volume of concrete. For severe exposure conditions, air content of the mortar alone (cement paste and sand) should be about 9%. Lower air content percentages for concrete with large aggregate reflect the fact that less mortar is needed for mixes that contain large gravel.

These levels of air-entrainment shown for different climate exposures are minimums. Higher air amounts are permissible as long as the design strength is maintained. Refer to ***Figure 1-31*** *for exposure locations.*

Slump

Air-Entrained Concrete

Figure 1–31. Weather Exposure Regions for Residential Concrete

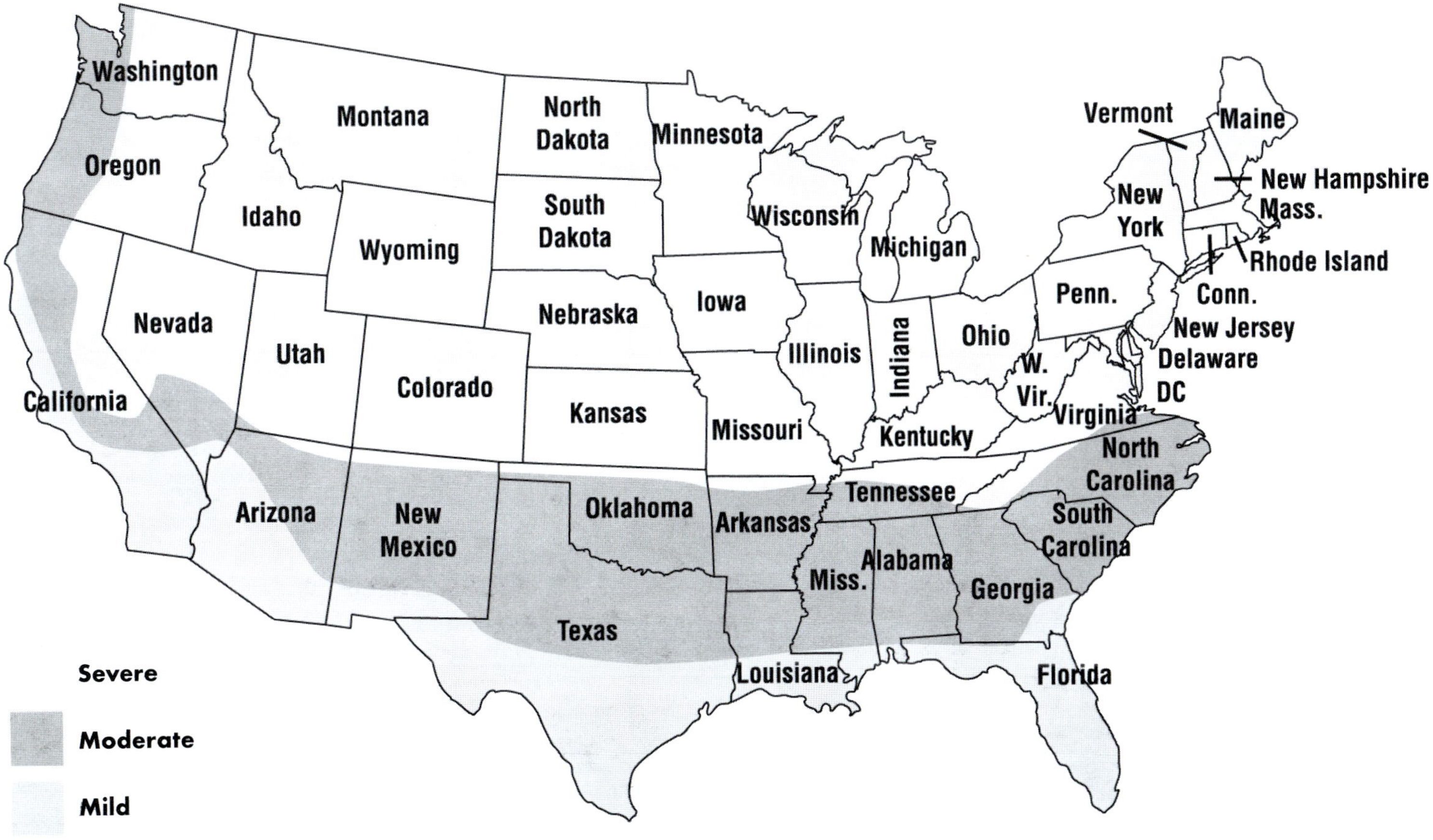

"Severe" and "Moderate" exposures are determined based on the likelihood that de-icing salts will be used at a given location. Check local climate data because icing conditions may vary locally with altitude.

Air entrainment is crucial for exposed concrete in cold climates, but it is recommended for almost all concrete, even in mild climates, because it reduces water demand, improves workability, reduces segregation of aggregate, and reduces bleeding of excess water. Recommended entrained-air percentages for different weather exposures are shown in **Figure 1-30**. Refer to the map in **Figure 1-31** for exposure regions throughout the continental United States.

Figure 1–32. Concrete Admixtures and Uses

Admixture	Common Active Ingredients	Use
Accelerator	Chloride: Calcium chloride Non-chloride: Triethanoline (TEA), Calcium formate, Calcium nitrite	To accelerate set and strength gain (primarily used in cold weather)
Retarder	Gypsum powder, Sugar*, Lignosulfonate, Hydroxycarbolic acid	To reduce setting time (primarily used in hot weather)
Water-reducer/ plasticizer	Lignosulfonate, Hydroxycarbolic acid	To produce a more workable mix
Superplasticizer	Sulfonated melamine formaldehyde (SMF), Sulfonated naphthlene formaldehyde (SNF), Modified lignosulfate (MLS)	To produce flowing plasticizer

* Large amounts of sugar will completely stop the setting reaction, so sugar is often used as a kill when a mixer breaks down and cannot be emptied.

Admixtures are chemical or mineral compounds used to alter the physical or chemical properties of concrete.

For more information: "Ordering Ready-Mix Concrete," JLC 3/89

Finishing air-entrained concrete. A concrete finisher may wait for bleed water to evaporate before starting to trowel the surface, but when concrete is air-entrained, bleed water may not appear. Excess water should still be allowed to evaporate from beneath the surface for a time before troweling begins — otherwise, water may be trapped just below a hard surface skin and cause later scaling or flaking.

ADMIXTURES

Small amounts of specialty admixtures can be used to modify concrete properties as needed. Common admixtures are shown in **Figure 1-32**. More detailed information on common admixtures used in residential concrete is given below:

Accelerators

Accelerating admixtures are used when rapid strength gain is required, such as when the risk of freezing or tight schedules requires faster curing. Accelerators increase the heat of hydration, shorten the set time, and increase the early strength. However, they can decrease the long-term strength of the concrete. The additional heat of hydration may also contribute to increased thermal shrinkage cracking when the concrete cools.

Accelerators do not stop concrete from freezing; they only let the concrete quickly develop its air-entrainment, which gives water in concrete pores a place to go to relieve the pressure of expanding ice. When temperatures are cold, concrete may need to be protected from freezing with insulation or within a heated enclosure (see "Cold-Weather Concrete," **page 35**).

Calcium Chloride

Calcium chloride is one of the most common accelerators. Chlorides can increase corrosion of reinforcing steel in concrete that's exposed to de-icing salts. (Non-chloride accelerators are available but seldom specified for residential work.) Calcium chloride can be added to the truck on-site, but it is better to have the ready-mix supplier add it at the plant. Accidentally varying the dose from one truckload to the next can cause color variations in finished slabs.

MAXIMUM DOSE = 2% CALCIUM CHLORIDE
(BY WEIGHT OF CEMENT)*

*About two 50-lb. bags of calcium chloride per 10-yd. load

Water Reducers and Plasticizers

Water reducers are used to increase slumps and make concrete more workable without adding water. This makes placement easier without compromising strength. High-range water reducers, or superplasticizers, create flowable concrete that can be readily worked into forms and around rebar, and is easily consolidated.

Superplasticizers

These can be tricky to work with. The effect from a superplasticizer typically lasts only about 30 minutes and wears off rapidly. Adding water to restore slump when the superplasticizer wears off creates inconsistent quality within the batch and should be avoided. It is possible to combine a superplasticizer with a lower-range water reducer to provide a longer working life; if you anticipate delays on site, consult with your supplier.

Set Retarders

Concrete sets up rapidly in warm weather, and it may be difficult to finish the surface before the concrete becomes too hard. Set retarders slow the rate of hydration and allow a longer working time. However, set retarders do not prevent evaporation and slump loss, so concrete may still dry out rapidly and quality can suffer. Plastic shrinkage cracking, which occurs because concrete dries out before setting, may actually be worse if a set retarder is used.

Admixtures

Pouring Concrete

PLACING CONCRETE

Concrete placement always happens under time pressure. Allow a half-hour to an hour per truckload of concrete — concrete is unusable after 90 minutes on the truck under normal conditions, and may go bad even faster in hot weather.

Pumping Concrete

Pumping calls for a special mix, and usually requires a specialty pumping subcontractor. Coordination is important, so a pre-construction meeting of all the contractors involved is recommended.

Pouring Concrete

Place walls in lifts of 1- to 2-ft., working around the perimeter of the building. This lets lower lifts stiffen before the pour gets deep, reducing pressure on the forms.

Place slabs in bands 4- to 6-ft.-wide, and strike off as you go so you won't have to disturb concrete that has already begun to set.

Place concrete close to its final position. Don't puddle it in one place and drag it along — the farther the concrete is pulled, the more the gravel separates from the paste (segregates), causing honeycombing and weakening the structure.

Do not drop the concrete more than 4 ft. Use a drop-chute with a 6-in.-diameter rubber or canvas hose to prevent segregation of aggregate.

Compact concrete as you place it by rodding, spading, or vibrating. This removes large air bubbles and prevents voids.

FINISHING CONCRETE

Strike off and bullfloat concrete as soon as it is placed, but wait for bleed water to evaporate before starting to trowel. Working concrete with water on the surface creates a weak layer and leads to dusting, crazing, and scaling. If concrete starts to set up before bleed water evaporates, sweep the bleed water off with a rubber hose or squeegee.

Edging is required to compact concrete near forms. Wait until bleed water evaporates to begin edging, but right after bullfloating, slip a mason's trowel down between the form and the concrete to cut the two apart in preparation for edging later.

Jointing. Joints can be made in the slab while it's still soft. The grooving tool must cut through the slab one-quarter its thickness to make the joint an effective control against shrinkage cracking (see "Control Joints," **page 55**).

Figure 1–33. Slab Curing Techniques

Method	Advantages	Disadvantages	Comments
Cover with plastic sheeting	Low cost, easy to handle, effective, does not leave film on slab	May mottle slab surface; can be torn or disturbed	Use large pieces to minimize joints, seal carefully, inspect and maintain, save for reuse
Apply spray curing compound	Cheap, simple, does not require maintenance or monitoring	Coverage may be incomplete if not applied with care; film may prevent bonding of finish floor; may not work in cold weather	Select opaque disappearing brand to monitor coverage; consult with suppliers concerning floor bonding
Continuous sprinkling or cover with wet burlap	Highly effective if sprinkled continuously, or burlap kept wet is used	Requires monitoring; may create runoff	Use clean water, provide drainage, monitor and maintain
Ponding	Best results if maintained continuously for prescribed period	Complicated; requires maintenance and monitoring; may cause runoff	May be practical for enclosed slabs (basements) and small areas

Special attention must be given to curing concrete slabs because their relatively large surface area gives up moisture and heat quickly. Proper curing keeps the moisture and temperature in the concrete as long as possible.

Floating drives aggregate just below the surface and removes surface imperfections. Floating should be done after edging and jointing, but before final troweling.

Troweling. Concrete is ready for hand-troweling when the weight of a worker makes a 1/4-in. heel-mark on the slab. For machine troweling (power-troweling), the concrete should be harder — a heel mark should be only 1/8-in.-deep.

Stripping forms. Allow concrete in forms to reach at least 500 psi compressive strength before stripping forms. This typically takes a day in mild weather, or three days in cold weather. Concrete suppliers can provide maturity curves that estimate the strength development of specific concrete mixes under specific field conditions — when in doubt, rely on the supplier's data.

CURING CONCRETE

Concrete hardens not by drying, but by hydration — a chemical reaction between cement and water. This reaction needs moisture and warmth. Curing is the technique of keeping the concrete moist and at the correct temperature (50°F to 90°F) for a period of at least three to seven days (the shorter period applies to concrete made with high-early-strength cement or with an accelerating admixture). If cured properly, concrete will be stronger, more abrasion-resistant, more durable, and less permeable.

Curing Problems

Drying out. If concrete dries out before curing is complete, hydration stops. Hardening will resume if the concrete becomes wet again, but in the field it is hard to resaturate concrete, so maintaining moisture for the minimum time is a better course.

Freezing. Concrete that freezes before reaching compressive strength of 500 psi may be ruined. If it freezes after reaching 500 psi, however, it will continue to harden when it warms up, as long as sufficient water is present. Again, maintaining good curing conditions for a few days is better than trying to reestablish them later.

Curing Conditions

Concrete is "comfortable" if people are comfortable. The best time to pour concrete work is when the weather is between 50°F and 70°F, humid, and not too sunny. In these conditions, little extra effort is required to cure concrete.

For walls in normal weather, leaving the forms in place will keep water trapped in the wall and allow curing to continue. The wall top should be kept wet with a soaker hose, or sealed with a spray-applied curing compound. If forms have to be stripped, spray the whole wall with curing compound or cover it with plastic sheeting (seal all seams and penetrations).

For slabs during mild weather, practical curing methods include ponding, sprinkling continuously, covering with plastic, covering with wet burlap, or sealing with a spray-applied curing compound. Each method has advantages and disadvantages (**Figure 1-33**). Spraying on a curing compound has become the most popular method because the one-time step is quick, simple, and cheap; but the coatings may prevent the adhesion of tile or other floor coverings, and care must be taken to get full coverage.

WEATHER CONDITIONS

Extreme weather presents challenges for placing, handling, and curing concrete. By slowing down hydration, cold

Finishing

Curing

Figure 1–34. Cold-Weather Guidelines for Concrete Walls and Footings

TYPE I CEMENT
Predicted low temperature prior to concrete reaching 500-psi compressive strength:

32°F to 20°F
Min. 470 lb. cement/yd. (5-sack mix)
Temperature of concrete: Min. of 60°F
Max. 6-in. slump
1% Calcium Chloride or equivalent accelerator
No protection required

19°F to 10°F
Min. 517 lb. cement/yd. (5 1/2-sack mix)
Temperature of concrete: Min. of 60°F
Max. 6-in. slump
2% Calcium Chloride or equivalent accelerator
No protection required

9°F to 0°F
Min. 564-lb. cement/yd. (6-sack mix)
Temperature of concrete: Min. of 60°F
Max. 6-in. slump
2% Calcium Chloride or equivalent accelerator
Cover top of wall with 6-ft.-wide insulated blanket

Below 0°F
Min. 564 lb. cement/yd. (6-sack mix)
Temperature of concrete: Min. of 60°F
Max. 6-in. slump
2% Calcium Chloride or equivalent accelerator
Cover entire wall with an insulated blanket or cover and provide auxiliary heat

TYPE III CEMENT
Predicted low temperature prior to concrete reaching 500-psi compressive strength:

32°F to 20°F
Min. 470 lb. cement/yd. (5-sack mix)
Temperature of concrete: Min. of 60°F
Max. 6-in. slump
No accelerator
No protection required

19°F to 10°F
Min. 517 lb. cement/yd. (5 1/2-sack mix)
Temperature of concrete: Min. of 60°F
Max. 6-in. slump
1% Calcium Chloride or equivalent accelerator

9°F to 0°F
Min. 564 lb. cement/yd. (6-sack mix)
Temperature of concrete: Min. of 60°F
Max. 6-in. slump
2% Calcium Chloride or equivalent accelerator
No protection required

Below 0°F
Min. 564 lb. cement/yd. (6-sack mix)
Temperature of concrete: Min. of 60°F
Max. 6-in. slump
2% Calcium Chloride or equivalent accelerator
Cover top of wall with 6-ft.-wide insulated blanket

For each type of cement used, follow these mix and protection guidelines to prevent freezing of plain, unreinforced concrete. Note: These recommendations are designed to prevent freezing damage to green concrete, but may not provide ideal curing conditions.

weather delays setting times and slows strength gain; plus, concrete may be damaged or ruined by freezing. Hot weather speeds up hydration and causes rapid drying, which can lead to various finish defects and structural weakness. Slabs are especially vulnerable to either cold or heat; if possible, slab work should be rescheduled for better weather.

COLD-WEATHER CONCRETE

With concrete, cold weather refers to temperatures averaging lower than 40°F, or dropping below 50°F for more than half the day.

When placing concrete in cold weather, take these precautions:

- **Thaw forms and subgrade.** Never place concrete on frozen ground or in icy or frosty forms. Ice and frost-swelled soil can fill the space meant to contain concrete, leaving future voids. Concrete may freeze instead of hardening, and could be damaged or weakened. Frost heave of soil after concrete is placed also may damage footings, walls, or slabs.

  To thaw ground and forms, cover them with straw or blankets, or build a heated enclosure. Make sure exhaust from heaters goes outside the enclosure; otherwise workers may be poisoned from carbon monoxide. Carbon dioxide in exhaust can also damage concrete surfaces.

- **Adjust mix for rapid strength gain.** Follow the supplier's recommendations for an appropriate mix for the placement and end-use weather conditions. To generate extra heat and speed up strength gain, options include adding extra cement to the mix, using Type III cement, or using an accelerating admixture (in very cold weather, all three strategies may be combined). General recommendations for cold-weather mixes and curing protection are shown in **Figure 1-34**.

Cold-Weather Concrete

- **Check concrete temperature.** Concrete coming down the chute should be at 60°F. Make your supplier aware of your expectations and be ready to reject loads that are cooler than 55°F.
- **Protect subgrade from freezing** before and after pour. Even after concrete is cured, frost heaves in soil can crack structures. To prevent this, insulate slabs, footings, and walls, cap foundation quickly, backfill as soon as possible, and heat structure.

Cold-Weather Finishing

In cold weather, set times are delayed (**Figure 1-35**) so schedule the pour early in the day. In cold temperatures

Figure 1–35. Concrete Temperature vs. Set Time

Concrete Temperature (F°)	Set Time (hrs.)
70	6
60	8
50	11
40	14
30	19

Set time (the time it takes for concrete to harden enough so that finishing can begin) increases by a third for every 10°F drop in concrete temperature.

(below 40°F), don't be surprised if finishing has to be done at night, long after regular quitting time. Bleed water may not evaporate; be ready to squeegee it off the slab. Slabs cool rapidly, so be ready to protect the slab with insulating blankets and straw as soon as finish troweling is complete.

Cold-Weather Curing

Ponding and sprinkling are impractical in freezing weather. Use curing compounds or plastic sheeting to hold moisture in fresh concrete. Provide enough insulation to keep concrete temperature above 50°F for at least three days (seven days is preferable). See **Figure 1-36** for the insulation level needed to maintain adequate warmth and **Figure 1-37** for insulation materials to meet R-value requirements.

After curing, exposed slabs need to dry thoroughly before they can withstand winter freezing. For that reason, pouring slabs in late fall is risky in colder regions.

HOT-WEATHER CONCRETE

Rapid evaporation causes most of the hot-weather problems for concrete. If sweat is evaporating quickly off workers, chances are that moisture is evaporating quickly from the concrete as well. Take these precautions:

- **Wet forms and subgrade.** Use a spray hose or sprinkler to dampen work area before beginning. Soak area well, but do not create standing water. Soaking the subgrade the night before the pour may be most practical.

A temperature drop greater than 50°F over 24 hours risks thermal shrinkage cracking. Insulation thrown over poured concrete, however, will trap the heat of hydration and allow the concrete to cool slowly. As outside air temperature drops, it may become harder to keep concrete at the correct temperature using insulation alone. Below a certain point for any given mix, provide a heated enclosure to prevent damage to the concrete from freezing.

Figure 1–36. Insulation Required to Keep Concrete at 55°F

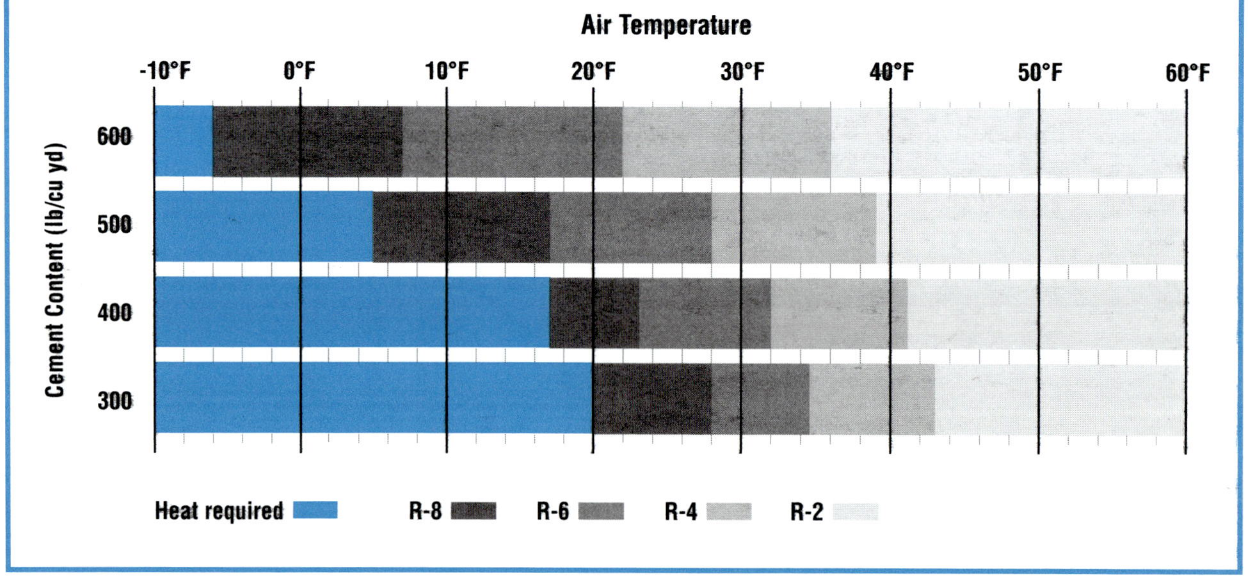

- **Chill the mix.** You can order concrete made with chilled water or even shaved ice. Some suppliers can refrigerate their aggregate also. Concrete should not be hotter than 60°F coming off the truck.
- **Provide shade and wind protection.** Sun screens, movable awnings, and wind barriers can reduce evaporation rates.
- **Fog the work.** Rent a fogging sprayer to elevate the humidity around the slab.
- **Retemper with care.** It's okay to replace water lost to evaporation, but not to add water to a partially hydrated load. In practice, add no more than a gallon or two per yard. Add this at the beginning, not midway through the load.
- **Adjust schedule.** Start work at sunup and finish before noon, or work in late evening.
- **Adjust manpower.** Have extra help ready to place and finish concrete quickly. If your normal crew is four people, add a fifth or sixth person.
- **Schedule work in stages.** Don't pour too far ahead of yourself — place small sections of slab that you can finish with the available labor.

Hot-Weather Finishing

Start to trowel as soon as bleed water has evaporated. Do not trowel water back into slab or sprinkle water onto slab (this causes crazing and flaking). However, if concrete begins to harden before water has evaporated, drag the water off with a hose or squeegee.

Hot-Weather Concrete

Hot-Weather Curing

For walls, apply curing compound to wall top right away, and to wall sides as soon as forms are stripped, or cover completely with plastic. Seal all plastic seams and penetrations, and ensure complete coverage with curing compound.

For slabs, start curing (cover with plastic or ponded water, spray on curing compound, or begin sprinkling) as soon as troweling is complete. Concrete must not be allowed to dry out.

Do not cover with black plastic, which absorbs solar heat and can cause too-rapid hydration or drying. Clear plastic is better, and reflective white plastic is best.

Figure 1–37. R-Values of Common Material

Material	R-Value
1/2-in. plywood	0.5
1-in. straw	2.0
1-in. mineral fiber blanket	3.0
1-in. expanded polyurethane	6.0

FOOTINGS SERVE TO SPREAD THE LOAD ONTO THE SOIL FOR adequate bearing, and to provide a smooth, level surface for wall forms or for block and mortar work. Footings also resist lateral soil pressure at the base of foundation walls.

2000 International Residential Code

Figure 1–38. Minimum Width of Concrete or Masonry Footings (in.)

	Loadbearing Value of Soil (psf)					
	1,500	2,000	2,500	3,000	3,500	4,000
Conventional Wood Frame Construction						
1-story	16 in.	12 in.	10 in.	8 in.	7 in.	6 in.
2-story	19	15	12	10	8	7
3-story	22	17	14	11	10	9
4-in. Brick Veneer over Wood Frame or 8-in. Hollow Concrete Masonry						
1-story	19 in.	15 in.	12 in.	10 in.	8 in.	7 in.
2-story	25	19	15	13	11	10
3-story	31	23	19	16	13	12
8-in. Solid or Fully Grouted Masonry						
1-story	22 in.	17 in.	13 in.	11 in.	10 in.	9 in.
2-story	31	23	19	16	13	12
3-story	40	30	24	20	17	15

The width of a foundation footing is based on the loadbearing value of the soil. Load-values assume undisturbed, native soils of known type or tested values of compacted soils.

For more information: "Footing Fundamentals," JLC 10/00

FOOTING LAYOUT

Footings can be larger than required dimensions, but should not be smaller. Footings should be straight and level to within 1/4 in. over 20 ft., and should be no more than 1/2-in. out of square in 20 ft.

Footing Width

The bearing width of footings varies according to soil strengths and loading conditions (**Figure 1-38**).

Footing Tie-In

Footings should be struck off level, but should never be troweled smooth. In addition, footings must be tied firmly to the wall above using either a keyway poured into the footing or reinforcing bar projecting from the footing into the concrete (**Figure 1-39**).

Figure 1–39. Footing Tie-In

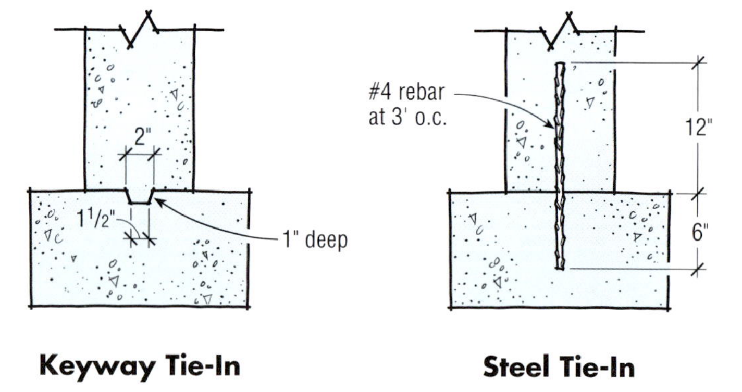

To tie a footing to the wall above, a keyway should be about 1 in. deep by 2 in. wide. When rebar is used, the typical rebar layout is one #4 bar every 3 ft. o.c.

Footing Layout

Footing Thickness

Footing proportions for plain concrete (unreinforced) footings are set by code (**Figure 1-40, left and center**). Generally, footings should be at least as deep as the thickness of the wall they support. The wall should be centered on the footing so that the projection of the footing on each side equals half the wall or footing thickness.

If a wider footing is required, the footing must be reinforced (**Figure 1-40, right**). Typically, 1/2-in. or 5/8-in. rebar will be required on 1-ft. centers, set about 3 in. up from the bottom of the footing.

Frost Depth

The footing should be placed at least 1 ft. below the frost line so it will not heave when the soil freezes. Frost depth varies depending on local climate; check with local codes for the local frost depth. On a hillside foundation (such as a basement walkout), plan on building a frostwall on the downslope side (**Figure 1-41**).

As an alternative to building a frostwall, a walk-out basement may be insulated to protect against frost damage (see "Insulation for a Walkout Basement," **page 89**).

Shallow foundations may offer an alternative to deep

Figure 1–40. Footing Dimensions

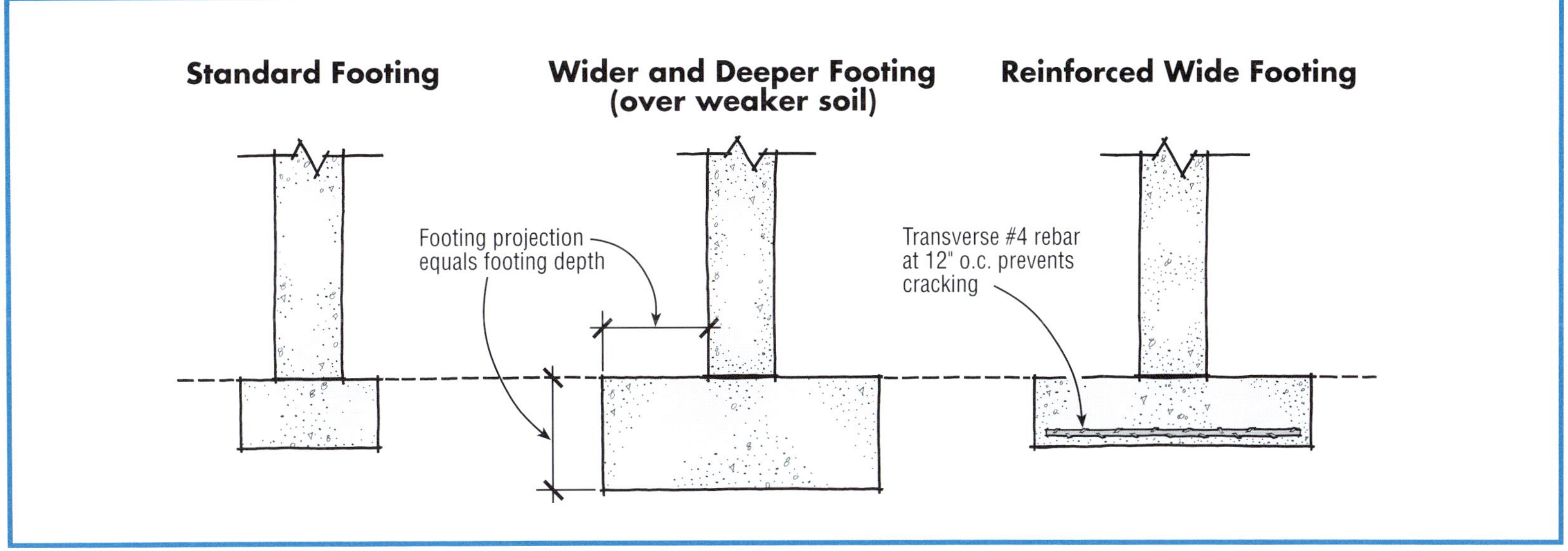

If a particular soil bearing capacity requires a wider footing, code allows the footing projection to increase, but the footing thickness must also be increased so that the footing depth equals the distance it projects from the wall (center). An alternative is to reinforce the wider footing (right), if local code allows.

Figure 1-41. Frostwall for Basement Walkout

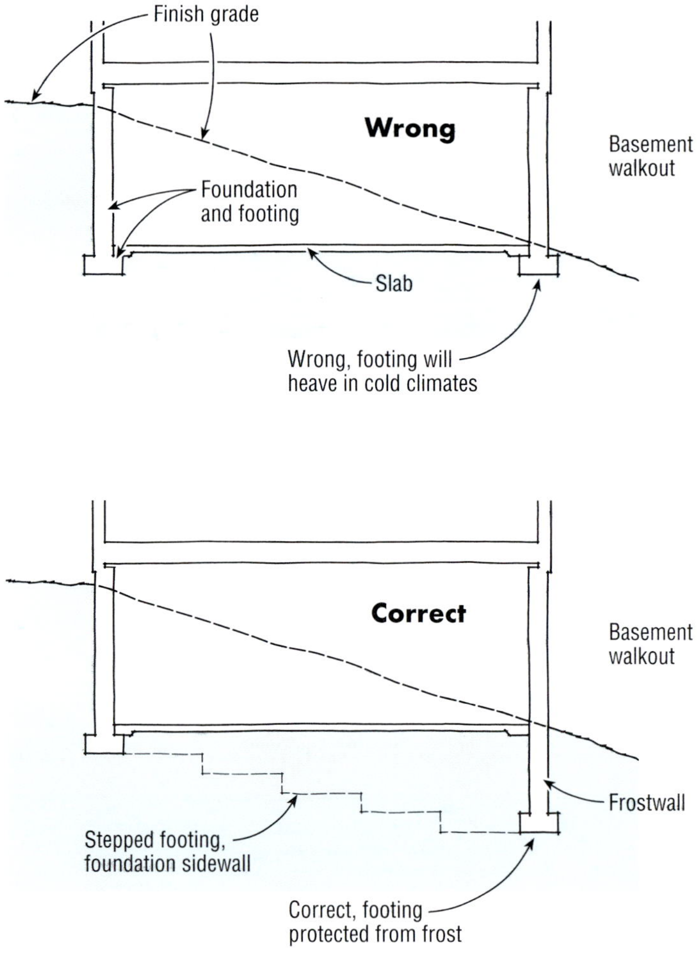

On a hillside foundation, a frostwall should be added on the downslope side to extend the footing below the frostline.

Figure 1-42. Stepped Footings

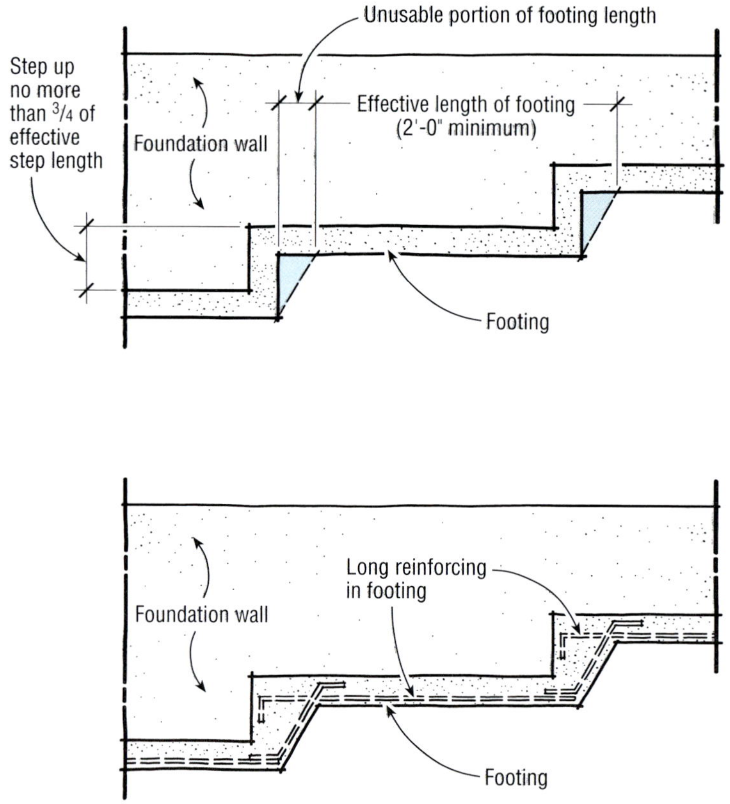

The rise of a stepped footing should not exceed 2 ft., and the footing should run at least 2 ft. horizontally between steps (at top). Typically, the corners of a stepped excavation are unstable. The footing should be sloped and reinforced, so the effective horizontal length of the footing is supported on well-compacted soil (at bottom).

For more information: "Fast Forming of Stepped Foundations," JLC 3/94 excavations if properly insulated and detailed. See "Frost-Protected Shallow Foundations," **page 86.**

STEPPED FOOTINGS

On sloped sites, footings must be kept level, so the footings will have to be stepped. Step lengths must be at least 2 ft., step heights must be no greater than three-quarters of step length, and vertical footing sections must be at least 6 in. thick (**Figure 1-42**).

On slopes too steep for stepped footings of these dimensions, consider a pier-and-grade-beam foundation (see "Pier Foundations," **page 82**).

JUMP FOOTINGS

When a short foundation wall ties into a tall wall (such as where a house with a full basement jumps to a stem wall

Frost Walls

Stepped Footings

Figure 1–43. Jump Footings

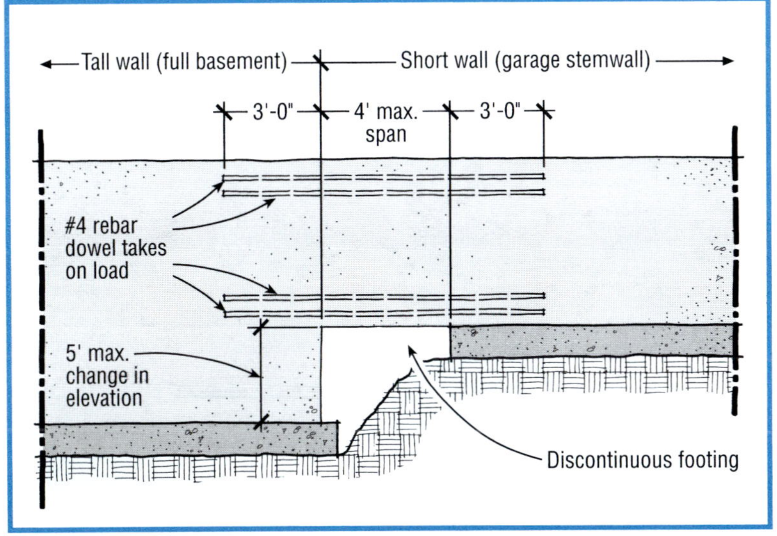

For poured concrete wall only: *Where a garage stemwall abuts a main basement wall, the short section of wall can be reinforced to span the gap. Use two #4 bars at the top and bottom of the short wall, extending 3 ft. into each adjoining section of wall above the footings. This solution is limited to a 4-ft. maximum span and a 5-ft. maximum change in elevation. If the walls are at right angles, the rebar can be bent accordingly.*

Figure 1–44. Pouring Footings in a Wet Trench

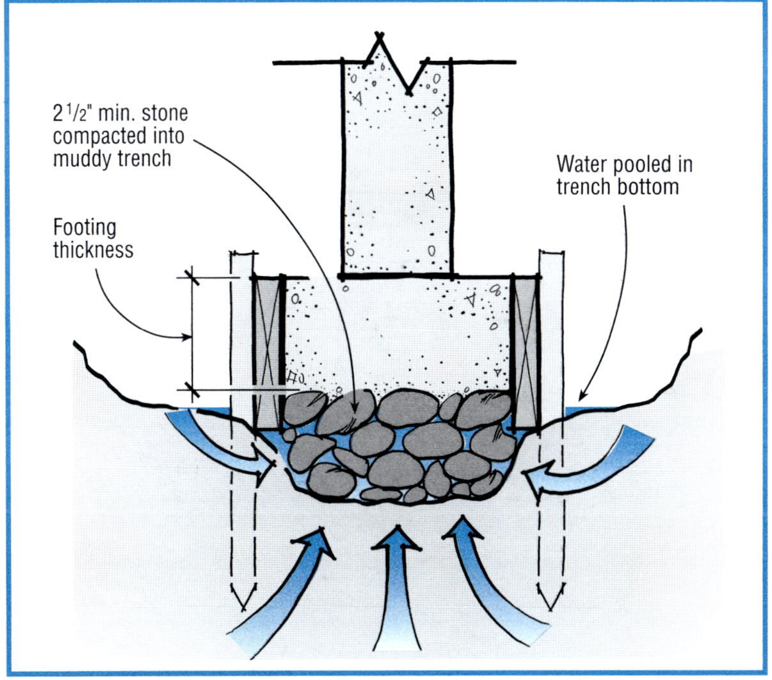

When water pools in a footing trench, place large cobbles in the bottom of the form, compacting them down into the mud. Muck and water may fill the spaces between the stones, but contact between stones will provide bearing. Use a stiff concrete mix when casting the footings.

for the garage), the footing can be interrupted and the foundation wall tied together with a reinforced foundation wall section. Masonry foundation walls have no real spanning capability, so the footing must be stepped down when elevations change, but poured concrete walls can be reinforced with steel to span openings (**Figure 1-43**, previous page). Check with a local code official first; the code on the books may have been written for masonry foundations.

FOOTING PROBLEMS

Foundation walls should be centered on the footing. In the event of a layout mistake, the footing can be repaired. The repair depends on the soil conditions.

A well-compacted gravel support may be allowed, but usually must be based on a good soil analysis and engineer's approval. Otherwise, an augmented footing must be poured using rebar dowels epoxied into the existing footing (**Figure 1-45**).

When footing trenches fill up with water in wet conditions, place a layer of well-tamped 3- to 4-in. cobbles (**Figure 1-44**, previous page) and use a stiff concrete mix when casting the footings.

Figure 1–45. Repairing Off-Center Footings

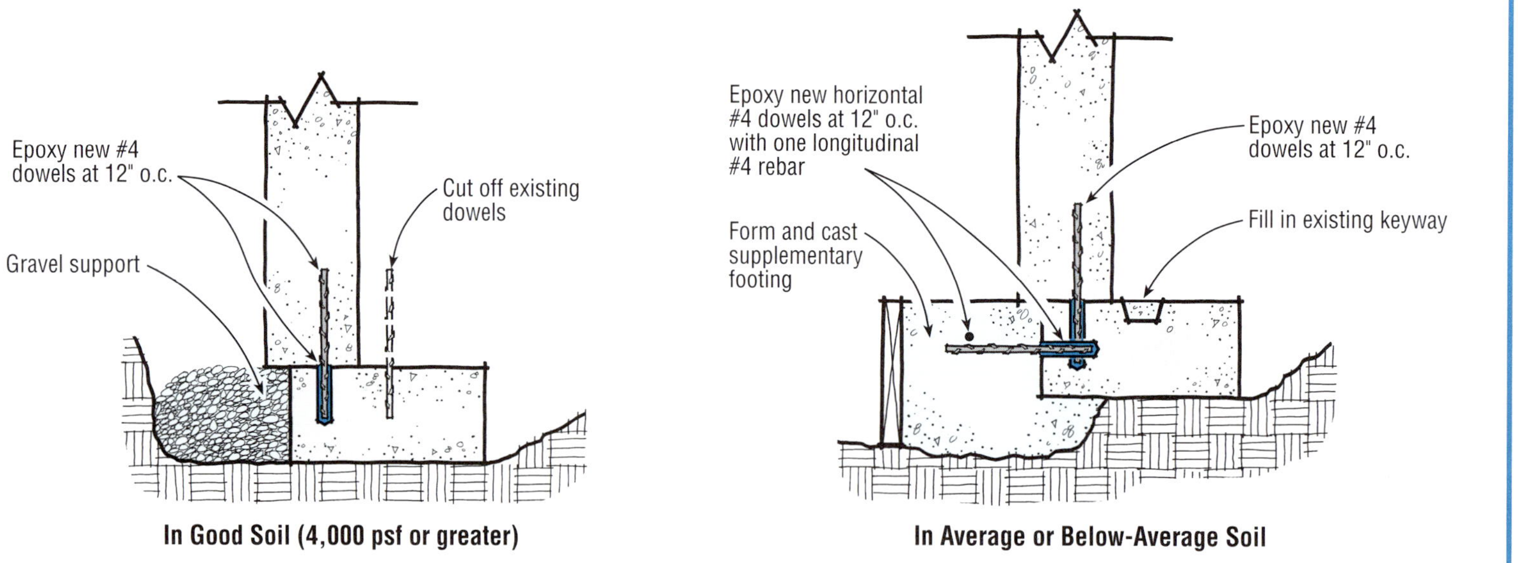

A mistake in footing layout can be corrected in strong soils by placing compacted gravel to support the wall (at left). In weaker soils, an augmented footing will need to be cast alongside the existing footing (at right).

HOW REBAR WORKS

When a sideways load is applied to a concrete wall, *tension* develops on the pressure-bearing side and *compression* develops on the side away from the load. In a free-standing retaining wall, the tension side is on the earth side of the wall (**Figure 1-103, page 92**). In a foundation wall that is held in place by floor framing and a basement slab, the tension side is on the inside (**Figure 1-46**, next page, and **Figure 1-66, page 63**). Concrete resists the compression, while the steel carries the tension and prevents the structure from breaking and toppling.

REBAR SIZES

Rebar comes in a range of diameters, numbered 3 through 11. The numbers denote the diameter of the bar in 1/8-in. increments. Thus, #3 bar is 3/8 in. in diameter, #4 bar is 4/8 (or 1/2) in., #5 bar is 5/8 in., and so on. The size most commonly used in residential construction is #4, though #5 and #6 bar are used often in hillside construction and tall concrete walls in seismic zones. Walkways, pool decks, steps, and simple landings often use #3 bar.

REBAR GRADES

Rebar is graded in primary classifications, commonly known as grade 40 and grade 60. Grade 40 is more malleable and easier to bend. Grade 60 is stiffer and does not bend as easily. Typically, Grade 40 is found in #3 and #4 bar, and Grade 60 in #5 and larger.

SIZE AND SPACING IN WALLS

Figure 1-47 shows proper reinforcement size and spacing for standard 8-in.-thick concrete and masonry walls at various heights. "Standard" construction implies work on well-drained sites with stable soils and the use of granular backfills and perimeter drains.

Repairing Footing Mistakes

Rebar Grades and Spacing

Reinforcement for Seismic Forces

In seismic zones, code requires extra reinforcement. For foundations supporting more than 4 ft. of unbalanced fill, follow the schedules shown in **Figure 1-48** and **Figure 1-49, pages 46-47**. For best results, err on the side of placing more rebar, and always backfill the foundation with a free-draining granular soil to reduce lateral forces (see "Backfill," **page 72**).

Reinforcement for Wet Sites

On sites with high seasonal water tables that will exert increased hydrostatic pressure on foundations, all codes require engineering.

PLACING REBAR IN WALLS

Rebar should always be placed near the *tension side* of the concrete.

In a full-height foundation wall, which is held in place by floor framing at the top and by the footing at the bottom, the tension side is toward the inside (**Figure 1-46**, next page, and **Figure 1-66, page 63**).

In a free-standing retaining wall or a half-height foundation wall, the tension would be on the side nearest the ground load (see "Stepped Foundation Walls," **page 53**).

Horizontal rebar is most effective along the top and bottom of the foundation elevation.

Concrete Coverage Over Rebar

To prevent corrosion, rebar must be covered with concrete:

- In pads and footings: 3-in. minimum coverage
- In walls and slabs, #6 or larger requires 2-in. minimum coverage; #5 or smaller needs 1 1/2-in. minimum coverage

Splicing Rebar

Tie all rebar with wire at splices. Overlap splices by 24 bar diameters (12-in. overlap for #4 bar, 15-in. overlap for #5 bar, 18-in. overlap for #6 bar), or as specified by an engineer.

> **Caution**
> Exposed rebar is dangerous. Cover exposed ends with rubber caps.

Figure 1–46. Rebar for Poured Foundation Walls

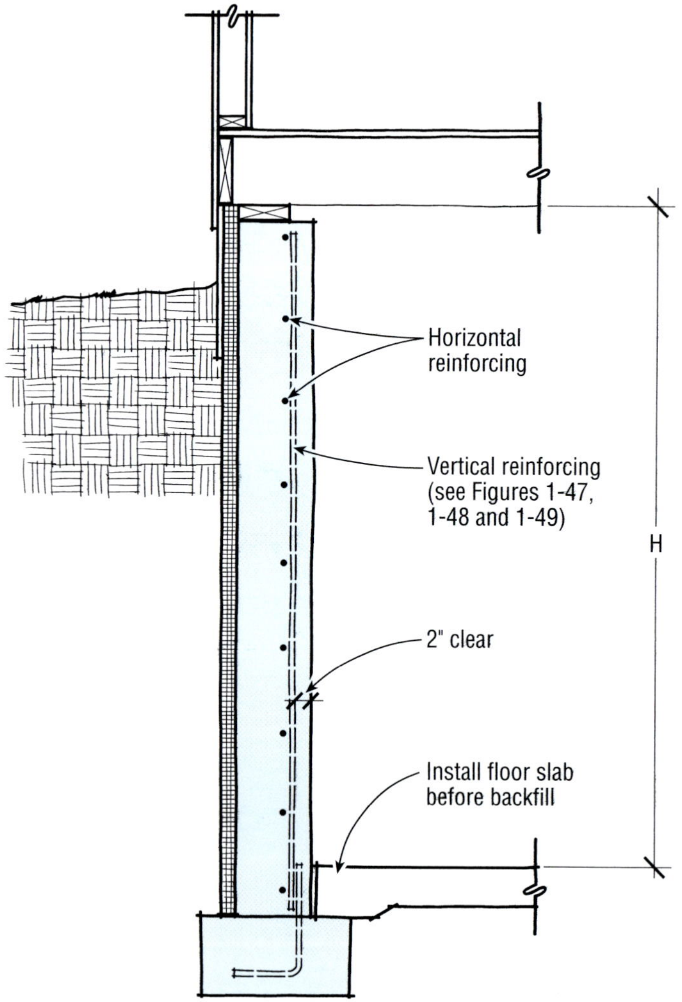

Place rebar towards the tension side of a concrete foundation wall that is pinned in place by floor framing and a basement slab. If the concrete wall is thicker than 10 in., a second layer of reinforcing, placed near the outside face, is recommended.

For more information: "Rebar Tips," JLC 7/94

Figure 1–47. Foundation Wall Reinforcements for Well-Drained Sites (8-in. wall thickness)

Maximum Wall Height (ft.)	Maximum Unbalanced Backfill Height[5] (ft.)	Minimum Vertical Reinforcement Size and Spacing[1], [2] for 8-in. Nominal Wall Thickness[3] — Soil Classes[4]		
		GW, GP, SW and SP soils	GM, GC, SM, SM-SC and ML soils	SC, MH, ML-CL and inorganic CL soils
6	5	#4 at 48″ o.c.	#4 at 48″ o.c.	#4 at 48″ o.c.
	6	#4 at 48″ o.c.	#4 at 40″ o.c.	#5 at 48″ o.c.
7	4	#4 at 48″ o.c.	#4 at 48″ o.c.	#4 at 48″ o.c.
	5	#4 at 48″ o.c.	#4 at 48″ o.c.	#4 at 40″ o.c.
	6	#4 at 48″ o.c.	#5 at 48″ o.c.	#5 at 40″ o.c.
	7	#4 at 40″ o.c.	#5 at 40″ o.c.	#6 at 48″ o.c.
8	5	#4 at 48″ o.c.	#4 at 48″ o.c.	#4 at 40″ o.c.
	6	#4 at 48″ o.c.	#5 at 48″ o.c.	#5 at 40″ o.c.
	7	#5 at 48″ o.c.	#6 at 48″ o.c.	#6 at 40″ o.c.
	8	#5 at 40″ o.c.	#6 at 40″ o.c.	#6 at 24″ o.c.
9	5	#4 at 48″ o.c.	#4 at 48″ o.c.	#5 at 48″ o.c.
	6	#4 at 48″ o.c.	#5 at 48″ o.c.	#6 at 48″ o.c.
	7	#5 at 48″ o.c.	#6 at 48″ o.c.	#6 at 32″ o.c.
	8	#5 at 40″ o.c.	#6 at 32″ o.c.	#6 at 24″ o.c.
	9	#6 at 40″ o.c.	#6 at 24″ o.c.	#6 at 16″ o.c.

(1) Alternative reinforcing bar sizes and spacings having an equivalent cross-sectional area of reinforcement per lineal foot of wall are permitted by code, provided the spacing of the reinforcement does not exceed 72 in.

(2) Use minimum Grade 60 rebar. The distance from the face of the soil side of the wall to the center of vertical reinforcement must be at least 5 in.

(3) For masonry walls, use Type M or S mortar and lay masonry in a running bond. See "Concrete Block Foundations," **page 56**.

(4) Soil classes refer to symbols explained in **Figure 1-5, page 4.**

(5) Unbalanced backfill height is the difference in height of the exterior and interior finish ground levels. Where an interior concrete slab is provided, the unbalanced backfill height is measured from the exterior finish ground level to the top of the interior concrete slab.

Adapted from 2000 International Residential Code

Splicing Rebar

Rebar Requirements in Foundation Walls

*The allowable designs shown here apply only to block and poured concrete foundations that sit above the seasonal water table and are not located in a seismic zone (see **Figures 1-48 and 1-49** for reinforcing schedules to meet these other site conditions).*

Adapted from 2000 International Residential Code

Figure 1–48. Foundation Wall Reinforcement for Seismic Zones (12-in. wall thickness)

Maximum Wall Height (ft.)	Maximum Unbalanced Backfill Height[5] (ft.)	Vertical Reinforcement Size and Spacing[1], [2] for 12-in. Nominal Wall Thickness[3] — Soil Classes[4]: GW, GP, SW and SP soils	GM, GC, SM, SM-SC and ML soils	SC, MH, ML-CL and inorganic CL soils
7	4	#4 at 72″ o.c.	#4 at 72″ o.c.	#4 at 72″ o.c.
	5	#4 at 72″ o.c.	#4 at 72″ o.c.	#4 at 72″ o.c.
	6	#4 at 72″ o.c.	#4 at 64″ o.c.	#4 at 48″ o.c.
	7	#4 at 72″ o.c.	#4 at 48″ o.c.	#5 at 56″ o.c.
8	5	#4 at 72″ o.c.	#4 at 72″ o.c.	#4 at 72″ o.c.
	6	#4 at 72″ o.c.	#4 at 56″ o.c.	#5 at 72″ o.c.
	7	#4 at 64″ o.c.	#5 at 64″ o.c.	#4 at 32″ o.c.
	8	#4 at 48″ o.c.	#4 at 32″ o.c.	#5 at 40″ o.c.
9	5	#4 at 72″ o.c.	#4 at 72″ o.c.	#4 at 72″ o.c.
	6	#4 at 72″ o.c.	#4 at 56″ o.c.	#5 at 64″ o.c.
	7	#4 at 56″ o.c.	#4 at 40″ o.c.	#6 at 64″ o.c.
	8	#4 at 64″ o.c.	#6 at 64″ o.c.	#6 at 48″ o.c.
	9	#5 at 56″ o.c.	#7 at 72″ o.c.	#6 at 40″ o.c.

(1) Alternative reinforcing bar sizes and spacings having an equivalent cross-sectional area of reinforcement per lineal foot of wall are permitted by code, provided the spacing of the reinforcement does not exceed 72 in.

(2) Use minimum Grade 60 rebar. The distance from the face of the soil side of the wall to the center of vertical reinforcement must be at least 8.75 in.

(3) For masonry walls, use Type M or S mortar and lay masonry in a running bond. See "Concrete Block Foundations," **page 56**.

(4) Soil classes refer to symbols explained in **Figure 1-5, page 4.**

(5) Unbalanced backfill height is the difference in height of the exterior and interior finish ground levels. Where an interior concrete slab is provided, the unbalanced backfill height is measured from the exterior finish ground level to the top of the interior concrete slab.

Rebar Requirements in Foundation Walls

Figure 1–49. Foundation Wall Thickness for Seismic Zones (10-in. wall thickness)

Maximum Wall Height (ft.)	Maximum Unbalanced Backfill Height(5) (ft.)	Vertical Reinforcement Size and Spacing(1), (2) for 10-inch Nominal Wall Thickness(3) — Soil classes(4): GW, GP, SW and SP soils	GM, GC, SM, SM-SC and ML soils	SC, MH, ML-CL and inorganic CL soils
7	4	#4 at 56″ o.c.	#4 at 56″ o.c.	#4 at 56″ o.c.
	5	#4 at 56″ o.c.	#4 at 56″ o.c.	#4 at 56″ o.c.
	6	#4 at 56″ o.c.	#4 at 48″ o.c.	#4 at 40″ o.c.
	7	#4 at 56″ o.c.	#5 at 56″ o.c.	#5 at 40″ o.c.
8	5	#4 at 56″ o.c.	#4 at 56″ o.c.	#4 at 48″ o.c.
	6	#4 at 56″ o.c.	#4 at 48″ o.c.	#5 at 56″ o.c.
	7	#4 at 48″ o.c.	#4 at 32″ o.c.	#6 at 56″ o.c.
	8	#5 at 56″ o.c.	#5 at 40″ o.c.	#7 at 56″ o.c.
9	5	#4 at 56″ o.c.	#4 at 56″ o.c.	#4 at 48″ o.c.
	6	#4 at 56″ o.c.	#4 at 40″ o.c.	#4 at 32″ o.c.
	7	#4 at 56″ o.c.	#5 at 48″ o.c.	#6 at 48″ o.c.
	8	#4 at 32″ o.c.	#6 at 48″ o.c.	#4 at 16″ o.c.
	9	#5 at 40″ o.c.	#6 at 40″ o.c.	#7 at 40″ o.c.

(1) Alternative reinforcing bar sizes and spacings having an equivalent cross-sectional area of reinforcement per lineal foot of wall are permitted by code, provided the spacing of the reinforcement does not exceed 72 in.

(2) Use minimum Grade 60 rebar. The distance from the face of the soil side of the wall to the center of vertical reinforcement must be at least 6.75 in.

(3) For masonry walls, use Type M or S mortar and lay masonry in a running bond. See "Concrete Block Foundations," **page 56**.

(4) Soil classes refer to symbols explained in **Figure 1-5, page 4.**

(5) Unbalanced backfill height is the difference in height of the exterior and interior finish ground levels. Where an interior concrete slab is provided, the unbalanced backfill height is measured from the exterior finish ground level to the top of the interior concrete slab.

Adapted from 2000 International Residential Code

CODE REQUIREMENTS

By code, all exterior sill plates must be anchored with minimum 1/2-in.-diameter anchor bolts.

Spacing

When casting foundations, place anchors at a maximum of 6 ft. o.c., and within 12 in. from the ends of each plate section (see **Figure 2-28, page 116**).

Embedment Depth

Foundation anchors must extend a minimum of 7 in. into concrete.

Figure 1–50. Anchor Bolt Layout Jig

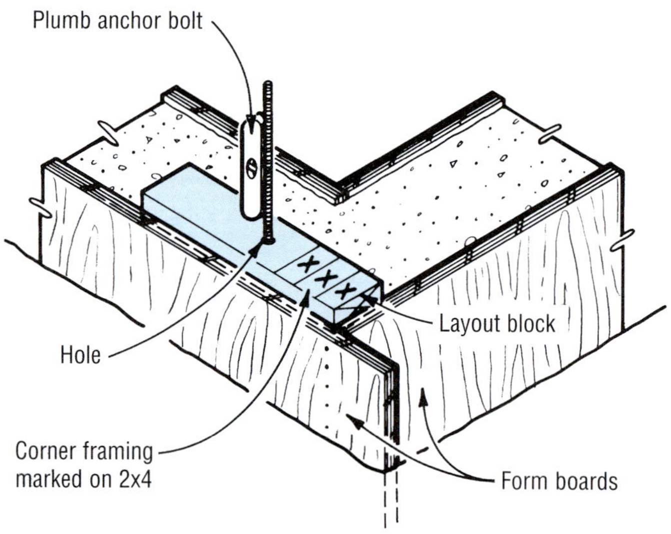

To ensure the exact position of an anchor bolt at corners or king posts (around openings), use a layout block attached to the form boards.

For more information: "Installing Seismic Framing Connectors," JLC 7/93

"Tying Down the House," JLC 9/97

PLACING ANCHOR BOLTS

Where anchor posts attach to a corner stud or at one side of a wall opening, they must be located precisely. A layout jig can be very helpful (**Figure 1-50**). If possible, wire the anchor posts directly to the rebar.

SEISMIC ANCHORS

In seismic zones, all shear walls must be mechanically fastened to the foundation with metal hold-downs, metal straps, closed-space anchor bolts, or some combination of these (**Figure 1-51**). Hold-downs require anchor bolts that extend up though the floor framing (**Figure 1-52**).

Code Requirements in Seismic Zones

In addition to following the anchor spacing requirements described above, code specifies:

- A minimum anchor spacing of 4 ft. o.c. for two-story buildings
- Anchors located within 12 in. of the ends of each plate section for all interior bearing walls and interior braced wall lines, as well as all exterior walls
- Plate washers measuring at least 2x2x3/16 in.

Refer to "Seismic and Wind Bracing," **pages 165-170** in Chapter 2, for more detail on securing seismic anchors to framing.

RETROFIT FOUNDATION ANCHORS

A variety of foundation anchors can be used to retrofit a missing anchor bolt or fasten framing into foundation materials (**Figure 1-53**). While many of these anchors are more expensive than shields or drop-in anchors, they don't require laborious spotting, which is difficult and time-consuming.

Selecting Retrofit Anchors

Follow the guidelines shown in **Figure 1-53** when retrofitting concrete anchors.

Tips for Placing Concrete Anchors

Usually the most important detail in installing any type of mechanical or epoxy anchor is drilling the hole according to the manufacturer's specs.

Do not drill too close to a concrete or masonry edge, or the anchor is likely to break out the substrate when inserted, and holding power may be compromised.

If the diameter of the hole is too large, concrete screws won't tap and hold properly, while sleeve and wedge anchors may spin in the hole.

Concrete screws will require a hole 1/4 in. deeper than the screw to leave room for displaced material, and drop-in anchors may fail if not placed in holes drilled to a required depth.

Placing Anchor Bolts

Seismic Anchors

Retrofit Foundation Anchors

Figure 1–51. Seismic Anchor Types

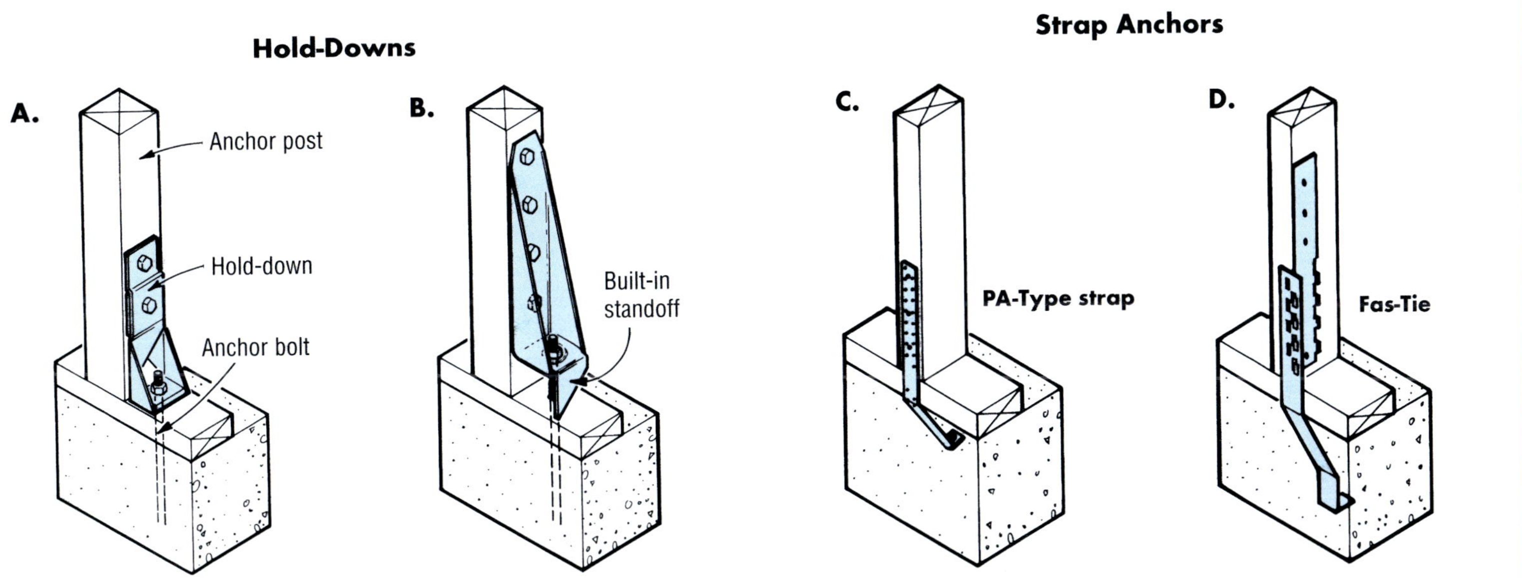

Hold-downs *either sit directly on the sill or bottom plate (A), or have a built-in standoff that lifts them off the plate (B) so the bolt-holes won't split out the end of the post.* ***Strap anchors*** *nail either directly to the stud (C), or through the sheathing into the stud. A two-part anchor (D) has two parts that lock together — one piece embedded in the concrete and a second piece nailed to the wide face of the stud.*

Overtightening may compromise the holding power of sleeve and wedge anchors. Take care to torque them down to manufacturer's specifications. Some anchors are also sensitive to the depth of the hole.

Retrofit Anchors in Unstable Substrates

For best results, place anchors in stable concrete. If you are uncertain if the concrete or masonry is stable, choose an epoxy anchor, which will help bind the base material together. Check with the manufacturer or catalog for the appropriate embedment depth and design load of each fastener — especially for any structural connections.

Figure 1–52. Hold-Down Anchor Bolts

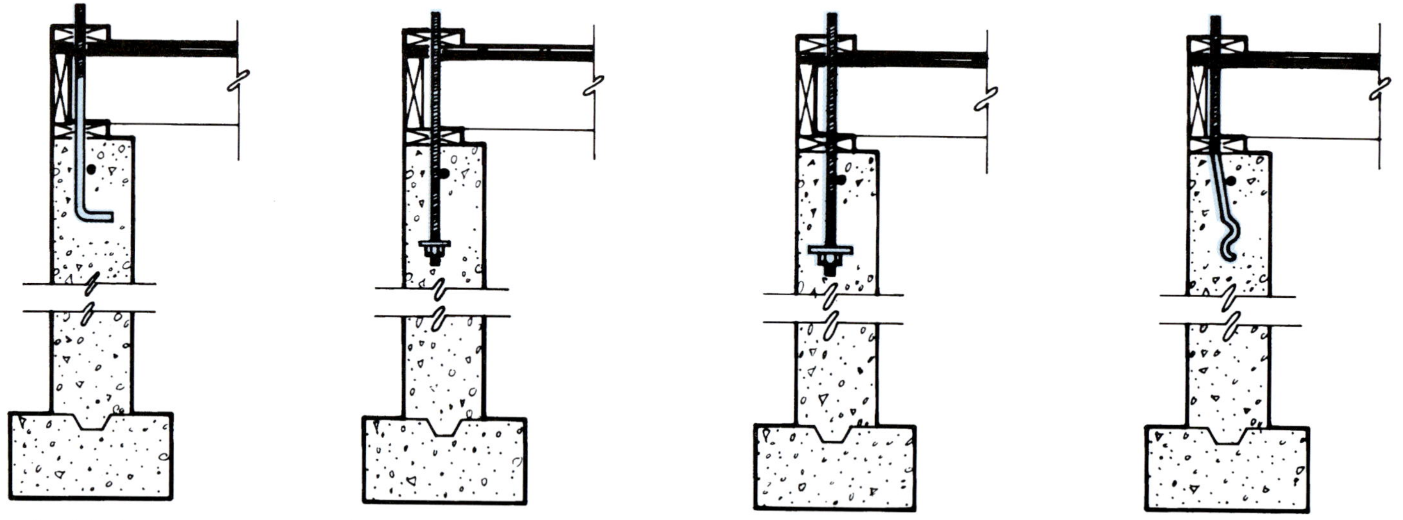

There are four types of anchor bolts for use with foundation hold-downs: (A) A threaded rod with a 90-degree bend in the end, like a large J-bolt; (B) a threaded rod or long bolt with a nut and washer on the end; (C) a threaded rod with a 1/4-in.-thick steel plate bolted to the end; and (D) a manufactured anchor bolt, such as Simpson Strong-Tie's STAB anchor.

For more information: "Seismic Support for Old Foundations," JLC 4/97

Selecting and Placing Concrete Anchors

Sill plate for bearing wall. If you've left out one or two anchor bolts, a few powder-actuated fasteners (PAFs) should do the job. If the slab is uneven and you need to draw the plate down, use a sleeve anchor. If you've left out all the anchor bolts for a bearing wall, use sleeve, wedge, or epoxy anchors.

Sole plate to slab for nonloadbearing partition. A PAF is the fastest way to connect interior partition plates (wood or steel) to a slab. Hammer anchors are a good alternative.

Steel column to a slab. To secure the bearing plate of a structural post to a concrete slab, use a wedge or epoxy anchor. If the concrete is at all soft, an epoxy anchor is more reliable.

Ledger for exterior deck or porch. To attach a framing ledger to a poured concrete or masonry wall, use a wedge anchor for strength. You may want to use a PAF to hold the ledger level while you're drilling for the wedge anchors.

Stair stringer in concrete bulkhead. A PAF is sufficient if you can fasten every 16 in. or so along the stringer. If you have to hang the stairs from one end of the stairwell, you're better off depending on a sleeve or wedge anchor.

Sleepers for wood flooring over a slab. A PAF will work well, unless the slab is way out of level (if it slopes towards a drain, for example). In this case, with shims beneath the sleepers, you may exceed the length of a PAF and need to use a wedge or sleeve anchor in the deepest areas.

Wood frame around basement openings. To box out a door or window with pressure-treated wood, PAFs are the quickest option; concrete screws or hammer anchors will also work.

Strapping to foundation wall. When fastening nailers for drywall or siding to a poured concrete or masonry wall, PAFs are the fastest method. Drive the pin into the horizontal mortar joint of a masonry wall. Concrete screws or hammer anchors are an alternative.

Figure 1-53. Typical Retrofit Concrete Anchors

Anchor Type	Size (in.)	Minimum Embedment Depth (in.)	Pullout Strength (lbs.) (3,000-psi concrete except as noted)
Hammer (sleeve type)	1/4 x 2	3/4	800
Hammer (nail type)	1/4 x 2 1/4	1 1/4	1,100
PAF	.140 x 2 1/2	1 1/4	1,400
Concrete screw	1/4 x 2 3/4	1 1/2	1,500
Sleeve	3/8 x 3	1 5/8	2,400
Wedge	3/8 x 4	3	4,000-5,000 (4,000-psi concrete)
Epoxy (dispenser type)	3/8 x 4	3 1/2	5,000-7,000 (4,000-psi concrete)
Epoxy (capsule type)	3/8 x 4	3 1/2	5,000-7,000 (4,000-psi concrete)

The sizes shown are typical for the type of fastener, although other sizes are available. Valves are typical only for the size listed and should be checked for the specific fastener and substrate. Values listed are "ultimate" pullout strength. In actual design, a safety factor of 4:1 to 8:1 is typically used.

Selecting Concrete Anchors

THE CONTROLLING REQUIREMENTS FOR FOUNDATION DESIGN are resistance to soil loads and protection from frost. Concrete walls easily support downward compressive loads, but lateral loads applied by soil pressure, and the pressures of expanding ice in the soil, can crack concrete walls. Foundation walls must be designed to resist these lateral loads.

CONCRETE WALL DIMENSIONS

By code, foundation walls must always extend above the finished grade at least 6 in. However, 18 in. is recommended to prevent splashback from deteriorating wood siding (see "Board Siding," **page 318**). The height above grade may be lowered to 4 in. if a masonry cladding, such as brick veneer, is used.

Maximum heights of foundation walls should be based on the required wall thickness and reinforcement required, as specified below.

Concrete Wall Thickness

The design thickness of foundation walls depends on soil conditions, the amount of reinforcement in the wall, the height of the wall, and the height of unbalanced fill. *Unbalanced fill* refers to the difference in ground-level height between the inside and outside of the foundation wall.

For more information: "Designing Concrete Basement Walls," JLC 4/02

While code will allow *plain,* or unreinforced, concrete and masonry block foundation walls (refer to Figure R404.1.1 (1), p. 67 of the 2000 International Residential Code), this is not recommended. Code sets minimums for safety, not necessarily for quality. To avoid the callback nightmare of a cracked foundation, all foundations should include reinforcing steel. An exception can be made for a plain concrete or masonry foundation wall less than 5 ft. tall with less than 4 ft. of unbalanced fill. In this case, use minimum 4,000-psi concrete (see "Specifying Ready-Mix," **page 26**).

Typically, a foundation wall should be thicker than the wall it supports. An 8-in.-thick wall is standard for supporting wood-framed walls, but only if the unbalanced fill height is less than 6 ft., the foundation is located on a well-drained site, and the foundation is properly reinforced.

Where poor soils exist, in seismic zones, and on hillside sites, a thicker 10- or 12-in. poured concrete wall may be required. In each of these cases, an engineer's review is recommended and may be required by local code.

REINFORCING CONCRETE WALLS

Follow general guidelines for reinforcing foundations (see "Rebar," **pages 43-47**).

For appropriate sizing and spacing of rebar in concrete foundations, follow the minimum code requirements shown in **Figures 1-47, 1-48,** and **1-49, pages 45-47**.

Rebar should always be placed near the tension side of the concrete (see "Placing Rebar in Walls," **page 43**). On a full-height foundation wall, which is held in place by floor framing at the top and by the footing at the bottom, the tension side is toward the inside (**Figure 1-46, page 44**).

Horizontal rebar is most effective along the top and bottom of the foundation elevation.

STEPPED FOUNDATION WALLS

Foundation walls on sloped sites are often built as a half-wall with a wood-frame cripple wall (**Figure 1-54**). In a half-height foundation wall, the hinge joint between the concrete and the cripple, or pony, wall creates a structural weakness. The half-height portion of the wall must be

Designing Foundation Walls

Figure 1–54. Full-Height vs. Half-Height Foundation Walls

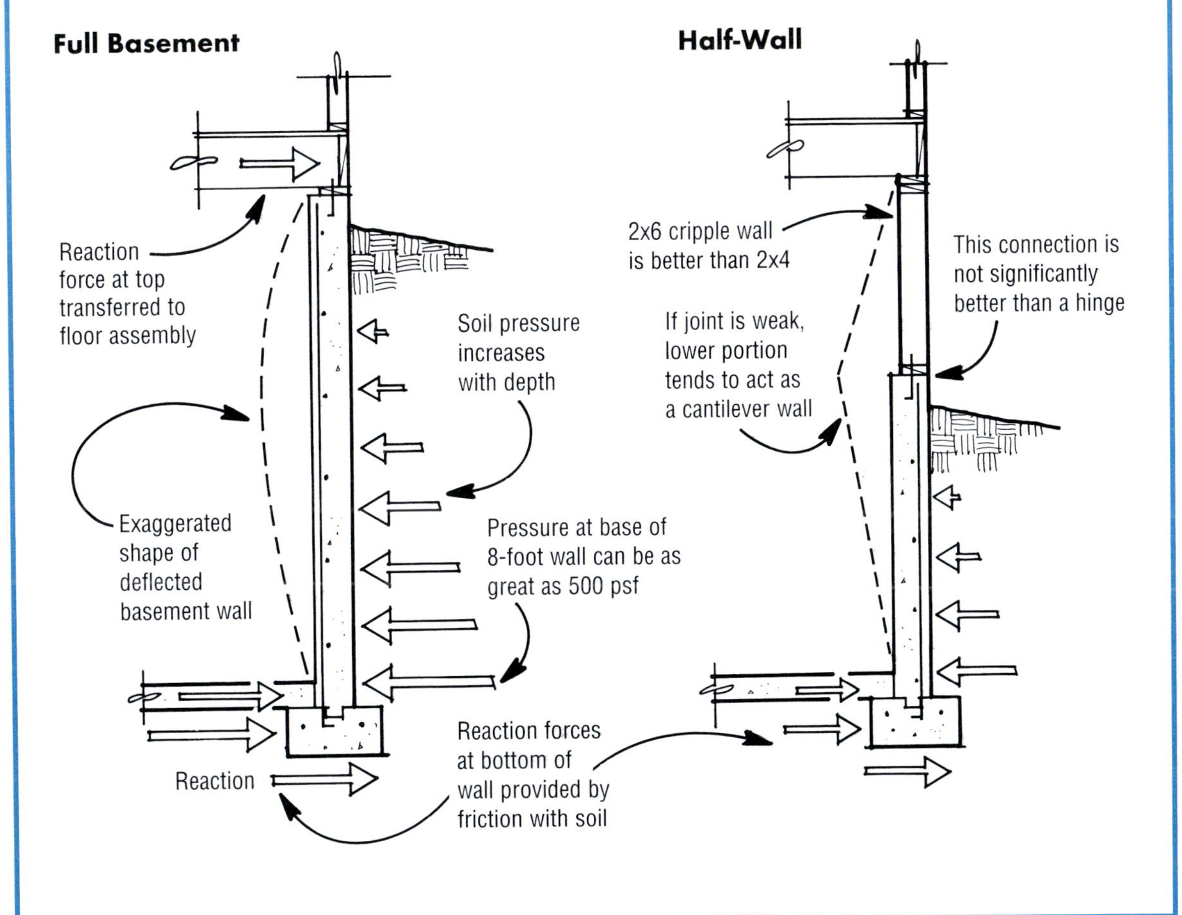

*A full-height foundation wall is braced at the top by floor framing and at the bottom by the concrete footing and slab, so it is likely to bow inward. Reinforcement placed towards the inside face will resist cracking caused by this bowing (**far left**). A concrete/cripple wall combination, however, is braced only at the bottom; the connection between the concrete and the wood wall is little more than a hinge (**near left**). To resist toppling inward, the concrete wall must be detailed like a free-standing retaining wall, with rebar placed towards the outside face.*

treated as a retaining wall, and reinforced, buttressed, or otherwise supported to resist the lateral loading of unbalanced soil (see "Retaining Walls," **page 91**).

Stepped Foundation Corners

When any foundation wall turns a corner, maintain the height for a distance equal to half the height of the higher wall before stepping it down (**Figure 1-55**).

STRUCTURAL BRACING FOR FOUNDATION WALLS

Always frame the floor deck before placing backfill. If this is not possible, install temporary bracing before placing backfill (see "Bracing Before Backfill," **page 72**).

Where soils impose an excessive lateral load, additional intermediate structural support for foundation walls can be provided by the cast foundation elements (**Figure 1-56**).

Figure 1–55. Stepped Foundation Corners

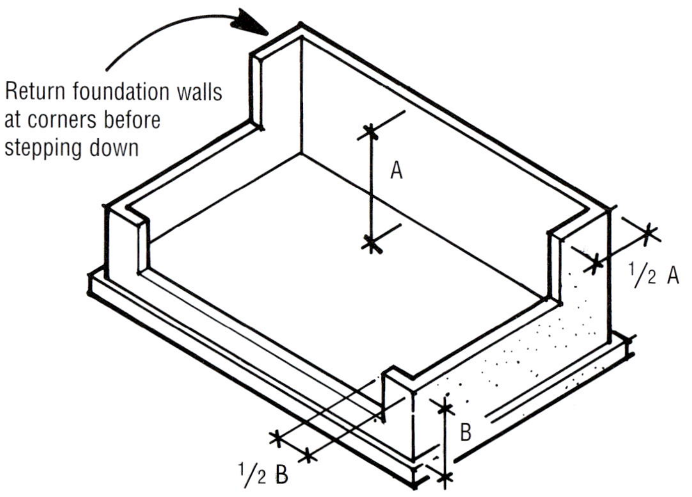

Regardless of height, every concrete wall should be braced at each corner. Maintain the higher wall height for a distance equal to half that wall height.

Figure 1–56. Cast-In Bracing for Foundation Walls

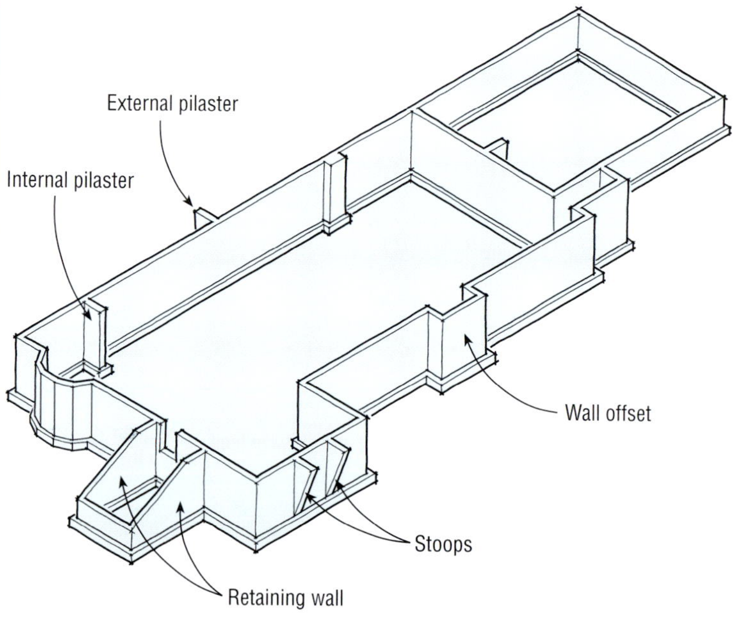

Any of the pop-outs or cast elements of a foundation wall will add lateral support, and they can be engineered to provide critical bracing of foundations in marginal soils.

CONTROL JOINTS

Some surface cracking is inevitable in the face of concrete walls. Rebar and wire mesh are placed to keep cracks from widening but will not prevent them from occurring. If minor, random cracking is unacceptable, use control joints to confine cracks to intended locations.

Place control joints no more than 20 ft. apart, and within 10 ft. from corners. Locate control joints at natural points of weakness, such as door and window openings, corners, and changes in elevation or section (**Figure 1-57**). In a stepped foundation, control joints should also be placed along a vertical in line with the point where the foundation changes elevations.

Stepped Walls

Control Joints for Walls and Slabs

Figure 1-57. Control Joints for Concrete Walls and Slabs

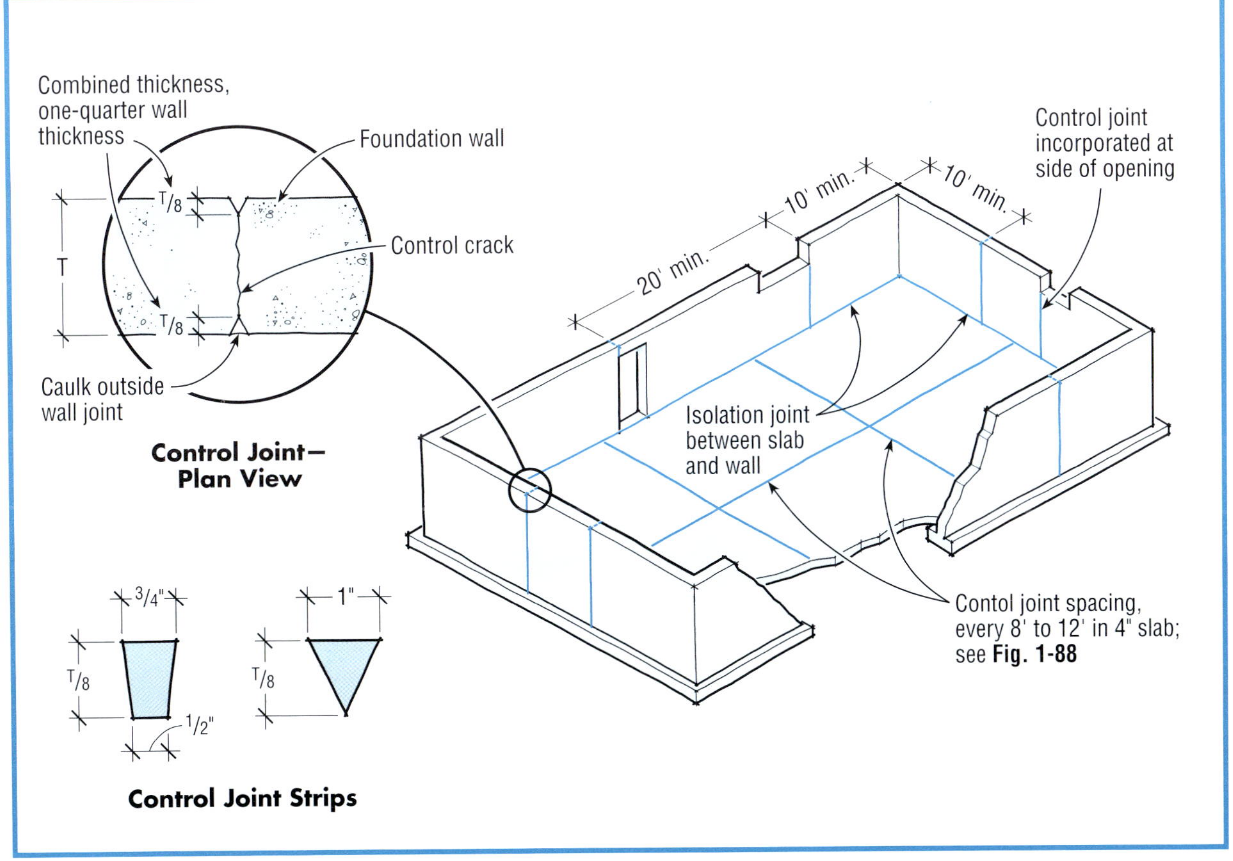

Control joints are formed into the concrete using triangular or keystone-shaped strips attached to the inside faces of the form (both sides). To be effective, the combined thickness of the strips should equal one-quarter the wall width. So, for an 8-in.-thick foundation wall, use 1-in.-deep strips. Place strips at minimum distances shown, at door and window openings.

FOUNDATIONS BUILT WITH CONCRETE BLOCK MAY COST LESS than poured concrete, depending on local material and labor costs. Block may also be more practical than poured concrete for small jobs, crawlspace foundations, or jobs where access by concrete trucks is limited.

Block foundation performance depends on proper design and detailing. Effective perimeter drainage is a necessity to reduce lateral soil pressures, especially for full basements (see "Perimeter Foundation Drains," **page 68**). Block is also very porous, so waterproofing or dampproofing is usually required to control moisture intrusion (see "Waterproofing and Dampproofing," **page 66**).

Figure 1-58. Commonly Used Concrete Masonry Units

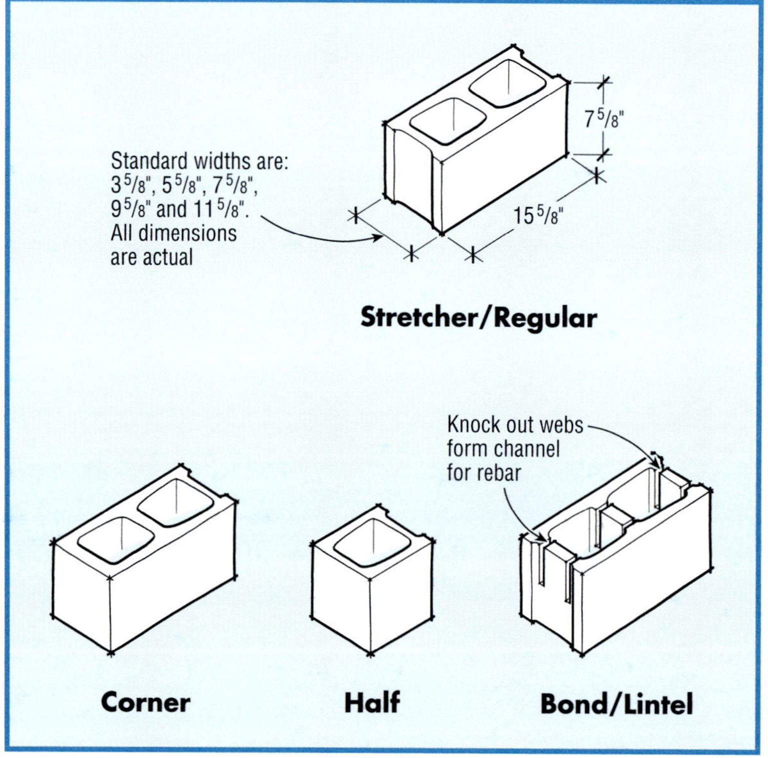

For more information: "Concrete Block Foundation," JLC 10/93

BLOCK TYPES

Most block basements are built with two-core or three-core hollow, loadbearing concrete masonry units (CMUs). Solid blocks are used to carry point loads and to cap foundations. **Figure 1-58** shows some of the common types of block likely to be needed for building a block basement. Other specially shaped blocks are available for use in bond beams, joist and girder pockets, window and door openings, pilasters, and piers.

Block Grades

When selecting block, choose blocks graded according to ASTM standards. Use only ASTM Grade N for foundations, not Grade S. Grade N has greater bearing strength and better resistance to frost action and moisture.

Block Types

Masonry units of ASTM Type I (moisture-controlled), rather than Type II, are recommended; Type I units are less susceptible to drying shrinkage and cracking.

Block Dimensions

Concrete block generally conforms to a modular system based on 4 in. or 8 in. Depth, height, and length of blocks are stated in full inches, but the actual dimensions are reduced by 3/8 in. or 5/8 in. to allow for the mortar joint. That way, walls can be designed in full lengths based on feet and inches (see "Estimating Block," **page 2**), and block laid with mortar joints will fall out correctly when installed.

In describing block, the wall or block width is always named first, then the course or block height, and finally the length of the block. For example, an 8x8x16-in. unit is actually 7⅝ in. thick, 7⅝ in. high, and 15⅝ in. long. Most block is available in half sizes, but odd lengths or heights may have to be cut on site.

MORTAR

Mortar serves to bond the units together, accommodate minor size variations between units, and create a seal against air and moisture. Mortar may also bond with reinforcing wire placed between courses.

Grout is different from mortar. Grout is used to fill block cores for reinforcement (see "Grouting Block Foundations," **page 60**).

Setting Block in Hot Weather

Hot, dry, windy conditions speed up the setting of mortar and also cause it to dry out from evaporation. Maintaining cool and moist conditions helps reduce these problems. For improved results in hot, dry weather, take these steps:

- Adjust mortar mix. Use more cement per batch to speed strength gain so mortar will set before water evaporates out.
- Reduce mortar exposure time. Lay mortar beds no more than 4 ft. ahead, and set block within one minute.
- Keep tools and materials cool. Store block, cement, lime, and sand in the shade; if sand becomes dry, dampen it by sprinkling.
- Use cool water to mix mortar.
- Set up windscreens to deflect drying breezes.
- Cover walls with plastic as soon as possible to hold moisture in, or spray-cure walls.

Mortar Ingredients

Mortar is made of Portland cement, hydrated lime, sand, and water. Portland cement gives mortar its compressive strength. The lime weakens the mix, but improves its workability and helps form a better seal. Sand provides strength and also reduces drying shrinkage. Water provides workability and enhances the bond to the block or brick.

Block Types and Grades

Mortar Specs

Mortar Types

Mortars are graded by type, based on the proportions of the ingredients (see **Figure 1-59**). The five traditional mortar types are known as Type M, S, N, O, and K, in descending order of compressive strength (using every other letter from the phrase "Mason Work"). Types with higher

Figure 1–59. Standard Mortar Mixes

Type	Cement	Lime	Sand	Compressive Strength (psi)
M	1	¼	3	2,500
S	1	½	4	1,800
N	1	1	6	750
O	1	2	9	350
K	1	3	12	75

Use these ratios of ingredients for mixing different mortar types. Typically, one bag of materials (cement and lime) is equivalent to 1 cu. ft. of loose material (sand). The ratios of ingredients alter the compressive strength of the mortars, but also affect workability and bond strengths. As a rule of thumb, tensile strength is a little less than one-tenth the compressive strength.

proportions of cement and lower proportions of lime have greater compressive strengths but are less workable.

For block joints: Many skilled masons prefer Type N (a.k.a. "6-1-1") for bonding block because it has better workability, allowing for a better-made joint and a better bond. The higher lime content also makes this mortar more "extensible," or flexible, in service.

For parging: Use Type M or S mortar for parging block prior to dampproofing (see "Parging Concrete Block," **page 66**).

Mortar Strength

Type M or Type S mortar is often specified for basement foundations because of its higher compressive strengths (**Figure 4-86, page 364**). However, the compressive strength of the mortar is less important than the quality of the joint and strength of the bond. (Cutting the mortar's compressive strength in half lowers the compressive strength of the total wall assembly by only about 10%.) If lateral soil loads are a concern, no mortar will provide sufficient strength. Instead, the block cores should be filled with grout and rebar (see "Reinforcing Block Foundations," **page 63**).

Mortar Water Content

Unlike concrete, which should be made with as little water as practical, mortar performs best when made with as much water as possible for a workable mix. Wetter mortar makes a stronger bond with the block, increasing the performance of the total wall system.

Site-Mixed Mortar

Mortar is usually mixed on site with water, Portland cement, lime, and sand. Site-mixing allows skilled masons to fine-tune the quality of the mortar, easing the labor and improving the quality of the job.

Pre-Bagged Mortar

Pre-bagged masonry cement mixes are available, with the cement and lime pre-proportioned so that only sand and water need to be added. If unskilled help is mixing the mortar, pre-bagged cements make it easier to get a uniform mix from batch to batch.

Setting Block in Cold Weather

Cool temperatures (below 40°F) slow the strength gain of mortar, and freezing can ruin the compressive and bond strength of mortar.

Below 40°F, take these steps:

- Heat mixing water. Maintain mortar temperature between 40°F and 120°F until placed.
- Cover work and materials. Use tarps to protect mortar, block, sand, and masonry work from wetting and freezing.

Below 32°F, add these precautions:

- Heat sand and block. Pile sand over a section of steel culvert pipe, and thaw frozen sand by directing the exhaust from a salamander heater into the pipe. Also thaw frozen block.
- Set up windbreaks or an enclosure.
- Protect work and materials. Maintain masonry and materials above 32°F with auxiliary heat or insulated blankets for 24 hours after laying block.

Below 20°F, take the following additional steps:

- Heat masonry units to 20°F or warmer.
- Provide heated enclosure for work. Keep masonry enclosure above 32°F for 24 hours after laying block.

Bagged Material Weights

Bags of Portland cement, masonry cement, and lime all contain 1 cu. ft. of material per bag, although they weigh different amounts. Portland cement weighs almost 100 lbs./cu. ft.; lime weighs about 50 lbs./cu. ft.; and masonry cements — which are mixtures of Portland and lime — weigh somewhere in between, depending on the type.

FOOTINGS FOR BLOCK FOUNDATIONS

Footings for block walls do not differ much from cast footings for poured concrete walls (see "Footings," **pages 38-42**).

If block and other materials will be staged inside the building perimeter before footings are poured, take care that stored materials do not prevent you from checking cross-diagonals and measuring foundation widths.

Figure 1–60. Running Bond

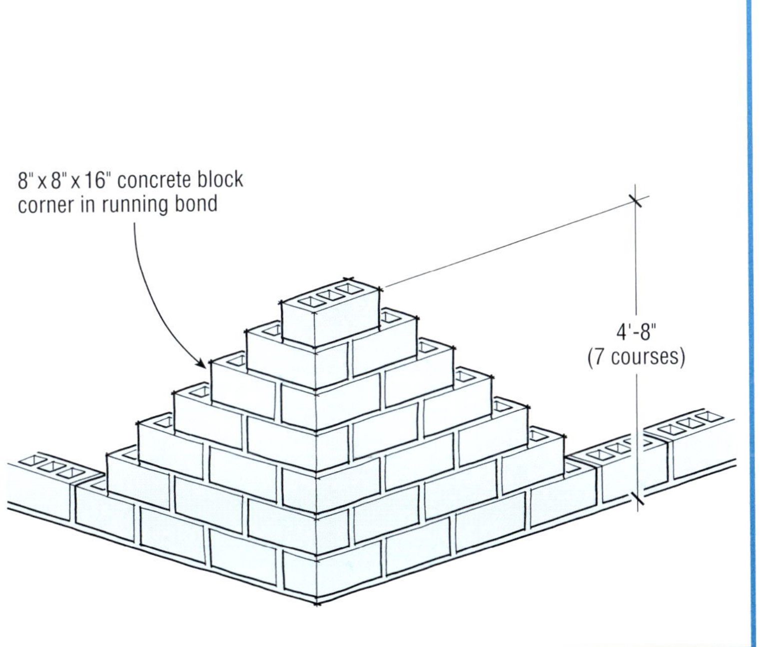

A running bond starts from the corner with each course offset by half a block from the one below it. By building up the corner first, a mason can establish the course heights and the plumb of the wall.

Figure 1–61. Cut Block Placement

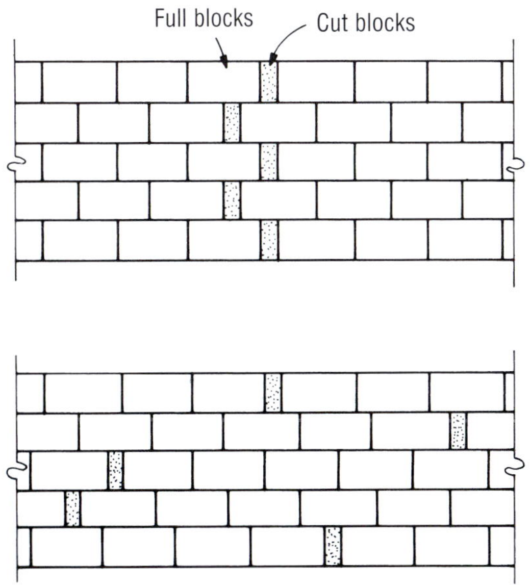

*If the length of a wall doesn't divide evenly into full blocks, align cut blocks in succeeding courses (**at top**). This keeps the cores lined up and looks much better than random placement (**at bottom**).*

SETTING BLOCK

As with concrete, mortar sets up best in humid climates with temperatures averaging above 40°F and below 70°F. In climate conditions above and below this, follow similar procedures described in "Hot-Weather Concrete," (**page 36**) and "Cold-Weather Concrete," (**page 35**).

GROUTING BLOCK FOUNDATIONS

Grout is used to fill cores and bond beams for strength, along with appropriately sized and placed steel rebar. Grout is not the same as mortar. It is really a type of concrete, made with Portland cement, water, sand, and small crushed rock or pea gravel, but little or no lime. Grout typically reaches a compressive strength of 2,500 to 3,000 psi.

Figure 1-62. Corner Layout for Block

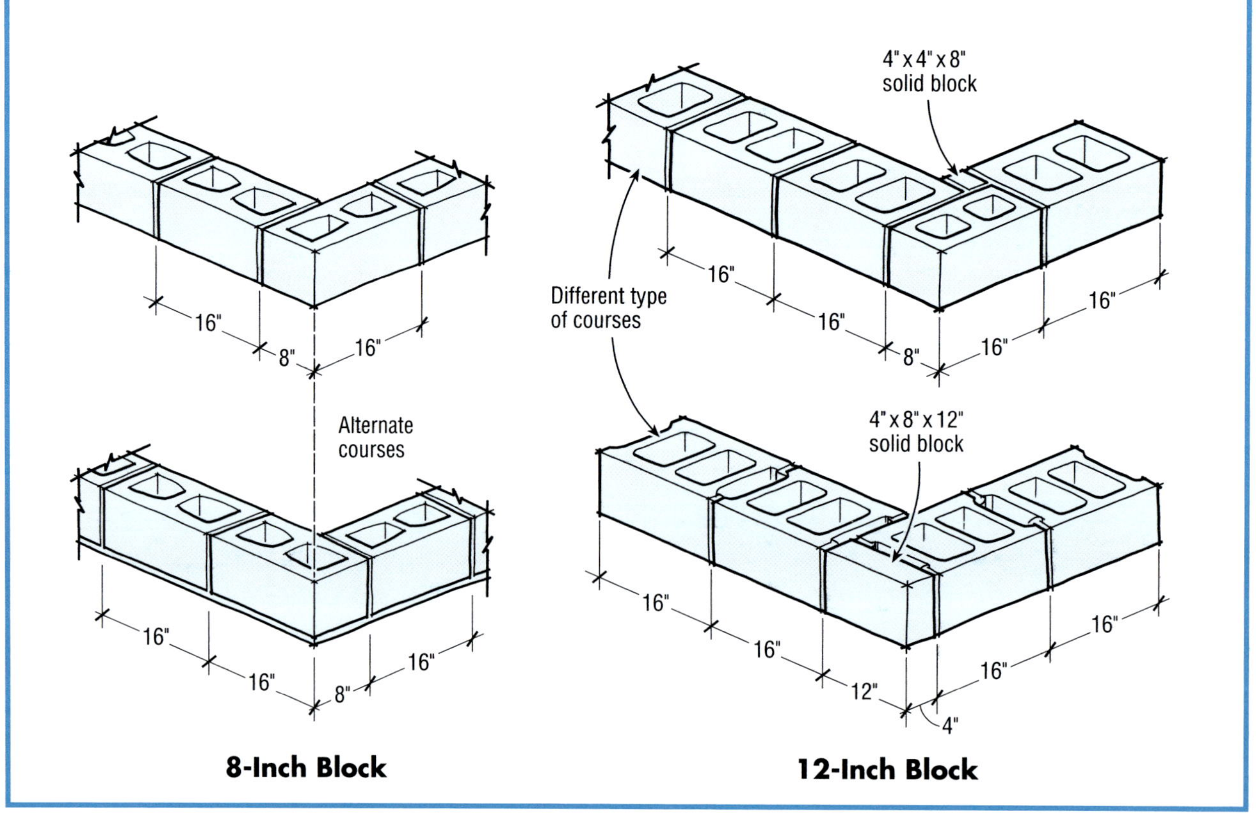

Maintain a running bond with standard block by weaving corners in alternate courses.

Ordering Ready-Mix Grout

Grout is best ordered from a ready-mix plant. A grout mix delivered by truck typically will be about an 8-bag concrete mix with maximum 1/2-in. gravel.

Site Mixed Grout

If mixing small batches of grout on site, make sure to mix the batch for at least five minutes. Place the grout soon after mixing. Discard any batch of grout that is not placed within 90 minutes.

BLOCK CONSTRUCTION DETAILS

If you have control over design decisions, lay out the distances between corners and bump-outs, as well as the dimensions of door and window openings, in increments of 8- or 16-in. to match full- and half-block lengths.

Running Bond

Concrete blocks are designed to be laid in a running bond (sometimes called half-bond) pattern in which the vertical joint of two adjoining blocks falls over the center of the block below (**Figure 1-60, page 59**).

Cut Block Placement

If blocks must be cut, maintain a running bond by staggering cut blocks symmetrically (**Figure 1-61, page 59**). Joints can be as close as 4 in. from a joint in the courses above and below and not affect the strength of the wall; but the appearance will suffer if the joints are off-center. Where

Grouting Block

Corner Layout

Figure 1–63. Corner Layout for 10-in. Block

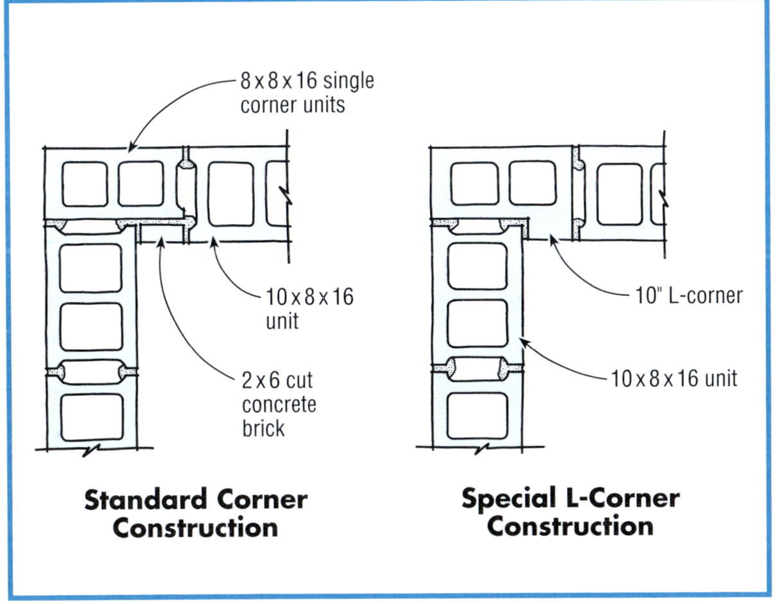

Figure 1–64. Corner Layout for 12-in. Block

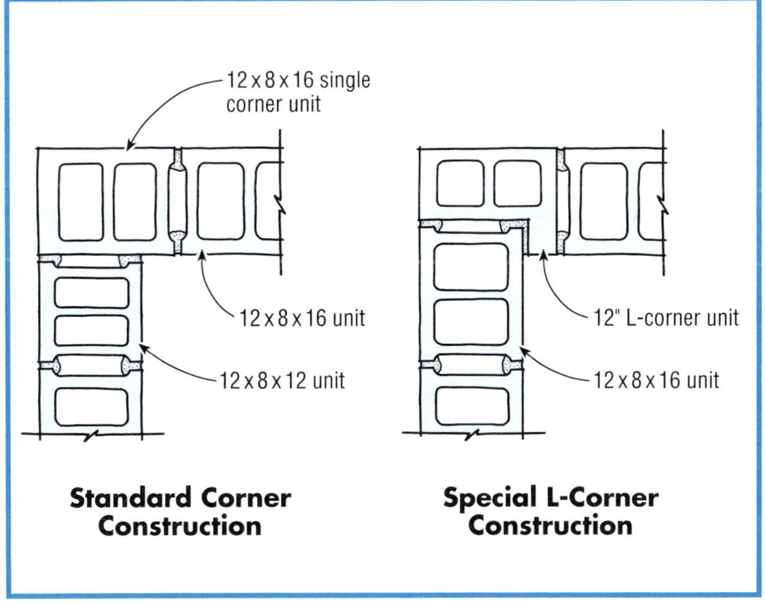

portions of the wall will have cores filled with concrete, keep the half-bond dead accurate.

Building Block Corners

Corners are a key detail in a block basement, and must be built to maintain a modular layout. With 8-in. block, weaving block at corners establishes the 8-in. running bond and keeps the cores in two-core block in line (**Figure 1-62, page 60**). For 10- and 12-in. block, a

Figure 1–65. Intersecting Block Walls

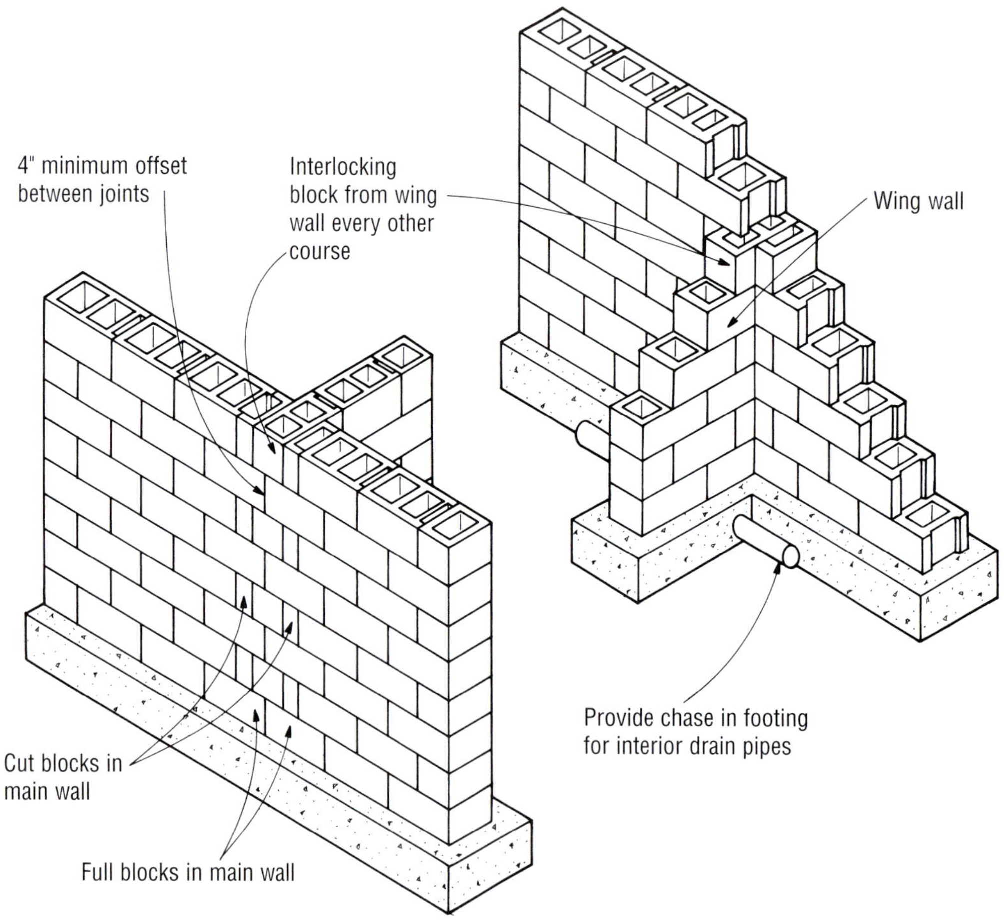

Where block walls join, the blocks must intersect on alternating courses. Fill the cores with grout and rebar running through the intersecting blocks to pin the walls together.

number of techniques can be used to lay out corners (**Figures 1-63** and **1-64, page 61**); the simplest method uses special corner block.

Concrete Block Intersections

Wall intersections add strength to help resist soil pressure. Buttresses or "wing walls" can be built to stiffen main walls even if no partition is called for (**Figure 1-65**).

Weave alternate courses together just as in building a corner, and grout all cores and reinforce with steel.

REINFORCING BLOCK FOUNDATIONS

Follow general guidelines for reinforcing foundations (see "Rebar," **pages 43-47**).

Rebar placed in grouted cores adds significant stiffness and strength to masonry basement walls. Place rebar towards the inside face of full-height foundation walls, and tie it securely into the footing (**Figure 1-66**).

For appropriate sizing and spacing of rebar in masonry foundations, follow the minimum code requirements shown in **Figures 1-47, 1-48,** and **1-49** (**pages 45-47**).

Reinforcing Openings in Block Foundations

To prevent cracking at stress points near openings, place reinforcing steel, as shown in **Figure 1-67**.

Anchoring Floor Framing to Block Foundations

In most block foundations, the floor framing acts as a critical brace to stabilize the top of the block wall. First-floor decks must be securely tied to the block:

- Anchor bolts should be embedded at least 12 in. into grouted cores. Ideally, the core with the anchor bolt should be fully grouted down to the footing, contain

Figure 1–66. Reinforcing Block Walls

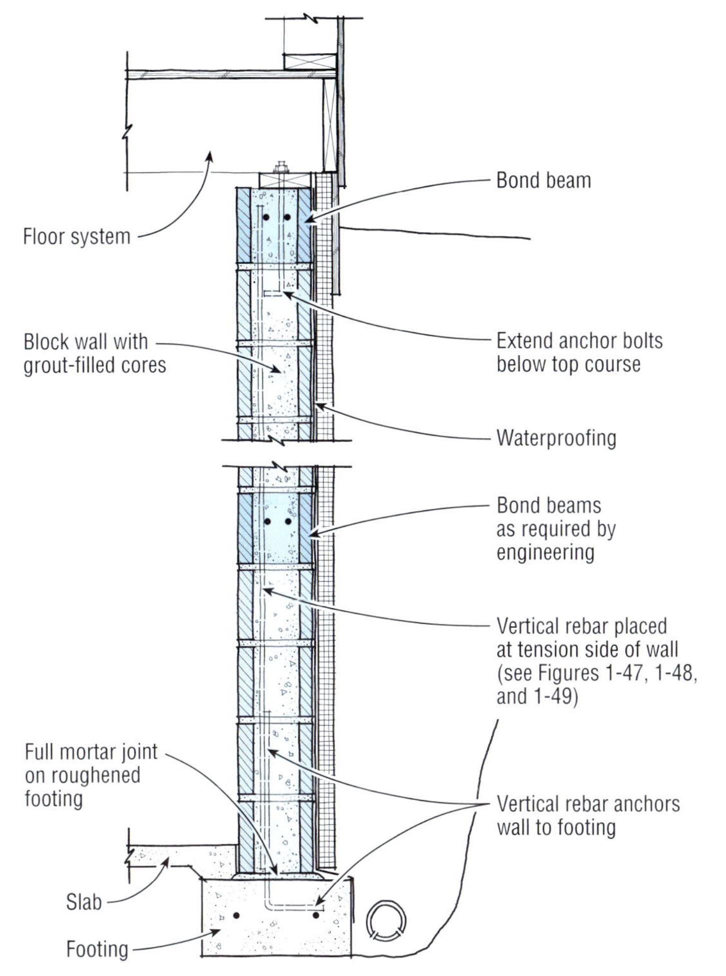

Reinforcing for block walls consists primarily of vertical rebar embedded in columns of grout-filled cores to the full-height of the foundation wall. Rebar must tie to the footing, and anchor bolts must extend below the top course of block.

Block Intersections

Reinforcing Block Foundations

Figure 1–67. Reinforced Opening in a Block Foundation

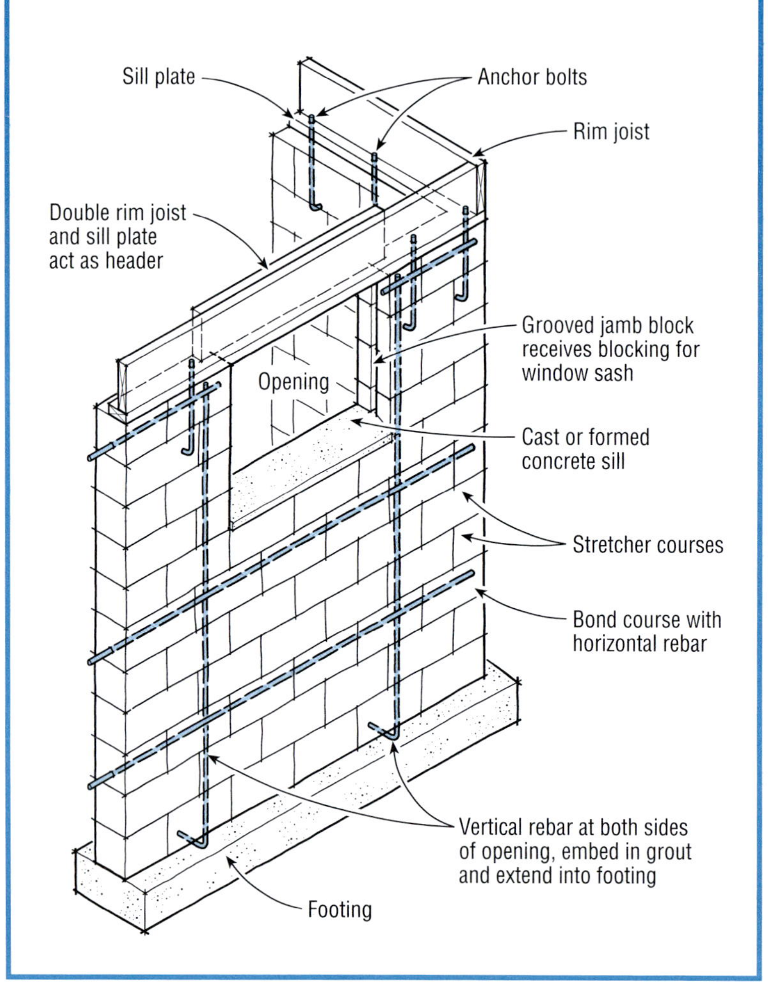

*In addition to the vertical rebar specified by code (**Figures 1-47, 1-48, and 1-49, pages 45-47**), additional rebar embedded in grout should surround window openings. In an opening at the top of a foundation wall, the sill plate and floor framing act as a header and tie together the top sections of the block.*

vertical rebar, and be tied to the footing with a rebar dowel (**Figure 1-66, previous page**).

- Floor joists must be securely fastened to the sill plate with toenails or clips (**Figure 1-68**).
- Bracing or blocking on joists parallel to foundation wall must extend three joists back from the wall (**Figure 1-69**).

Bond Beams

In tall foundation walls (8 ft. or higher) a bond beam running along the top course is recommended to tie walls

Figure 1–68. Tying Block to Floor Framing

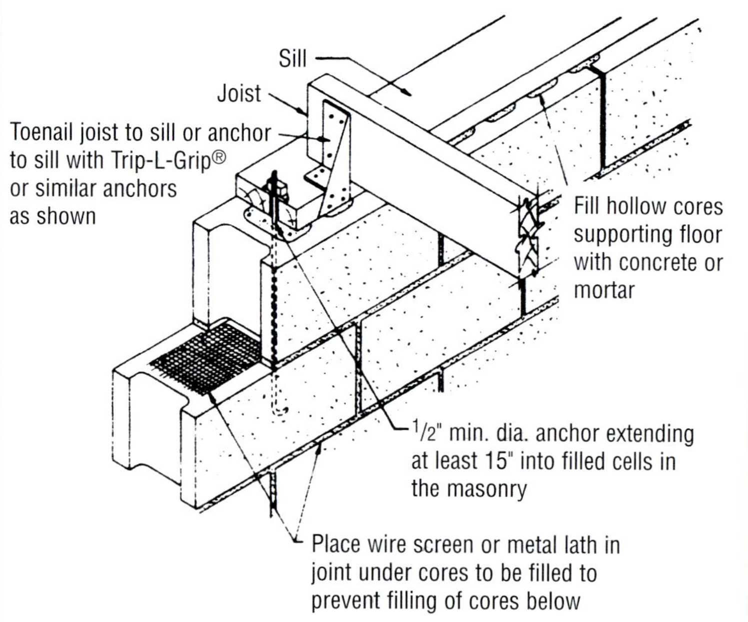

In order to distribute the loads from joists and beams, the bearing course of block should either be capped by 4-in.-thick solid masonry, or anchored into concrete-filled (grouted) block cores.

together, particularly in poor soil conditions. On wet sites and in seismic zones, an engineer may also specify bond beams for lateral support midway in the foundation wall elevation (**Figure 1-70**).

To form bond beams, place rebar and pour grout in the channel made by knocking out the webs in special bond-beam block. Use two #4 bars in 8-ft. walls and two #5 bars in 10- or 12-ft. walls. Lap bars at joints, and bend bars around corners by at least 24 bar diameters.

Joint Reinforcement

Wire-mesh joint reinforcement is used to help reduce the size of cracks, but it does not add appreciable structural strength. Typically, mesh is placed in the mortar joints every second or third course (**Figure 1-70**).

Openings in Block Walls

CONTROL JOINTS

Bond Beams

Avoid using expansion and control joints in basement construction when possible because they weaken the walls in the horizontal span and increase the potential for foundation leaks.

Figure 1–69. Bracing Block Walls

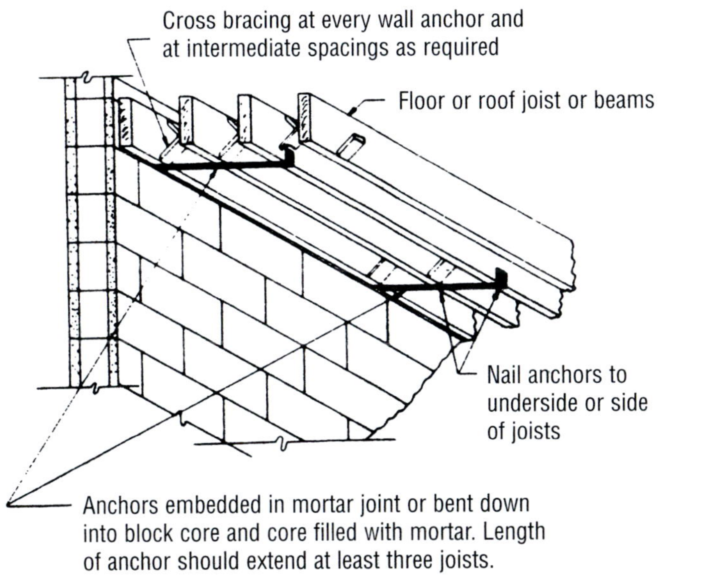

Joists running parallel to a block wall should be braced every 8 ft. The bracing (or joist blocking) should span at least three joists.

Figure 1–70. Bond Beam

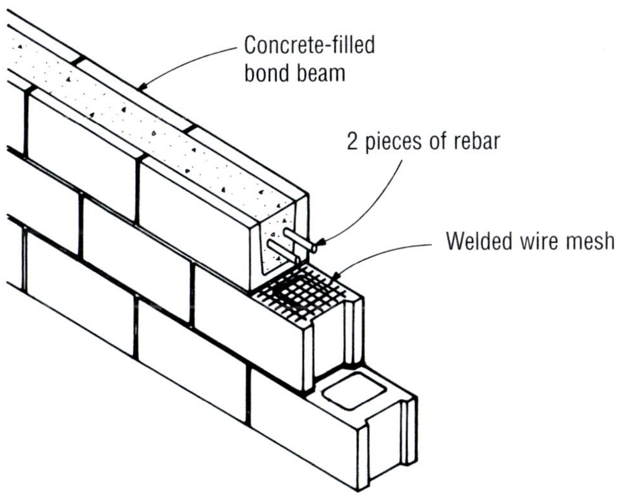

To strengthen a block wall against lateral pressure, place a horizontal bond beam halfway up the wall.

WATERPROOFING REFERS TO A CONTINUOUS MEMBRANE covering that will resist water under hydrostatic pressure (i.e., standing water). *Dampproofing* is a water-resistant coating designed to shed water and resist soil moisture. Dampproofing is not effective against water under pressure.

DAMPPROOFING

A simple dampproof coating is sufficient where positive drainage is reliable, the seasonal water table never rises above the footing, or the basement is not living space.

Figure 1–71. Parging for Concrete Block Foundation

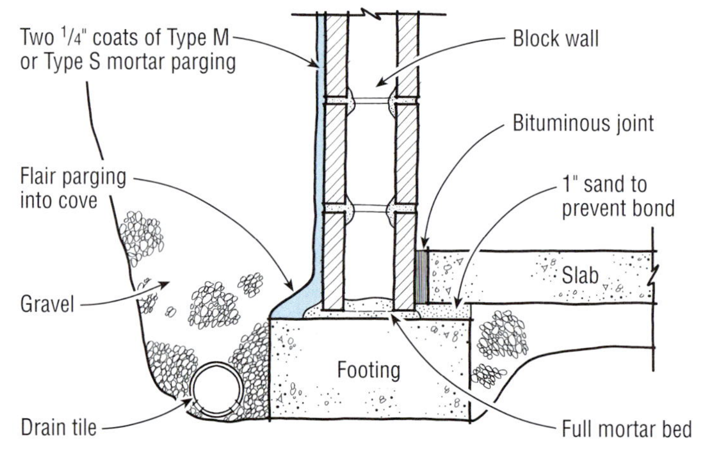

Parging should be applied to block foundations before applying a dampproof coating. At the bottom of the wall, flare the parging into a cove to shed water from the footing/wall joint.

For more information: "Waterproofing Basement Walls," JLC 9/02

"Spray-Applied Foundation Water-Proofing," JLC 7/99

"Foundation Waterproofing Options," JLC 3/95

Dampproofing Concrete Block Basements

Spray-applied waterproofing and some waterproofing membranes can be applied directly to concrete block. However, dampproofing should be applied over parging.

Parging Concrete Block

Before parging walls, brush off any dirt and dampen wall with a water spray (do not soak).

To parge walls, trowel on two coats of Type M or S mortar. Each coat should be at least 1/4 in. thick, for a total minimum thickness of 1/2 in. Roughen the first coat with a brush when the wall is partially dry to ensure good bonding.

Allow the first coat to harden for 24 hours before applying the second coat. Cure the topcoat for 48 hours before applying dampproofing.

Form the parging into a cove where the wall meets the footing to direct water away from the joint (**Figure 1-71**).

WATERPROOFING

Apply waterproofing products in the following situations: When soils are slow draining; basement walls are below the water table; drainage is unreliable; or the basement encloses habitable space (**Figure 1-72**).

Drainage Panels

Some waterproofing systems employ a dimpled or expanded-fiber geomat fastened against the foundation, either with or without an additional coating material applied to the foundation wall. These drainage panels provide an immediate break in hydrostatic pressure, carrying water from the backfill into the perimeter drains.

Figure 1–72. Types of Waterproofing

Type	Suppliers and Brands	Pros	Cons	Comments
Liquid Membranes	Tremco Barrier Solutions' Tuff-N-Dri (spray-applied polymer-modified asphalt); Tremco's Tremproof 60 and Pecora's Duramem 500 (polyurethane liquid membranes for trowel, roller, or spray application)	Quick application, low in-place cost	Possible inconsistency in coverage, 60-mil min. required	Follow manufacturer's recommended procedures carefully for voids and joints. Cement cove or fillet may be needed at footing-wall corner. Monitor coverage.
Sheet Membranes	W.R. Grace's Bituthene, Pecora's Duramem 700-SM, and W.R. Meadows' Sealtight and Mel-Rol	Consistent thickness, easy patching of holes, "fish-mouths," puckers, and wrinkles	High cost, tricky to use (sticks to everything)	Many details to learn – surface preparation, priming, patching, joint treatment, terminations, lap joints, penetrations, and corners
Cementitious Waterproofing	Thoroseal from Thoro System Products	Easy to use, readily available	No elongation – will not accommodate joint or crack movement	Use acrylic additive for better bonding and durability
Bentonite Clay	Volclay panels from Colloid Environmental Technologies	Nonhazardous, nonpolluting, easy and quick to apply, can go on at low temperatures	Cannot be inspected for integrity (seal forms after backfilling)	"Pumped-in" bentonite retrofits are of dubious value

Availability and installation methods of waterproofing materials vary widely. While waterproofing will resist standing water in occasional extreme situations, no waterproofing system is designed to work without proper foundation drainage.

PERIMETER FOUNDATION DRAINS

Every foundation below grade should be equipped with an exterior perimeter drain. Always work out a drainage plan before digging to locate cleanout and daylight locations of perimeter footing drains (**Figure 1-73**).

Perimeter foundation drainage must always be planned in conjunction with well-draining backfill (see "Soil Drainage," **page 8**, and "Backfill," **page 72**).

Placing Footing Drains

Place drain "tile" — typically 4-in. Schedule-C PVC or ABS pipe — with the holes down on a bed of gravel.

Figure 1–73. Perimeter Drainage Plan

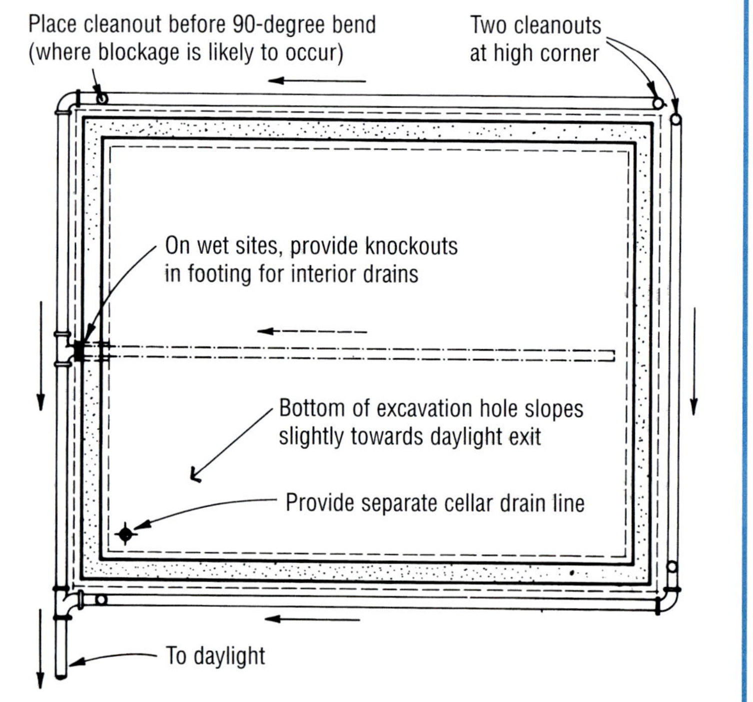

Plan drainage, including cleanout and daylight locations, before digging. On wet sites, plan on knockouts through the footing to connect to a drain under the slab.

Figure 1–74. Interior Sump Basket

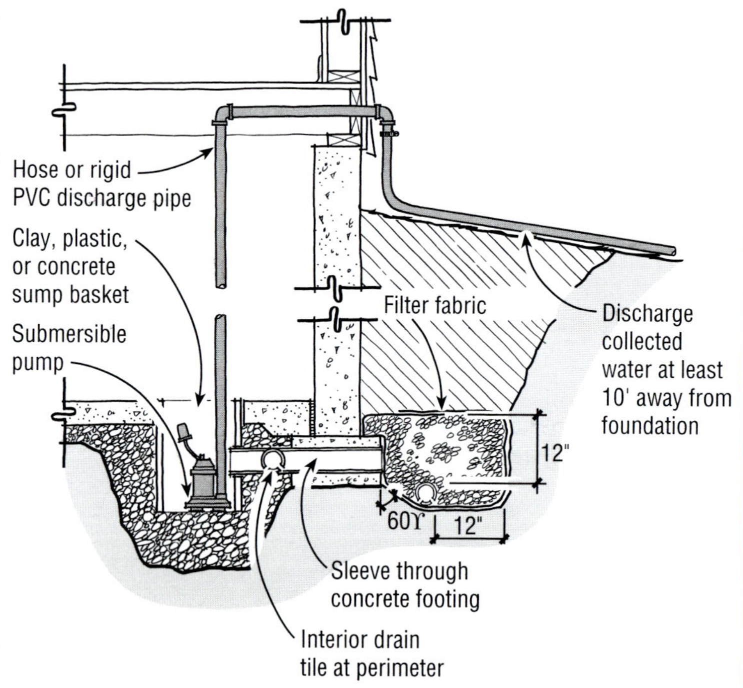

An interior sump basket picks up excess water flowing through sleeves in the footing. A submersible pump discharges the water at ground level away from the foundation.

For more information: "Foundation Drainage," JLC 3/99

Figure 1–75. Placing Footing Drains

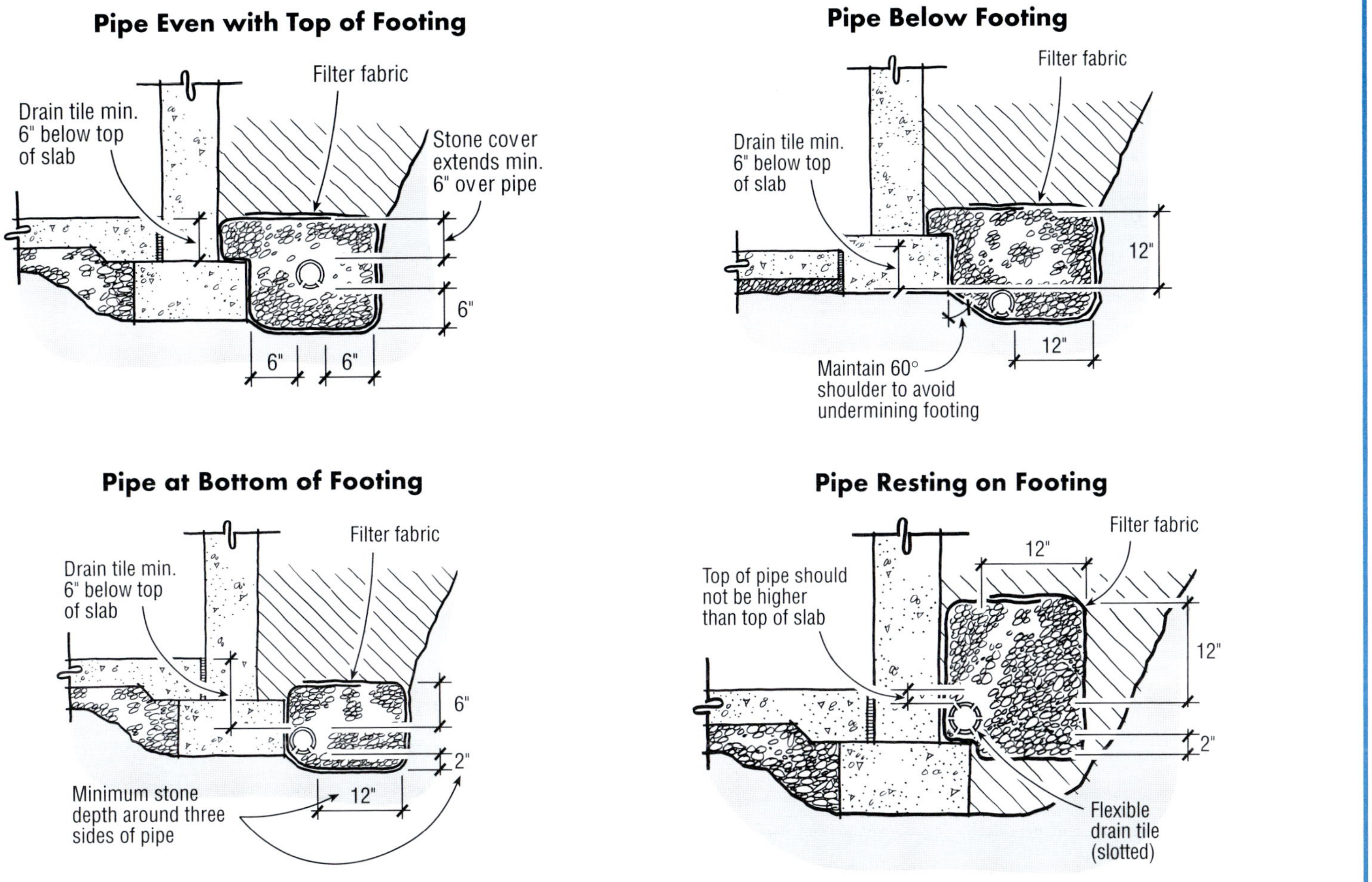

*The best location for rigid drain pipe is alongside the footing (**at left, top and bottom**). Ideally, the drain should be at least 6 in. below the top of the slab and always covered by at least 6 in. of stone. If tile is placed below the footing, do not place it too close to the footing or water may undermine the footing (**at right, top**). To keep flexible drain pipe from developing low spots, place it on top of the footings, making sure that the top of the pipe is not higher than the top of the interior slab and that it is covered by a 12-in.-deep bed of stone (**at right, bottom**).*

Perimeter Drains

Sump Drains

Figure 1–76. Surface Drainage Without Gutters

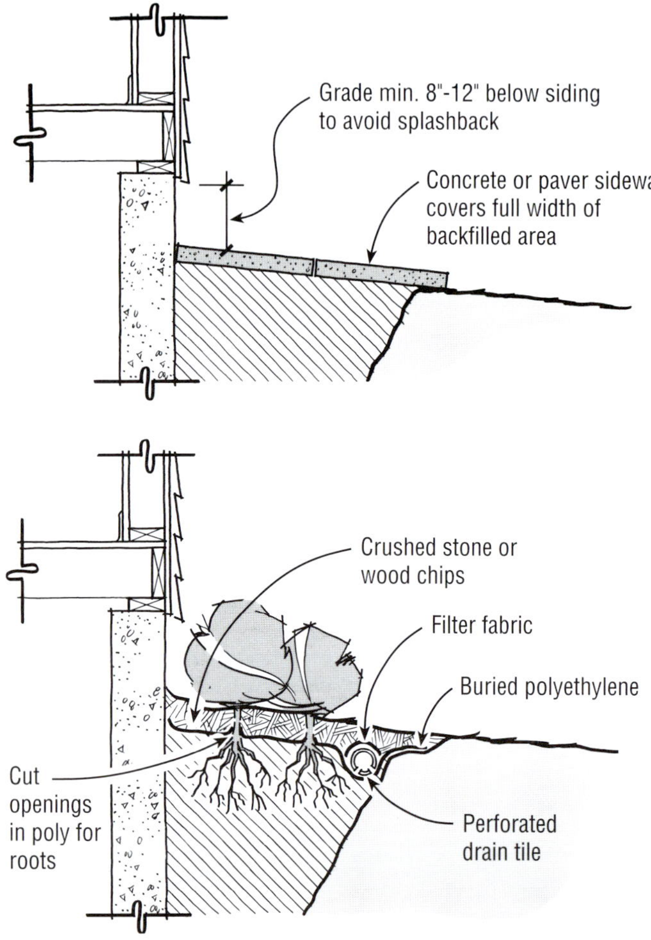

*A properly sloped concrete or paver sidewalk will reduce the amount of runoff that percolates through the backfill (**at top**). Where perimeter plantings are used to landscape, improve drainage by burying a sheet of polyethylene below the plant bed, with openings cut out for roots (**at bottom**).*

Figure 1–77. Channeling Gutter Discharge

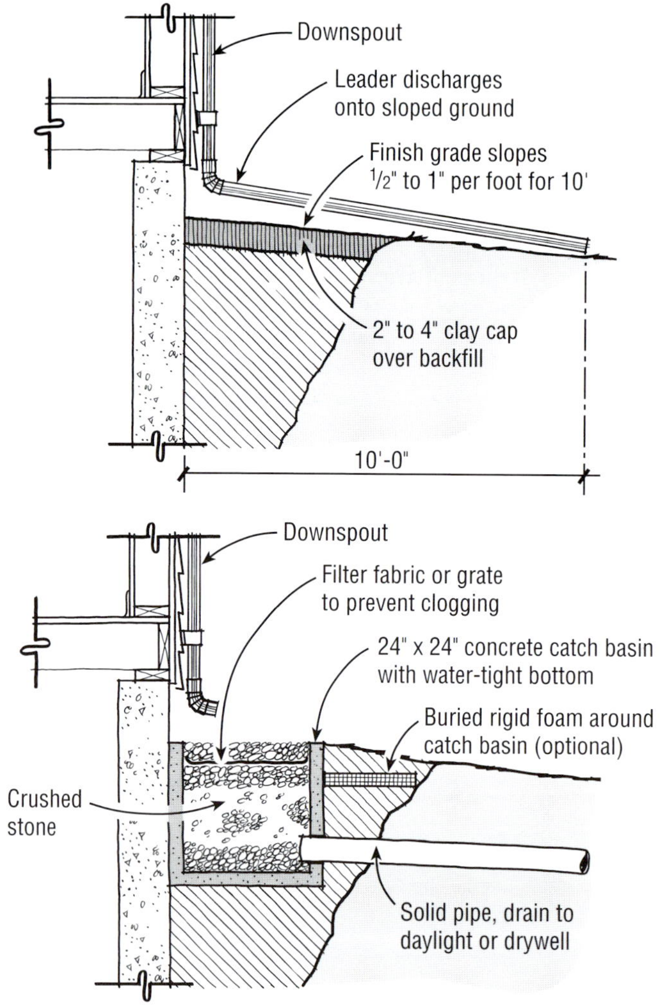

*Sloped downspout leaders should discharge at least 10 ft. away from the foundation wall (**at top**). Otherwise, downspouts should discharge into a catch basin (**at bottom**).*

The gravel surrounding the drain should be wrapped with a geotextile, or filter fabric, to keep fine particles from clogging the drain tile.

Place drain tile straight and level, or with a slight pitch toward the exit. Do not attempt to create a pitch when using flexible drain tile, because undulations in the level can cause clogging. Flexible drain pipe should be placed on the footing shelf to prevent dips and sags (**Figure 1-75, page 69**). In either case, perimeter drain pipe should never be above the surface of the slab.

Tile should drain to daylight. If this is not possible, install a sump basket and a sump pump that discharges above ground away from the building (**Figure 1-74, page 68**).

Drainage Forms

Perimeter drainage can be provided with stay-in-place footing forms, such as Form-A-Drain® — to ensure a level perimeter drain. These systems usually provide a larger capacity than typical foundation pipe systems.

SURFACE DRAINAGE

While perimeter drainage can handle water migrating through the soil, it may not be enough to handle a heavy run-off of surface water. The first line of defense against basement water problems should be handled at the surface.

Sloped Grade

Surface Drainage

The finish grade around the house should slope away from the foundation at the rate of $^1/_2$ to 1 in. per ft. for 6 to 10 ft. A 2- to 4-in. cap of silty-clay material — sometimes called a *ground cap* — will keep runoff from percolating down through the backfill. Better yet, use concrete or paver sidewalks sloped away from the house, or shallow subsurface drain pipes (**Figure 1-76**).

Make sure to properly compact backfill as you place it; otherwise, the soil will settle over time and create a slope draining toward the house (see "Backfill," **page 72**).

Gutter Downspouts

Use gutters and downspouts to divert roof runoff away from the foundation perimeter (see "Gutters" in Chapter 3, **page 283**). If leaders dump out right next to the house they will concentrate the problem in a smaller area. Extend downspout leaders 10 ft. away from foundation, or provide a sump basket and subsurface drain at the bottom of the downspout (**Figure 1-77**).

BRACING BEFORE BACKFILL

Backfill should be placed only after the floor framing has been completed to brace the top of the foundation wall against the backfill load. If this is not possible, provide temporary foundation bracing, as shown in **Figure 1-78**.

PLACING BACKFILL

Careful backfilling helps prevent foundation damage and soil settlement. Backfill only with well-draining granular fill (see "Drainage," **pages 68-71**).

Minimum Backfill Amount

Place the granular backfill in the zone from footing to near grade, widening to a minimum 30-degree angle from the footing up (**Figure 1-79**).

Figure 1–78. Temporary Foundation Bracing

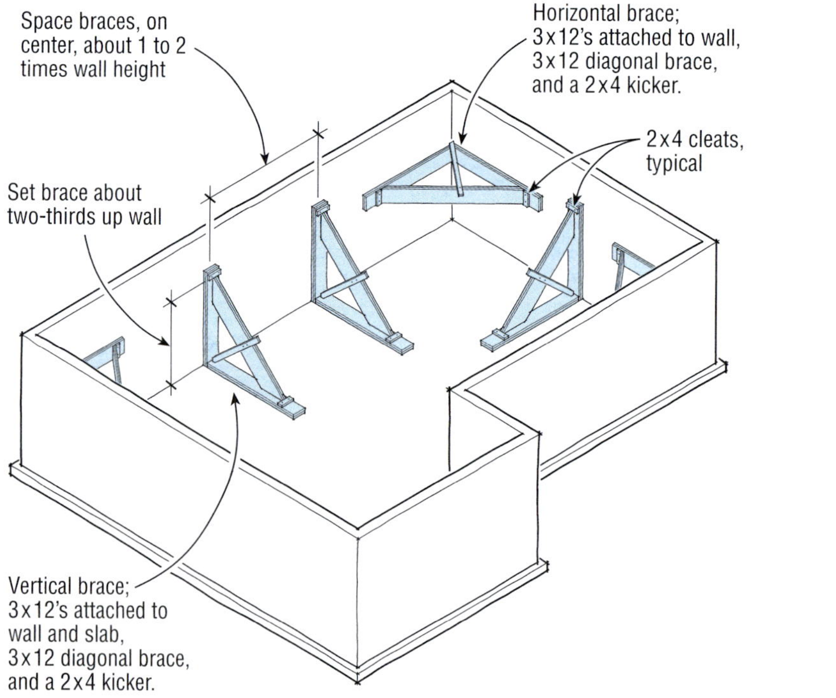

If backfilling must be done before the floor deck has been framed, install temporary foundation bracing. Set braces at about two-thirds the wall height and space braces apart about one to two times the wall height.

For more information: "Backfilling Basics," JLC 10/94

Foundation Backfill Checklist

- **Use coarse granular material.** Sandy or gravelly soils make the best backfill.
- **Compact the backfill.** Place and compact the backfill in 6-in. lifts. Use hand compactors near the foundation, not machinery (see "Compacting Soil," **page 9**). Uniform-graded gravel is largely self-compacting, but sandy, silty, or clay soils will settle if not compacted.
- **Slope final grade away from foundation.** A minimum pitch of 6 in. in 10 ft. will direct surface water away from the foundation. For added insurance against settling, build final grade up deeper and at a steeper slope so it will still slope away from foundation after settlement. For additional drainage details, see "Drainage," **pages 68-71**.

Figure 1–79. Backfilling Foundation Walls

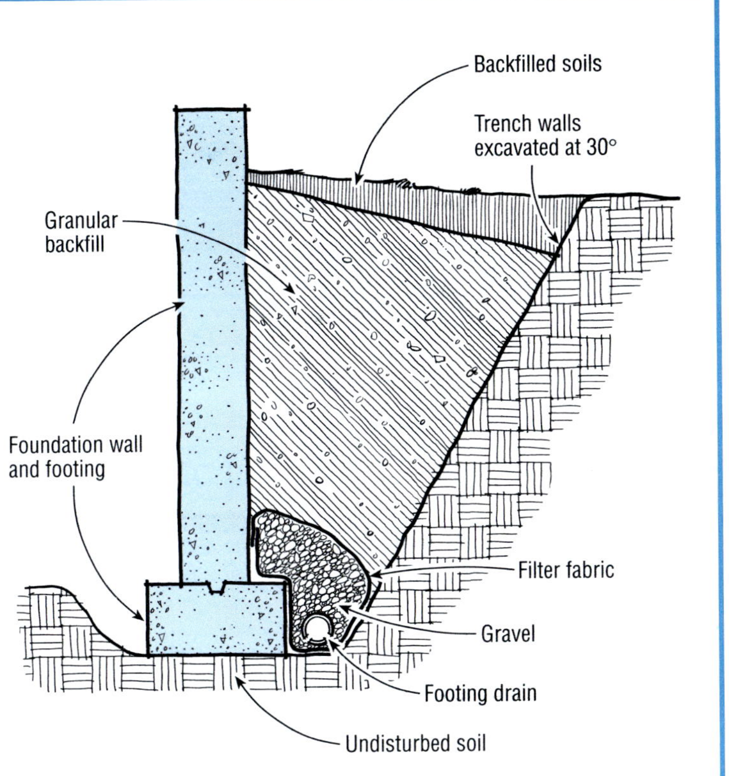

*If a foundation has not been excavated at least 3- to 4-ft. beyond the footing (see "Excavating Foundation Holes," **page 12**), widen the cut to at least 30 degrees from the base of the footing before placing backfill.*

Temporary
Foundation
Bracing

Backfill

SITE LAYOUT FOR SLAB FOUNDATIONS

When framing a floor system, there is a little latitude in both the timing and installation of the utilities. But when building on a slab, the utilities must be defined precisely before any concrete is poured.

Plumbers First

When starting layout, make sure the plumber is the first trade on site. Let other subs work around him — it's a lot easier for an electrician to work around obstructions than it is for the plumber.

Figure 1–80. Forming a Structural Slab

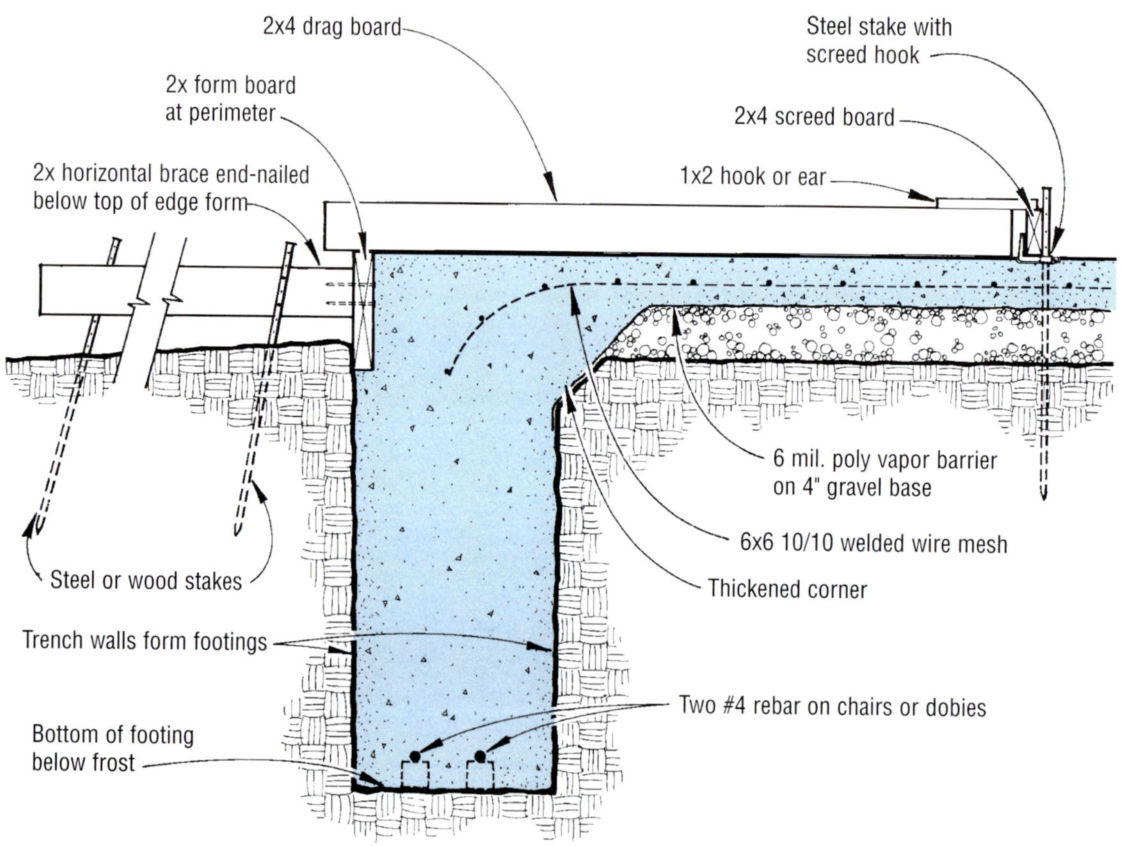

For more information: "Slab-On-Grade Foundations," JLC 4/95

In slab-on-grade construction, excavate accurately; the trench walls typically act as footing forms. The slab's subbase should be a minimum 4-in. compacted gravel base.

Before excavation begins, work through the plumbing plans to define all drain/waste and supply lines and vent stack locations, and review plans for all utilities running below the slab, such as electric, gas, and phone lines.

Subslab Utilities

After the service lines have been installed, all plastic or copper lines should be properly bedded (clean fill, no rocks) and tested for leaks prior to backfilling. All copper should be checked for dents or abrasions and wrapped in split-foam insulation, or something similar, to protect it.

Install caution tape: Make sure every trade buries color-coded caution tape 12 in. below grade and in line with their work (**Figure 1-83, page** 77). These plastic caution tapes are color coded: red for electric, blue for water, green for sewer, yellow for gas, and orange for phone.

Mark location of new work: Use spray paint to color-code the location of new work after backfilling trenches. It's easy to forget where lines run a few days after a trench has been backfilled. Don't leave anything to memory: Before the concrete trucks show up, make notes either on the drawings or in a site log.

SUBGRADE AND SUBBASE

Subgrade material should consist of well-drained native soil. If soils are not suitable, excavate and replace with a deep subbase layer of compacted gravel.

Granular fill, such as gravel or a mixture of sand and gravel, works best as a subbase under a slab because it compacts well (**Figure 1-80**). The fill should be completely thawed and free of organic material — melted ice and decomposed organic material will leave voids (see "Soils," **pages 3-10**).

Base Compaction

Slab Layout

Residential slabs, even with steel reinforcement, cannot span over voids or soft spots (the amount of steel used is not enough for structural reinforcement unless designed for that purpose by an engineer). Compact subgrade and

Prepping Subgrade

Slab Layout Checklist

Keep careful track of these critical layout details for structural slabs:

- **Benchmark:** A slab excavation should be as level as possible. Set an elevation benchmark prior to excavation. Place the benchmark somewhere convenient and make sure everybody on the site knows where it is (**Figure 1-81**, next page).
- **Concrete formwork:** Form monolithic slabs as shown in **Figure 1-80**. Double-check forms, verifying that they are located properly, both in relation to the corner points the surveyor set prior to excavation and in relation to the elevation benchmark.
- **Reference lines:** Mark on the edge of forms one reference point from which each subtrade can take dimensions and work. A string line across the forms is ideal. Locate this reference line at some feature of the building that cannot change, such as the longest bearing wall. Make sure that all trades clearly understand how this line corresponds to their work.
- **Edge of slab/wall:** Take time to understand the relationship between the edge of the form and the edge of the wall. Explain it to the plumber and other subs (**Figure 1-82, page** 77).
- **Penetrations:** Double-check all penetrations before the slab is poured. These include toilet flanges, drain boxes, turn-ups for water, electric, gas, phone, and floor drains.

Figure 1–81. Structural Slab Layout and Formwork

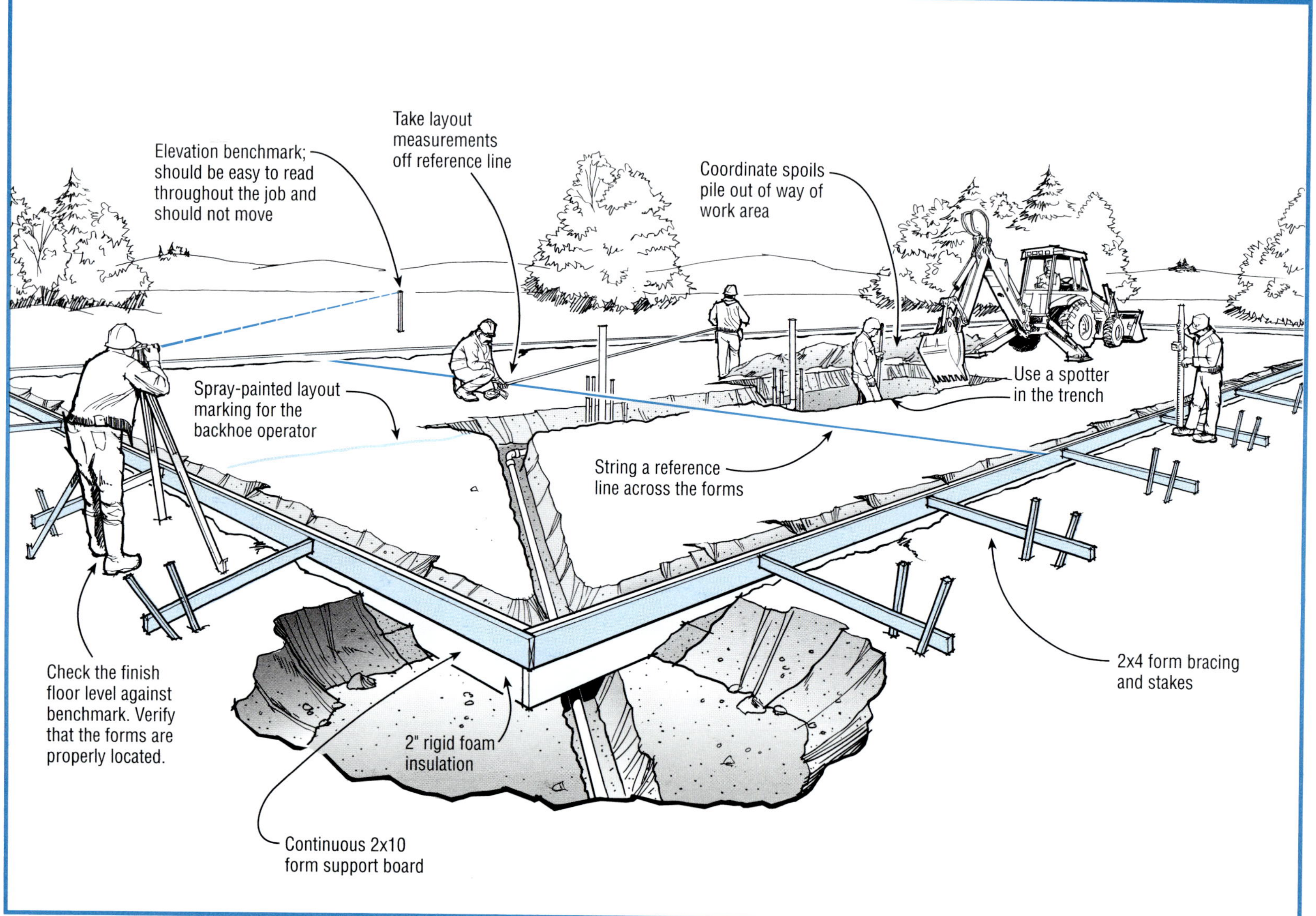

Check the elevation of forms against a benchmark. Layout lines for slabs should be clearly marked with spray paint for the excavator.

For more information: "Layout Under the Slab," JLC 12/00

"Foundation Layout by Transit," JLC 3/91

subbase after utility trenching to prevent settlement, and under footings to provide even bearing strength.

Compaction is usually expressed as a percentage at optimal moisture content — usually 95% or greater (see "Compacting Soil," **page 9**). Run a plate tamper or jumping jack until there is very little impression left with each successive pass.

All fill material should be reasonably free of moisture, but if it's too dry, you can pound on it all day and it won't compact much. In this case, spray it sparingly with water.

Level Base

The subbase must be graded to the same level across the full width of the slab. Level the subbase to within ½ in. (less than 12.4% of a 4-in. slab). Where subgrade elevation

Figure 1–82. Slab/Wall Relationship

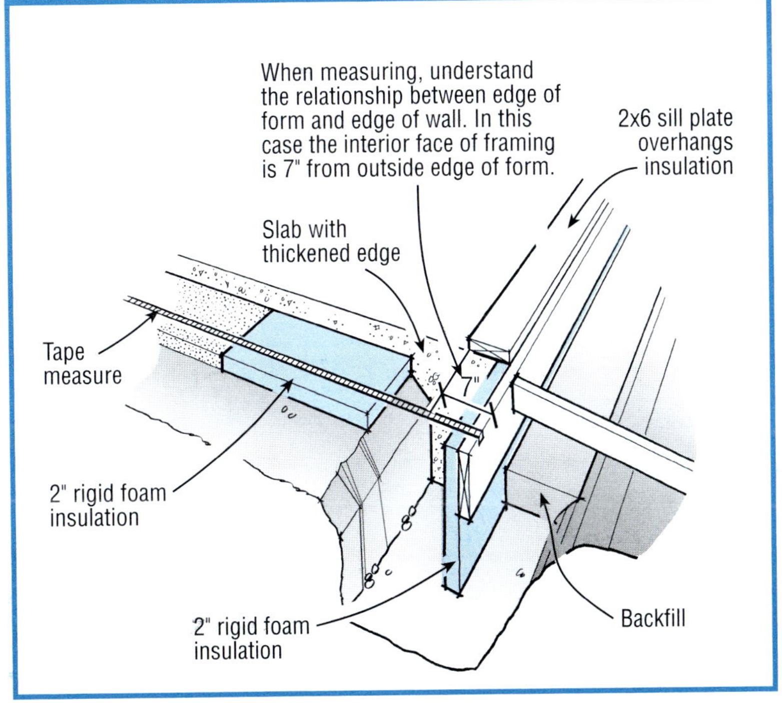

When measuring formwork, the layout crew must understand the relationship between the edge of the form and the edge of the wall. In this case, the interior face of framing is 7 in. from the outside edge of the form.

Figure 1–83. Utility Trench Infill

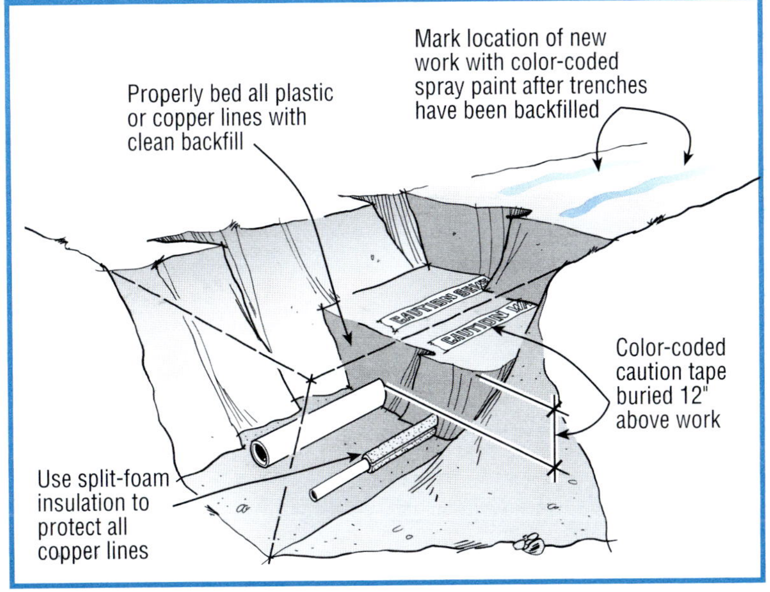

Water supply lines must be below the frostline outside the foundation perimeter, and should be protected with split foam insulation throughout. Clearly mark all utility lines with caution tape 12 in. below grade, and transfer these markings to the surface with spray paint using color-coded caution tape: Red = Electric; Blue = Water; Green = Sewer; Yellow = Gas; Orange = Phone.

is inconsistent, the slab will vary in thickness. This variation will cause it to cure unevenly, which will stress the concrete, increasing the likelihood of cracking.

Subslab Vapor Barriers

Install a subslab vapor barrier on interior slabs to keep moisture and soil gases from entering the home. Use a minimum 4-mil to 6-mil poly. For best results, use a cross-woven poly. Lap seams by at least 6 in.

Capillary Break

An optional 3- to 4-in. layer of sand or gravel over the poly, directly under the slab, helps reduce concrete curling and cracking by allowing water to escape from the slab at the bottom as well as the top (**Figure 1-84**). Damp sand is typical, but gravel or a sand-gravel mix can also be used.

Figure 1–84. Capillary Break Beneath Slabs

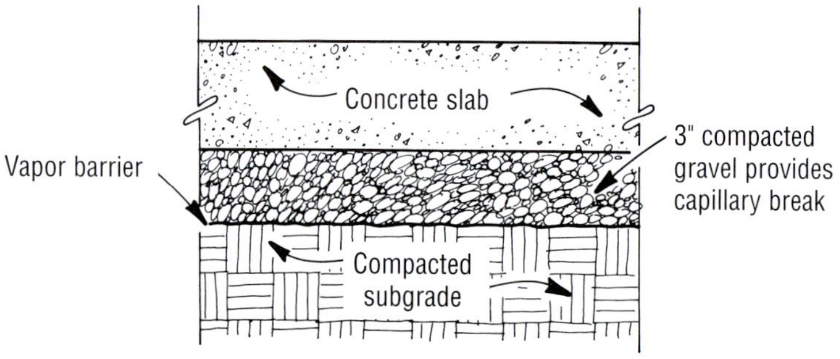

To prevent a slab from curling, use a 3-in.-thick layer of sand or gravel over the poly vapor barrier. Compact the sand before pouring the slab. While sand is more commonly used, gravel works better than sand because it is less likely to mix with the concrete or to create thin spots in the slab.

CONCRETE FOR STRUCTURAL SLABS

Depending on the slab's purpose, use a 3,000-psi to 4,000-psi mix (**Figure 1-85**) at a 4-in. slump (for increased workability, consider a water-reducing admixture). For a harder surface, use a 6-bag mix. For more information on concrete mixes, see "Concrete," **pages 26-37**).

Use air-entrained concrete for slabs exposed to freezing.

STRUCTURAL REINFORCEMENT FOR SLABS

Increase the thickness of a slab to 8 in. under bearing walls and structural columns (**Figure 1-86**).

In thickened bearing pads and exterior wall footings, lay two #4 rebars (see "Rebar," **page 43**).

Figure 1–85. Concrete for Slabs

Floor Class	Application	Minimum Thickness	28-day Strength	Slump
1	Residential or tile covered	4"	3,000 psi	4"
2	Offices, churches, hospitals, schools, ornamental residential	4"	3,500 psi	4"
3	Drives, sidewalks for residences, garage floors	4"	3,500 psi	4"
4	Business or commercial walks	5"	3,500 psi	4"
5	Light industrial and commercial	5"	4,000 psi	3"

Figure 1–86. Slab Bearing Support

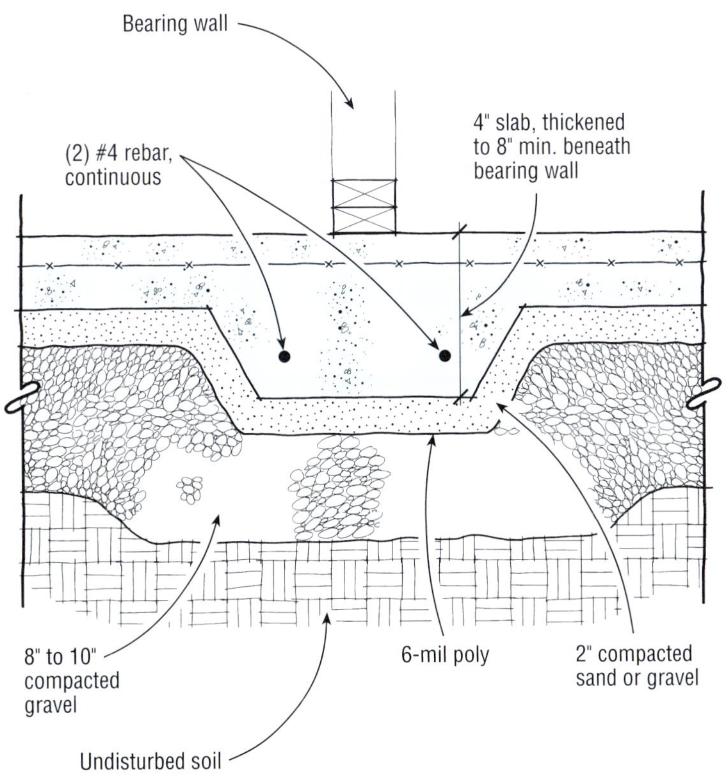

Thicken slabs beneath interior bearing walls and columns to a minimum of 8 in., using two #4 rebar about 2 in. from the bottom surface. The bearing width of all footings varies according to soil strengths and loading conditions, as shown in ***Figure 1-40, page 39****.*

Figure 1–87. Crack Control Joints

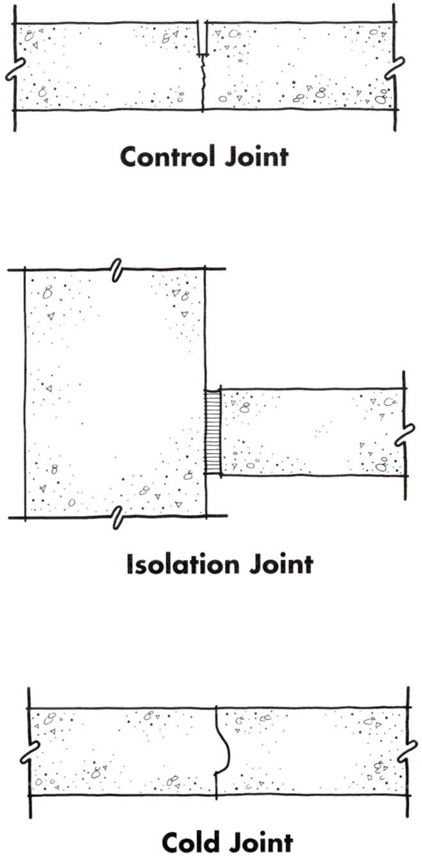

*Control joints (**center**) extend only partially into the slab surface and determine where cracks will form. Isolation, or expansion, joints (**top**) extend all the way through the slab and allow sections to slide back and forth without cracking. Cold joints (**bottom**) are necessary when a pour must be interrupted, and should coincide with control joints.*

Reinforcing for Slabs

Crack Control

Figure 1-88. Control Joints for Slabs

Slab Thickness (in.)	Joint Spacing (ft.)	Joint Depth (in.)
4	8 to 12	3/4 to 1
5	10 to 15	1 to 1 1/4
6	12 to 18	1 1/4 to 1 1/2
7	14 to 21	1 3/8 to 1 3/4
8	16 to 24	1 5/8 to 2
9	18 to 27	1 3/4 to 2 1/4

Figure 1-89. Isolating Basement Slabs

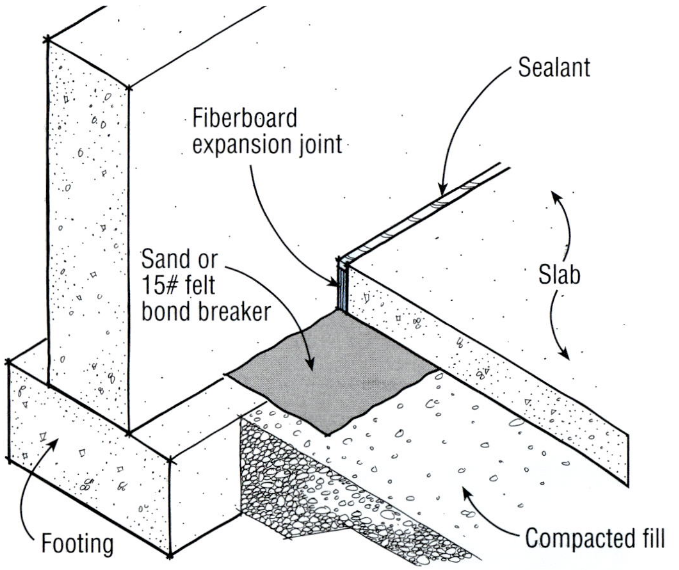

An expansion joint should be installed at the edges of a basement slab, and a bond breaker should be installed between the footing and the slab to help prevent cracking as the slab moves.

Under a fireplace, thicken the slab to 12 in. and use #5 reinforcing bars, placed 12 in. o.c. each way to form a gridwork.

STEEL REINFORCEMENT FOR SLABS

To limit cracking, residential slabs require steel reinforcement — either rebar or welded wire. This steel does not increase the load-carrying strength unless purposely engineered to do so.

Rebar

In the field of a slab, use #3 or #4 rebar at 16 in. o.c. Place rebar in the center of the slab section. Use dobies or wire chairs to hold the steel off the ground. Do not use brick to support rebar, because it will pull moisture out of the surrounding concrete too quickly, creating a stress point.

Pipe or conduit laid directly on the ground will dramatically weaken the slab by effectively creating a score line on the bottom (tension) surface.

Welded-Wire Mesh

To effectively limit cracks, 6x6 wire mesh in slabs should be placed in the upper part of the slab, where cracks are noticeable, but no closer than 2 in. to the top surface. Use dobies or wire chairs to hold the steel off the ground.

At slab edges, curve the mesh down into the perimeter footings (**Figure 1-80, page 74**). Do not lap wire sections at joints. Instead, stop the mesh several inches back from the joint location.

If control joints are spaced close together, and a low-slump mix is used, wire mesh is probably not necessary.

CONTROL JOINTS FOR SLABS

Three kinds of joints are used in concrete slabs (**Figure 1-87**, previous page):

- **Control joints** confine cracks to intended locations
- **Isolation joints** allow structural and non-structural elements to move separately
- **Cold, or construction, joints** define where a concrete pour has been interrupted

Cutting Control Joints

Control joints can be tooled in the slab surface with a grooving trowel or cut with an abrasive or diamond circular-saw blade at recommended spacing and depths (**Figure 1-88**). Make tooled joints during finishing, as soon as bleed water has evaporated. Cut sawn joints as soon as possible after finishing the slab. Random cracks may appear as rapidly as within six hours in hot weather.

Isolation Joints

Provide isolation, or expansion, joints at garage and basement slab edges (**Figure 1-89**) and at column footings (**Figure 1-90**). This prevents movement of one element from damaging the other. Expansion joints can be made of asphalt-impregnated fiberboard or a plastic strip that can be zipped off after concrete is placed.

For a finished appearance in slabs, score fiberboard expansion material prior to placing, so it can be snapped out and the joint sealed with caulk (**Figure 1-91**). This same detail should be employed with exterior driveway expansion joints; in this case, the caulk provides an important protection from potential freeze-thaw damage (see "Concrete Driveways," **page 24**).

Basement slabs must also be isolated from the footing, either with sand (**Figure 1-71, page 66**) or with 15# felt (**Figure 1-89**).

Figure 1-90. Isolating Column Footings

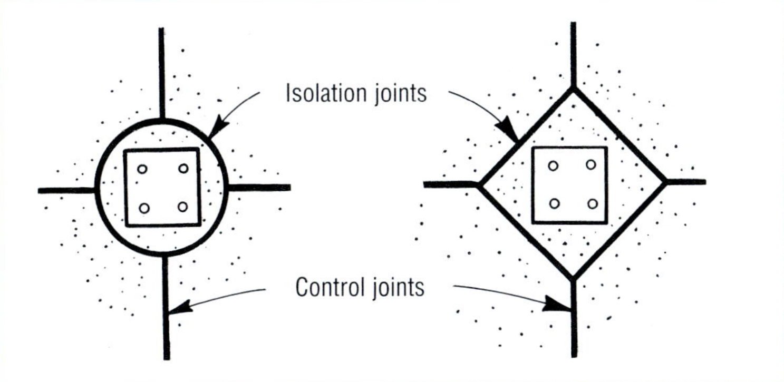

*Structural columns must be isolated from the slab using block-outs that extend through the slab thickness (**Figure 1-87**, page 79). When forming square block-outs, align the points of the square to meet control joints.*

Control Joints

Figure 1-91. Finished Expansion Joint

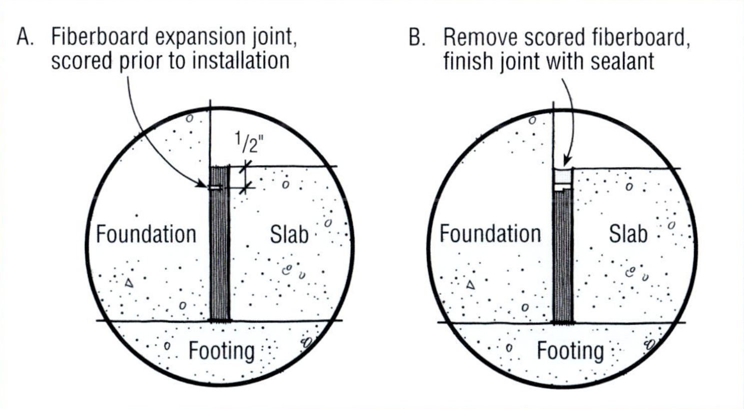

For a finished appearance at the slab edge, score the fiberboard on a table saw so the top 1/2 in. can be easily snapped off after the pour. Then finish the joint with sealant.

PIER-AND-GRADE-BEAM FOUNDATIONS CAN BE AS SIMPLE AS A sonotube foundation for a porch or small addition, or as complex as an engineered foundation for expansive soils.

SIMPLE PIER FOUNDATIONS

Simple pier-and-grade-beam foundations can support decks, porches, pole barns, and small home additions, providing good insulation and ventilation (**Figure 1-92**).

Figure 1–92. Simple Pier Foundations

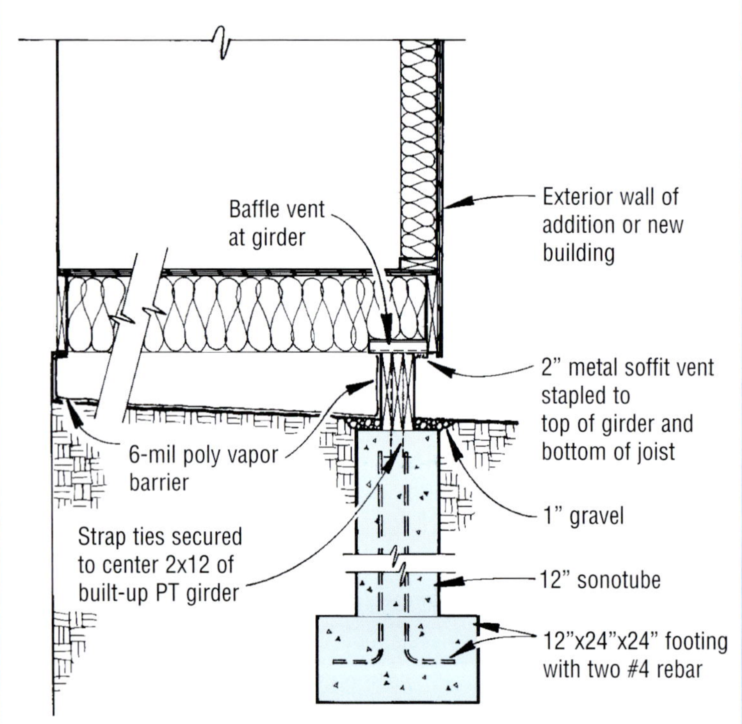

Sonotubes, placed on pier footings with a pressure-treated grade beam, can be installed easily without disturbing existing utility lines.

For more information: "Post and Pier Room Additions," JLC 7/94

Sizing Piers

For sizing pier diameters for small foundations, the rule of thumb is "1 inch per foot of span." Thus, a deck that spans 8 ft. will stand comfortably on 8-in.-diameter piers, while a deck that spans 10 ft. requires 10-in.-diameter piers. For spans longer than 12 ft., add a second row of piers and a second girder at the center of the joist span.

For calculating concrete amounts, see "Estimating Concrete for Piers," **page 1**.

Pier Footings

Footings for shallow piers (less than 6 ft. deep) will help prevent the pier from settling. The pier footing should be as thick as the pier's diameter, with sides that measure twice that much. So an 8-in. pier, for example, should rest on a footing that's 8 in. thick and 16 in. square, while a 12-in. pier should rest on a footing that's 12 in. thick and 24 in. square.

Simple Pier Foundation Details

- Use strap ties to anchor piers to a triple pressure-treated 2-by girder. The ties wrap around the center stick in the built-up girder and lap over the top. (The girder looks better if the ties aren't exposed on its face.)
- Install a poly vapor barrier over the ground beneath the joists, and run it up and staple it to the inside of the girder. This helps prevent water vapor from building up in the shallow space beneath the joists.
- Additional moisture control can be added by cantilevering joists 2 in. beyond the rim of the girder, and installing a 2-in.-wide standard metal soffit vent on the underside of the overhanging joists.

Figure 1–93. Engineered Pier Foundation

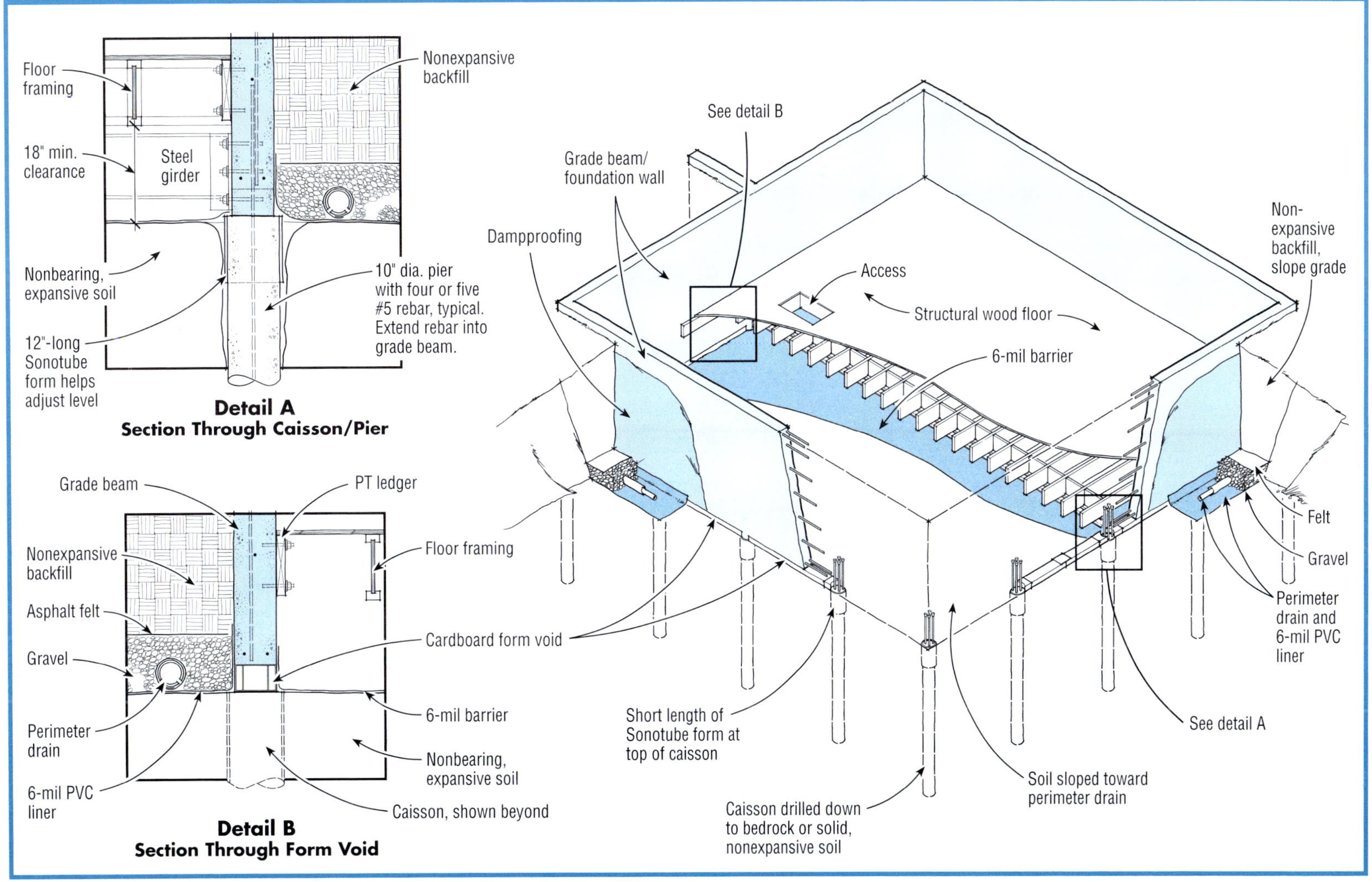

Unlike a continuous poured foundation, the heavily reinforced walls (the grade-beams) of this engineered pier foundation are isolated from the underlying soil, except where they bear on caissons that extend far below expansive soils to bedrock or more stable soils. Cardboard form voids (inset ***at top left****) provide temporary support when pouring the walls, but give way without stressing the foundation when expansive soils swell.*

Simple Pier Foundations

Grade Beam

- Finished grade should slope well away from a pier foundation — about a slope of 4 in. in 10 ft.

ENGINEERED PIER FOUNDATIONS

Pier-and-grade-beam foundations work where slopes are too steep for conventional stepped footings, where soils are weak, or where expansive soils would move conventional footings (**Figure 1-93**, previous page). Piers can penetrate deep through unstable soils and reach stronger soil or rock, where a combination of end bearing and soil friction against the sides of the pier supports the weight of the house. Grade beams transfer house loads onto the piers.

Figure 1-94. Grade-Beam Section

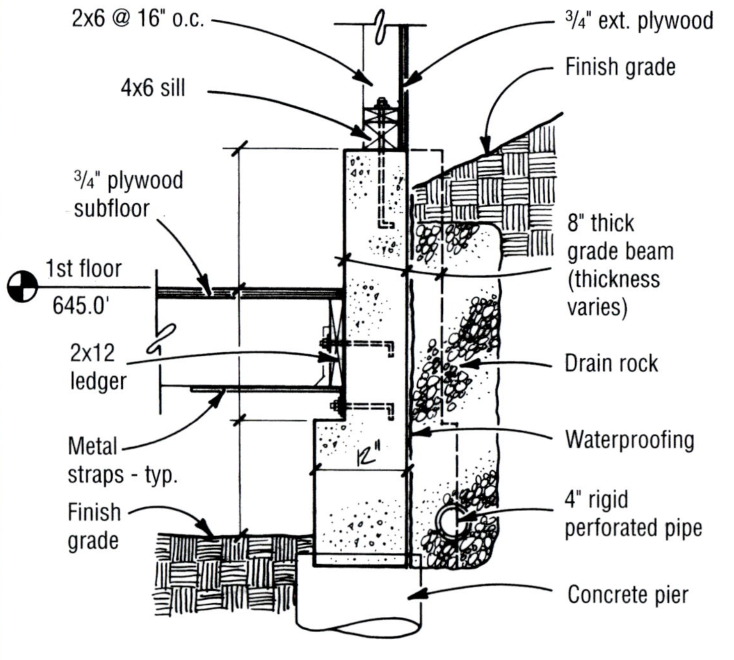

Grade-beams should be centered on piers to transfer house loads directly to the piers. Provide well-draining backfill and perimeter drains to relieve soil pressure on the beam between piers.

For more information: "Trouble-Free Foundations for Expansive Soils," JLC 11/02

Engineering

A soils engineer and a structural engineer are generally necessary to design this type of foundation and to supervise or inspect construction.

Pier Dimensions

Depth and diameter of piers are calculated based on soil characteristics and building loads. Piers may be anywhere from 5 to 20 ft. deep or deeper, depending on site conditions. Pier diameters are typically 10 to 12 in. Piers are generally spaced from 5- to 12-ft. apart. Of course, in seismic zones, on steep sites, and in areas with extreme soil conditions, these dimensions can be much greater.

Grade-Beam Dimensions

Beams are generally a minimum of 6 in. wide and 12 in. deep (**Figure 1-94**), but can be much larger (**Figure 1-93**), depending on loads and spans. A grade-beam may slope to conform to site contours, in which case a pony wall is typically framed over the beam (see **Figure 1-54, page 53**).

Reinforcement

Reinforcing steel must also be sized and installed in accordance with engineering calculations. Large (12-in.-diameter) piers should have at least four pieces of vertical rebar, creating rebar cages that tie into horizontal rebar in the beams (**Figure 1-95**).

Void Forms

If soils are expansive, space must be created under the beams to allow soil to expand without stressing the structure. This is accomplished with cardboard boxes placed at the bottom of the form (**Figure 1-93, page 83**). In less extreme cases, expanded foam (not extruded) can be used to form beams; the spaces in the foam will give a little to absorb slight expansions.

Anchor Bolts

Anchor bolts should be cast into the beams. In places where cripple walls are needed above grade-beams to create a level elevation for floor framing, the spacing between anchor bolts may need to be closer than usual (consult the engineer on this detail). In addition, horizontal anchor bolts may be required to secure the rim joist of floor framing (**Figure 1-94**).

Backfill

Grade-beams should be backfilled with a well-compacted granular fill, and perimeter drainage should be installed that drains to daylight (see "Drainage," **page 68**).

PIER FOUNDATION DETAILS

- **Pier tops must be clean when beams are cast.** Dirt or debris between the pier and the beam can weaken the joint.
- **Soil variations can affect the bearing strength of individual piers.** If soil color or consistency changes during drilling, alert the engineer: Pier depth may have to be increased if weaker soils are encountered. Piers that are too shallow are a common source of failure.
- **Locate beams accurately on centers of piers.** Off-center beams can crack piers.
- **Maintain coverage of rebar within piers and beams.** Rebar in piers must be protected by at least 3 in. of concrete coverage. Use dobies or wire spacers to maintain distance between piers and sides of hole (engineer should check this detail). Coverage in beams must be as specified in engineer's drawings (minimum 2 in.).

Engineered Pier Foundations

Figure 1-95. Pier-to-Grade-Beam

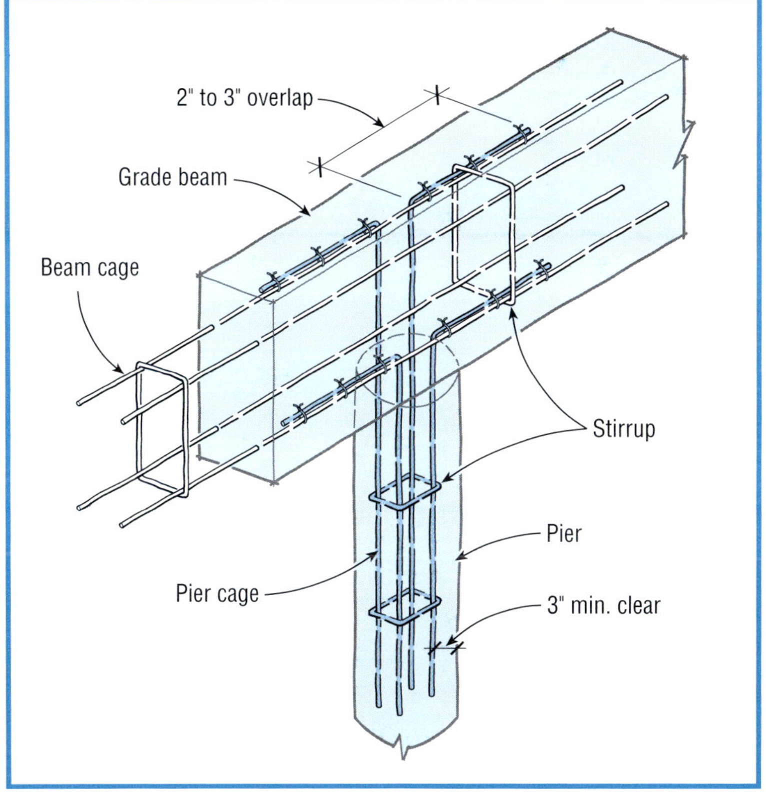

Bend rebar to overlap rebar cages in the piers and the grade beam. Each overlap should be at least 24 bar diameters at splices and spliced tightly with wire.

INSULATING FOUNDATION WALLS

Foundation walls give up a significant amount of heat. Insulate poured-concrete and block foundation walls with a minimum of 1 in. of extruded polystyrene insulation. Expanded polystyrene (EPS), polyisocyanurate, and polyurethane rigid insulations are not recommended for below-grade applications.

Cold-Climate Foundation Insulation

Install insulation running the full height of the foundation wall. In cold-climates, "taper" exterior foundation insulation as follows:

- Install 1-in.-thick 2x8-ft. tongue-and-groove sheets vertically from footing to sill plate.
- Next, place a second layer of half-sheets (2x4-ft.) extending down 4 ft. with joints offset.
- Follow with a third layer of 2x8 sheets running horizontally across the top of the foundation.

Insulation Protection

Protect vertical insulation on the exterior of foundation walls with 3/8- or 1/2-in. pressure-treated plywood or a parging coat of Type M or S mortar applied over expanded-wire metal lath (see "Stucco," **page 346**).

Figure 1-96. Shallow Foundation Insulation

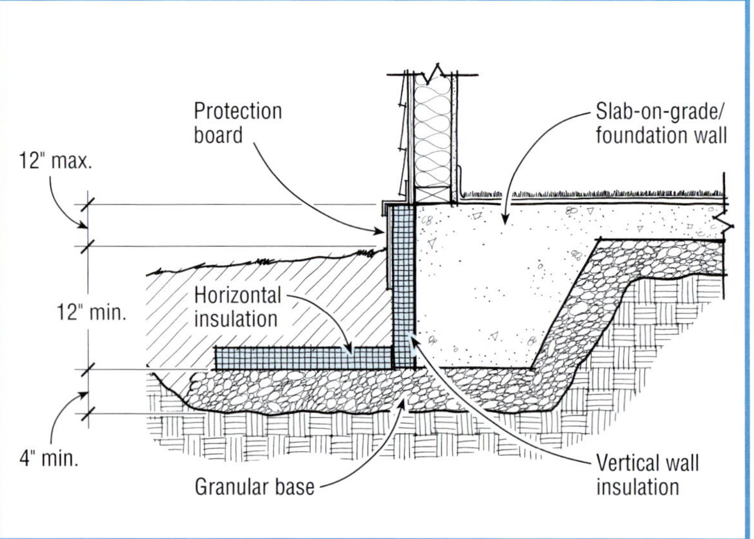

In a frost-protected foundation, foam insulation, rather than soil depth, keeps the soil at the footing from freezing. In mild climates, code requires only perimeter insulation projecting at least 24 in. to protect subslab soils from freezing and heaving.

For more information: "Frost-Protected Shallow Foundations," JLC 9/96

"Shallow Foundations," JLC 3/90

FROST-PROTECTED SHALLOW FOUNDATIONS

In cold climates, footings traditionally have been built below the maximum frost penetration depth (as deep as 5 ft. in the far north). However, codes now allow builders to use shallow footings or slabs-on-grade in any climate, as long as the soil under and around the foundation is protected from freezing with rigid insulating foam.

Horizontal Insulation for Structural Slabs

Insulation placement varies depending on whether the structure is heated or unheated:

- In mild climates, heated structures will need only vertical insulation along the edge of a structural slab.
- In colder areas, wing insulation is required, especially at the corners (**Figure 1-96**). Vertical and wing insulation specifications for the heated frost-protected shallow foundation are shown at right in **Figure 1-97**.

Protection for Horizontal Insulation

By code, horizontal insulation placed less than 12 in. below

Figure 1–97. Minimum Insulation Requirements for Frost-Protected Foundations Under Heated Buildings

Adapted from 2000 International Residential Code

Heating Degree Days (°F days)	Vertical Insulation Min. R-Value*	Horizontal Insulation Min. R-Value*		Horizontal Insulation Dimensions per plan view (in.)		
		along walls	at corners	A	B	C
1,500 or less	4.5	not required	not required	not required	not required	not required
2,000	5.6	not required	not required	not required	not required	not required
2,500	6.7	1.7	4.9	12	24	40
3,000	7.9	6.5	8.6	12	24	40
3,500	9.0	8.0	11.2	24	30	60
4,000	10.1	10.5	13.1	24	36	60

* **Note:** Use the following R-values to determine insulation thicknesses required for this application:

Type II expanded polystyrene – 2.4R per inch;

Type IV, Type VI, and Type X extruded polystyrene – 4.5R per inch;

Type IX expanded polystyrene – 3.2R per inch;

Type X extruded polystyrene – 4.5R per inch.

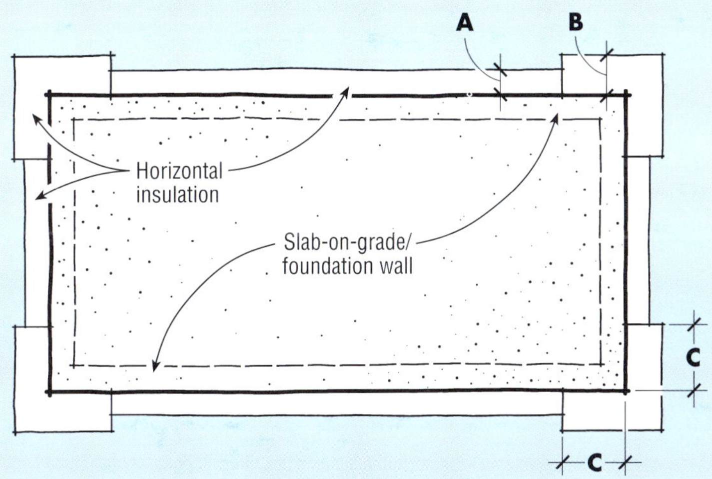

Plan View

Shallow foundations for heated buildings require insulation for protection against frost damage. Some state energy conservation standards may require greater values than those shown in the table above. Interpolation between values is permissible.

grade must be protected by a layer of concrete, asphalt paving, cementitious board, or pressure-treated plywood.

Perimeter Drainage
Slope finish grade at least 4 in. per 10 ft. In clay and silt-laden native soils, the soil beneath horizontal insulation should be backfilled with a well-compacted granular fill and perimeter drainage should be installed that drains to daylight (see "Drainage," **page 68**).

FROST-PROTECTED SHALLOW GARAGE FOUNDATIONS

An unheated structure, such as a detached garage, needs continuous insulation under the entire slab, extending out beyond the perimeter (**Figure 1-98**). Since there is no building heat to capture, this insulation is intended to prevent earth heat from escaping.

For more information: "Design Guide for Frost-protected Shallow Foundations," NAHB Research Center at 301/249-4000, www.nahbrc.org

Figure 1–98. Insulation for Unheated Garage

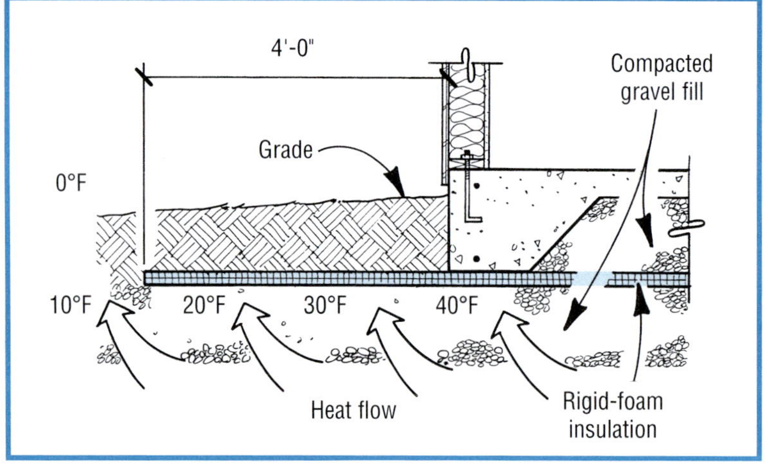

For unheated slabs-on-grade, high-compression foam is placed under the entire slab to retain ground heat. The insulation must extend horizontally 4 ft. beyond the footing.

Foam for Frost-Protected Shallow Foundations
In this design, the footing and slab are bearing on the foam. By code, the foam used for frost-protected slabs must have a density of 2 lbs. per cu. ft. (See "Note" in **Figure 1-97** for allowable Types.) Most expanded foam stocked by lumberyards is only 3/4-lb. or 1-lb. density and lacks sufficient compressive strength. High-compression foam usually needs to be special-ordered.

Figure 1–99. Insulating Walk-Out Basements

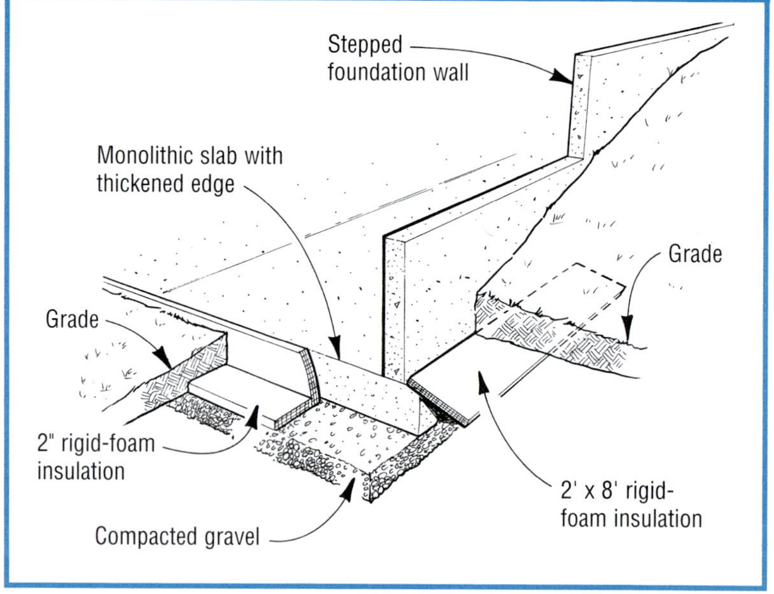

Insulate walk-out basement foundations with foam to avoid having to step down the excavation. A 2x8-ft. length of foam board protects the footing of the main wall at the corner where the backfill is shallow.

INSULATION FOR WALKOUT BASEMENTS

The principle of frost protection can also be applied to special cases where foundation elements are vulnerable to frost, such as walkout basements (**Figure 1-99**). In these cases, carefully placed insulation can prevent frost action from damaging or moving structural elements.

INSULATION FOR PIERS

In some soils, even a 48-in.-deep post footing will sometimes be heaved up by frost. Post-and-pier foundations for decks and porches can be frost-proofed, using a 4-ft. strip of foam around post footings to keep the ground below from freezing (**Figure 1-100**).

Figure 1–100. Insulating Piers

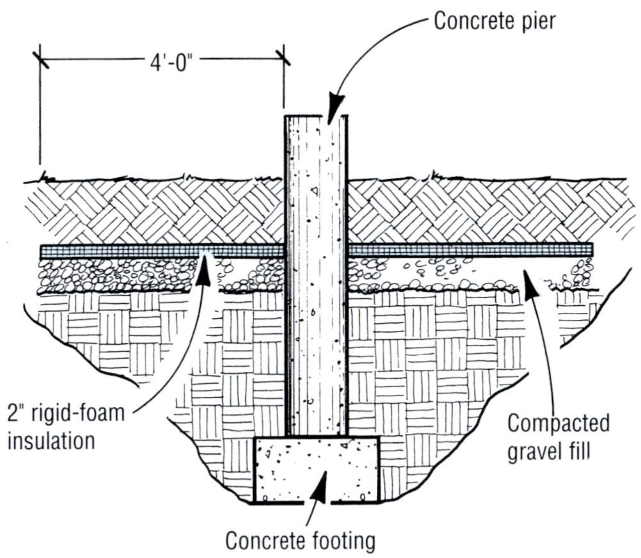

To prevent frost heaving of pier foundations, place 2-in. foam horizontally around the piers about 1 ft. below grade, as shown.

Insulation for Walkout Basements

Insulation for Piers

Permanent Wood Foundations

1 - FOUNDATIONS: Permanent Wood Foundations

PERMANENT WOOD FOUNDATIONS IS THE TERM USED BY THE wood-treating industry, and adopted by residential building codes, for basements built with pressure-treated wood that is rated for ground contact (**Figure 1-101, next page**).

APPROVED PRESSURE-TREATED WOOD

When buying wood for a permanent wood foundation, check the wood treater's stamp and be sure it bears the notation "C-22," which denotes a standard created by the American Wood-Preservers' Assoc. (817/326-6300, www.awpa.com) to cover foundation-grade lumber. This kind of wood typically must be special-ordered, and can be as much as twice the price of in-stock, 0.4-pcf pressure-treated lumber found in most lumberyards.

Figure 1–101. Permanent Wood Foundation

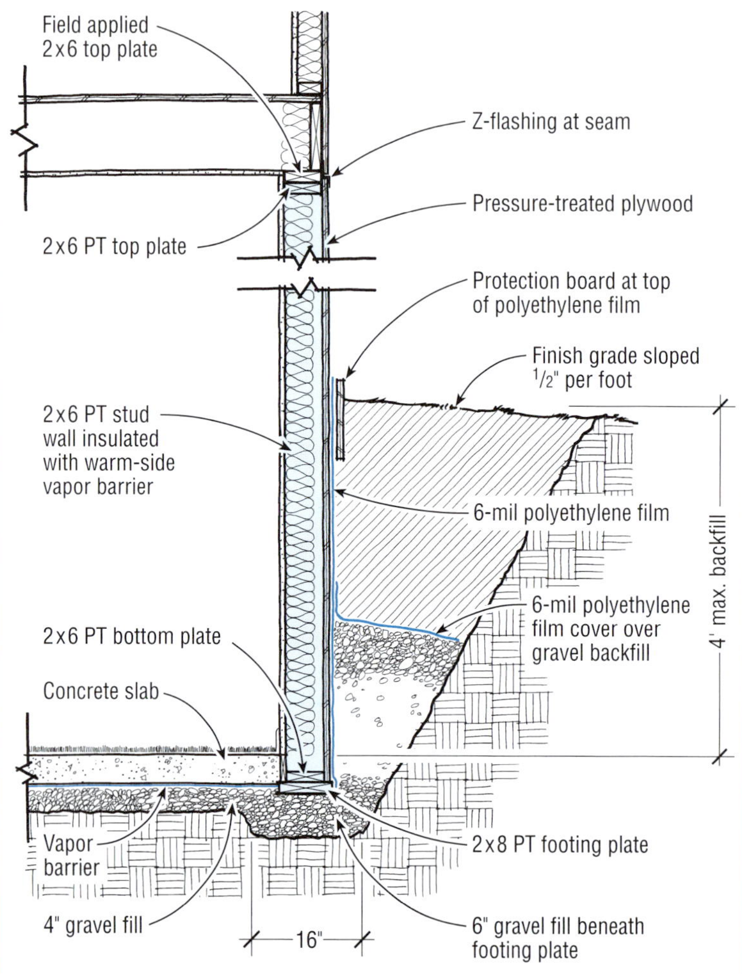

Where soil conditions permit, a permanent wood foundation may be an inexpensive foundation alternative. Frame the foundation only with "C-22" pressure-treated wood approved for ground contact.

For more information: "Permanent Wood Foundations: Design and Construction Guide," Southern Forest Products Assoc., 504/443-4464; download at http://store.southernpine.com/images/ref400.pdf

CRITICAL INSTALLATION DETAILS

- **Subbase:** Subbase should extend below frost line. Use gravel or crushed stone, properly compacted, under wall and slab. Subbase under perimeter walls should be at least 16 in. wide and 16 in. deep. The subbase under the slab should be at least 4 in. deep.
- **Footing plate:** Use "C-22" pressure-treated 2x8, rated for this use. All cuts in footing plates or any other lumber must be field-treated with preservative.
- **Bottom plate:** Use "C-22" pressure-treated 2x6.
- **Wall framing:** Use "C-22" pressure-treated 2x6 studs, framed 16 in. o.c. or 12 in. o.c., depending on load.
- **Sheathing:** Use pressure-treated plywood.
- **Insulation:** Use fiberglass batts, leaving a 2-in. gap between insulation and exterior wall, and install a vapor barrier on the warm side.
- **Waterproofing:** Use a 6-mil polyethylene film on sheathing exterior, terminated at grade. Protect termination with treated plywood or 1-in. treated lumber. Place poly over sheathing and perimeter-drainage gravel before placing soil backfill.
- **Floor slab:** Pour 4-in. concrete, placed before backfilling.
- **First floor deck:** Build in accordance with code, framed before backfilling. (Note: Backfilling before floor deck and slab are in place is likely to cause damage or collapse.)
- **Backfill:** Use only gravel at perimeter (approximately 16 in. deep above footing plate and brought up to within 12 in. of the finish grade); remaining backfill may be native soil. Total depth of backfill must not exceed 4 ft.

RETAINING WALLS HOLD BACK THE PRESSURE OF EARTH embankments. Unlike basement walls, they are not braced at the top by a floor system, and must be designed so they will not topple or fail in bending at the bottom.

Engineering

Doubling the soil height quadruples the pressure, so any retaining wall higher than 4 ft. should be designed by an engineer. If vehicle loads or other heavy loads will be placed above a wall, it should also be engineered.

Railings

Some codes require safety railings on any retaining wall higher than 30 in., whether there is a walkway or not. Typically, railings must be 42 in. high.

FORCES IN A RETAINING WALL

Soil pressure on a retaining wall increases with increasing depth (**Figure 1-102**). Walls can fail in bending, by sliding, or by toppling. When engineering a wall, all three failure modes must be analyzed.

Drainage for Retaining Walls

Soil pressures on a retaining wall increase drastically when the soil is saturated (see "Soils," **page** 3). At the same time, the chance of soils sliding or overturning increases when soil is wet. Backfilling with poorly draining material, or failing to provide positive drainage, greatly increases the odds of wall failure. Always backfill retaining walls with free-draining granular material (sand or gravel), and provide drains that allow water to escape from behind walls (see "Drainage," **page** 68).

REINFORCING RETAINING WALLS

Follow general guidelines for reinforcing foundations (see "Rebar," **page** 43).

Earth Pressure

Figure 1–102. Forces in a Retaining Wall

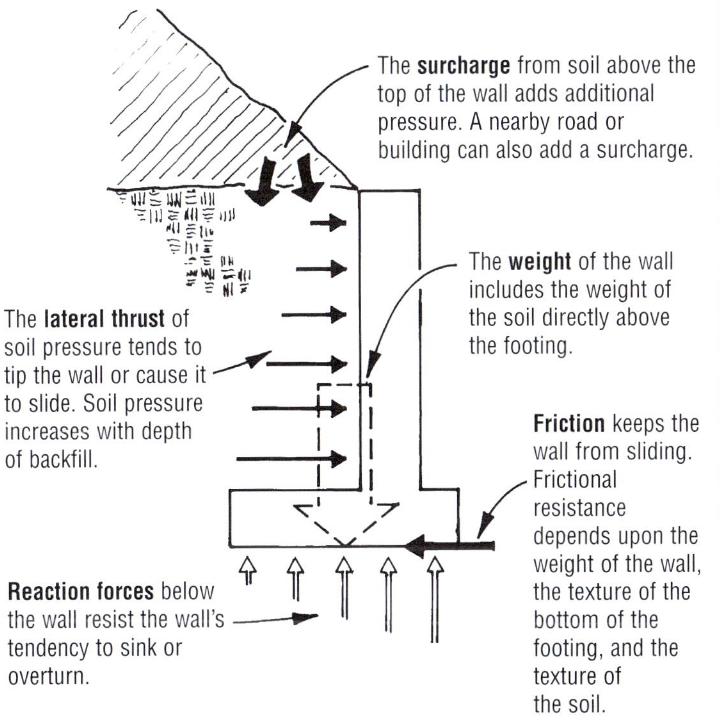

The greater the depth of the wall, the greater the total lateral force of the soil. This exerts an overturning force that is resisted by the weight of the soil over the footing and the weight of the wall itself. Friction at the base of the footing keeps the wall from sliding.

In a retaining wall, which is not braced at the top like a foundation wall, steel goes on the side of the wall close to the soil load, where tensile stresses occur (**Figure 1-103**).

Figure 1–103. Rebar in Concrete Retaining Walls

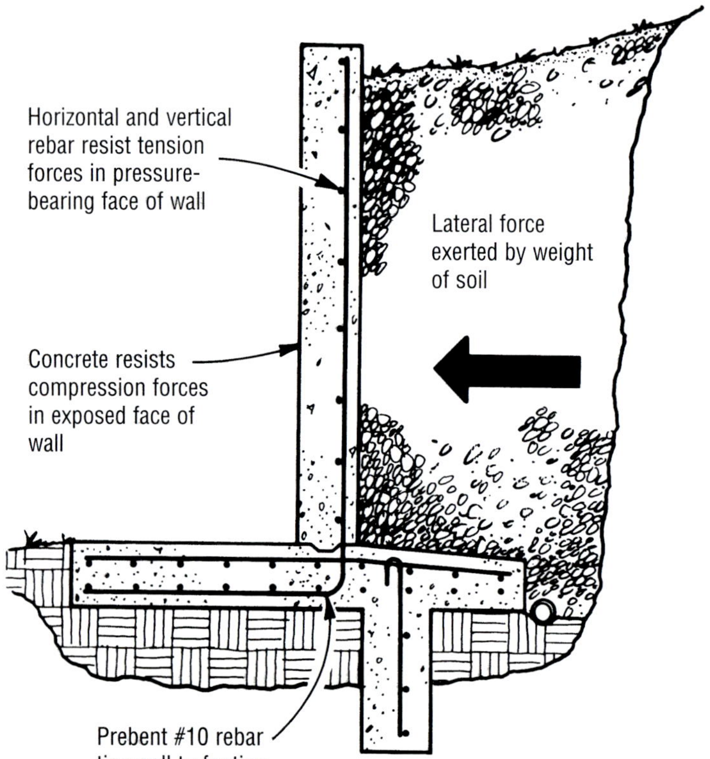

The compressive strength of concrete and the tensile strength of steel work together to resist lateral pressures placed on a retaining wall.

For more information: "Trouble-Shooting Low Retaining Walls," JLC 5/93

POURED CONCRETE AND MASONRY RETAINING WALLS

For concrete and masonry retaining walls less than 5-ft.-tall, follow guidelines for steel placement and dimensions as shown in **Figure 1-104**. Taller walls should be designed by a qualified engineer.

TIMBER RETAINING WALLS

Crib walls built out of landscaping ties function like a gravity wall. The mass of earth in the crib structure holds back the pressure of the soil behind it. Wood must be treated against rot. Timbers can be joined with 10- or 11-in.-long, hot-dipped, galvanized spikes or with rebar dowels in pre-drilled holes.

Crib walls can be as high as 30 ft., but a qualified engineer should design walls higher than 5 ft. For walls 5 ft. or shorter, follow the guidelines shown in **Figure 1-105**.

Tiebacks for Low Walls

For very low timber retaining walls in granular soil, you may be able to use tiebacks instead of an actual crib design (see **Figure 1-106**). Tiebacks spaced 16 ft. o.c., staggered on alternate courses, are typical. (For greater holding power, space tiebacks closer.) Tiebacks should extend into soil a distance equal to the wall height.

INTERLOCKING BLOCK RETAINING WALLS

Geogrid is a plastic grid material used to reinforce soil banks. In combination with interlocking block systems, it can be used to create a block-faced earth retaining wall that functions similarly to a concrete gravity retaining wall (**Figure 1-107, page 95**).

Figure 1-104. Masonry Retaining Walls

Poured Concrete

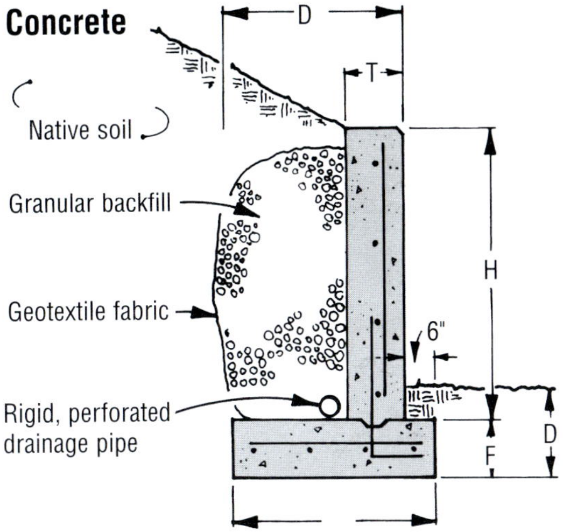

Concrete Block

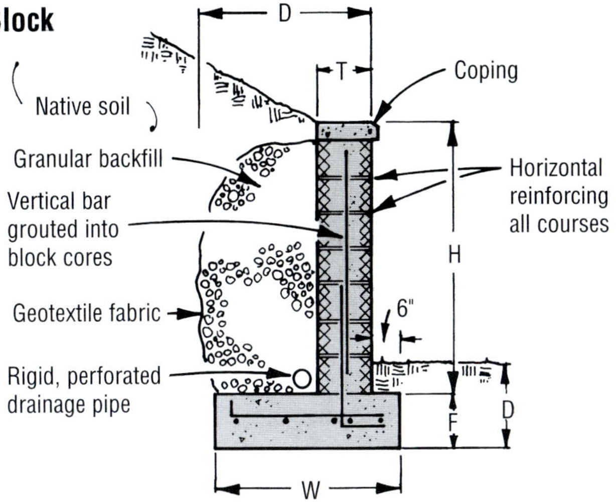

Mortared or Dry-Laid Stone

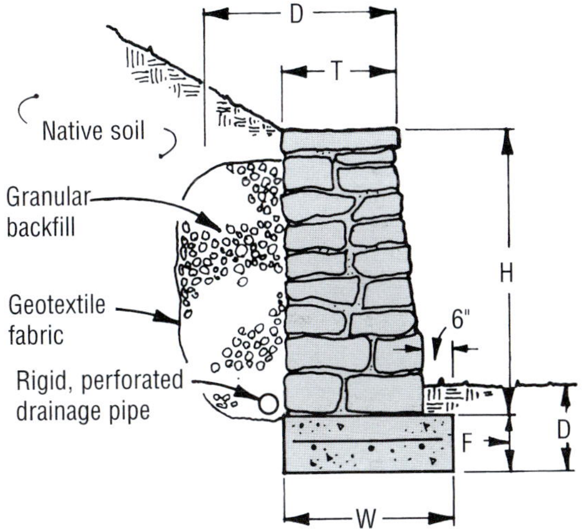

Recommended Dimensions for Low Masonry Retaining Walls*

H	W	Steel Rebar	Bar Spacing	F	T	D
2'	20"	#3 (3/8")	2'-0" o.c.	9"	8"	Local frost depth or 12" to 18"
3'	25"	#4 (1/2")	2'-0" o.c.	10"	8"	
4'	32"	#5 (5/8")	2'-0" o.c.	11"	10"	
5'	42"	#5 (5/8")	1'-6" o.c.	12"	12"	

*Suggested details for walls no higher than 5 ft. where dense, coarse-grain soil exists below footings. Not for loose or soft sand, peat, or clay.

Figure 1–105. Timber Retaining Walls

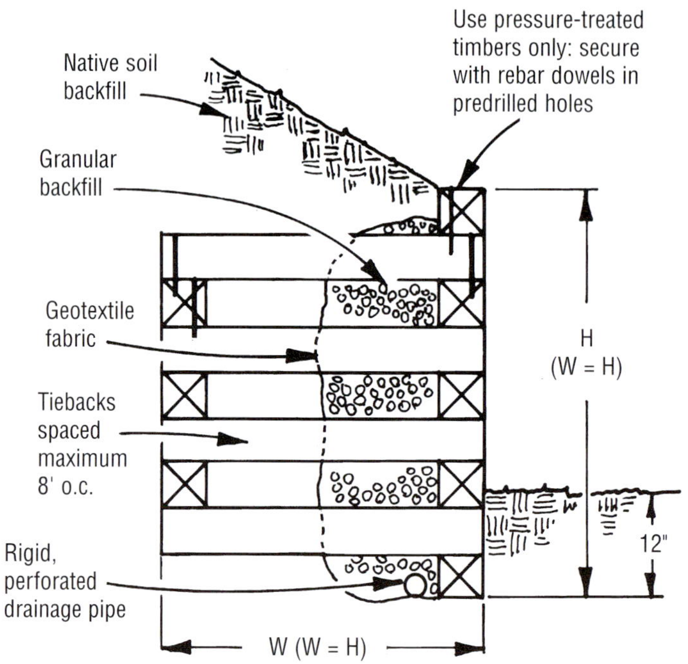

Recommended Dimensions for Low Timber Crib Walls*

Timber Size	Dowel Size	Spacing of Tiebacks
6x6	1/2" (#4 bar)**	6'-0" (max.)
8x8	3/4" (#6 bar)**	8'-0" (max.)

* Details apply to walls no higher than 5 ft.

**In acidic soils, increase by 1/4 in. or use hot-dipped galvanized.

For more information: "Low Retaining Walls," JLC 3/88

"Trouble-shooting Low Retaining Walls," JLC 5/93

"Practical Engineering: Interlocking Block Retaining Walls," JLC 5/94

Figure 1–106. Tiebacks for Low Retaining Walls

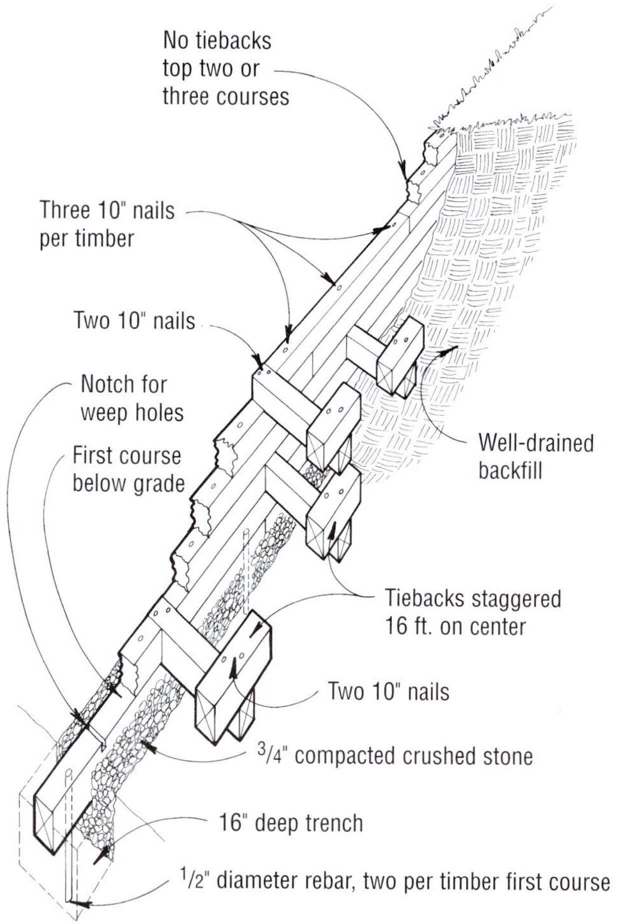

T-shaped timber tiebacks every 16 ft. on each course help to stabilize very low timber walls. In heavy soils, backfill with sand or crushed stone. With new, surfaced timbers that fit snugly together, weep holes or perforated drain pipe is needed to relieve hydrostatic pressure.

Retaining Walls continued from page 92

Engineered Tall Walls

A qualified engineer should design block walls higher than 5 ft. Usually, the vertical spacing of the geogrid is about twice the depth of the masonry units, and the geogrid extends into the soil horizontally a distance equal to 60% to 80% the total wall height, depending on the surcharge above the wall.

Figure 1–107. Interlocking Block Retaining walls

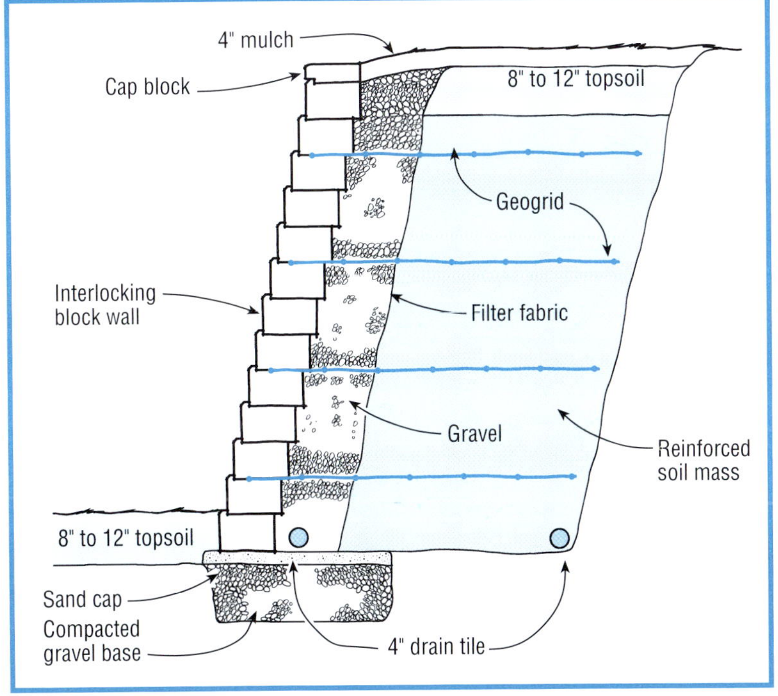

In an interlocking block wall, geogrid is typically placed on soil layers at 1- to 2-ft. intervals, and the soil is compacted in lifts. Block without geogrid may be sufficient for walls up to 4 ft. in height; with geogrid, walls can be as high as 16 ft. or more.

Radon Testing

Radon is a naturally occurring radioactive gas that is known to cause cancer in high doses. The risk associated with radon in residences remains uncertain, but the Environmental Protection Agency has recommended that every existing home be tested for radon and that action be taken if the test shows radon concentrations greater than 4.0 picocuries/liter of air.

Check Local Sources

The EPA also recommends that new homes be built with radon resistant elements (**Figure 1-108**). This recommendation has been adopted as a requirement by some codes. These precautions are more important in areas where higher concentrations of radioactive materials are present in local rock. Check with state public health and environmental agencies about testing and risks in your area.

Subslab Poly

The EPA-recommended details for radon abatement call for poly to be placed above the gravel subbase and below the slab. Placing concrete directly on plastic can contribute to slab cracking due to uneven shrinkage and curling. For that reason, consider putting a layer of damp compacted sand above the poly before placing the concrete (see "Subslab Vapor Barriers," **page** 78).

CRITICAL CONSTRUCTION DETAILS

The EPA recommends the following system to prevent radon from entering basements and to conduct radon from below the slab to the outdoors:

- **Subslab:** Place gas-permeable layer (4-in. layer of clean gravel).
- **In basements:** Plastic sheeting should be placed on top of gravel sub-base and under slab.
- **In crawlspaces:** The sheeting is placed over the crawl-space floor.
- **Seal and caulk** all cracks, joints, or penetrations.
- **Use 3- or 4-in. PVC vent stack** from sub-slab to above roof. Mark this clearly so plumbers won't mistake it for a drain vent.
- **In attics:** Supply a junction box for power in case homeowners choose to install a powered vent fan.

Figure 1–108. Radon Elimination Strategies

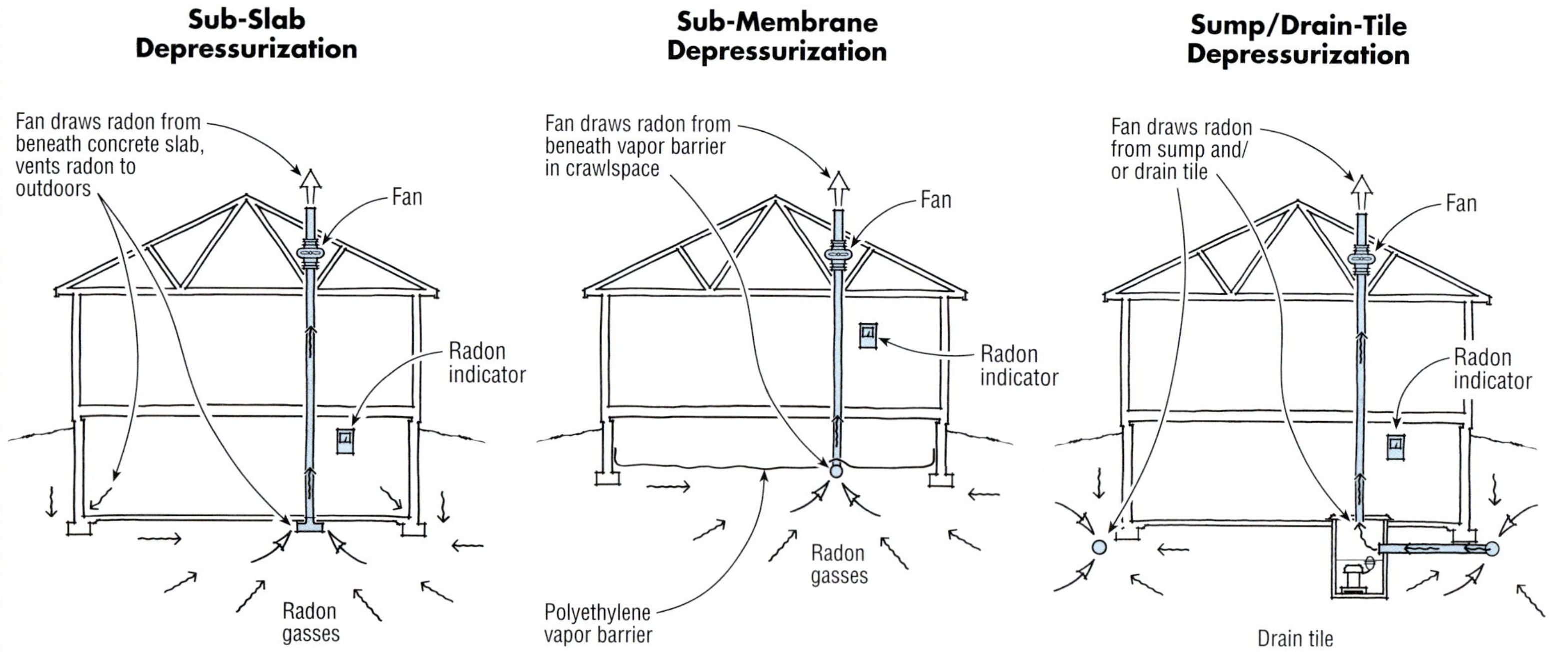

*Different strategies for venting radon are required for different foundations: Houses with basements (**left**), houses with crawlspaces (**center**), and basements or slabs with sump chambers (**right**).*

For more information: "Radon Vent Retrofit," JLC 12/01

CONVERTING LINEAR FEET TO BOARD-FEET

Formula method. To find board-feet, multiply the total length (in ft.) by the nominal lumber thickness by the width (in in.), and then divide the total by 12. For example: (10) 8-ft. 2x4s = (80 × 2 × 4) ÷ 12 = 53.3 board-feet

$$(L \times T \times W) \div 12$$

Factor method. Alternatively, use the conversion factors in Figure 2-1.

Figure 2–1. Linear Feet to Board-Feet Conversion

Nominal Lumber Size	Conversion Factor
1x4	.33333
1x6	.50000
1x8	.66667
1x10	.833333
1x12	1.00000
2x3	.50000
2x4	.666667
2x6	1.00000
2x8	1.33333
2x10	1.66667
4x4	1.33333
4x6	2.00000
6x6	3.00000

As an alternative method of calculating board-feet, multiply the lin. ft. of each lumber size you are using by the corresponding conversion factor. Example: (10) 8-ft. 2x4s = 80 × .666667 = 53.3 board-feet

FLOORS AND CEILINGS

Estimating Joists

Estimating Number of Joists

To calculate the number of joists, use the formulas shown in Figure 2-2 for the appropriate on-center spacing.

Figure 2–2. Estimating Joists and Rafters

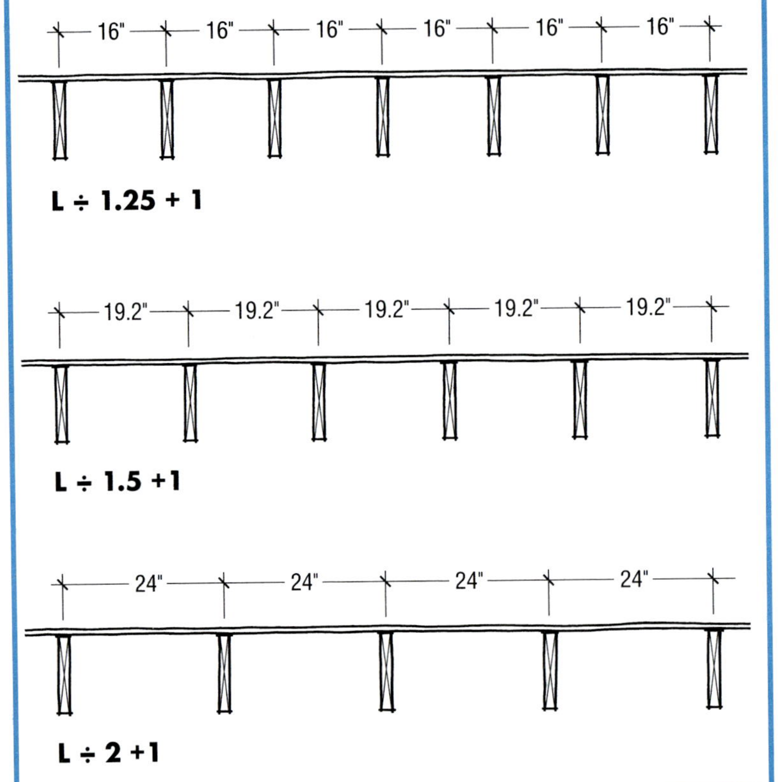

To calculate the number of joists needed: 1) Measure the width of the room (ft.); 2) divide by the appropriate o.c. spacing (ft.); and 3) add one to start.

Estimating Rim Joists

Remember to include rim stock in the total joist count. Rim joist stock is calculated as follows:

BUILDING LENGTH (FT.) $\times$ 2 $\div$ LUMBER LENGTH

Lumber lengths should be multiples of the o.c. spacing. For example, use 12s, 16s, 20s, etc., for 16-in. o.c. joist spacing. Depending on the length of the building, it may be more efficient to count these individually, mixing lumber lengths that break evenly on the joists.

Estimating Subflooring

The most accurate method of estimating subflooring is to graph scaled lines over the floor plans in 4x8-ft. increments. For a quick alternative, calculate as follows:

- For each section of subfloor, square off any jogs or cantilevers from the outside perimeter on the plans to form one large perimeter rectangle.
- For each section, multiply the length by the width of the perimeter rectangle.
- Divide by 32 (for 4x8 panels).

FLOOR AREA = LENGTH $\times$ WIDTH

SHEATHING (4X8 PANELS) = FLOOR AREA $\div$ 32

Remember to add a couple of sheets to the total panel count to accommodate cutting error and waste, especially if there are jogs in the floor plan.

For more information: "Estimating Rules of Thumb," JLC 3/96

WALLS

Estimating Plate Stock

Order wall plates in a quantity that's at least four times the total length of the walls. On walls that run in the same direction as the trusses or joists, an additional plate is needed for drywall backing at the ceiling. More will be needed to cover waste, miscellaneous backing, and continuous fire blocking for walls over 8 ft. high.

Estimating Studs

For a small house (less than 2,000 sq. ft.) with 16-in. o.c. framing, order 1 stud for each lin. ft. of wall framing — both interior and exterior. For a larger house framed 16 in. o.c., order 1.25 studs for each lin. ft. of wall framing.

Estimating Headers

In a house with windows under 36 in. wide, use the following shortcuts:

For solid-sawn headers:

1) Count the number of windows and doors; count French doors or sliders as 2;

2) Divide by 3 and order that many 10-footers.

For doubled-2x headers:

1) Count the number of windows and doors; count French doors or sliders as 2;

2) Multiply by 2;

3) Divide by 3 and order that many 10-footers.

For extra large or extra narrow windows, calculate headers individually.

Estimating Subflooring

Estimating Plates, Studs, and Headers

Figure 2–3. Rafter Line Length Ratios

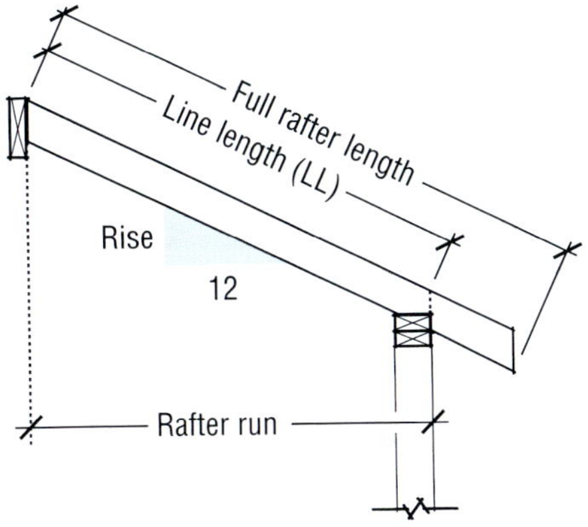

To convert Rafter Run to Line Length:

1) Select LL ratio for given slope in table above;

2) Multiply rafter run by LL ratio.

To find Full Rafter Length (add for any overhang at the eaves):

1) Multiply the horizontal length of overhang by the same LL ratio;

2) Add this overhang length to the rafter LL.

Rafter Line-Length (LL) Ratios

Roof Pitch/12	COM LL Ratio	H/V LL Ratio
1	1.0035	1.4167
1½	1.0078	1.4197
2	1.0138	1.4240
2½	1.0215	1.4295
3	1.0308	1.4361
3½	1.0417	1.4440
4	1.0541	1.4530
4½	1.0680	1.4631
5	1.0833	1.4743
5½	1.1000	1.4866
6	1.1180	1.5000
6½	1.1373	1.5144
7	1.1577	1.5298
7½	1.1792	1.5462
8	1.2019	1.5635
8½	1.2254	1.5817
9	1.2500	1.6008
9½	1.2754	1.6207
10	1.3017	1.6415
10½	1.3288	1.6630
11	1.3566	1.6853
11½	1.3851	1.7083
12	1.4142	1.7321
14	1.5366	1.8333
16	1.6667	1.9437
18	1.8028	2.0616
24	2.2361	2.4495

When estimating rafter lengths and calculating spans, convert the rafter run (horizontal distance) to a sloped distance (line length) using these ratios.

Examples: On a 6/12 roof with a run of 15 ft., find the net rafter length for the common rafters: 15 ft. × 1.118 = 16.77 ft. or 16 ft. 9¼ in. Round up to the next 2-ft. increment for the lumber length needed (18-footers).*

If the roof has a 16-in. overhang, measured along the rafter, the job will require 20-ft. rafters: 16 in. × 1.118 = 1.491 ft. + 16.77 ft. = 18.261 ft. or 18 ft. 3¼ in. Round up to 20-footers.

* For more precise calculations of rafters, use an *effective run*, which is shortened for the ridge thickness. For more information, see *A Roof Cutter's Secrets to Framing the Custom Home* (JLC Books)

Estimating Wall Sheathing

For rectangular wall areas:

WALL AREA = TOTAL WALL LENGTH × WALL HEIGHT

SHEATHING (4x8 PANELS) = WALL AREA ÷ 32

1) Multiply the total length of exterior walls by the wall height to get total wall area;

2) Subtract the areas of major openings such as sliders or large windows;

3) Divide by 32 (for 4x8 panels).

Remember to account for gable-end walls.

For regular gables (same pitch both sides of ridge):

Multiply span by total rise. This gives you the total for two gable-ends.

For irregular gables or single gable-end walls:

1) Multiply total run by total rise for each gable end;

2) Divide by 32.

ROOFS

Estimating Rafter Length

To find the full rafter length, multiply the correct line length ratio in **Figure 2-3** by the horizontal run of the rafter. This gives the net rafter length, or *line length*, from top plate to ridge. Then add for the plumb cut at the ridge and overhang to find the full rafter length.

For more information: "Unit-Price Estimating," JLC 4/94

"Unit-Pricing Pitfalls," JLC 7/99

Estimating Number of Common Rafters

To calculate the number of common rafters, use the formulas provided in **Figure 2-2 (page 97)** for the appropriate on-center spacing.

Remember to include ridge stock.

Estimating Hip and Valley Rafter Lengths

To calculate the length of each hip rafter, use the Hip/Valley LL ratio shown in **Figure 2-3**.

Estimating Jack Rafters

For a simple hip or valley, each pair of jack rafters is equal in length to one common (one long jack + one short jack = one common).

Estimating Roof Sheathing

ROOF AREA = PERIMETER WALL LENGTHS (INCL. OVERHANGS) × RAFTER LINE LENGTH

SHEATHING (4x8 PANELS) = ROOF AREA ÷ 32

1) Add overhang distance to exterior wall lengths;

2) Multiply the total length of the perimeter (wall length + overhang) by the rafter line length (**Figure 2-3**) to get total roof area (on gable roofs, don't include the gable-end wall in the total length);

3) Divide by 32 (for 4x8 panels).

Calculate roofs of different pitches individually.

Remember to add a couple of sheets to the total to account for error and waste.

SELECTING DIMENSIONAL LUMBER

Dimensional lumber is differentiated by several species groups. The main species groups in the U.S. are Spruce-Pine-Fir (Canadian), Douglas Fir-Larch, Hem-Fir, and Southern Pine (typically pressure-treated). Each species group is available in a number of grades, but unless otherwise specified, most framing lumber is #2. Lower grades may be allowed for studs (Stud-Grade) and top plates (Utility-Grade).

Strength values for these and other North American species and species groups are shown in **Figure 2-4**. For convenience, the span tables in this book (**pages 185-208**) are listed by species group. If a species or group is not listed in the charts, use the spans for a species or group with the same or higher Fb value.

Estimating Rafters

Design Values for Lumber

Figure 2–4. Strength of Wood

Design Values[1] for 2x8 (Nominal) Lumber

Species Group	Grade	Extreme Fiber Stress in Bending (Fb)[2]	Modulus of Elasticity (E)[3]
D-Fir-L	Select Strength	1,620	1.9
	No. 1/ No. 2	1,020	1.6
	No.3	570	1.4
SPF	Select Strength	1,500	1.5
	No. 1/ No. 2	1,050	1.4
	No. 3	600	1.2
Hem-Fir	Select Strength	1,560	1.7
	No. 1/ No. 2	1,200	1.6
	No.3	690	1.4
SYP	Select Strength	2,300	1.8
	No. 1	1,500	1.7
	No. 2	1,200	1.6
	No. 3	600	1.4

(1) These values include a size factor for 8-inch-wide nominal members used in normal conditions (lumber with a moisture content <= 19% placed on edge). Wet lumber or flat members require higher values.
(2) psi
(3) million psi

*This table shows comparative design values for the four main wood species used in this manual. The values assume the lumber will be loaded on the edge. Extreme fiber stress in bending (Fb) is a measure of the lumber's strength to resist loads applied perpendicular to the grain. This load produces **tension** in the wood fibers along the edge farthest from the applied load, and **compression** in the fibers along the edge nearest to the load. The Modulus of Elasticity (E) is a ratio of the amount the wood will **deflect** in proportion to the applied load. **E is a measure of stiffness, whereas Fb is a measure of strength.***

Shrinkage

Kiln-dried lumber is stamped K-D (kiln-dried) or S-Dry (surface dry), and is shipped with a moisture content of about 19%. Anything larger than a 6x6 is generally not available K-D.

In a completed building, framing eventually dries to an average of 6% to 11% moisture content, depending on climate. This drying causes the lumber to shrink **across** the grain; shrinkage **along** the grain is negligible. **Figure 2-5** shows the degree of shrinkage in flat-sawn framing lumber.

For more information: "Framing Details for Wood Shrinkage," JLC 4/96

For more information on design values: The U.S. Span Book for Major Wood Species (available free from Canadian Wood Council; 800/463-5091; http://www.cwc.ca/publications/us_span_book/)

Figure 2–5. Predicted Shrinkage of Dimension Lumber

Lumber Size	Actual Width	Width @ 19% MC (at Delivery)	Width @ 11% MC (Humid Climates)	Width @ 8% MC (Average Climates)	Width @ 6% MC (Arid Climates)
2x4	3 1/2"	3 1/2"	3 7/16"	3 3/8"	3 3/8"
2x6	5 1/2"	5 1/2"	5 3/8"	5 5/16"	5 5/16"
2x8	7 1/2"	7 1/4"	7 1/8"	7 1/16"	7"
2x10	9 1/4"	9 1/4"	9 1/16"	9"	8 15/16"
2x12	11 1/4"	11 1/4"	11"	10 15/16"	10 7/8"

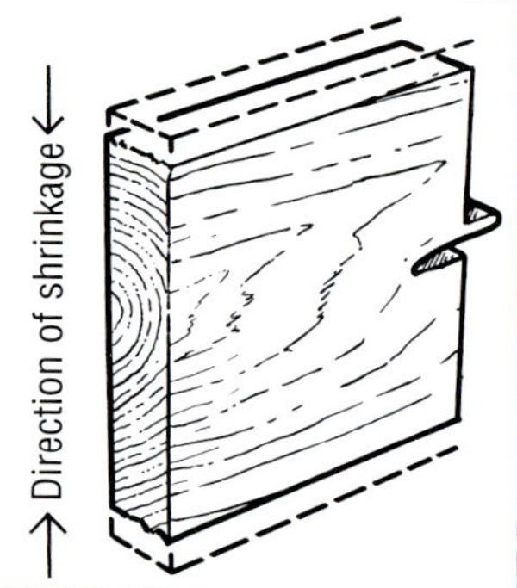

Framing lumber shrinks primarily across its width; shrinkage from end to end is insignificant. Actual shrinkage varies depending on the lumber's moisture content when delivered and the area's climate.

Shrinkage in large carrying beams can cause one part of a house to settle more than others, causing drywall cracks and other problems (**Figure 2-6**).

Figure 2–6. Avoiding Cumulative Shrinkage

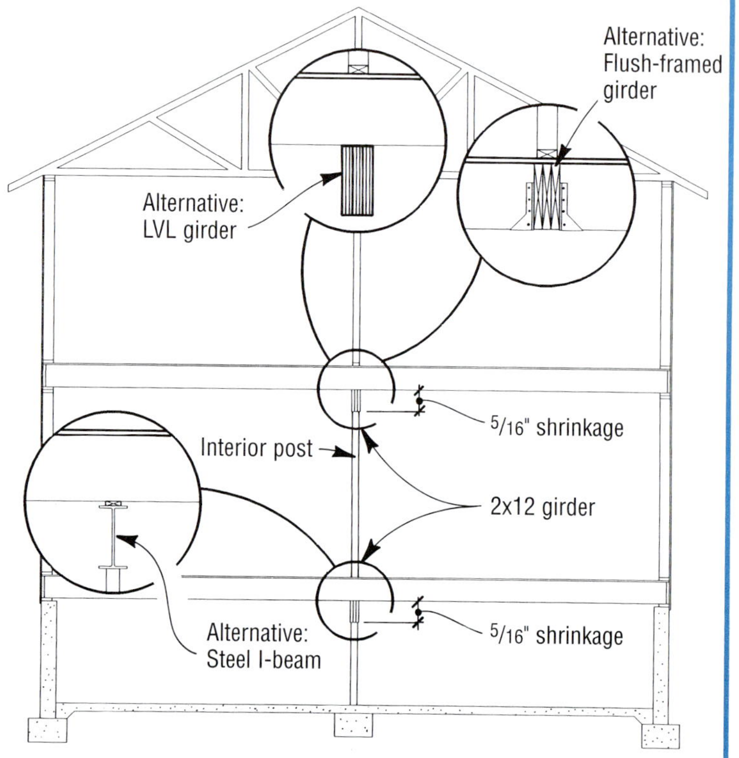

The two 2x12 girders in this building will shrink enough to cause a 1/2-in. drop in the second-floor level — enough to cause nail pops and cracks in the finishes. Use steel, engineered lumber, or flush framing to eliminate the problem.

However, using flush beams with hangers, or engineered lumber or steel, can reduce the potential for shrinkage problems. If dimensional lumber is hung flush from a beam made of steel or engineered lumber, the result can be a bulge at the beam (see **Figure 2-7**).

Figure 2–7. Flush-Framed Floor Joists

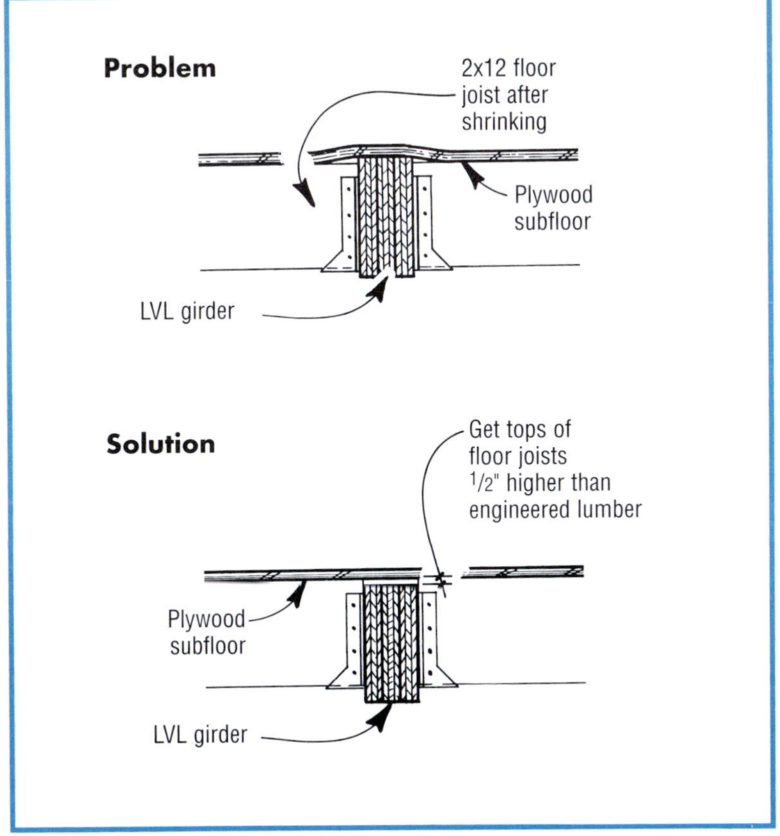

When installing dimensional-lumber floor joists flush with the top of engineered or steel beams, install the joists 1/2 in. higher than the girder to accommodate shrinkage.

SELECTING SHEATHING AND SUBFLOOR PANELS

Sheathing and subflooring should consist of performance-rated structural panels. These are made from plywood or oriented-strand board (OSB), and should carry a stamp (**Figure 2-8**) from APA-The Engineered Wood Association (formerly the American Plywood Association).

Lumber Shrinkage

Reading Plywood Ratings

Panel Use Ratings

For framing, choose only from the following panel grades:

- **Sheathing** panels are designed for use on roof decks and walls;
- **Structural 1** panels are designed for use on shear walls;
- **Sturdi-Floor®** tongue-and-groove panels are designed for use as subflooring.

Figure 2–8. Reading Sheathing Grade Stamps

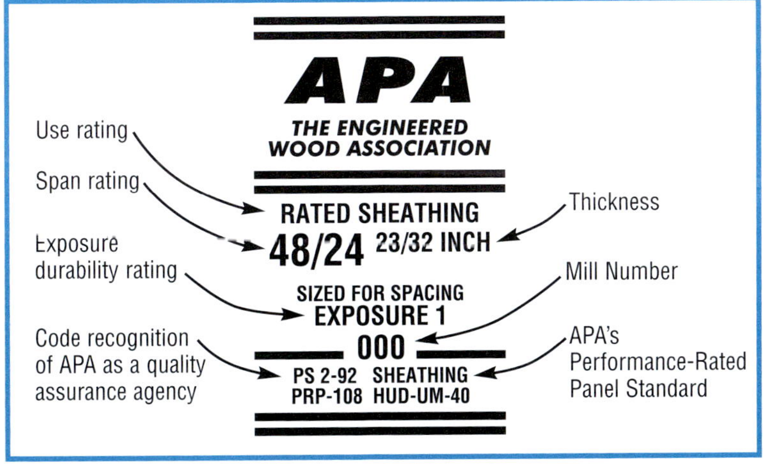

The key information found on a grade stamp for sheathing-rated plywood or OSB panels includes the panel's use rating, span rating, exposure durability, and thickness.

For more information: "Rating OSB Performance," JLC 2/94

Load-Span Tables for APA Structural-Use Panels (APA publications: http://www.apawood.org/pdfs/managed/Q225D.pdf)

Panel Handbook and Grade Glossary (APA publications: http://www.apawood.org/pdfs/managed/X505-R.pdf)

Panel Exposure Ratings

For framing, select panels with only the following exposure ratings:

- **Exterior** panels can be used outdoors;
- **Exposure 1** panels can withstand moisture during normal construction delays — up to a full season, according to APA. However, the panels should be covered up as soon as possible.

Other exposure ratings for wood panels include:

- **Exposure 2 panels** are typically used for protected construction and industrial applications where the potential for high humidity and water leakage may exist for limited times;
- **Interior panels** are made with water-soluble interior-grade glues and are intended for interior applications only, not for framing.

Panel Span Ratings

Span ratings on sheathing grade stamps may have two numbers such as 24/16. The first indicates the allowable roof span (in this case, 24 in. o.c.) and the second indicates the subflooring span (16 in. o.c.). Span ratings for multi-purpose panels are: 24/0, 24/16, 32/16, 40/20, and 48/24.

If a sheathing panel stamp includes only one span number, the panel is for use on walls only.

Sturdi-Floor® panels are engineered for use as subflooring. They have span ratings of 16-, 20-, 24-, 32-, and 48-in.

Panel Thickness

Plywood and OSB are always 1/32 in. smaller than their nominal sizes. For example, 1/2-in. plywood actually measures only 15/32 in.

Roof Panel Span

Roof panels should be a minimum of 1/2 in. thick for 16-in. o.c. framing, and 5/8 in. thick for 24-in. o.c. framing.

For roofs framed 24-in. o.c., use H-clips to support the edges of the sheathing at mid-span. Clips create the proper gap and help stiffen panel edges.

OSB

While the strength and nail-holding ability of OSB and plywood are equal for similarly rated panels, OSB swells more when wet. Because of this, some builders avoid OSB in subflooring, especially in kitchens or baths, where exposure to wet conditions is likely. Some flooring contractors will not install hardwood flooring over OSB subflooring, although tests have shown that it performs the same as plywood in this application. OSB is not considered a suitable underlayment for use directly under ceramic tile or resilient flooring, however.

Unused OSB should be stored under cover on the job site. If an installed OSB subfloor is left uncovered for a long period after installation, consider treating it with a water repellent.

SELECTING ENGINEERED LUMBER

Engineered studs and beams are not direct substitutes for solid lumber. Each engineered product has special installation requirements. Be sure to consult the product's technical literature before using any new engineered product.

Most engineered lumber is made with exterior glue. The glue will hold up under normal construction delays, but it is not waterproof. Therefore, all engineered lumber should be protected from the weather on-site and stored flat.

Wood I-Joists

Using wood I-joists places the strongest material (typically LVL) where the bending stresses are greatest — in the *flanges* along the top and bottom of the beam. This lets the lightweight beams span great distances with minimal deflection, providing a very stiff floor.

I-joist *webs* consist of plywood or OSB, making them more susceptible to water damage than sawn lumber. Therefore, I-joists must be stored under cover — particularly those with OSB webs. Care must be taken when drilling or notching the webs to allow mechanicals to pass through (see "Boring and Notching I-Joists," **page 123**).

Engineered Studs

For tall walls and kitchen walls that must remain straight for countertop and cabinet runs, engineered studs are a cost-effective alternative to dimensional lumber. Follow manufacturers' guidelines for installing specific products. General use precautions are shown in **Figure 2-9**.

Finger-jointed studs aren't any stronger than dimensional lumber of the same species, but because the defects have been cut out, they are more dimensionally stable. In addition, the grain is interrupted by glue joints, making finger-jointed studs less likely to twist and bow.

The glue that holds the joints together is water-resistant, but these studs should be protected from prolonged exposure to the weather.

Laminated Strand Lumber (LSL). LSL is made from strands of wood bonded with a polyurethane adhesive, and it is used mostly for rim boards, studs, and headers. LSL studs are straight and true, but heavier than solid lumber. They have sharp edges so wear gloves when handling. LSL studs take air nails about the same way dimensional lumber does, but there are fewer problems with splits if nails are kept at least 1/2 in. from the edges.

OSB vs. Plywood

Selecting Engineered Lumber

Figure 2–9. Engineered Studs for Tall Walls

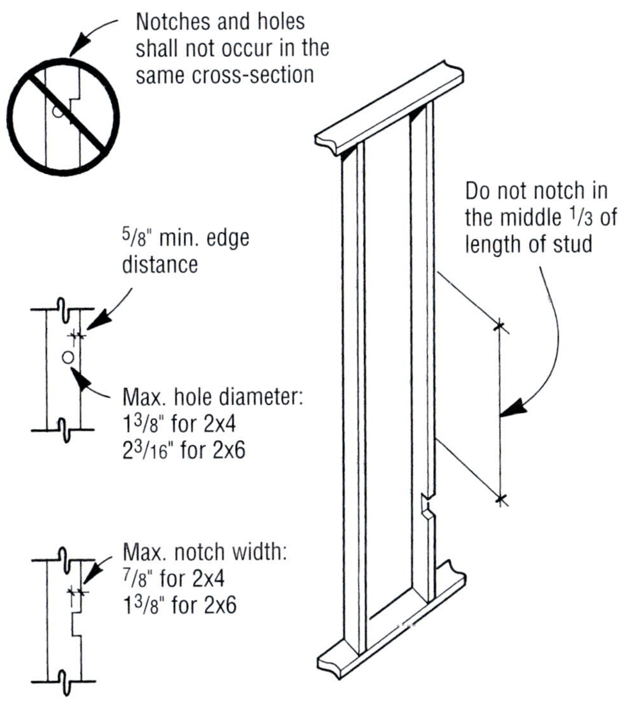

Certain restrictions may apply to drilling holes and cutting notches in tall engineered studs. Whenever using engineered lumber, consult the manufacturer's specs for design values, load tables, and nailing requirements.

Engineered Beams

Engineered beams may consist of laminated veneer lumber (LVL), glue-laminated timbers (glulams), or parallel-strand lumber (Parallam®). Be aware that the large loads carried by an engineered beam will concentrate at the bearing points. To keep the beam's fibers from crushing under the weight, be sure to allow sufficient bearing surface. Installation of large engineered beams may require a crane or lift.

Laminated-Veneer Lumber (LVL). LVL is a strong, versatile, engineered product that resembles plywood. LVL beams are typically built up from several LVL members nailed or bolted together. Bolting is preferable where the beam is side-loaded (e.g., a floor system with flush framing) or where it will be exposed to the weather. Side loading with joist hangers can cause the outer LVL members to "peel off" if they are not bolted together.

Before ordering LVL, it's important to note the grade, species, and manufacturer specified by the designer. For example, Douglas-fir LVL is somewhat weaker than yellow-pine LVL. While most LVL has a water-resistant coating, it is still vulnerable to cupping when exposed to moisture. So store it dry.

Glue-Laminated Timbers (glulams). A glulam consists of dimension lumber face-laminated with structural adhesives. Three appearance grades are available — industrial, architectural, and premium. Design values are the same for all three grades, but the better grades have fewer surface defects. This is important in residential construction where most glulams are left exposed.

Glulams come in widths ranging from $2^{1}/2$ to $10^{3}/4$ in. Depths are multiples of the lamination thickness. For example, a glulam with eight $1^{1}/2$-in. laminations will be 12 in. deep.

Most glulams are delivered in a wrapping that protects the surface of any exposed beam during shipping and installation. If possible, don't remove this wrapping until all interior finishing is complete. Protect wrapping so that water doesn't leak in and become trapped.

Never notch or drill a glulam without consulting either the fabricator or a structural engineer. Connections should be made so that the beam can swell and shrink with moisture swings.

Parallel-Strand Lumber (Parallam). Sold under the brand name Parallam®, this product consists of strands of veneer that are aligned, coated with glue, and cured with a combination of pressure and microwaves. The resulting beam is stronger and stiffer than solid-sawn lumber and is equivalent in strength to multiple LVLs. However, you save the expense of bolting and don't need to worry about side loading or warping from moisture. The material is somewhat harder to cut and nail than solid lumber.

HOLDING POWER OF NAILS

A nail's holding power is a function of its diameter, how far it penetrates the wood, and the type of wood it penetrates. In framing, nails should never be loaded in withdrawal (where the load acts parallel to the nail shank and tries to pull it out). Instead, framing nails should be loaded only laterally (where the load acts perpendicular to the shank). Also, nails have more holding power when driven into the side grain rather than into the end grain. In fact, nailing into the end grain will reduce a nail's lateral load capacity by approximately one-third.

Box vs. Common Nails

Lateral strength is largely a function of a nail's diameter and the density of the type of wood into which the nail is driven. For example, 10d and 12d nails have the same diameter and the same lateral strength in each type of wood. Common nails are stronger than box nails because of their greater diameter (**Figure 2-10**).

Selecting Engineered Beams

Selecting Nails

Figure 2–10. Lateral Strength of Common vs. Box Nails

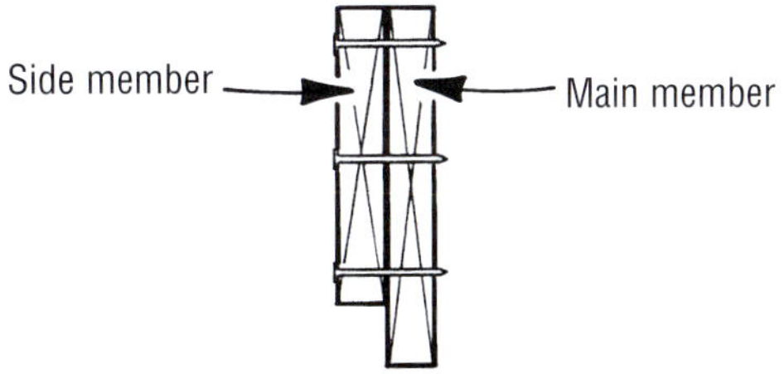

Penny Wt.	**8d**		**10d**		**12d**		**16d**		**20d**	
Nail Type	Common	Box	Common	Box	Common	Box	Common	Box	Common	Box
Diameter (in.)	0.131	0.113	0.148	0.128	0.148	0.128	0.162	0.135	0.192	0.148
Side Member Thickness (in.)	0-3/4		0-3/4		0-3/4		1 1/2		1 1/2	
	Lateral Rating (lb.)									
SPF	70	57	83	68	83	68	120	88	144	100
D-F-L	90	72	105	87	105	87	141	103	170	118
Hem-Fir	73	58	85	70	85	70	122	89	147	102
SYP	104	79	121	101	121	101	154	113	185	128

Nails are rated for "lateral capacity," which is greatly affected by the nail's diameter. Since box nails are skinnier than commons, they have lower strength values. The values shown vary with the type of framing lumber used, and they assume the nail will penetrate the main member (illustration at top) at least 12 diameters.

When substituting box for common nails, calculate the nail size needed using the conversion ratios shown in **Figure 2-11**.

Withdrawal Strength of Nails

Nails are much stronger when loaded laterally (across the nail) than when loaded in withdrawal (along the length of the nail). Withdrawal from end grain is particularly weak and not accepted as a structural connection by most codes. Withdrawal values from sidegrain are given in **Figure 2-12**.

Figure 2-11. Conversion Ratio for Common to Box Nails

To convert required number of common nails of a given size to box nails, multiply by the appropriate ratio and round up.

Penny Wt.	8d	10d	12d	16d	20d
Ratio	1.23	1.22	1.22	1.36	1.44

S-P-F lumber assumed.

Do not substitute a common nail that's specified on the plans for an equal number of box nails. Instead, multiply the specified number of common nails by the conversion ratio shown in the chart and round up to find the equivalent number of box nails.

Toenails

For toenails loaded in withdrawal, multiply the values in **Figure 2-12** by .67. For toenails loaded laterally, multiply the values in **Figure 2-10** by .83. Toenails should be driven at an angle of about 30 degrees from the face of the stud or other member being attached (**Figure 2-13**).

PNEUMATIC NAILS

Pneumatic nails are typically sold by a specified shank diameter in inches (.120, .131, and .148 are common for framing nails). These are typically skinnier than common nails of equal length and, therefore, have lower lateral strength (**Figure 2-14**) and withdrawal values (**Figure 2-15**). Consult manufacturers for lateral strength of specific nail types.

For more information on air-nail design values: National Evaluation Report (NER-272): Power-Driven Staples and Nails For Use in All Types of Building Construction (available on the Senco web-site: http://www.senco.com/pdf/facts/ner272.pdf)

Figure 2-12. Withdrawal Strength Design Values of Common vs. Box Nails

Penny Wt.	8d		10d		12d		16d		20d	
Nail Type	Common	Box	Common	Box	Common	Box	Common	Box	Common	Box
Diameter (in.)	0.131	0.113	0.148	0.128	0.148	0.128	0.162	0.135	0.192	0.148
	Withdrawal Value*									
SPF	21	18	23	20	23	20	26	21	30	23
D-F-L	32	28	36	31	36	31	40	33	47	36
Hem-Fir	22	19	25	21	25	21	27	23	32	25
SYP	41	35	46	40	46	40	50	42	59	46

* psi per inch of penetration into side grain of main member

This table shows the allowable load values for typical common and box nails. These values represent the pounds per square inch of load applied per penetration into the side grain of the main member.

GALVANIZED NAILS

Galvanized nails are made from zinc-coated steel. The zinc protects the steel from rust. How well a nail resists rust depends on the amount of zinc and how it's applied.

Hot-dipped nails (coated by immersion in molten zinc) are the most durable.

Hot-galvanized (HG) nails are coated with zinc chips in a hot tumbler, a process that can leave an uneven coating. A certain percentage of these nails will rust when used outside.

Electroplated nails are smooth and shiny, but the thin zinc coating will oxidize away quickly when exposed. This makes them a poor choice for exterior work.

Figure 2–13. Proper Toenailing

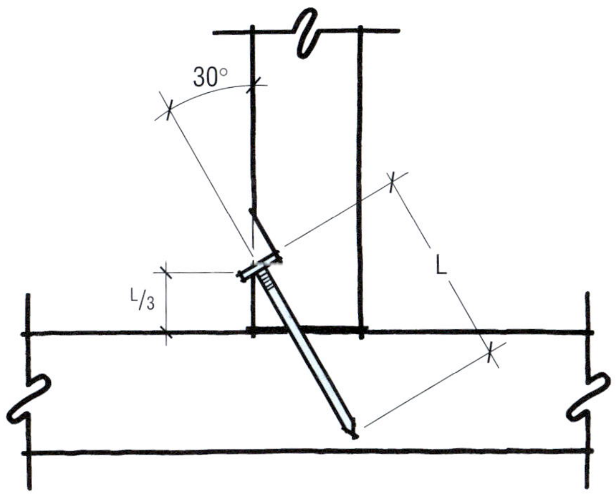

Place toenails away from the end of the board a distance that's equal to one-third the length of the nail. Angle the nail so it comes through at the center of the board's thickness.

Selecting Galvanized Nails

Figure 2–14. Lateral Strength of Air Nails

	Wood Species			
Nail Diameter	SPF	DFL	Hem-Fir	SYP
0.12"	19	29	20	37
0.131"	21	32	22	41
0.148"	23	36	25	46
0.162"	26	40	27	50

Values are based on a 10-year "normal" load duration, and assume: 1) smooth-shank nails driven perpendicular to the wood grain; and 2) both side and main members are of the same wood species. When toenailing, multiply the values by .83, as described in "Toenails," opposite page.

Nail Design Values

Figure 2–15. Withdrawal Strength Load Values for Air Nails

		Wood Species			
Nail Length	Nail Dia.	SPF	DFL	Hem-Fir	SYP
$2^1/2$"	0.131"	52	62	54	67
3"	0.12"	69	81	71	89
3"	0.131"	79	93	80	101
3"	0.148"	84	99	86	109
$3^1/4$"	0.12"	69	81	71	89
$3^1/4$"	0.131"	79	93	80	101
$3^1/2$"	0.162"	92	109	94	119

Withdrawal values (psi per inch of penetration into the main member) are based on a 10-year "normal" load duration, and assume: 1) smooth-shank nails driven perpendicular to the wood grain; and 2) both side and main members are of the same wood species. When toenailing, multiply the values by .83, as described in "Toenails," opposite page.

Mechanical plated, or "peen" plated, nails have a zinc layer that isn't as thick as that used on a hot-dipped nail, so it is less corrosion-resistant. However, mechanical plating is appropriate for screws as the thin coating won't clog the threads.

ALUMINUM AND STAINLESS-STEEL NAILS

Aluminum nails are sometimes used to fasten exterior siding. They're very resistant to rust and corrosion, but may corrode when used with some flashing metals. (See **Figures 3-51** and **3-52, pages 256-257.**)

Stainless steel. While expensive, stainless-steel nails are the most rust-resistant under nearly all conditions. They are highly recommended below grade or in homes that will be exposed to salt air. They're also recommended for cedar or redwood trim, and for siding that will be left to weather without stain or paint. Stainless-steel nails are typically available as Type 304 and 316. The 316 type are more durable.

NAILING RULES OF THUMB

Typical nailing schedules appear in **Figure 2-16.**

Two is better than one. In general, never rely on a single nail. Use at least two.

Nail spacing. Don't space nails closer than one-quarter their length to the edge of the board.

Nail penetration. To hold at full strength, nails should penetrate the wood a depth that's at least 11 times their diameter — 1 1/2 in. for 8d nails and 1 3/4 in. for 16d nails.

Figure 2–16. Nailing Schedules

Roof Framing	**Number of Common Nails**	**Number of Air Nails**
Rafter to top plate (toe-nailed)[1]	3-8d per rafter	3-3 x .131 per rafter
Ceiling joist to top plate (toe-nailed)[1]	3-16d per joist	5-3 x .131 per rafter
Ceiling joist to parallel rafter (face-nailed)[1][2]	3-16d per joist	4-3 x .131[1] per rafter
Ceiling joist laps over partitions (faced-nailed)	3-16d per lap	4-3 x .131 per rafter
Collar tie to rafter (face-nailed)[1]	3-16d per joist[1][2]	4-3 x .131 per rafter
Blocking to rafter (toe-nailed)	2-8d each end	2-3 x .131 each end
Roof rafter to ridge beam[1]	3-10d per rafter	3-3 x .131 per rafter
Jack rafter to hip (toe-nailed)	3-10d per rafter	3-3 x .131 per rafter
Roof Sheathing[1]		
Structural panels	1-8d every 6 in. edge, 12 in. field	1-2 1/2 x .131 every 6 in. edge, 12 in. field

Nailing Schedules

Figure 2–16. Nailing Schedules, continued

Wall Framing	**Number of Common Nails**	**Number of Air Nails**
Top or bottom plate to stud (end-nailed)[1]	2-16d per stud	3-3 x .131 per stud
	3-16d per 2x6 stud	4-3 x .131 per stud
Top or bottom plate to stud (toe-nailed)[1]	3-16d per stud	4-3 x .131 per stud
	4-16d per 2x6 stud	
	5-16d per stud	3-3 x .131 per stud
Top plate to top plate (face-nailed)[1]	2-16d every 16 in.	2-3 x .131 every 12 in.
Top plates at intersections (face-nailed)	2-16d each side of joint	4-3 x .131 each side of joint
Stud to stud (face-nailed)	2-16d every 24 in.	2-3 x .131 every 16 in.
Header to header (face-nailed)	1-16d every 16 in. along edges	1-3 x .131 every 12 in. along edges
Bottom plate to floor joist, band joist, end joist, or blocking (face-nailed)[3][4]	1-16d every 16 in.	1-3 x .131 every 8 in. along edges
Wall Sheathing[1]		
Structural panels	1-8d every 6 in. edge, 12 in. field	1-2 1/2 x .131 every 6 in. edge, 12 in. field
Floor Framing		
Joist to sill, top plate or girder (toe-nailed)	4-8d per joist	4-3 x .131 per joist
Bridging to joist (toe-nailed)	2-8d each end	2-2 1/2 x .131 each end
Blocking to joist (toe-nailed)	2-8d each end	2-2 1/2 x .131 each end
Blocking to sill or top plate (toe-nailed)	3-16d each block	4-3 x .131 each block
Ledger strip to beam (face-nailed)	3-16d below each joist	4-3 x .131 below each joist
Joist on ledger to beam (toe-nailed)	3-8d per joist	3-3 x .131 per joist
Rim joist to joist (end-nailed)	3-16d per joist	4-3 x .131 per joist
Rim joist to sill or top plate (toe-nailed)[1]	2-16d every 12 in.	2-3 x .131 every 8 in.
Floor Sheathing		
Structural panels		
1 in. or less	1-8d every 6 in. edge, 12 in. field	1-2 1/2 x .131 every 6 in. edge, 12 in. field
Greater than 1 in.	1-10d every 6 in. edge, 6 in. field	1-3 1/4 x .131 every 6 in. edge, 6 in. field

(1) For high wind areas, consult local codes.
(2) See **Figure 2-87, page 145**, for detailed schedules to accommodate specific roof loading conditions.
(3) Nailing requirements are based on wall sheathing nailed 6 in. o.c. at the panel edge. If wall sheathing is nailed 3 in. o.c. at the panel edge to obtain higher shear capacities, nailing requirements for structural members must be doubled, or alternate connectors, such as shear plates, must be used to maintain the load path.
(4) When wall sheathing is continuous over connected members the tabulated number of nails can be reduced to 1-16d nail per foot.

HANGERS FOR SOLID WOOD JOISTS

When using joist hangers, it's critical that you use the proper size hanger and fill all holes with nails.

Do not split ends. A nail that splits the wood will reduce the joist's carrying capacity. In critical situations, you can predrill nail holes a depth of up to three-quarters the nail's diameter.

Double-shear connectors provide greater strength with fewer nails (**Figure 2-17**).

Joist-Hanger Nails

A particular hanger may call for 8d, 10d, 12d, or 16d common nails. Special joist-hanger nails also may be used. Though they have 8d or 10d shank diameters, they are only 1 1/4- to 1 1/2-in. long, so they can be driven into the joist without protruding through the other side. Using them where the hanger calls for full-length 10d or 12d nails will reduce the hanger's load capacity by 20% to 30%. Hanger catalogs give adjustment factors for specific nail sizes. Don't use joist-hanger nails in place of longer nails in double-shear hangers or as face nails on skewed hangers.

Effect of Lumber Shrinkage on Hangers

Joist hangers are designed for use with dry lumber. If lumber is wet, shrinkage may leave a gap between the bottom of the joist and the hanger's saddle or between the top of the joist and the floor decking.

WOOD I-JOIST HANGERS

Wood I-joists require special hangers. Do not try to adapt conventional hangers.

Figure 2–17. Double Shear Connectors

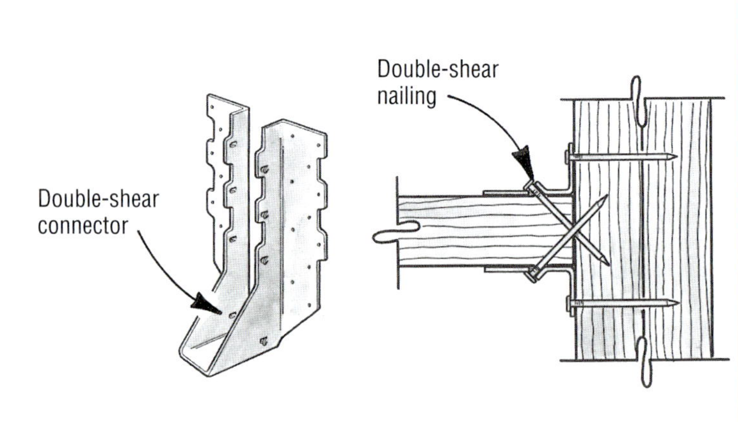

The strongest joist hangers are those that offer double shear nailing that crosses the joist ends. The angle also makes it easier to nail them to the joist in a joist bay.

Figure 2–18. I-Joist Hanger Height

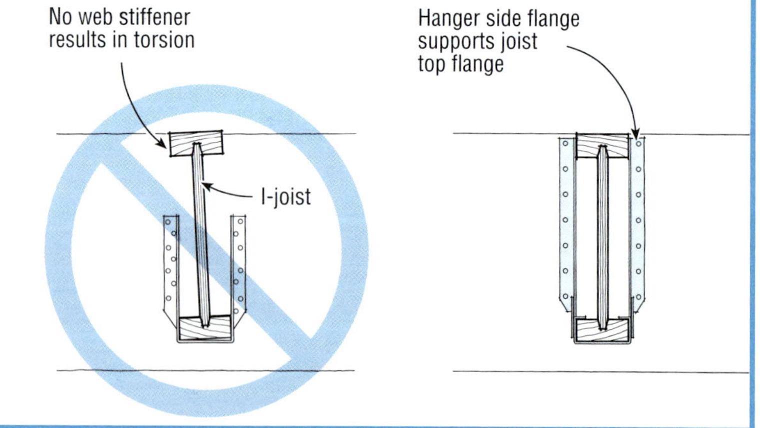

Joist hangers must be the full height of the I-joist (right), unless web stiffeners are used. Otherwise the joist may twist (left).

For more information: "Practical Engineering: Using Metal Connectors," JLC 5/97

I-Joist Hanger Height
Hangers for wood I-joists should be tall enough to catch the I-joist's top flange unless a web stiffener is used; otherwise, the joist may roll in the hanger (**Figure 2-18**). If a web stiffener is used, the hanger side flange should be at least 60% of the joist depth.

I-Joist Hanger Width
Hangers for wood I-joists should be the same width as the joist flange. Do not trim the flange to fit a narrower hanger; this would reduce the joist's strength. A wider hanger will either leave a gap, potentially causing a squeak, or will require filling with plywood, which would unevenly load the hanger.

Nailing I-Joist Hangers
Nail in every hole with correct sized nails; leaving nails out greatly reduces the hanger's capacity. Nails that are too long can hit the bottom of the hanger and curl under the joist, causing a squeak.

Web Stiffeners
For I-joist to I-joist connections, install web stiffeners against the bottom of the top flange of the carrying joist (**Figure 2-19**).

Wood Nailers for I-Joists
For hangers that bear on wood nailers, the nailer should not overhang its support more than 1/4 in. or sit more than 1/8 in. in from the edge (**Figure 2-20**).

Figure 2–19. Hangers on I-Joists

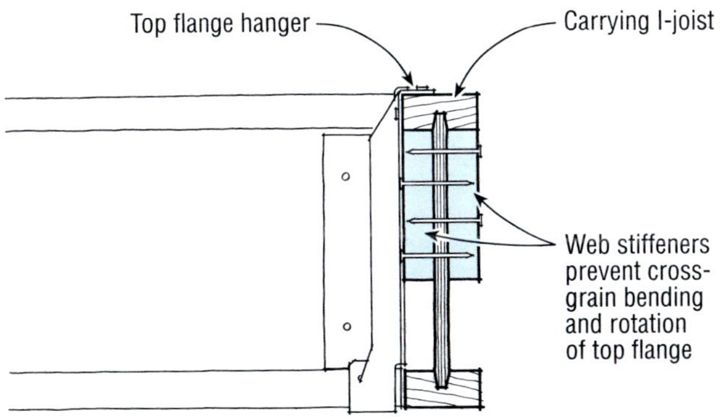

Before installing hangers to an I-joist, install web stiffeners against the bottom of the top flange of the carrying joist so that every nail in the hanger can be filled.

Figure 2–20. I-Joist Hangers on Concrete and Steel

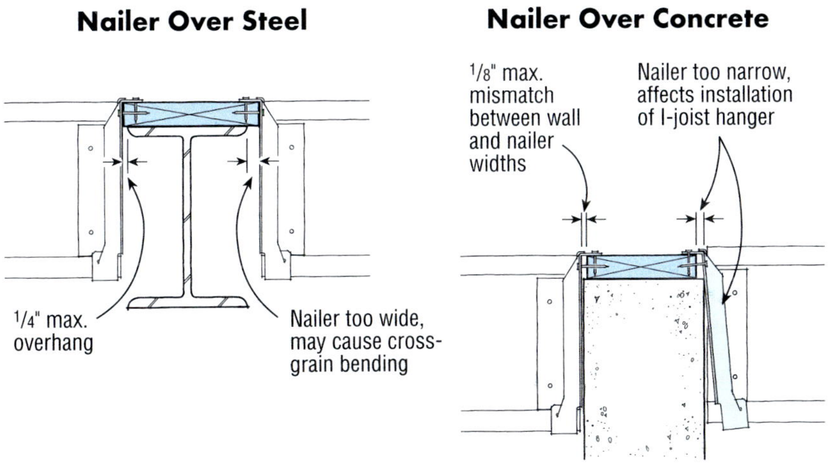

When applying a wood nailer over concrete or steel, make sure the nailer is not too wide (left) or too narrow (right), or this will compromise the strength of the joist hanger.

FLOOR SYSTEMS GET THEIR STRENGTH FROM THE DIAPHRAGM action of the subflooring, which is typically T&G plywood glued and either nailed or screwed over uniformly spaced joists.

JOIST DEPTH

Rule of thumb. Most codes assume residential floor loads to be 50 pounds per sq. ft. — 40 pounds live load plus 10 pounds dead load. If load conditions do not exceed this value, you can use the following rule of thumb for sizing uniformly loaded residential floor joists:

HALF THE SPAN + 2 = JOIST DEPTH

First, round the clear span of the floor joist up to the nearest foot and divide by 2. Then add 2 to the answer. For example: For a clear span of 15 ft. 6 in., round up to 16 and divide this span by 2, giving you 8. Next, add 2 to get the required depth in inches (8 + 2 = 10). It's important to use the *actual* lumber dimensions, not the nominal dimensions, so a 2x12 floor joist will be required.

Figure 2-21. Live Loads for Floors and Ceilings

Component	Live Load (lbs./sq. ft.)
Residential rooms	40
Decks	60
Balconies	60
Fire escapes	40
Stairs	40
Ceiling joists (limited attic storage)	20
Ceiling joists (no attic storage)	10

Live loads include all loads that are not part of the building itself (dead loads), such as people, furnishings, wind, and snow.

DESIGN LOADS FOR FLOOR JOISTS

Live loads. A 40-psf live load will meet most codes for residential living spaces (some require only 30 psf in bedrooms). For typical live loads, see **Figure 2-21**.

Dead loads. For dead loads, 10 psf is adequate for standard wood-frame construction. For mortar-set ceramic tile floors, frame for a 20-psf dead load. For other special situations, calculate loads based on weights of building materials (**Figure 2-22**).

JOIST SPACING

Floor and ceiling joists are usually framed 16 in. o.c. However they may be framed at 12 in. o.c. to increase stiffness, or to increase the span in one section of a floor system without changing lumber depth.

JOIST SPAN TABLES

Joist span means the clear, horizontal distance between supports (**Figure 2-23**). The actual joist length is longer

Figure 2-23. Joist Span

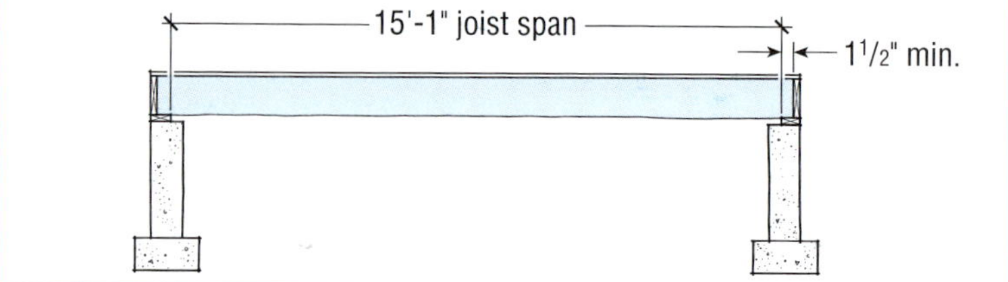

Span charts provide the clear distance between supports; actual joist length must include an additional bearing length on each end.

Figure 2–22. Weights of Common Building Materials

Materials	Pounds per Sq. Ft.
Studs, Joists & Rafters	
2x4s, 16 in. o.c.	9
2x6s, 16 in. o.c.	1.4
2x6s, 24 in. o.c.	1
2x8s, 16 in. o.c.	1.9
2x8s, 24 in. o.c.	1.3
2x10s, 16 in. o.c.	2.4
2x10s, 24 in. o.c.	1.6
2x12s, 16 in. o.c.	2.9
2x12s, 24 in. o.c.	2
Residential roof truss, 24 in. o.c.	4
Sheet Goods	
1/2 in. plywood	1.5
3/4 in. plywood	2.3
1/2 in. drywall	2
5/8 in. drywall	2.5

Materials	Pounds per Sq. Ft.
Flooring	
3/4 in. hardwood strip flooring	4
3/8 in. ceramic tile	2.5
1/2 in. quarry tile	6
1 in. mortar bed	12
Cement backerboard	3.5
Carpet and pad	3
Exterior Wall Cladding	
Wood siding	1.5
Three-coat stucco	10
Window unit	8
Roofing	
Asphalt shingles	2.5 - 4.5
Clay tiles	9 - 12
Spanish tile	19
Mortar bed for roof tile	10
Insulation (per in. thickness)	
Fiberglass batt insulation	.05
Rigid foam	.2

This table can be used as an approximate guide for calculating dead loads. For the exact weight of specific materials, it's best to check with the manufacturer.

to provide bearing at each end (see "I-Joist Bearing," **page 124**). Spans for dimensional lumber floor and ceiling joists are given in **Figures 2-24** to **2-27, pages 185-188.**

If the species or species group you are using is not listed, choose one with the same or slightly lower Fb values in **Figure 2-4, page 101**. If necessary, you can interpolate between two values.

Figure 2–28. Anchor Bolt Placement

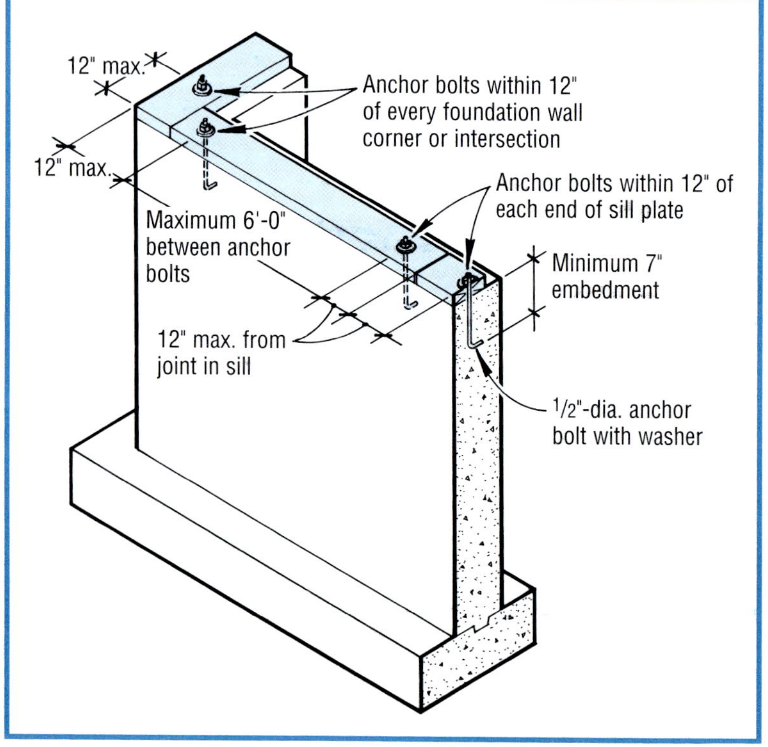

Each sill section requires at least two anchor bolts, placed as shown.

FRAMING DETAILS

Sill Plates

For sill plates in contact with concrete or masonry foundations, use pressure-treated wood to protect against decay and insect attack.

Anchor Bolts

Bolt the sill plate to the foundation with anchor bolts or strap anchors spaced no more than 6 ft. apart.

Most codes call for minimum 1/2-in. steel J-bolts embedded at least 7 in. in concrete, mortar joints, or solid-grouted masonry units. Each section of sill must have at least two bolts, with one bolt placed somewhere between 3 1/2 to 12 in. from each end (**Figure 2-28**). Each bolt needs a nut and adequately sized washer.

Wedge anchors. If anchor bolts are missing, use 1/2-in. wedge anchors or adhesive anchors that use hammer-in glass capsules (see "Retrofit Foundation Anchors," **page 48**).

Figure 2–29. Tapered Joist Ends

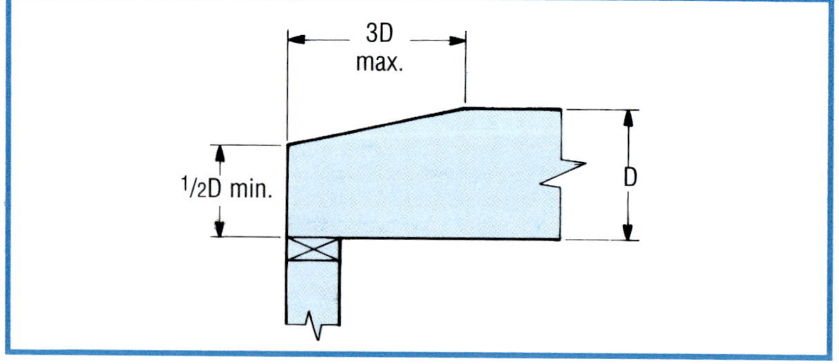

Where ceiling joists are taper-cut at the eaves, make sure the width of the cut does not exceed three times the joist depth and that the end of the joist is at least half the depth.

With a wedge anchor, tightening the nut expands the anchor until it's locked tight against the sides of the hole. A wedge anchor must be tightened to the manufacturer's specified torque. Tightening it too much can pulverize the surrounding concrete. Also, anchors can't be placed too close to the edge of a concrete slab or foundation wall.

Adhesive anchors. Adhesive anchors can be spaced closer to each other, and closer to the edge of a slab or wall, than wedge anchors. Drill a hole in the concrete and drop the capsule in. Then pound the anchor into the hole with a small sledgehammer; this mixes the chemicals in the capsule. **Note:** The capsules have a long shelf life, but should be protected from freezing.

Anchor Bolt Placement

Joist Bearing

Joists should have at least 1 1/2 in. of good bearing on wood or metal, or 3 in. on masonry (**Figure 2-23, page 114**). If necessary, you can safely notch a joist at its bearing point up to one-sixth the depth of the joist.

Ceiling joists are often tapered to fit beneath the roof rafters. But tapering them too much will weaken the joists at their bearing point. If you must taper the ends of a ceiling joist, follow the guideline shown in **Figure 2-29**.

Joist Bearing

Boring and Notching Joists

In general, do not make any notches in the middle third of a joist. Also, avoid vertical cuts in the bottom of a joist; angle the cuts instead to reduce the likelihood of splitting. For other rules, follow the recommendations in **Figure 2-30**.

Where mechanicals need to pass through a joist supporting a partition, double the joist with blocking in between to create a cavity (**Figure 2-31**).

Boring and Notching Joists

Rim Joists and Bridging

To prevent joists from twisting, the ends must be nailed to a solid band joist (rim joist) or header. If there is no band joist or header, nail solid blocking between the joist ends. Most codes don't require additional bridging unless the joist is at least six times as high as it is wide, in which case bridging is needed every 8 ft. However, solid or X-type bridging at mid-span will help stiffen any floor.

Figure 2-30. Cutting, Notching, and Boring Lumber Joists

Joist Size	Maximum Hole	Maximum Notch Depth	Maximum End Notch
2x4	None	None	None
2x6	1 1/2	7/8	1 3/8
2x8	2 3/8	1 1/4	1 7/8
2x10	3	1 1/2	2 3/8
2x12	3 3/4	1 7/8	2 7/8

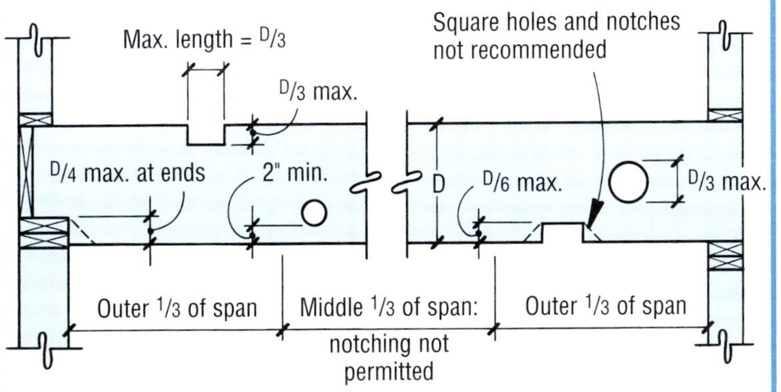

Do not notch a span's middle third where bending forces are greatest. For all calculations, use actual, not nominal, dimensions.

LOAD PATHS

All loads start at the roof and must transfer on an unbroken path through structural elements to the foundation. Many cracking problems, which are misinterpreted as "settling," are actually caused by broken load paths. These broken paths result in loads being carried by areas that were not designed to carry them.

Figure 2-31. Accommodating Mechanicals

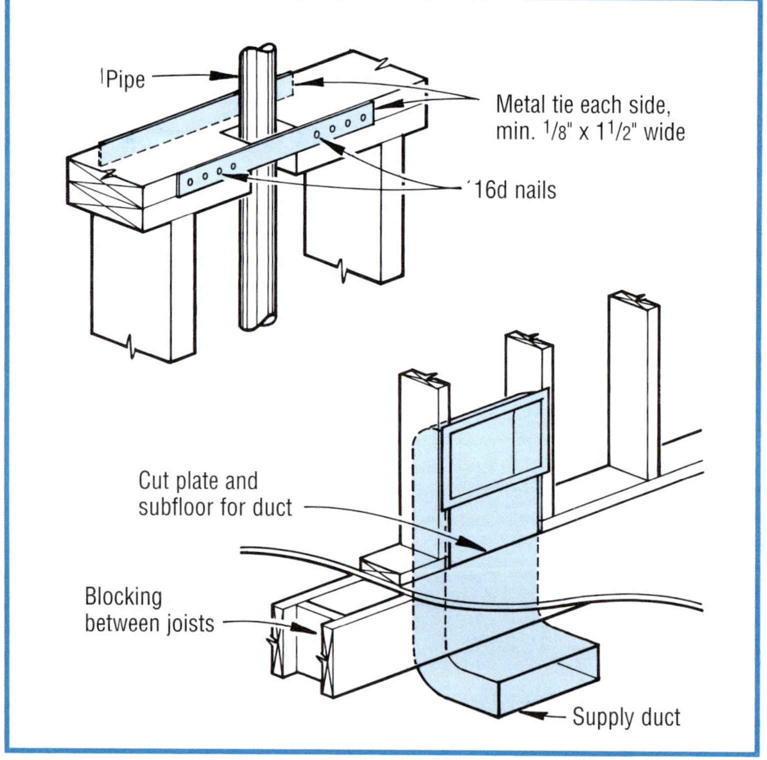

Joists supporting partitions should not be cut or notched to accommodate mechanicals. Instead, double the joist with blocking in between to create a cavity.

For more information: "Practical Engineering: Misplaced Load Paths," JLC 4/96

"Solutions to Common Framing Problems," JLC 6/98

Support for Loadbearing Walls

Parallel to joists. Ideally, if a loadbearing wall runs parallel to the floor joists, then it should sit directly over a beam or a joist supported by a loadbearing wall below. If the loadbearing wall sits between two joists, complete the load path by installing solid blocking between the two joists every 16 in. o.c.

Perpendicular to joists. If a loadbearing wall runs perpendicular to the floor joists, it should be offset from a supporting beam or loadbearing wall below by no more than the depth of the joists (**Figure 2-32**).

Figure 2-32. Aligning Bearing Walls

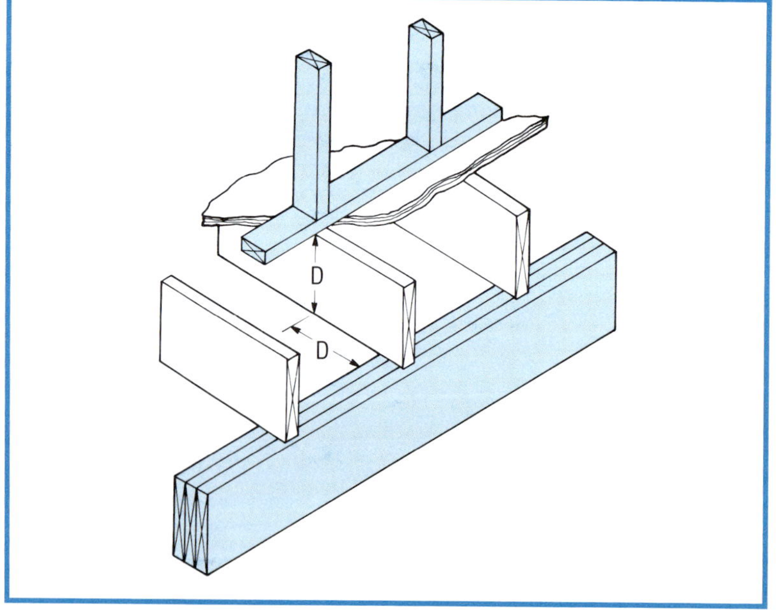

Never offset a loadbearing wall from a beam or loadbearing wall below by more than the depth of the joists.

Support for Non-Loadbearing Walls

A non-loadbearing wall that runs parallel to the joists doesn't need to be placed directly over a joist. However, when it's not directly over a joist, solid blocking should be installed between the joists to carry the load (**Figure 2-33**). If the wall is placed directly over a joist, that joist should be doubled.

Point Loads

A concentrated load that bears on a floor system — such as a structural post — must sit directly over a post or beam below. In addition, solid blocking (with a cross-section at least as large as the post) must carry the load through the floor system to the underlying post or beam.

Load Paths

FLOOR OPENINGS

Framing Floor Openings

Where a floor opening is more than 4 ft. wide, both the headers and trimmers should be doubled (**Figure 2-34**). Headers over 6 ft. long must be fastened with steel connectors and should be sized according to the load.

A tail joist in a framed opening that's over 12 ft. long should be connected with hangers. Otherwise, it should sit on a 2x2 (min.) ledger nailed to the face of the header.

Figure 2–33. Blocking under Non-Loadbearing Walls

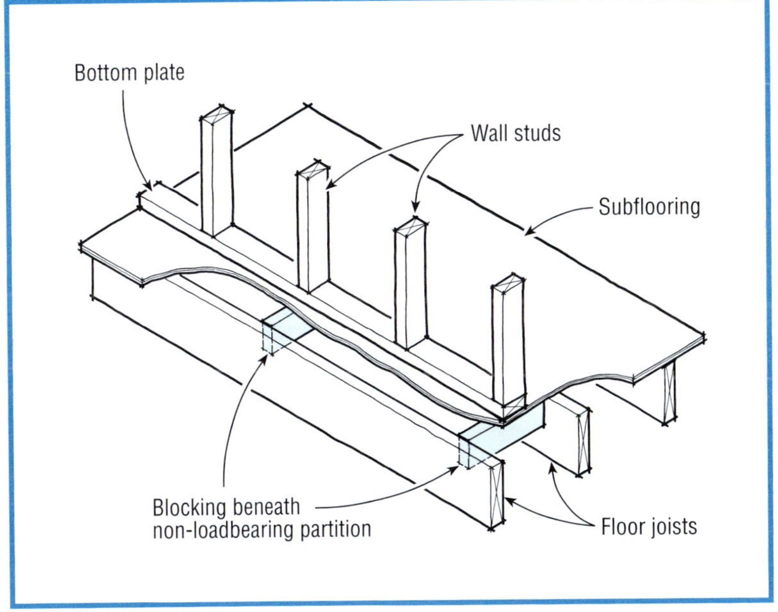

Install blocking where a non-loadbearing wall rests between two joists. Where it rests directly on a joist, the joist should be doubled.

Figure 2–34. Framed Openings

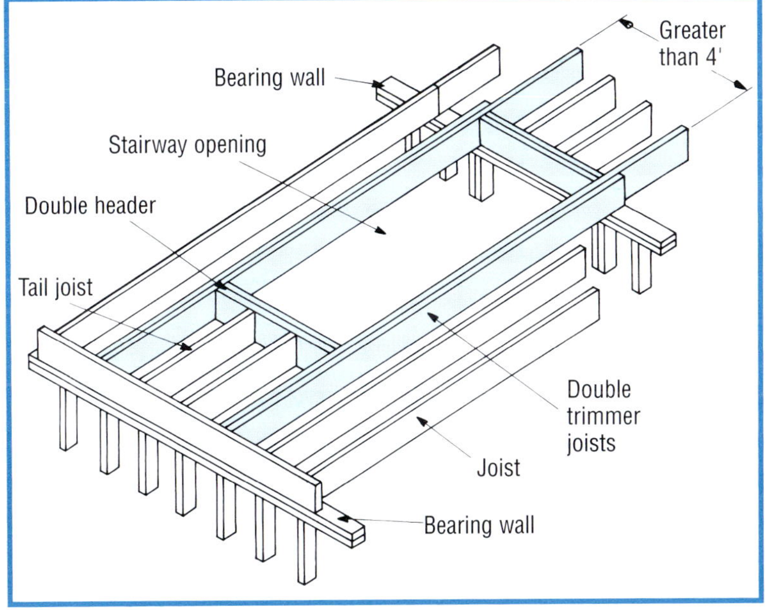

Where a header in a floor opening exceeds 4 ft., double the headers and trimmers. Where the header exceeds 6 ft., it should be installed with framing anchors.

Figure 2–35. Cantilever Details

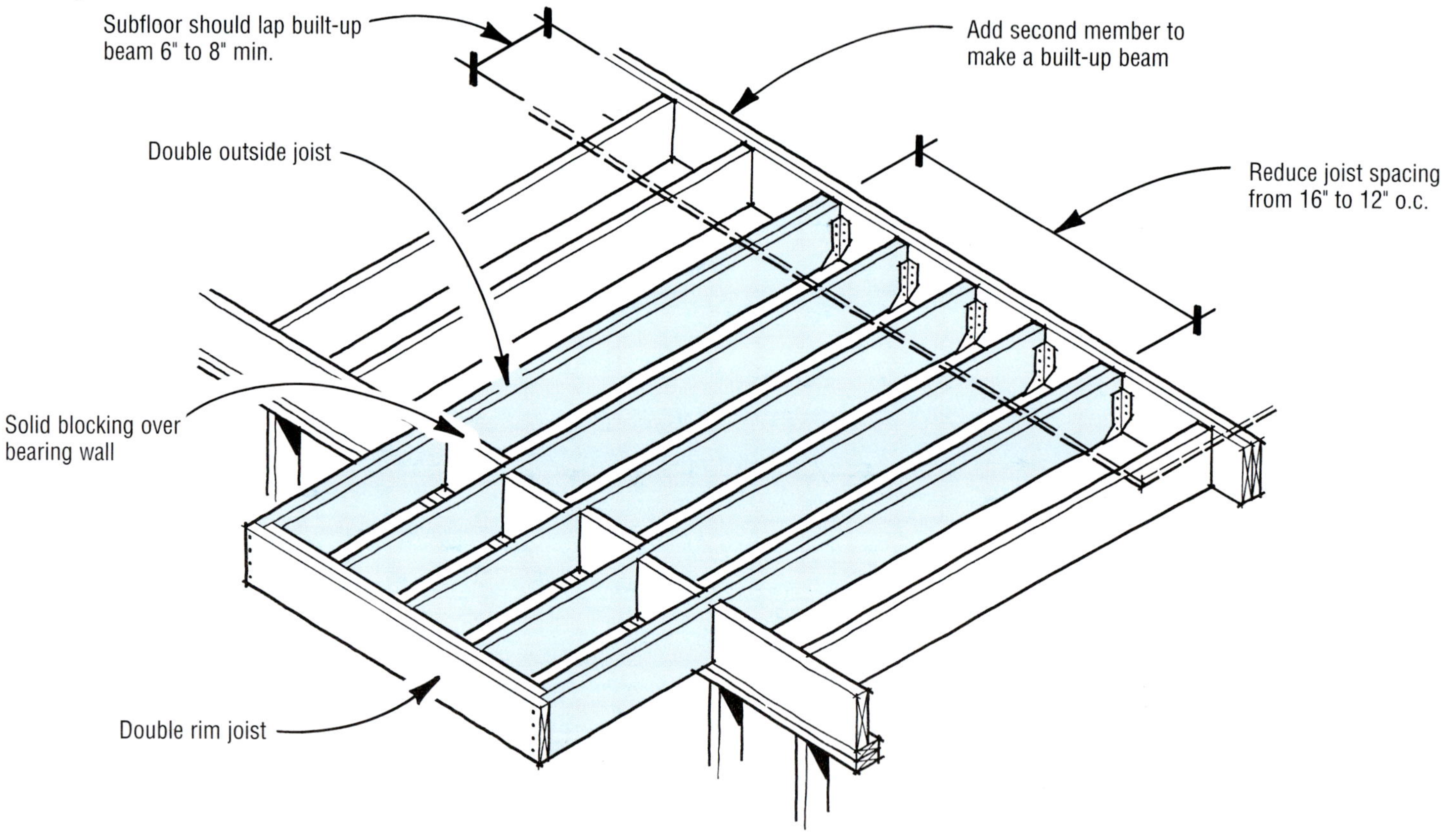

When framing any cantilevered floor area, pay strict attention to all connections and the sizes of supporting members.

CANTILEVERED FLOORS

Basic framing details for all cantilevers are shown in Figure 2-35.

Loadbearing Cantilevers

Where a cantilevered floor supports only a roof load with a total span of 28 ft. or less (such as on a Garrison-style home), the joists may cantilever up to one-eighth the length of the lumber span. Where the loads are greater, the cantilever should not exceed the depth of the joists unless the system is engineered.

Non-Loadbearing Cantilevers

A cantilever that supports a non-loadbearing wall should not exceed one-fourth the joist span. For non-loadbearing cantilevered decks (floor load only), the maximum overhang is one-third the span (Figure 2-36).

Cantilevers

Engineered Cantilevers

Cantilevers that exceed these limits should be engineered. Doubling joists or reducing joist spaces, along with upgrading connections, often permits greater overhangs.

Figure 2–36. Cantilever Rules of Thumb

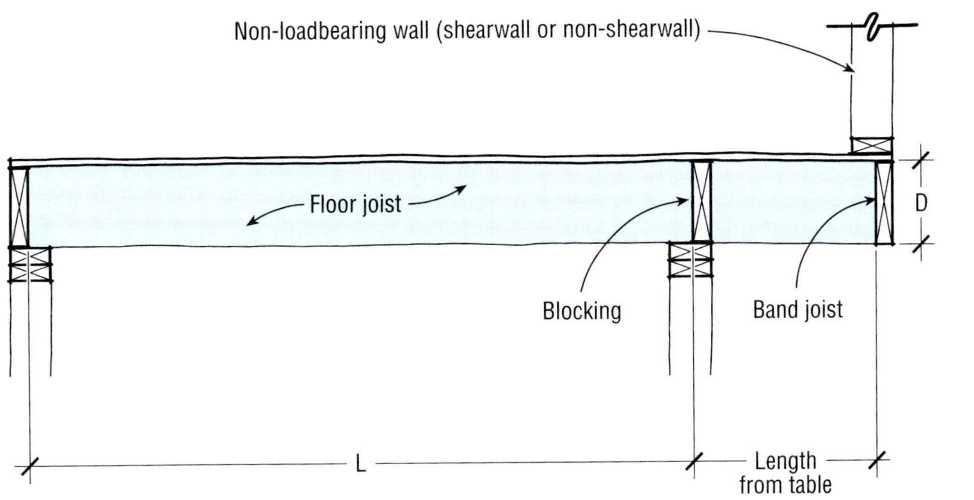

Maximum Length of Overhang	Load Conditions
D	Supports loadbearing wall
L/8	Supports only roof load with clear span of less than 28 ft.
L/4	Supports only non-loadbearing wall
L/3	Supports only floor load

The maximum distance cantilevered joists may extend past a bearing wall or other support depends on the load bearing on the end of the cantilever.

The best quality subfloor consists of tongue-and-groove (T&G) plywood that's glued and either nailed or screwed to the joists. T&G subflooring improves the perceived stiffness of the floor, and helps prevent squeaks. Subflooring that is not T&G must have solid blocking under all joints between panels.

SUBFLOOR SPANS

While T&G subflooring does not require solid blocking at joints, every sheet must rest on at least two joists. Panels are rated for spans of 16-, 20-, or 24-in., depending on thickness and grade. Span information is stamped on individual sheets (**Figure 2-8, page 103**).

FASTENING SUBFLOOR PANELS

Nail panels with minimum 6d common nails, spaced 6 in. o.c. at the edges of the panel and 12 in. o.c. in the field. Keep nails at least 3/8 in. from edges. (For shear wall nailing requirements, see "Shear Walls," **page 165**.)

For nailing schedules, see Figure 2-16, **pages 110-111**.

Gaps for Subfloor

APA recommends that subfloor panels be gapped 1/8 in. along the ends and edges. A 10d box nail may be used to gauge the spacing.

Gluing Subfloor

When nailing subfloor panels to joists under wet conditions, use a construction adhesive labeled AFG-01 and/or ASTM D 3498. Solvent-based construction adhesives typically dissipate the water better than latex- and urethane-based adhesives.

Roof and Wall Sheathing

Do not glue roof or wall sheathing to framing. Sheathing panels tend to be thinner than subflooring panels, so they're more likely to buckle when wet.

Subflooring

SPANS, INSTALLATION DETAILS, AND OTHER WOOD I-JOIST specifications vary from one manufacturer to the next. Therefore, it is important to consult the manufacturer's literature when using these products.

BORING AND NOTCHING I-JOISTS

Rules vary, depending on the I-joist manufacturer, but the following guidelines generally apply to cutting holes in wood I-joists.

No Holes Through Flanges

Never make holes or saw cuts in I-joist flanges — the bending stresses are too high. Even a shallow kerf in the lower flange can critically weaken an I-joist.

Holes in Webs

Boring and Notching I-Joists

Follow manufacturers' guidelines for minimum distances between holes and the bearing point (**Figure 2-37**). Use manufacturers' knockouts where practical. In general, you can cut a hole up to $1^1/2$ in. in diameter anywhere in the web.

Figure 2–37. Cutting and Notching Rules for Wood I-Joists

Min. Distance (ft.-in.) from Inside Face of Support to Near Edge of Hole

I-Joist Depth	TJI/Pro	Hole Diameter				
$9^1/2$"	150	1'-0"	1'-6"	3'-0"	5'-0"	6'-6"
	250	1'-0"	2'-6"	4'-0"	5'-6"	7'-6"
$11^7/8$"	150	1'-0"	1'-0"	1'-0"	2'-0"	3'-0"
	250	1'-0"	1'-0"	2'-0"	3'-0"	4'-6"
	350	1'-0"	2'-0"	3'-0"	4'-6"	5'-6"
	550	1'-0"	1'-6"	3'-0"	4'-6"	6'-0"
14"	250	1'-0"	1'-0"	1'-0"	1'-0"	1'-6"
	350	1'-0"	1'-0"	1'-0"	1'-6"	3'-0"
	550	1'-0"	1'-0"	1'-0"	2'-6"	4'-0"
16"	250	1'-0"	1'-0"	1'-0"	1'-0"	1'-0"
	350	1'-0"	1'-0"	1'-0"	1'-0"	1'-0"
	550	1'-0"	1'-0"	1'-0"	1'-0"	2'-0"

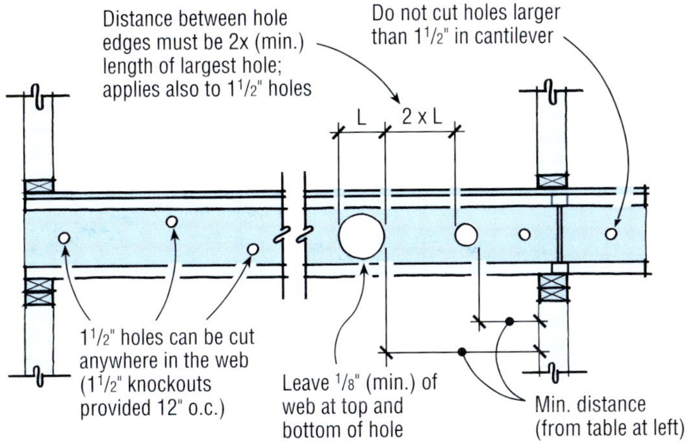

Note: Distances in the charts above are based on uniformly loaded joists using the maximum loads shown in Trus Joist's brochure. For other load conditions or hole configurations, contact a Trus Joist representative. For simple span (5-foot minimum) uniformly loaded joists, one maximum-size hole may be located at the center of the joist span provided no other holes occur in the joist. **DO NOT** cut into joist flanges when cutting out web.

Most manufacturers allow holes up to $1^1/2$ in. anywhere in the web, but require that the distance between any two holes be at least twice the diameter of the larger hole. In all cases, follow manufacturer's specifications for hole locations. The example provided here is courtesy of Trus Joist. Larger holes are allowed per manufacturer's specification.

Hole size. Permissible hole size increases as you near the center of the span where shear stresses are the smallest. Most manufacturers require that the distance between two holes be at least twice the diameter of the larger hole.

Avoid rectangular cutouts. With rectangular holes in webs, avoid overcutting the corners. Rounded corners are best. One technique is to drill a 1-in. hole in each corner and then cut between the holes.

Figure 2–38. I-Joist Bearing

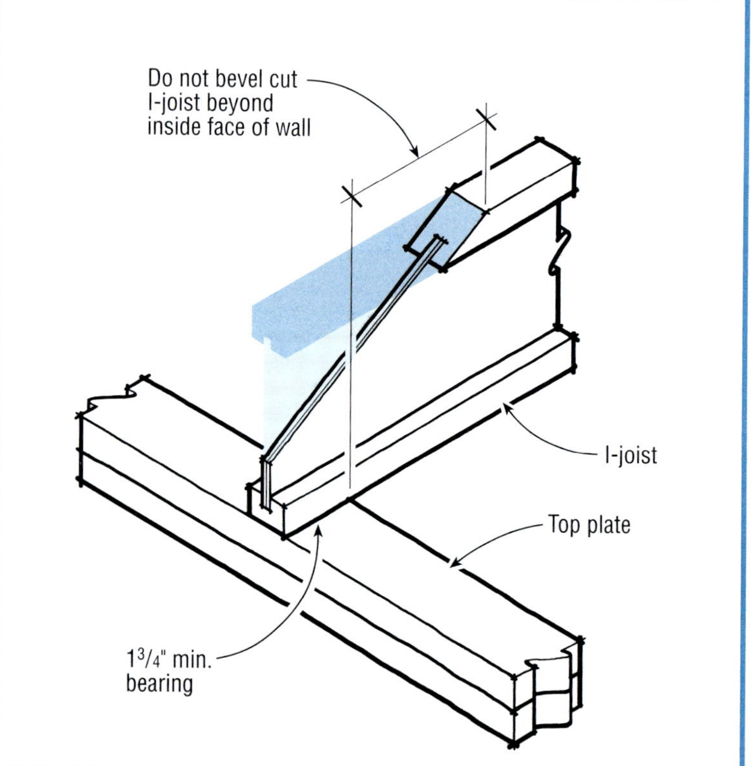

Provide adequate bearing for I-joists and do not taper the ends.

For more information: APA Performance Rated I-Joists (http://www.apawood.org/pdfs/managed/EWSZ725B.pdf)

I-JOIST BEARING

Most I-joists require at least 1 3/4 in. of bearing at each end. If the end of the joist is cut at a bevel, the top chord must also extend beyond the face of the wall (**Figure 2-38**).

TEMPORARY BRACING FOR I-JOISTS

Unbraced I-joists are flimsy and unstable until they are fully braced and sheathed. It doesn't take much to roll them over and damage them, so install all blocking and rim boards as soon as possible.

Sheathing and 1x4 Bracing

Stabilize the joists by sheathing the first 4 ft. of the floor system at one end of the joist bays and then bracing the rest of the joists to the sheathing area with 1x4 braces placed every 6 to 8 ft. Nail the braces flat to the top of the joists and lap them at the ends by at least two joists.

Avoid Overloading I-Joists

Be careful not to overload I-joists with heavy construction materials, such as stacks of plywood or pallets of bricks. If such loads are unavoidable, place them directly over load-bearing walls, posts, or other suitable supports.

I-JOIST LOAD PATHS

Loadbearing walls should be stacked over girders or load-bearing walls below. Use squash blocks or I-joist blocking to carry the load through the floor system (**Figure 2-39**).

Do not offset loadbearing walls unless the design has been engineered.

Rim Joists for I-Joist Floors

To keep joists from twisting — and to help transfer loads from the wall above to the wall or foundation below —

either attach the ends of I-joists to a full-height (band) joist or header or install blocking between the joists.

Rim Joist Materials

Acceptable rim joist materials include: the same-size I-joists as those used in the floor system; one or two layers of plywood or OSB; or an engineered lumber or metal product (Figure 2-40).

Don't mix wood I-joists with solid lumber rim joists. Not only are the two materials difficult to match in height, but also solid lumber will shrink after installation, leaving too much weight on the I-joists.

Figure 2–39. Support for Bearing Walls

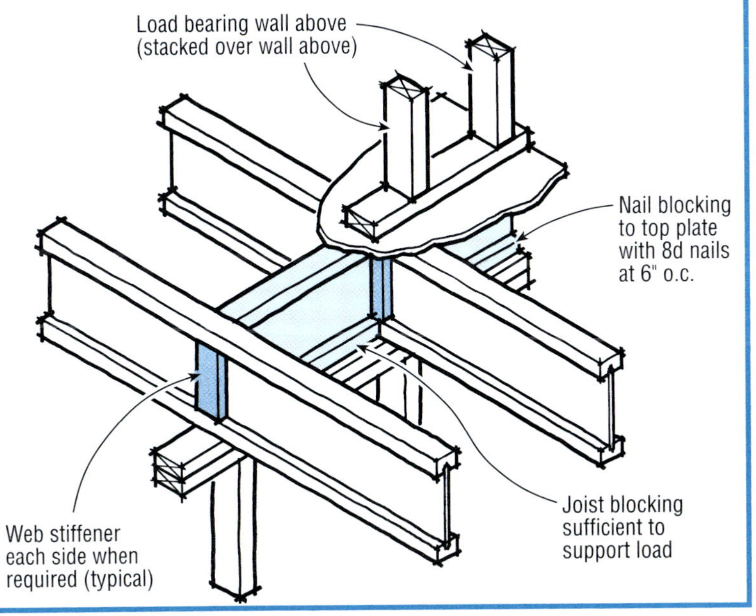

Where a loadbearing wall sits over a girder or another wall, full-height blocking should be installed to carry the load.

Web Stiffeners for I-Joists

Web stiffeners of plywood, OSB, or 2x are used to keep I-joist webs from buckling at bearing points or other intersections. (Thin web materials cannot transfer large shear forces.) With joist hangers, web stiffeners are often required to provide solid nailing.

Figure 2–40. Rim Joists

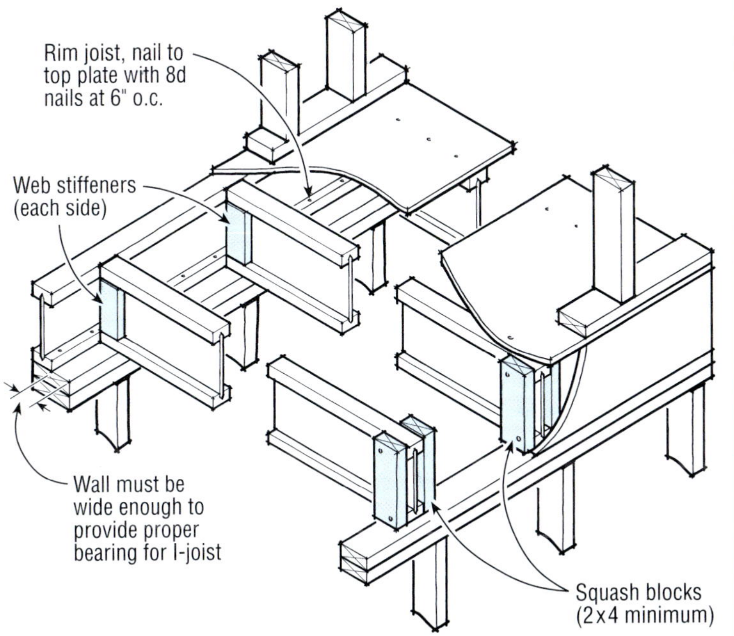

In I-joist construction, band joists typically consist of wood I-joists, plywood, or special composite rim joist materials. Web stiffeners may be required on long spans. Squash blocks may be required with thinner rim joist materials when there is a load-bearing wall above. Check manufacturer's specifications.

I-Joist Bearing
Temporary Bracing
I-Joist Load Paths
Rim Joists

Web stiffeners at bearing points should be installed tight to the top of the bottom flange. Leave a 1/8- to 1/4-in. space at the top flange to prevent the stiffener from prying the top flange off under load (**Figures 2-41**). One exception is where the joists support a load from above at mid-span. In this case, install the stiffener tight to the top flange with the space at the bottom.

Web stiffeners go on both sides of an I-joist with the face grain parallel to the length of the joist (for plywood and OSB stiffeners). Nail plywood or OSB stiffeners from each side with three or more 8d nails, staggered and clinched at the ends. For solid 2x stiffeners, use three 16d nails (two from one side, one from the other) and clinch the ends if they penetrate. In all cases, follow manufacturers' nailing schedules.

Figure 2–41. Web Stiffeners

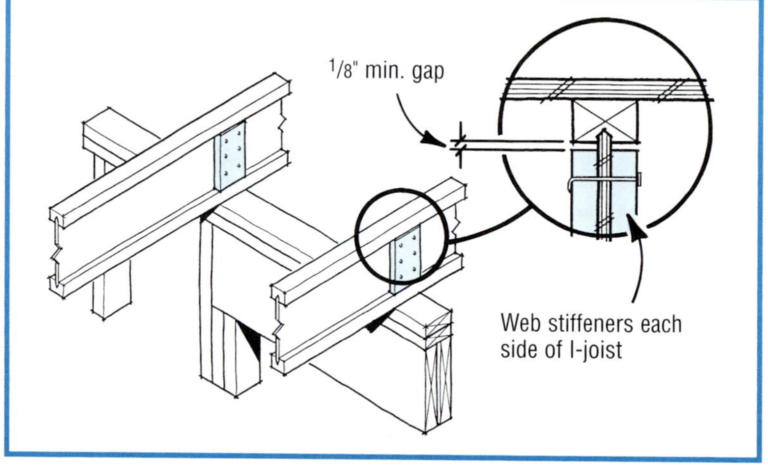

Web stiffeners of plywood, OSB, or 2x lumber (for large I-joists) are required at certain bearing points to transfer shear and prevent buckling of the web. Except where there is a concentrated load from above, web stiffeners should fit tight to the bottom flange with a space at the top. Web stiffeners also provide solid nailing at joist hangers. Follow manufacturers' nailing schedules.

For more information: "Framing Floors with I-Joists," JLC 4/02

Squash Blocks for I-Joists

Squash blocks are required under concentrated loads such as posts, and, in some cases, where the joists must transfer the load from a bearing wall above to another bearing wall below.

Squash blocks should be slightly larger than the depth of the joists to ensure that they pick up the full load. Fasten blocks to the top and bottom joist flanges with 8d nails (**Figure 2-42**).

Figure 2–42. Squash Blocks

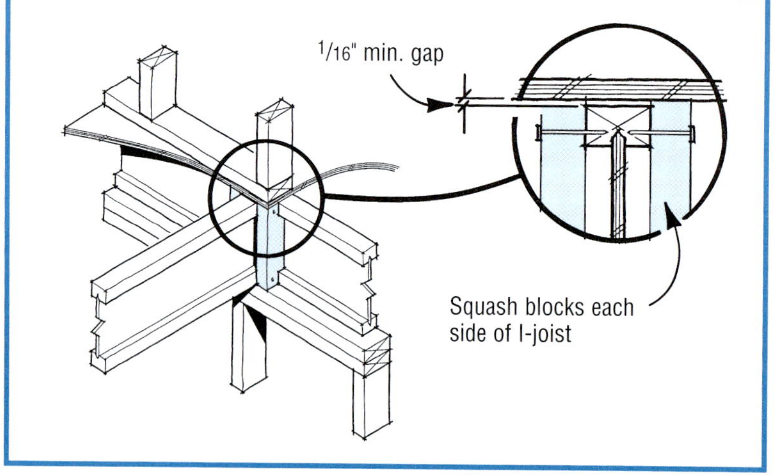

Squash blocks are required to transfer concentrated loads through the floor system at outside walls. They may be used under interior loadbearing walls instead of full-depth blocking between the I-joists.

Filler Blocks for I-Joists

Where two or more I-joists serve as a girder, the web area should be filled in with solid blocking of plywood, OSB, or dimensional lumber. The filler will ensure that both members carry the load.

Filler pieces should be a minimum of 4 ft. long, depending on the manufacturers' specs, and should be installed with a gap at the top (**Figure 2-43**). Nails should be clinched to prevent pullout.

FLOOR OPENINGS

Where a joist or stair stringer attaches to the side of an I-joist, nail backer blocks to the web to provide a nailing surface. These full-width blocks should extend 6 to 12 in. past either side of the intersecting joist or stringer and may be needed on both sides of the web. Refer to the manufacturer's literature for details (**Figure 2-44**).

Web Stiffeners
Squash Blocks
Backer Blocks

Figure 2–43. Filler Blocks

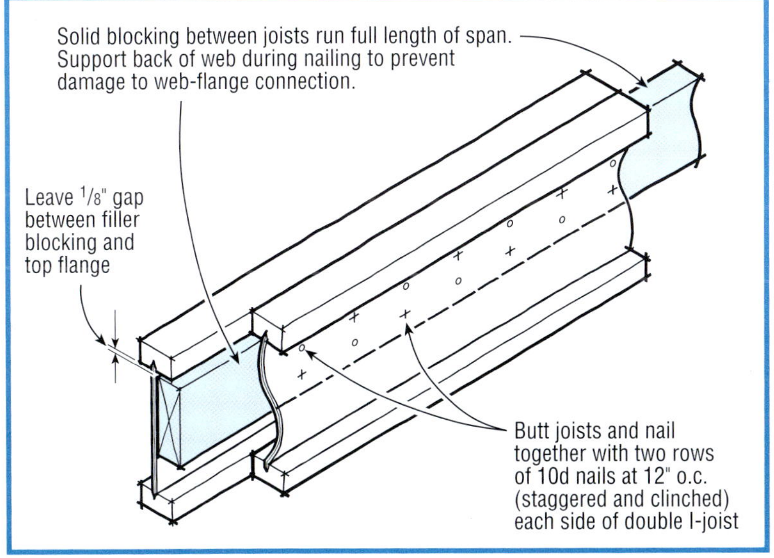

Leave a minimum 1/8-in. gap at the top of a filler block, which is required where two I-joists serve as a beam.

Figure 2–44. Backer Blocks

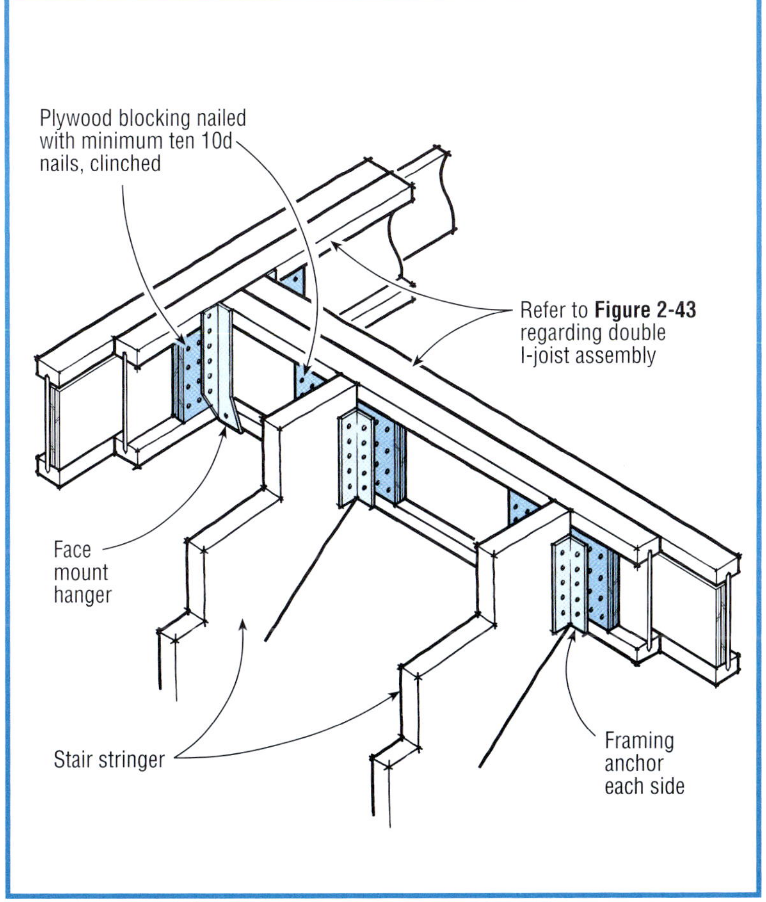

Backer blocks provide nailing for a joist or stair stringer that attaches at a right angle.

Figure 2–45. Cantilevered I-Joists

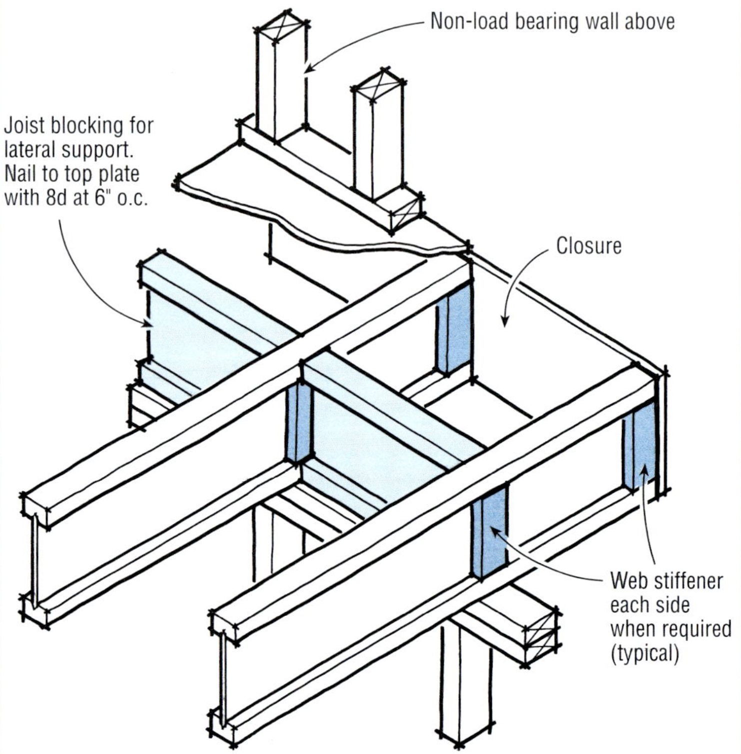

In general, manufacturers allow non-loadbearing cantilevers up to one-third the length of the joist span (to a maximum of 4 ft.). Loadbearing cantilevers are allowed up to 2 ft. However, manufacturers' guidelines for blocking and reinforcing must be strictly followed.

CANTILEVERED I-JOISTS

There are several ways to frame cantilevers with I-joists. Most manufacturers permit non-loadbearing cantilevers up to one-third the length of the joist span (to a maximum of 4 ft.). Loadbearing cantilevers may be up to 2 ft. long. Since a cantilever can easily over-stress an I-joist that's not reinforced, it's important to strictly follow the manufacturer's guidelines for allowable spans and required blocking. **Figure 2-45** shows some sample details.

STRUCTURALLY, AN OPEN-WEB FLOOR TRUSS RESEMBLES AN I-beam in that it puts most of its material along its top and bottom edges where stresses are greatest. To strengthen a truss, the fabricator may double its top and bottom chords, make side-by-side girder trusses, use larger truss plates or stronger wood, or use some combination of these techniques.

Caution: Loose Plates
Connector plates should be centered over the joint and firmly embedded in the wood. Never refasten a loose plate: The bond cannot be restored once it's been broken.

TRUSS TYPES

Floor trusses can be designed to sit on either their bottom chord or their top chord (**Figure 2-46**).

Cantilevered I-Joists

HANDLING TRUSSES

When receiving a truckload of trusses, reject those with excessive splits in chords or braces, those with knots close to metal plates, or those with loose or deformed plates. Also reject any that show evidence of having been damaged and repaired. Beware of warped or wet lumber, which can set up dangerous stresses as it shrinks and dries.

Handling Floor Trusses

Figure 2–46. Truss Types

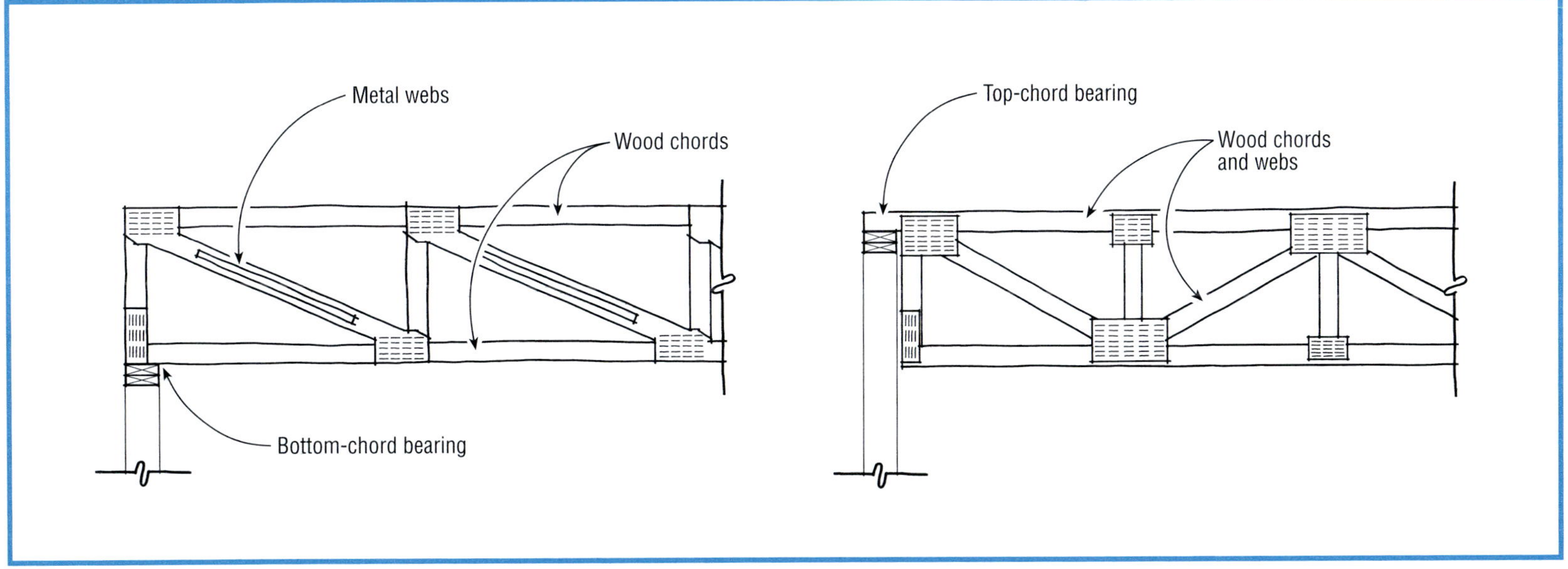

Floor trusses typically have either wood or metal webs. ***Bottom-chord-bearing*** *trusses (left) sit atop a wall plate or sill like a standard floor joist.* ***Top-chord-bearing*** *trusses (right) hang from the top plate.*

Figure 2–47. Lifting Floor Trusses by Crane

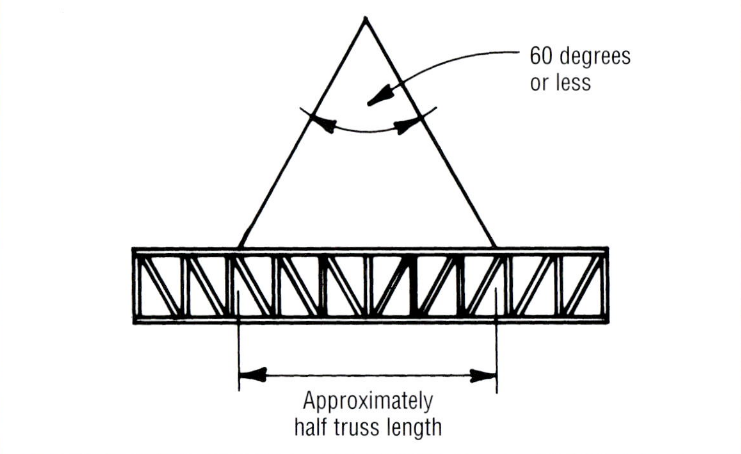

When lifting floor trusses with a crane, always lift from two points and never lift the truss sideways, which can loosen connector plates.

Lifting With a Crane

If you use a crane, always lift from two points (**Figure** 2-47) and never lift the truss sideways; the excess flexing can loosen the connector plates, causing eventual failure.

INSTALLING FLOOR TRUSSES

Floor trusses are usually spaced 24 in. o.c., and are typically lifted by hand, rolled into place, fastened, and braced.

Right Side Up

It is crucial to install each truss right side up, according to the label attached to the truss. Each web member is designed to be in compression or tension, but not both.

Bearing at Center Girder

A bottom-chord-bearing truss that crosses a girder or bearing wall may be designed to function as two simple beams. In this case, a tag attached to the truss will indicate that

Figure 2–48. Special Field Cuts

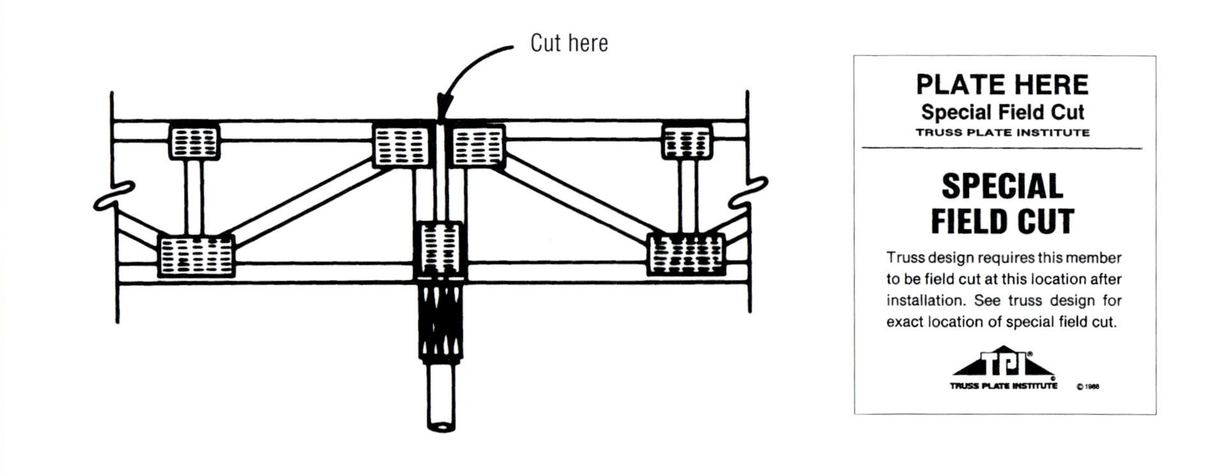

A bottom-chord-bearing truss that crosses a girder or bearing wall may be designed to function as two simple beams. A tag will instruct the builder to cut through the top chord after installation.

For more information: "Framing with Floor Trusses," JLC 4/93

the top chord should be cut after installation (**Figure** 2-48). Without this cut, a load applied to one end of the truss will lift the other, creating a seesaw action.

Stairwell Openings

Girder trusses are designed to support several standard floor trusses. These are often used around stairwell openings (**Figure** 2-49).

BRACING FLOOR TRUSSES

Bracing prevents trusses from bending, twisting, or otherwise deforming. On bottom-chord-bearing trusses, the ends are tied together with 2x4 ledgers that also serve as a nailing base for the perimeter of the plywood deck (**Figure** 2-50). Underneath, 2x6 strongbacks laid on edge should run continuously through the webs of all the trusses at 10-ft. intervals (one run for a 20-ft. truss, two runs for longer trusses).

Figure 2–49. Stairwell Openings

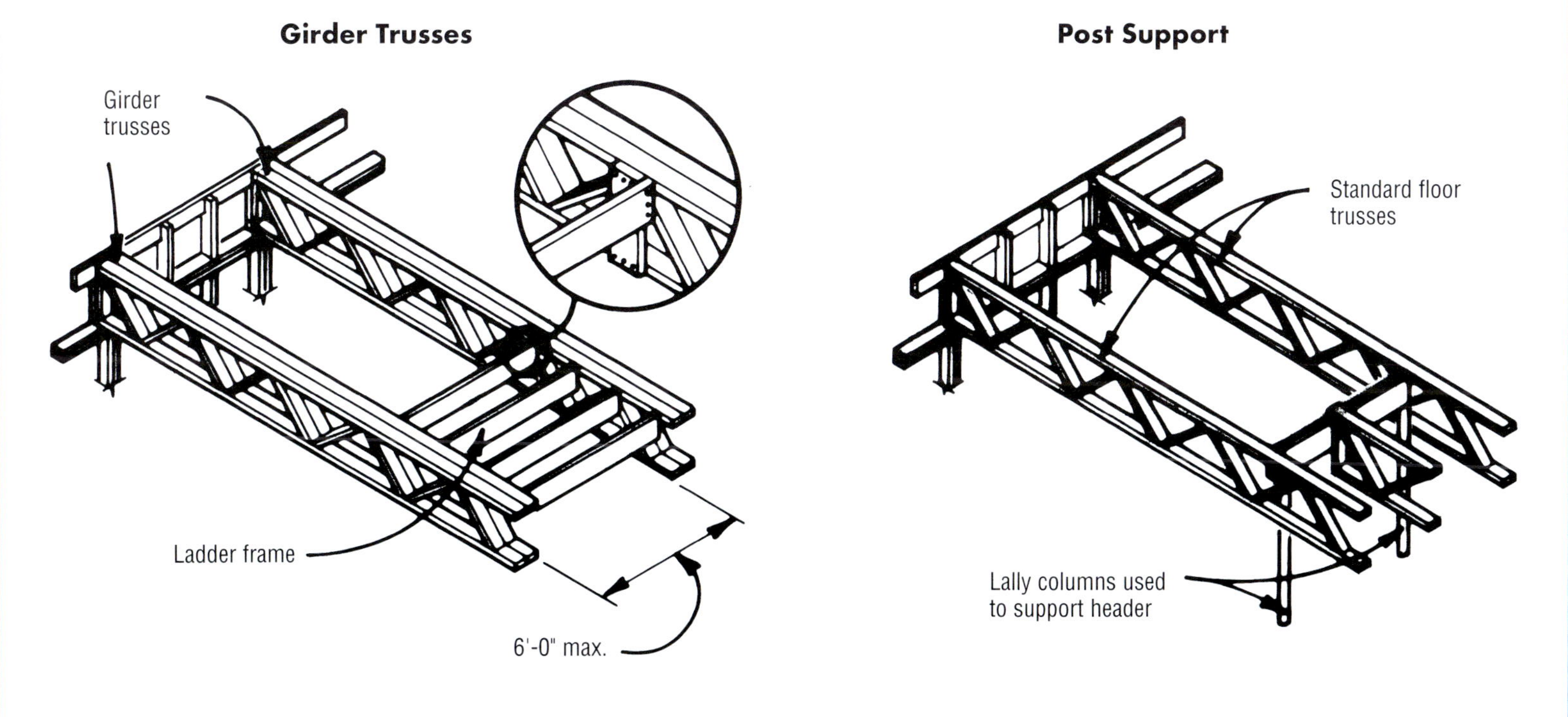

Double-width girder trusses are designed to support several standard floor trusses. These are often used around stairwell openings (left). Another option is to support staircase headers on posts or lally columns (right).

The 2x6 serves the same purpose as bridging in a standard floor, distributing concentrated loads over a wider area.

FIRE-STOPPING FOR FLOOR TRUSSES

Top-chord-bearing trusses short-circuit the fire-stopping ordinarily provided by the top plate. Common solutions are to extend the drywall past the trusses to the top plate, or to insert a separate 2x4 fire stop inside each stud bay just below the bottom chord of the truss (**Figure 2-51**). Check with local codes for requirements in your area.

The ends of bottom-chord-bearing trusses are designed to be tied together with 2x4 ledgers that also serve as a nailing base for the perimeter of the plywood deck. In addition, install horizontal 2x6 strongbacks at 10-ft. intervals to distribute loads and prevent twisting, which can loosen truss plates.

Figure 2–50. Bracing Bottom-Chord-Bearing Trusses

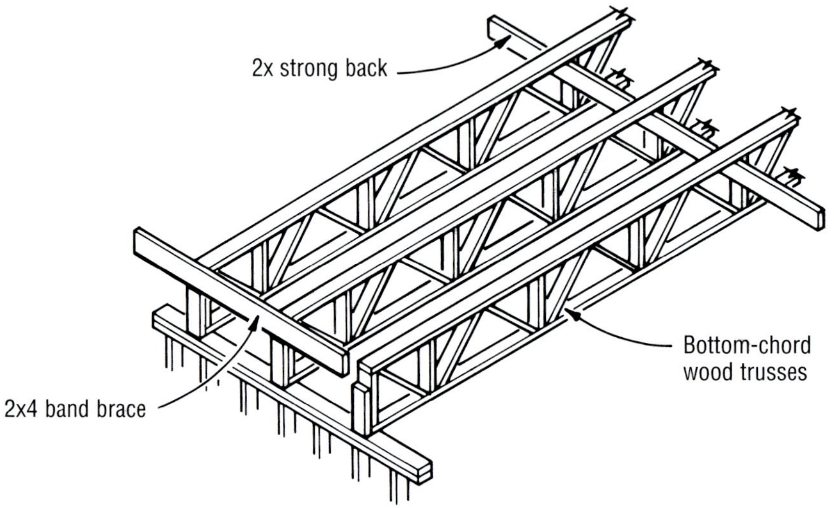

Many codes require fire-stopping with open-web trusses at the intersections of wall and ceiling spaces. Techniques include extending the drywall past the truss to the top plate (near right), or installing a 2x4 fire-stop inside each stud bay just below the truss's bottom chord (far right).

Figure 2–51. Fire-Stopping for Open-Web Trusses

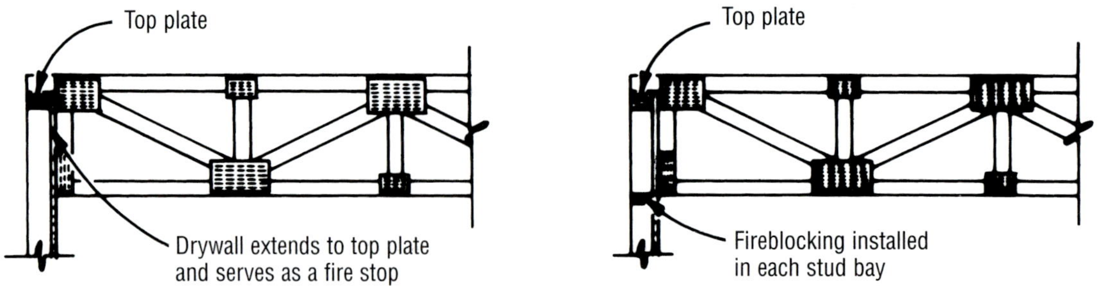

STUD SPACING

Maximum stud spacing for Standard or better-grade studs is shown in **Figure 2-52**. Lower-grade 2x4 studs cannot exceed 16 in. o.c.

TOP PLATES

Most model codes require loadbearing walls to have double top plates lapped so that the splices are at least 4 ft. apart. Two 16d nails are required on each side of the splice and additional nails are required every 24 in. in the overlap area (**Figure 2-53**). See nailing schedules (**Figure 2-16, pages 110-111**) for air-nail requirements.

Double vs. Single Top Plates

Single top plates are considerably weaker than a traditional double top plate. However, many codes permit single top plates as long as they are joined at splices, corners, and intersecting walls with .036-in.-thick galvanized steel plates, measuring 3x6 in. Plates should be fastened with six 8d nails on each side (six nails for ICBO).

Figure 2-52. Maximum Stud Spacing (in.)

Stud Size	Supporting Roof and Ceiling Only	Supporting One Floor Roof and Ceiling	Supporting Two Floors Roof and Ceiling
2x4	24*	16	–
2x6	24	24	16

* Shall be reduced to 16 inches if Utility grade studs are used.
Adapted from IRC Table R-602.3(5)

When using single top plates, rafters or joists must not be offset from studs by more than $1^{1}/2$ in. (**Figure 2-54**).

MAXIMUM STUD LENGTHS

Studs used in exterior loadbearing walls, whether 2x4 or 2x6, should not be more than 10 ft. high without being engineered.

Engineered solutions include reducing stud spacing, using wider studs, doubling the studs, and using studs made from engineered lumber.

Stud Length and Spacing

Double vs. Single Top Plate

Figure 2-53. Splicing Top Plates in Bearing Walls

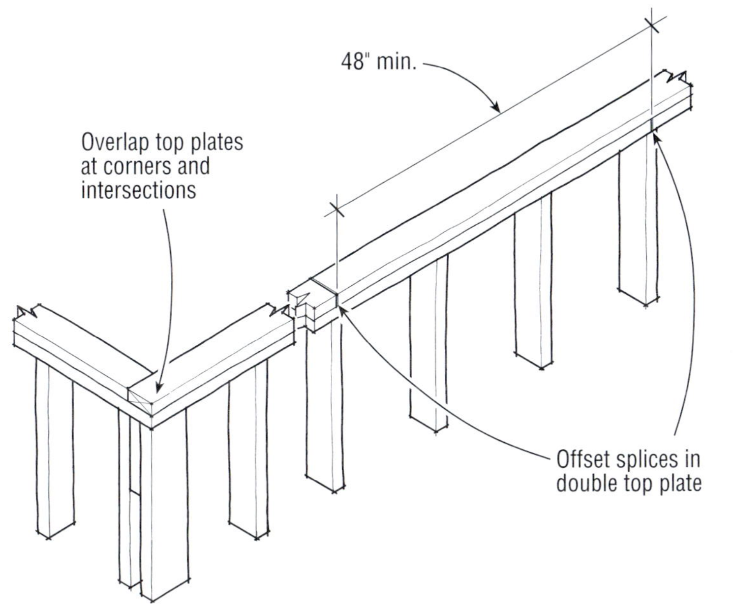

Double top plates must be lapped at corners and intersections and splices must be staggered at least 48 in. with at least four 16d nails spread across the lap.

Figure 2-54. Double vs. Single Top Plate

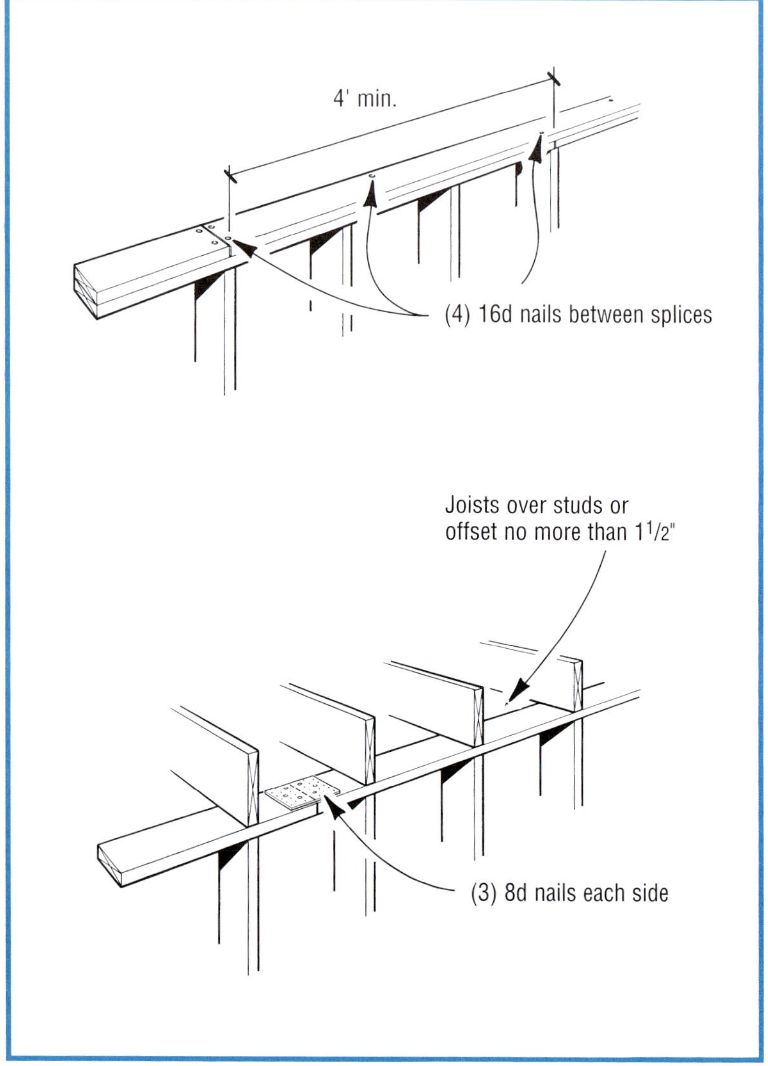

Although some model codes allow a single top plate with steel connectors and joists lined up over studs, the traditional double top plate produces a stronger wall.

For more information: Tall Walls Workbook (Canadian Wood Council; download at: http://www.cwc.ca/publications/workbooks/Tall_walls_en.pdf)

NOTCHING AND BORING STUDS AND PLATES

Studs

When notching or drilling through studs under 10 ft. tall, follow the guidelines shown in **Figure 2-55**. Never make notches and holes across from one another in the same section of a stud. For studs over 10 ft. in height, follow guidelines shown in **Figure 2-9, page 105**.

Plates

If more than 50% of the width of a top plate is notched or drilled away, then the edge of the plate must be reinforced with a 24-gauge steel plate that spans the distance between the adjoining studs (**Figure 2-31, page 118**).

Figure 2-55. Notching and Boring Regular Studs

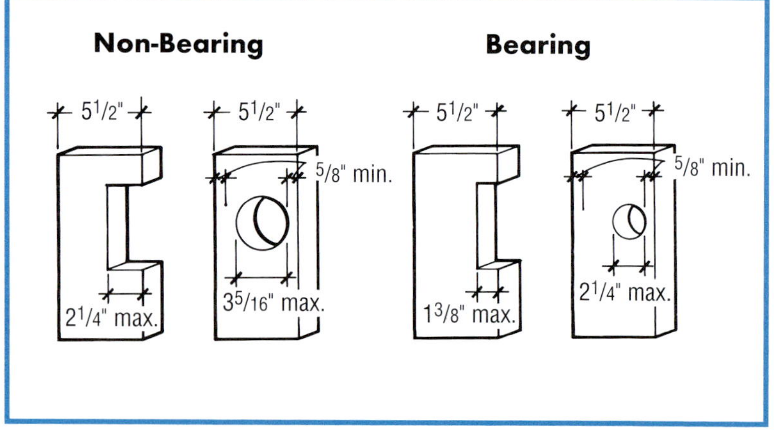

Notches in loadbearing studs under 10-ft.-tall should not exceed 25% of the stud depth and should not occur in the middle third of the stud. Bored holes should not exceed 40% of the stud depth and should be at least 5/8 in. from the edge. See **Figure 2-8, page 103** *for notching and drilling studs over 10 ft. tall.*

Notching and Boring Studs

Corner Studs

Wall Bracing

CORNER STUDS

On three-stud corners, builders often replace the middle stud with wood spacers. While that is acceptable in most cases, the three-stud corner is more than just a nailer for sheathing or drywall — it must also transfer shear loads between adjoining walls (**Figure 2-56**). For this reason, don't omit the third stud where shear resistance is critical.

WALL BRACING

In most residential construction, plywood or OSB sheathing panels are used for lateral bracing and are nailed 6 in. o.c. at the edges and 12 in. o.c. in the field. This provides excellent racking resistance. Bracing values for plywood can be improved by using larger nails or by spacing nails more closely together.

Bracing with Foam Sheathing

When using a non-structural sheathing such as foam board, install bracing at the building corners to provide resistance to racking. This bracing may consist of diagonal 1x4s or vertical sheets of plywood or OSB (**Figure 2-57**).

Diagonal bracing. Diagonal bracing is typically notched into the top and bottom plates and all intervening studs, and nailed with two 8d nails at each stud. Use 3/4-in. plywood instead of 1x4s for braces.

Figure 2–56. Corner Framing

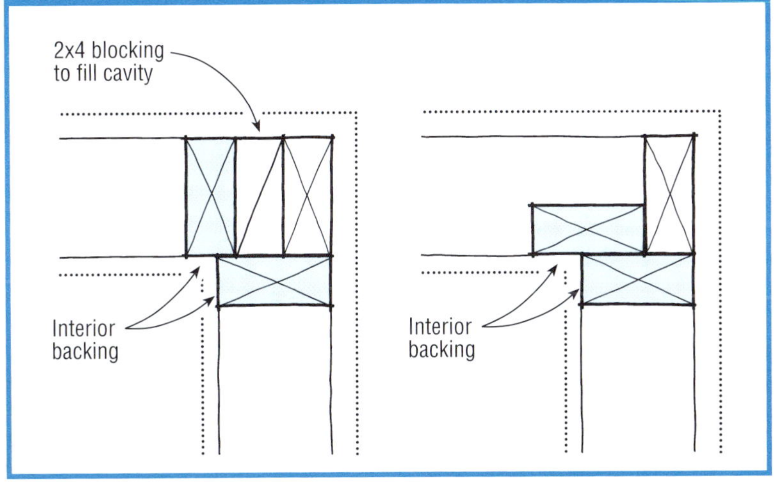

A three-stud "California" corner (right) uses less lumber and can be insulated better than a traditionally framed corner. Never replace the third stud of a three-stud corner with blocks; not only is it a nailer for drywall, but it also provides critical shear strength.

Figure 2–57. Code-Minimum Bracing

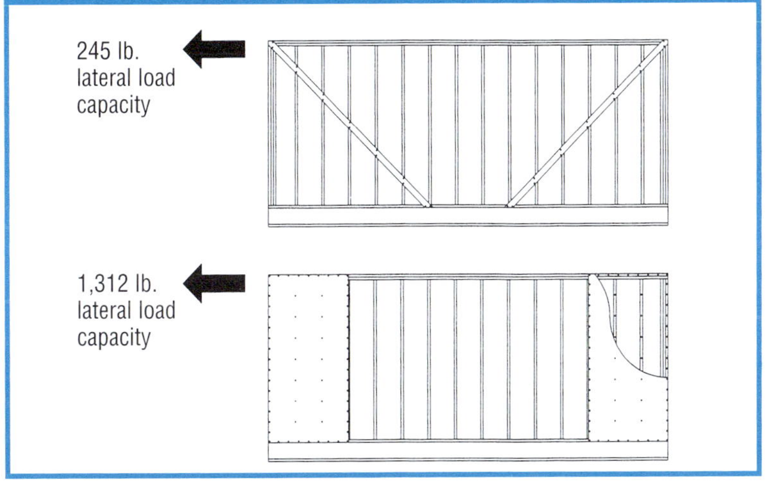

Foam sheathing does not provide adequate racking resistance. However, a pair of 7/16-in. plywood braces at each end of the wall will provide more than five times the strength of let-in 1x4s. The plywood should be nailed with 6d common nails 6 in. o.c. on the edges and 12 in. in the field.

Figure 2–58. Header Options

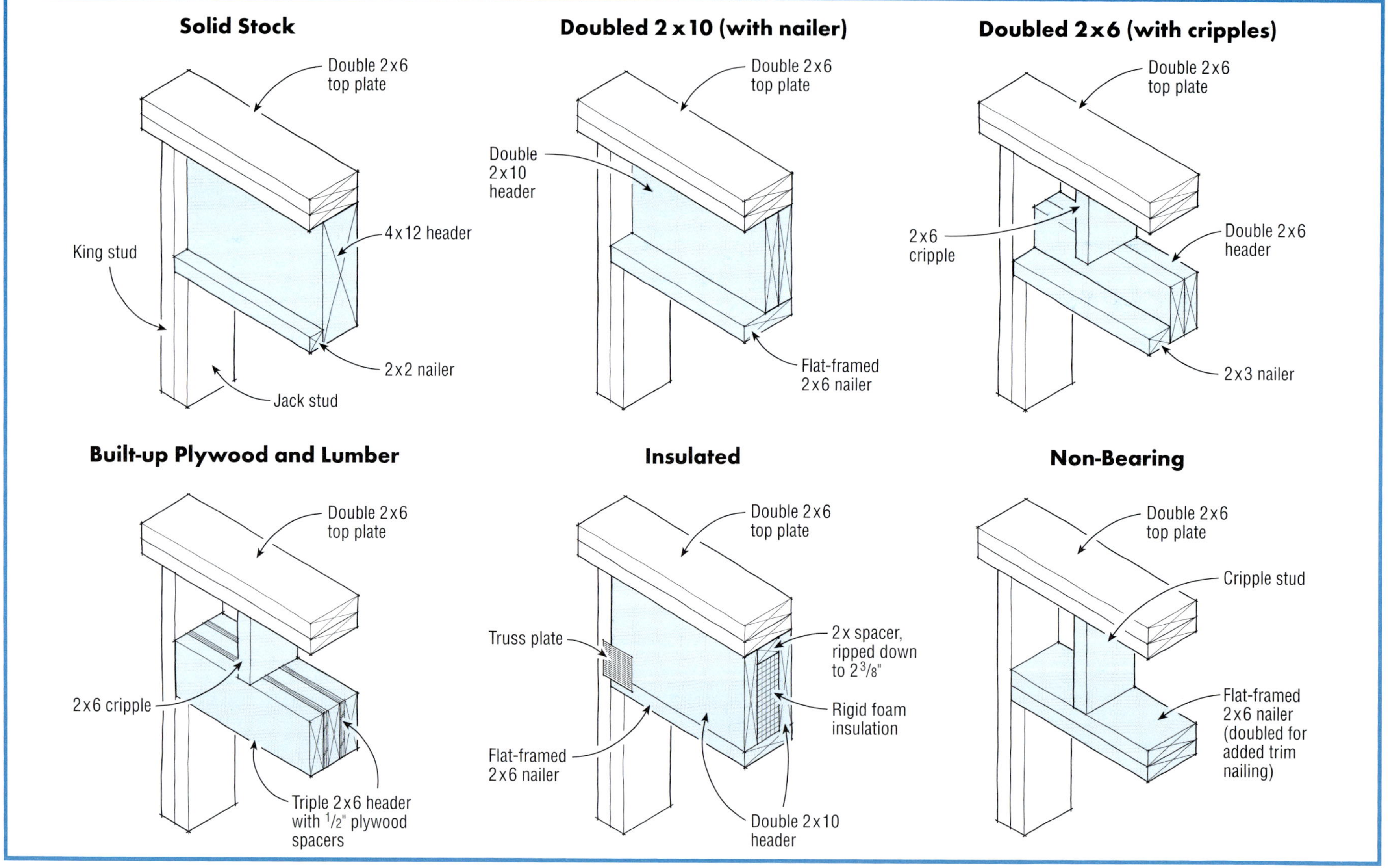

Using cripples above the header allows for smaller dimension lumber (depending on span requirements), but this option is more labor intensive. To avoid cripples, use 4x12s or 2x12s, or build up to the same depth using 2x10s with an added flat-framed member.

Metal Bracing

Depending on code, T-shaped metal bracing may sometimes be used in place of wood bracing — typically only as temporary bracing during construction.

Wind Bracing

In earthquake-prone or high-wind areas, pay special attention to the nailing of wall and roof sheathing, as well as to all the vertical framing connections that structurally join the building from foundation to roof. Consult local codes for nailing requirements (see also "Shear Walls," **page 165**).

HEADERS

Window and door headers can be framed in a variety of ways, depending on the lumber available and the insulation requirements, as shown in **Figure 2-58**.

Header Height

Set headers at the height of door rough openings. For a standard prehung door, the header height needs to be at least 6 ft. 10 in. (**Figure 2-59**).

Set window headers at the same height as doors so all openings align.

Rim-joist headers. In a two-story house, rough opening heights can be raised by doubling up on the rim joist, creating a girder to meet structural requirements for the span (see span tables, **pages 189-194**).

Sizing Headers

Rough openings in loadbearing walls require headers that are supported at either end by jack studs (trimmers). Header sizes for various house configurations and the number of jack studs required at each end are shown in **Figures 2-60 through 2-65, pages 189-194.**

Figure 2–59. Header Height

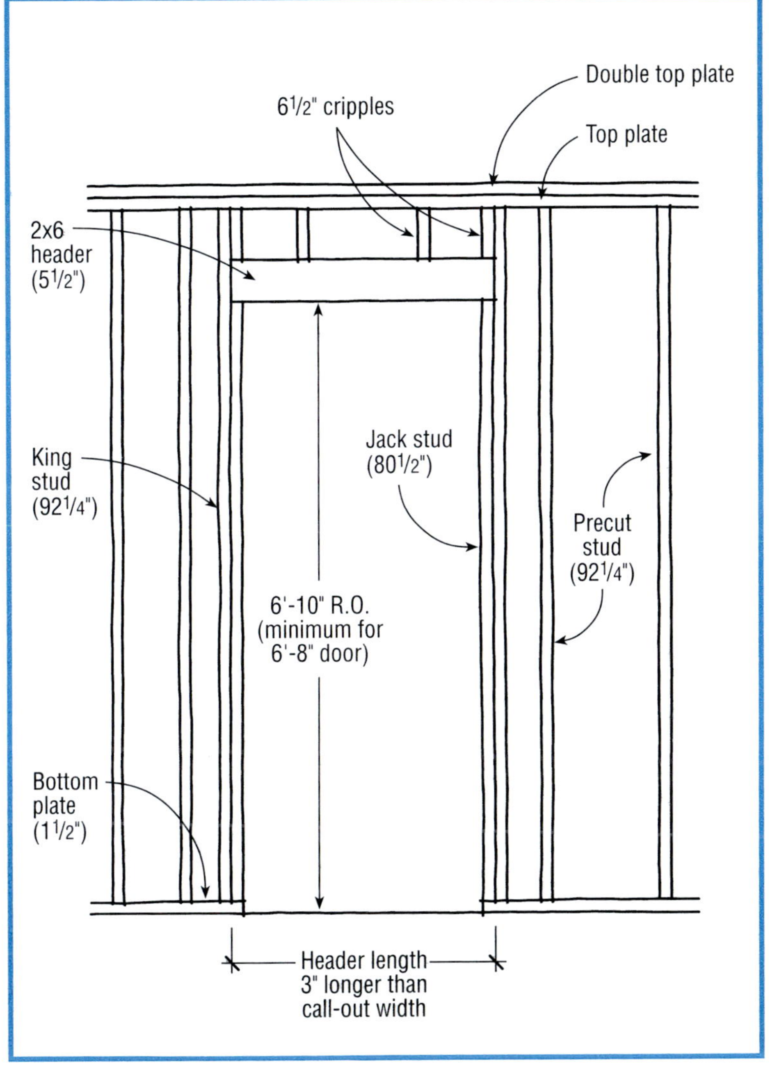

With prehung doors and 92 1/4-in. precut studs, use a standard 80 1/2-in. jack stud. (If using 92 5/8-in. precuts, lengthen the jack stud to 81 1/8 in.)

Header Options

Corner Window Headers

A corner window requires a cantilevered header. **Figure 2-66** shows the basic building details for such cantilevered headers. To meet actual load conditions, an engineer should specify header dimensions and connection criteria.

Even with proper header sizing, some deflection may occur. Do not nail the window directly to the header. Manufacturers of corner windows provide metal installation clips to isolate the windows from the rough opening. These clips should be installed using foam backer rod between the header jam and the rough opening. Secure trim to the jamb with a minimal number of light-gauge nails.

Figure 2-66. Cantilevered Headers

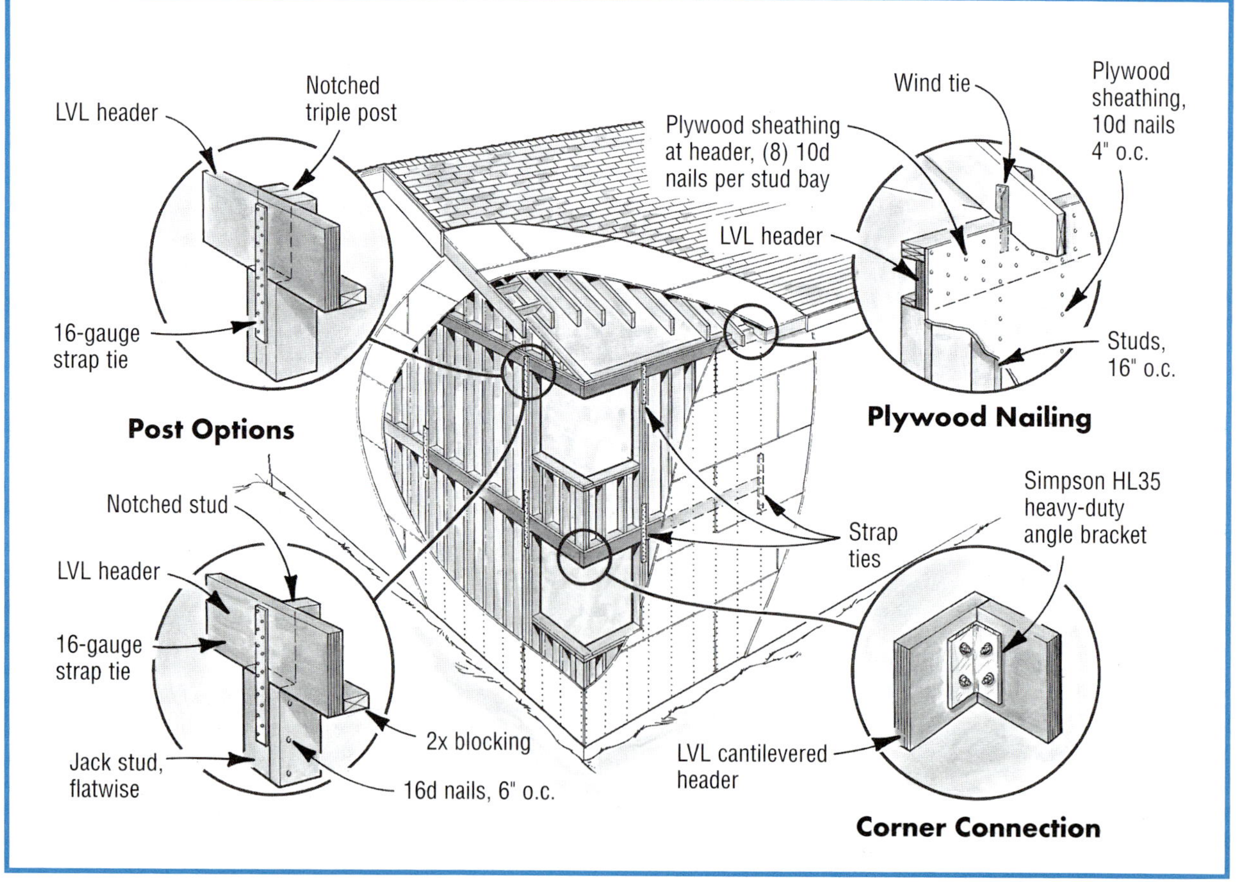

LVL headers, securely anchored to the wall framing by strap ties and thoroughly nailed sheathing, are cantilevered above the window opening. For a two-story structure, the lower header may be built-in as the rim joist.

For more information: "Framing Corner Windows," JLC 4/97

FIRE-STOPPING DETAILS FOR WALLS

In general, residential building codes call for fire stops at the following locations (**Figure 2-67**):

- In stud walls and partitions at every floor and ceiling level, including furred spaces;
- At every 10 ft. of vertical distance in tall walls or partitions;
- At all interconnections between vertical and horizontal concealed spaces, such as at soffits, cove ceilings, and drop ceilings;
- In the stud spaces along the side of stair stringers;
- At openings around pipes, ducts, chimneys, and fireplaces at floor and ceiling levels.

Fire-Stop Materials

Code-approved fire-stop materials include nominal 2-in.-thick lumber, double layers of 3/4-in. plywood or particleboard (with joints backed by the same material), 1/2-in.-thick drywall, and 1/2-in. cement board. Unfaced fiberglass batts can be used as fire-stopping under IRC R602.8.1.1 as long the batts fill the stud cavity for at least 16 in. of vertical space.

Corner Windows

Firestopping in Walls

Irregular Openings

All fire stops should be cut to fit tightly. Voids of more than 1/8 in. will undermine the material's fire-stopping ability. For irregularly shaped openings, use an "intumescent" fire-stopping compound. Because these products expand when heated, they will fill the void left by plastic piping or cable insulation that has melted away.

Figure 2-67. Fire-Stop Details

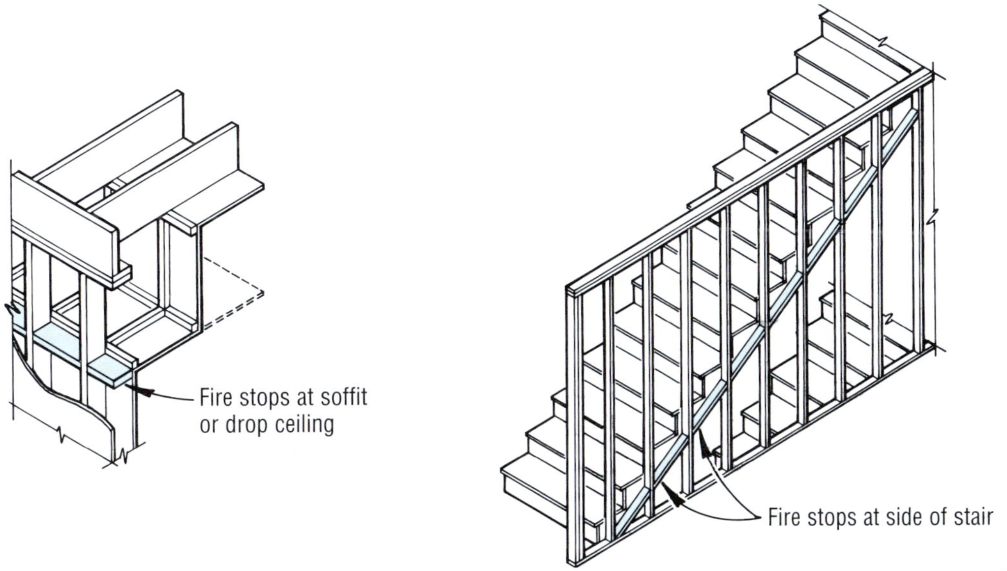

Fire stops made of approved materials are required in walls between floors and wherever concealed vertical and horizontal spaces intersect. Stairwell walls and any walls or partitions over 10 ft. tall also need fire stops.

RAFTER LOADS

Loads on residential rafters typically include snow and wind (the most common live loads), roofing materials, and interior finish materials (dead loads).

Live loading conditions for roofs vary widely with snow and wind loads. Be sure to check local codes for structural design requirements in your area.

For precise dead load calculations, refer to the material weights shown in **Figure 2-22, page 115.**

Figure 2–68. Rafter Span

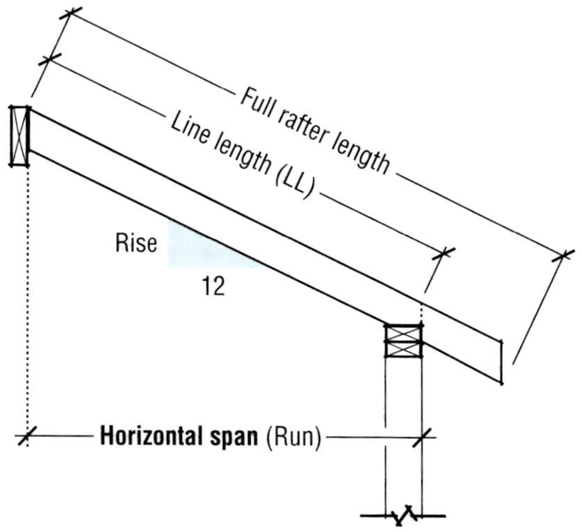

For more information: "Straight Talk About Hip and Valley Rafters," JLC 5/94

*Rafter span tables provide a horizontal rafter distance. These must be multiplied by a line length ratio (***Figure 2-3, page 99***) to calculate a line length (sloping distance). Length for plumb cuts and overhang must be added to find full rafter length. Note: The rafter spans in the span tables (***pages 195-208***) have been shortened for the thickness of the ridge.*

COMMON RAFTER SPANS

Rafter span is the horizontal distance from bearing at the ridge to bearing at the top plate (**Figure 2-68**).

To convert rafter span (horizontal distance) to rafter length (sloped distance), use the conversion factors in **Figure 2-3, page 99**.

Maximum rafter spans for common lumber species are given in **Figures 2-69 to 2-82, pages 195-208.**

HIP AND VALLEY RAFTERS

While hip and valley rafters carry significant loads, they are often undersized. To allow for full bearing, hip and valley rafters should be at least one size larger than commons and jacks.

Sizing Hip and Valley Rafters

The loads on hip and valley rafters are *tributary loads* transferred from a wide area of the roof (**Figure 2-83**). The hip and valley rafter sizing table (**Figure 2-84, page 142**) accounts for such tributary loads.

Apex Support

In addition to choosing the right size hip or valley rafter, the point load at the high (ridge) end of a hip or valley must be well supported. Typically, common rafters tied together by ceiling joists or collar ties support this load. Large cathedral ceilings should be engineered to provide adequate support. This may include tension rods along the wall plates or reinforced plate connections for hip and valley rafters (**Figure 2-85, page 143**).

Rafter Loads

Sizing Hip and Valley Rafters

Figure 2-83. Calculating Tributary Load (W) on Hip and Valley Rafters

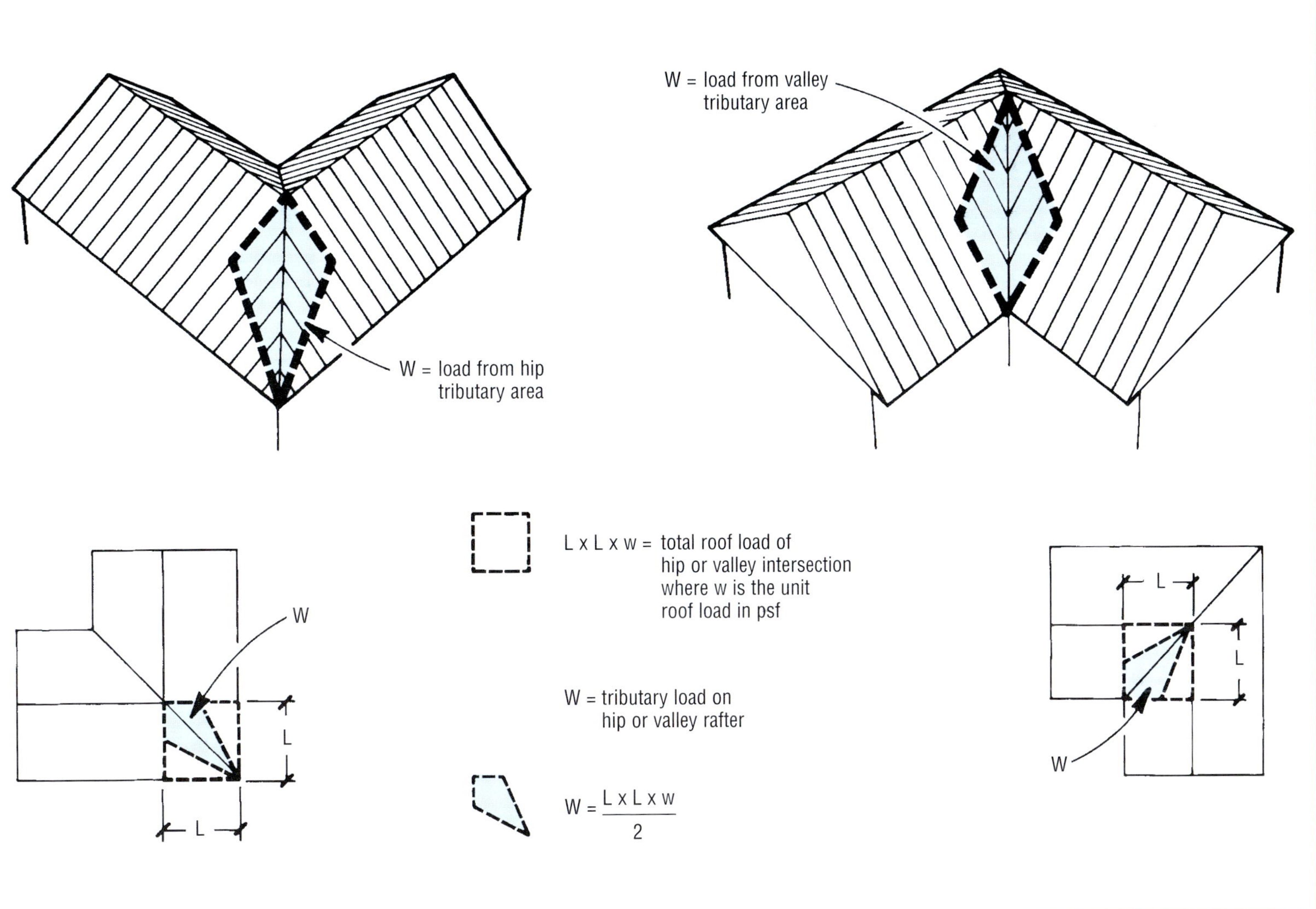

A hip or valley rafter picks up half the load of all the jack rafters feeding into it — its tributary load. *In plan view, this tributary load is kite-shaped. Note that the hip's tributary area is wider at the top and the valley's tributary area is wider at the bottom.*

Figure 2–84. Hip and Valley Rafter Sizing Table

L in feet	15 psf live load	30 psf live load	45 psf live load	60 psf live load
4	2x6	2x6	2x6	2x8*
5	2x6	2x6	2x6	2x8
6	2x6	2x8	2x8	2x10
7	2x8	2x10	2x10	2x12
8	2x10	2x12	2x12	(2) 2x12
9	2x10	(2) 2x12	(2) 2x12	(2) 2x12
10	2x12	(2) 2x12	(2) 2x12	LVL 11⅞
11	(2) 2x12	(2) 2x12	LVL 11⅞	LVL 11⅞
12	(2) 2x12	LVL 11⅞	LVL 11⅞	LVL 14
13	(2) 2x12	LVL 11⅞	LVL 14	(2) LVL 14
14	LVL 11⅞	LVL 14	(2) LVL 14	(2) LVL 14
15	LVL 11⅞	LVL 14	(2) LVL 14	(2) LVL 14
16	LVL 11⅞	(2) LVL 14	(2) LVL 14	(2) LVL 16*
17	LVL 14	(2) LVL 14	(2) LVL 16	(2) LVL 16
18	LVL 14	(2) LVL 14	(2) LVL 16	*
19	(2) LVL 14	(2) LVL 16*	*	–
20	(2) LVL 14	(2) LVL 16	–	–

*Deflection limited

Note: The calculations behind this table are based on the following:

10 psf dead load
L/240 deflection
No slope adjustments

F_b wood = 1,000 psi
E wood = 1,000,000 psi
F_b LVL = 2,800 psi
E LVL = 2,000,000 psi

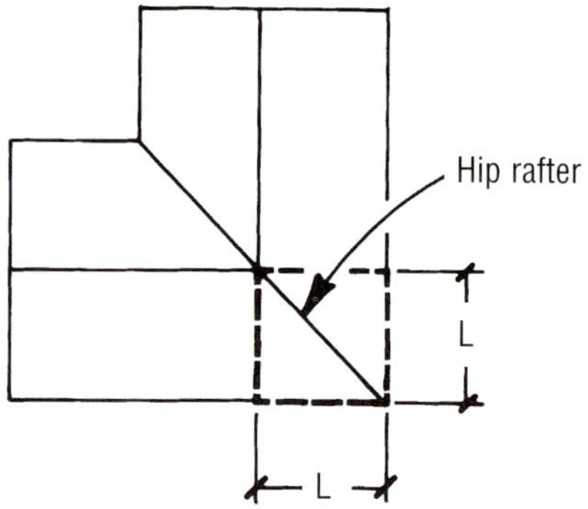

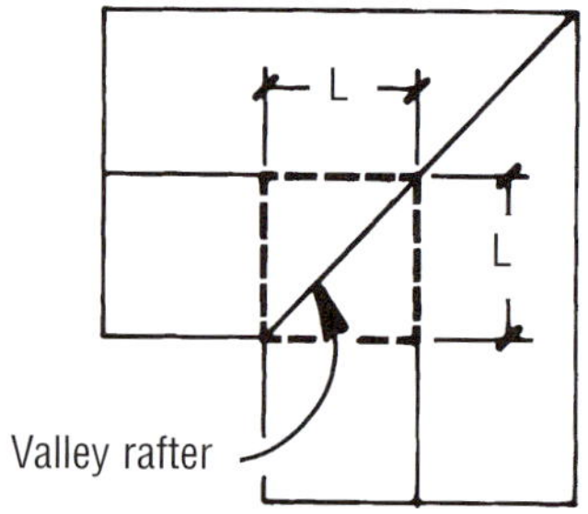

The table above summarizes a series of sizing calculations for various roof design loads. L represents the length of a hip or valley rafter for equally pitched roof intersections only.

Figure 2–85. Apex Support for Cathedral Ceiling Hips and Valleys

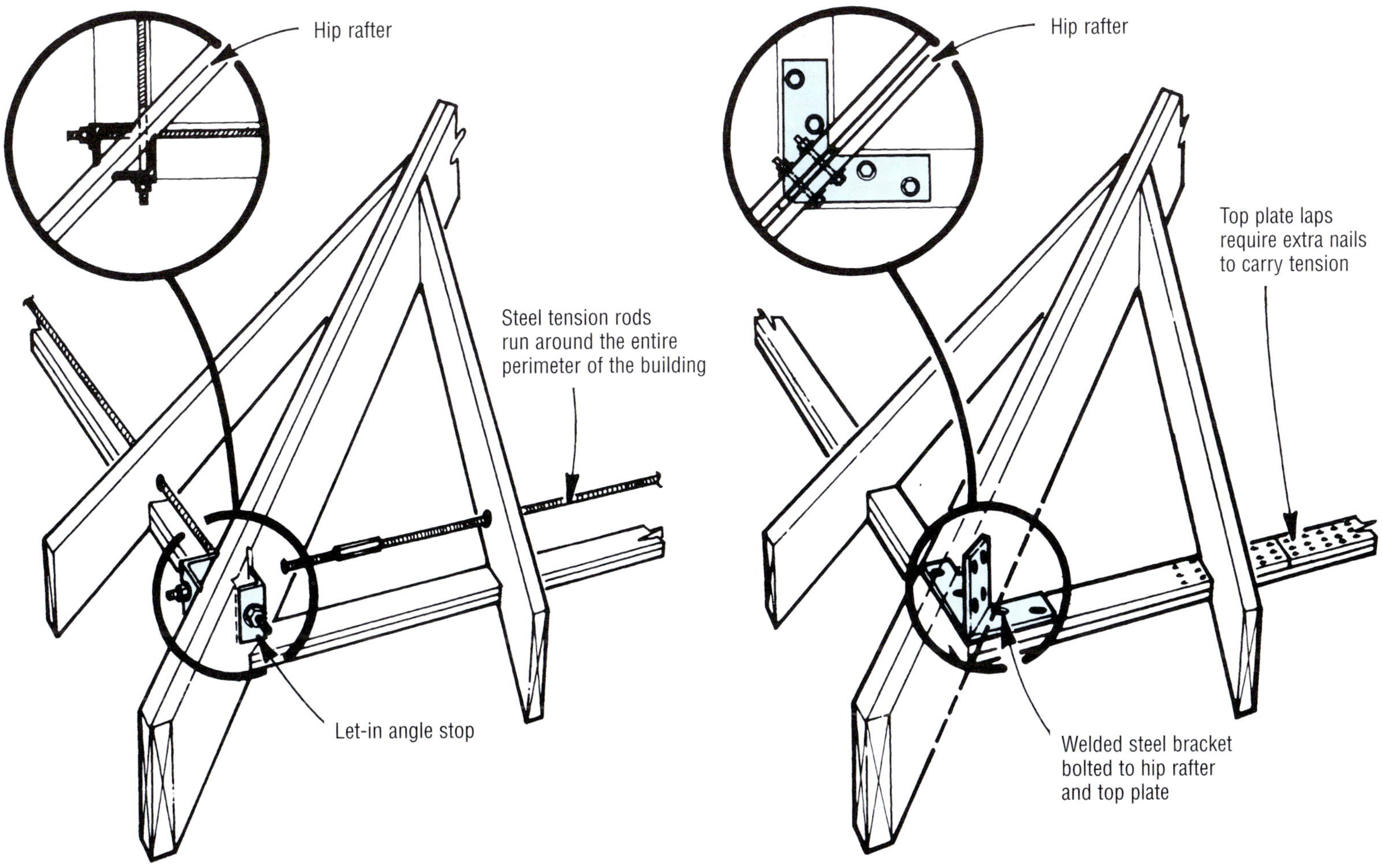

Hip and Valley Structure

In a large hip roof with full cathedral ceilings, there are no support posts or collar ties to resist the spread of the rafters. In such cases, install steel rod tension ties (left) around the perimeter of the building to prevent the bottom ends of the hip rafters from spreading. Alternatively, use welded steel brackets to secure the hip rafters at the corners, allowing double 2x6 top plates to act as tension members (right).

Figure 2-86. Rafter Bearing

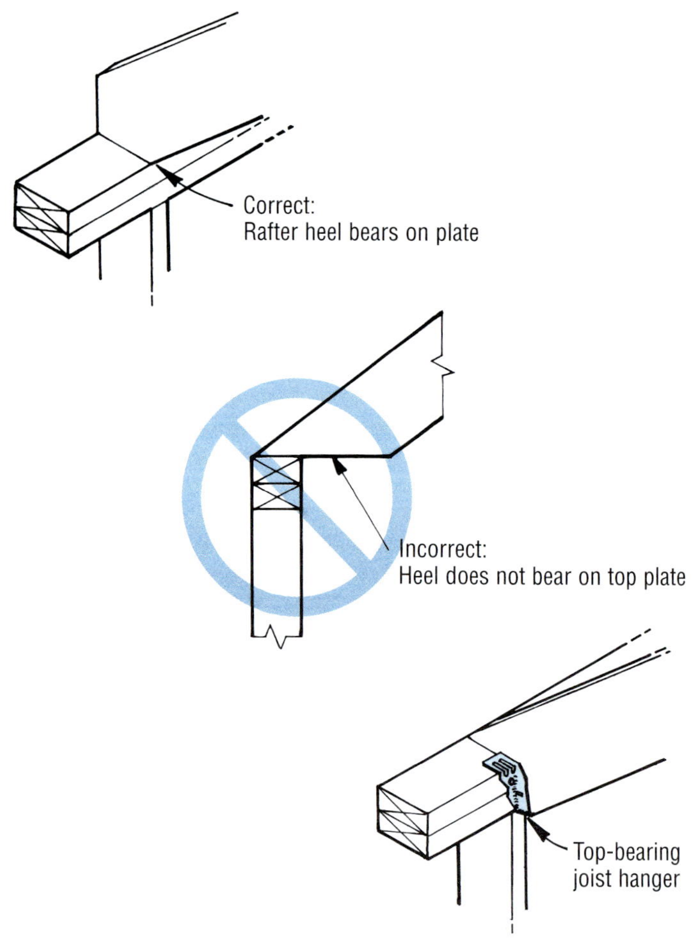

A rafter should rest on a top plate with bearing on its heel (top), not its toe (middle). Where this isn't possible, a top-bearing joist hanger can be used (bottom). The joist hanger also keeps the rafter from rotating, a job that normally requires ceiling joists or solid blocking.

RAFTER BEARING

Rafters should have bearing of at least 1 1/2 in. on the top plate or beam. Bearing should be on the heel of the rafter, not the toe. Bearing on the toe places most of the load on only a portion of the rafter and can lead to its splitting (**Figure 2-86**).

A non-structural ridge board should be at least one size deeper than the common rafters to provide full bearing for the rafter's plumb cut.

STRUCTURAL BRACING

Rafters must be restrained from spreading. This can be accomplished with adequate connections to ceiling joists, with collar ties, or with a structural ridge beam.

Rafter-Joist Connections

The connection between roof and wall at the eaves needs to resist uplift forces caused by wind (and earthquake activity in seismic zones), as well as horizontal thrust caused by roof loads. Low roof slopes, long rafter spans, and heavy snow loads increase horizontal thrust significantly.

Unless the roof has a structural ridge, the ceiling joists must be continuous across the building and must be fastened to the rafter tails with the number of nails shown in **Figure 2-87**. If the number of nails is excessive for the amount of nailing area, use bolts or framing clips instead.

A weak link in this system is where ceiling joists join at the center of a building. This connection requires the same nailing schedule as a collar tie-to-rafter connection (**Figure 2-16, pages 110-111**). When the ceiling joists are in the same plane, as with a flush-framed beam, steel strap ties can be used (**Figure 2-88**).

Figure 2–87. Nailing and Bolting Schedule for Eaves Tie Connections

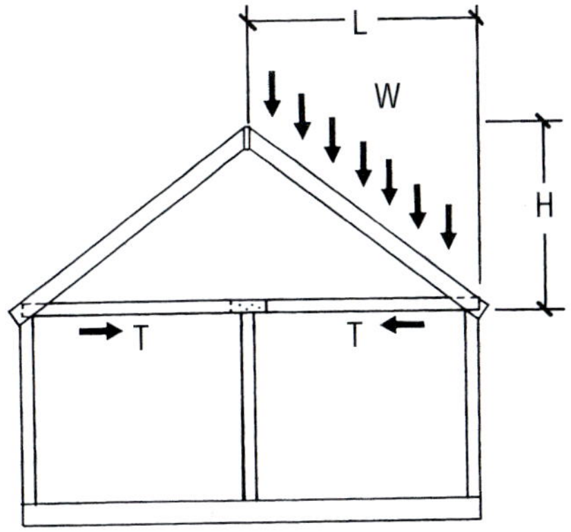

W = L (ft.) × rafter spacing (ft.)
× roof design load (L.L. + D.L. in psf)

$$T = \frac{W \times L}{2H}$$

Roof Pitch	Building Width (2L) 20'	24'	30'	36'
3/12	**T = 1,064 lb.** 11 nails or 3 bolts	**T = 1,276 lb.** 13 nails or 4 bolts	**T = 1,596 lb.** 16 nails or 4 bolts	**T = 1,915 lb.** 20 nails or 5 bolts
4/12	**T = 798 lb.** 8 nails or 2 bolts	**T = 958 lb.** 10 nails or 3 bolts	**T = 1,197 lb.** 12 nails or 3 bolts	**T = 1,437 lb.** 15 nails or 4 bolts
6/12	**T = 532 lb.** 6 nails or 2 bolts	**T = 638 lb.** 7 nails or 2 bolts	**T = 798 lb.** 8 nails or 2 bolts	**T = 957 lb.** 10 nails or 3 bolts
9/12	**T = 355 lb.** 4 nails or 1 bolt	**T = 425 lb.** 5 nails or 2 bolts	**T = 532 lb.** 6 nails or 2 bolts	**T = 638 lb.** 7 nails or 2 bolts
12/12	**T = 266 lb.** 3 nails or 1 bolt	**T = 319 lb.** 4 nails or 1 bolt	**T = 399 lb.** 4 nails or 1 bolt	**T = 479 lb.** 5 nails or 2 bolts

Note: This table is based on 30-pound psf live load plus 10-pound psf dead load. Nails are assumed to be 16d, at 100-pound capacity each; bolts are 1/2 inch, at 400-pound capacity each. Rafters are 16 inches on-center.

Figure 2–88. Connecting Flush-Framed Ceiling Joists

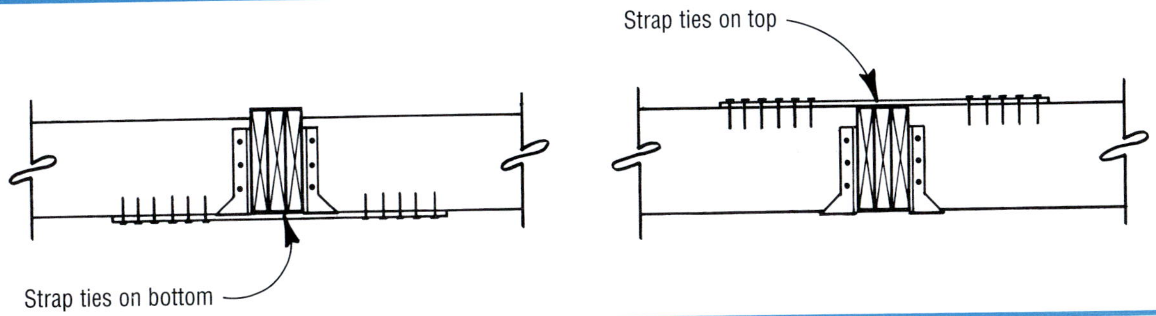

When a flush-framed header or girder interrupts the ceiling joists at midspan, use strap ties to connect the joists and carry tension created by the roof load.

Figure 2–89. Placing Collar Ties

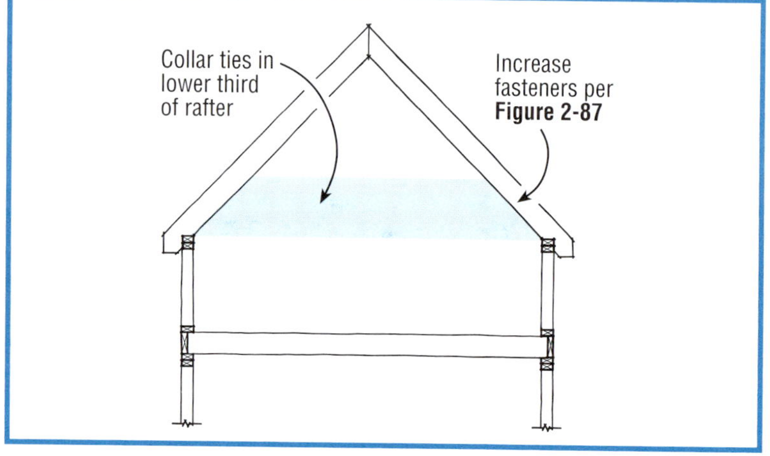

The purpose of traditional collar ties is to strengthen the rafter-to-ridge connection. However, if collar ties are needed to prevent the rafter ends from spreading, then they should remain in the lower third of the span (rule of thumb) or be engineered.

Figure 2–91. Structural Ridge Connections

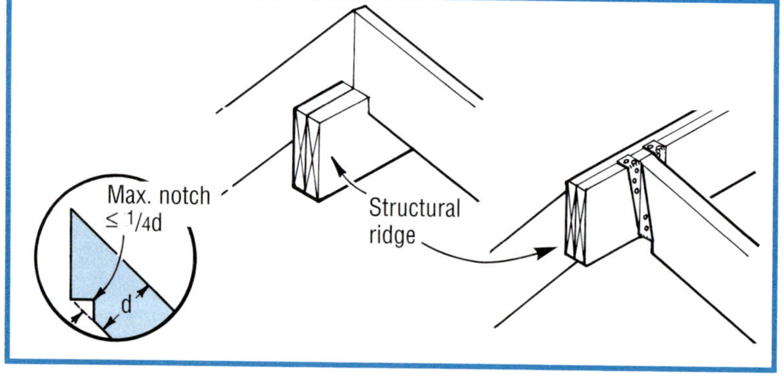

Either rest the rafters on top of a structural ridge or use joist hangers attached to the ridge beam. The notch in the rafter should be no deeper than one-fourth its width.

Collar Ties

Collar ties often are placed near the ridge, but this placement only serves to strengthen the rafter-to-ridge connection and to stiffen the rafters. To effectively keep rafters from spreading, collar ties must be in the lower third of the rafter (**Figure 2-89**). They are not necessary if the rafters are adequately nailed to ceiling joists.

Collar-tie connections should be treated like rafter-to-joist connections (**Figure 2-87**, previous page).

Structural Ridge Beams

In a cathedral ceiling, the outward thrust of rafters is not restrained by ceiling joists or rafter ties and the rafter tops must bear on a structural ridge beam. Sizing for structural ridge beams is shown in **Figure 2-90**.

Figure 2–92. Purlin Design

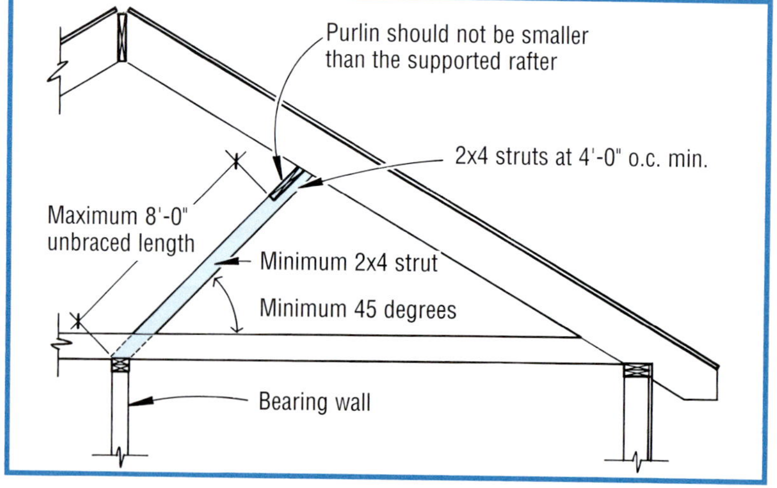

Purlins and struts can effectively stiffen rafter spans. Purlins should be at least the same dimension as rafters and a bearing wall should support the struts.

Rafters may sit on top of a structural ridge or may be supported by joist hangers or anchors (**Figure 2-91**).

Purlins and Roof Struts

Purlins and struts can be used to reduce rafter spans or to stiffen rafters (**Figure 2-92**). To properly support the roof, the purlins should be at least the same dimension as the rafters while the struts should transfer the load to a bearing wall at an angle not less than 45 degrees. The IRC 2000 sets the maximum spacing of 2x4 struts at 4 ft. o.c.

Collar Ties

Structural Ridge Beams

Purlins

Figure 2–90. Structural Ridge Beam Spans for 20 psf Roof Live Load

	Roof Dead Load = 10 psf			Roof Dead Load = 20 psf		
	Building Width (ft.)			Building Width (ft.)		
	20	28	36	20	28	36
Size	Maximum Ridge Beam Spans for Common Lumber Species (ft.-in.)*					
1-2x6	4-10	4-1	3-7	4-2	3-6	3-1
1-2x8	6-1	5-2	4-7	5-3	4-6	3-11
1-2x10	7-5	6-4	5-7	6-5	5-5	4-10
1-2x12	8-8	7-4	6-5	7-6	6-4	5-7
2-2x6	7-2	6-0	5-4	6-2	5-3	4-7
2-2x8	9-1	7-8	6-9	7-10	6-7	5-10
2-2x10	11-1	9-4	8-3	9-7	8-1	7-2
2-2x12	12-10	10-10	9-7	11-1	9-5	8-3
3-2x8	11-1	9-7	8-5	9-10	8-4	7-4
3-2x10	13-10	11-8	10-4	12-0	10-2	8-11
3-2x12	16-1	13-7	12-0	13-11	11-9	10-4
4-2x8	12-3	10-11	9-9	11-1	9-7	8-5
4-2x10	15-7	13-6	11-11	13-10	11-8	10-4

* Tabulated values assume #2 Grade Douglas Fir-Larch, Hem-Fir, Southern Pine, or Spruce-Pine-Fir lumber.

Cathedral ceilings require structural ridge beams. This table shows two dead loading conditions (10 and 20 psf) under live load conditions of 20 psf. In areas with heavier snow and wind live loads, stronger ridge beams will be required. For longer spans and heavier loading conditions, consider using LVL or other engineered materials.

RAISED RAFTER PLATES

To create added space for insulation at the eaves, many builders raise the rafters up and fasten them to a plate nailed to the top of the ceiling joists. While this reduces heat loss and ice dams, it also weakens the important connection between the ceiling joists and the rafter ends, thereby requiring reinforcement (**Figure 2-93**).

BLOCKING FOR END RESTRAINT

Rafter ends connected to ceiling joists are generally not at risk of rotating. But where rafters sit on a beam or plate by themselves or are raised above the joists, full-height blocking should be installed between the rafters or a rim joist (or subfascia) should be nailed to their ends.

Blocking is often needed to close off open eaves (no soffit), but not for structural reasons.

OVERHANGS

In general, roof overhangs should not exceed 2 ft. or one-third the horizontal span of the rafter, whichever is less (**Figure 2-94**).

Figure 2–93. Raised-Rafter-Plate Connections

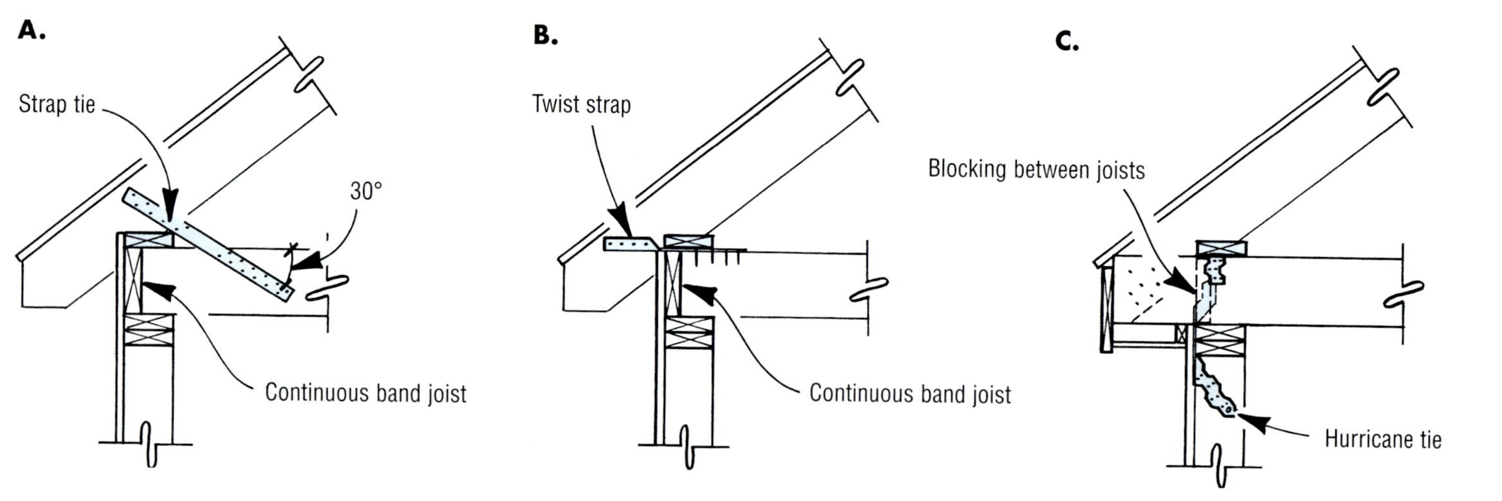

Using rated strap ties (A) is an effective way to resist roof thrust and uplift on raised rafters. When an attic floor is in the way, twist straps (B) will work, but may require additional wind ties in high wind areas. Extending the attic joist beyond the wall (C) allows for hurricane ties to help resist uplift.

Raised Rafters

Overhangs

Rake Overhangs

There are two common ways to build overhangs at rakes:

- For overhangs shorter than 1 ft., blocks nailed to the gable-end wall are adequate.
- For larger overhangs, short "outlookers," or purlins, are typically cantilevered (**Figure 2-95**). These should not cantilever more than 2 ft. or half the span of the purlin, whichever is less.

ROOF OPENINGS

In general, at roof openings more than 4 ft. wide, both headers and trimmers should be doubled. At openings more than 6 ft. wide, joist hangers should support headers unless they bear on a supporting beam or wall.

Figure 2–94. Roof Overhangs

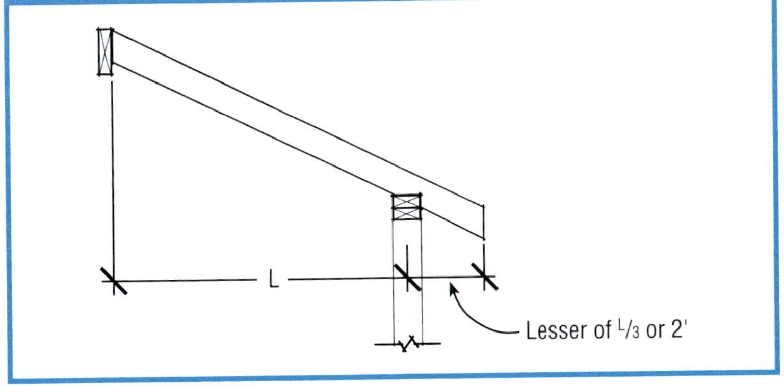

Roof overhangs should not exceed 2 ft. or one-third of the rafter span, whichever is less.

Figure 2–95. Rake Overhangs

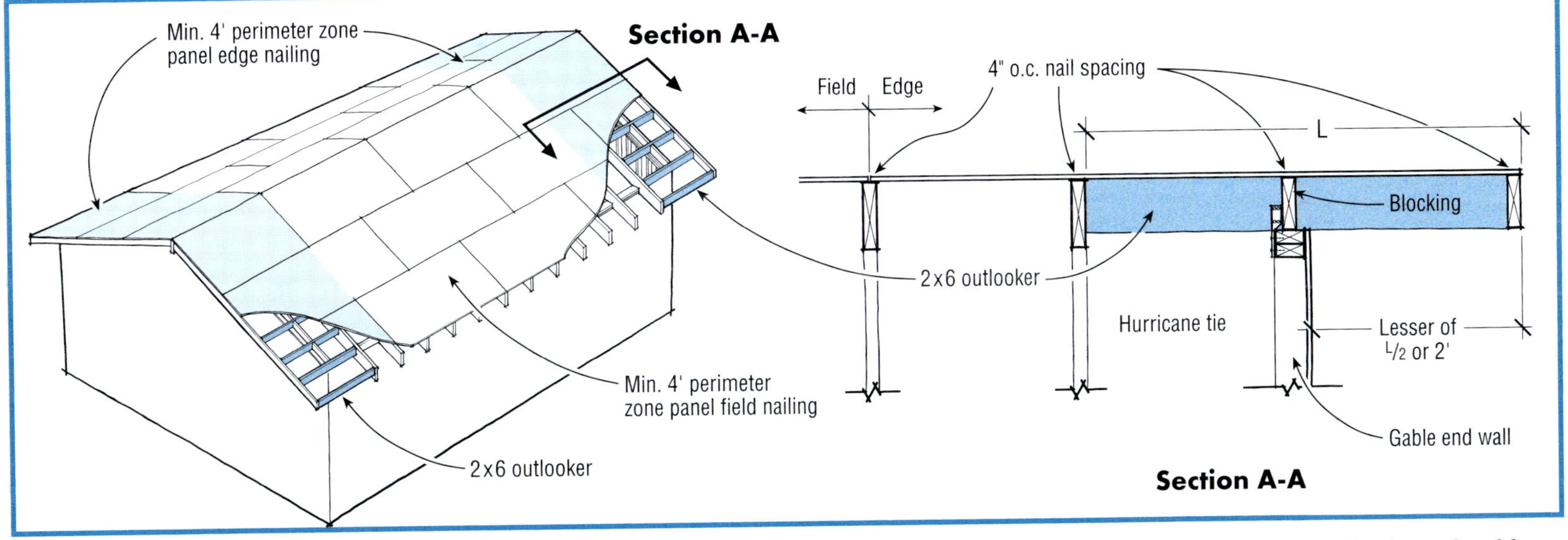

Outlookers at the rake should not exceed 2 ft. or half the span of the outlooker, whichever is less. Blocking-type outlookers should not exceed 1 ft. In high-wind areas, nailing of the sheathing around the roof's perimeter may need to be upgraded.

Figure 2–96. Center Support for Shed Dormers

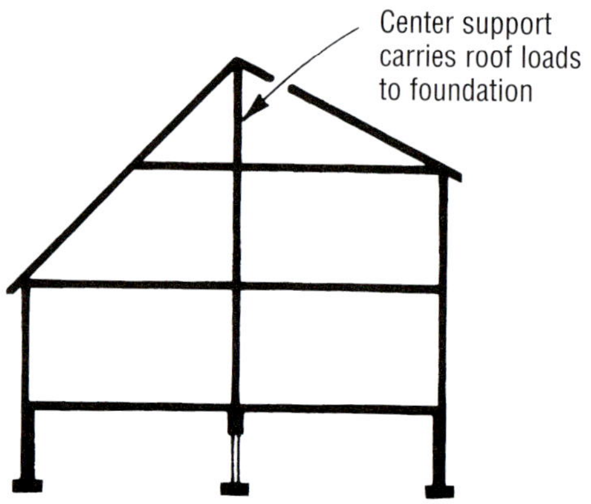

A center support wall, or a structural ridge in the same location, will handle the unbalanced loads caused by adding a shed dormer. The loads must follow a continuous path to the footing.

SHED DORMERS

Adding a shed dormer to a typical gable roof upsets the balance of forces by breaking the stable triangle formed by the rafters and ceiling joists.

Two common ways to support the dormer loads are with a center loadbearing wall (**Figure 2-96**) or a structural ridge. Either solution must provide a continuous load path to the foundation.

Shed Dormers With Structural Ridges

A structural ridge can replace the ridge or can be installed beneath it (**Figure 2-97**).

Where a structural dormer ridge is dropped, it should tie in securely to the structural ridge or center loadbearing wall (**Figure 2-98**).

With no structural ridge or center loadbearing walls, small shed dormers are still possible but a combination of structural elements must be added (**Figure 2-99, page 152**).

Figure 2–97. Retrofit Ridge Connections for Shed Dormers

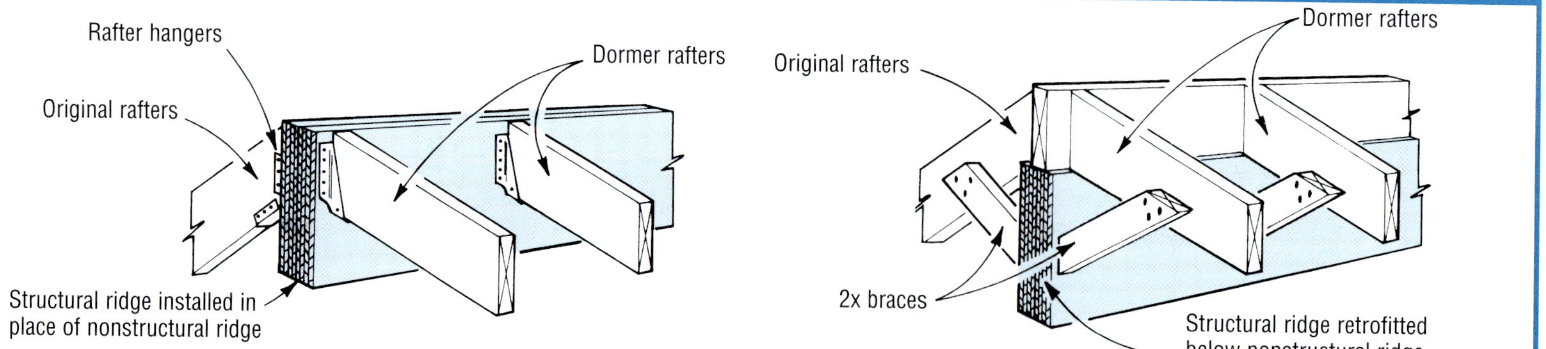

Whether a structural ridge replaces the existing ridge (left) or is installed below (right), it must be sized to carry the loads introduced by the addition.

Figure 2–98. Dropped-Ridge Dormer (With Structural Ridge)

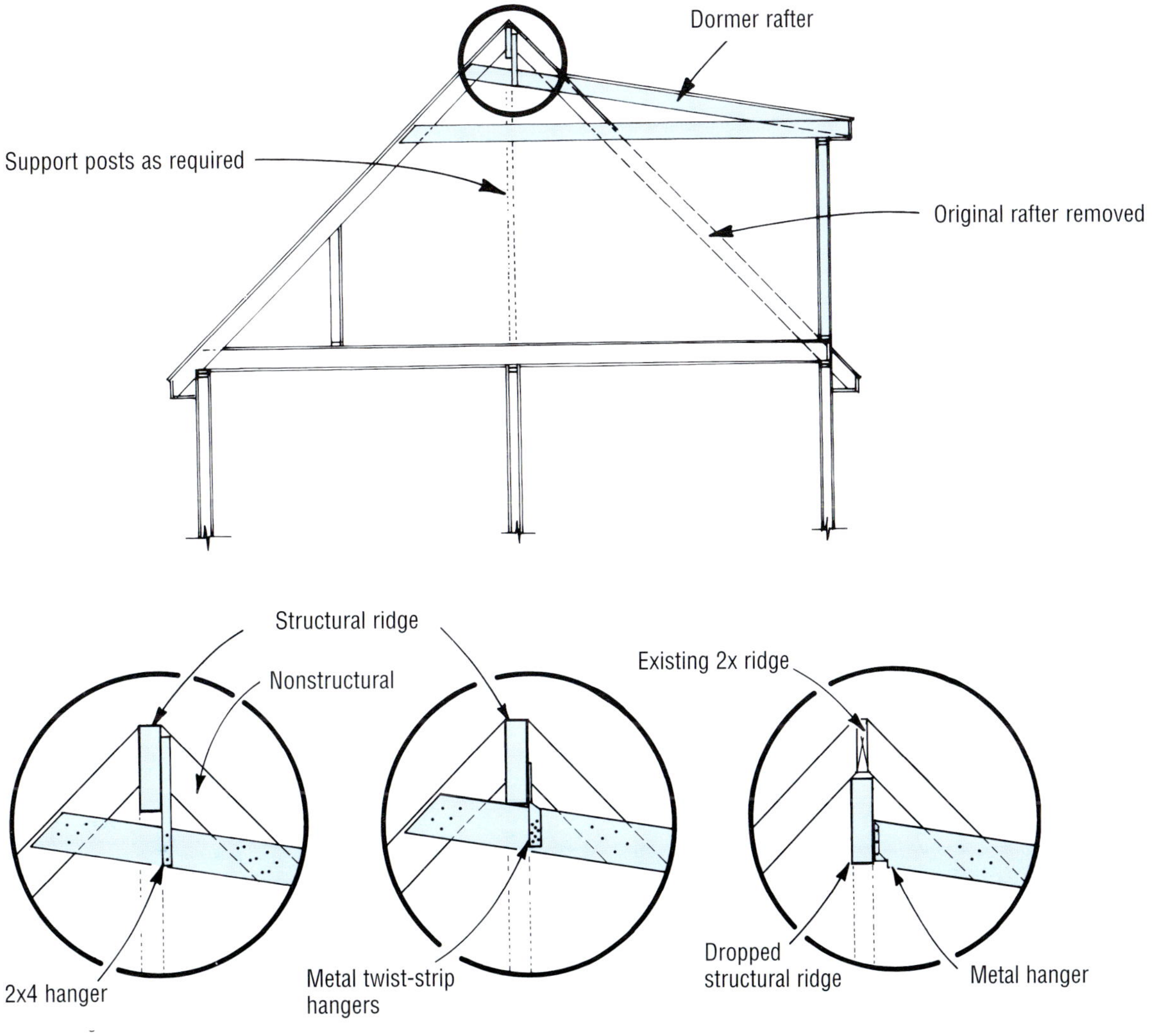

Where the dormer ridge is dropped below the main ridge, the dormer ridge must be tied securely to the structural ridge or center loadbearing wall. Several options are shown.

Figure 2–99. Precautions for Small Shed Dormers

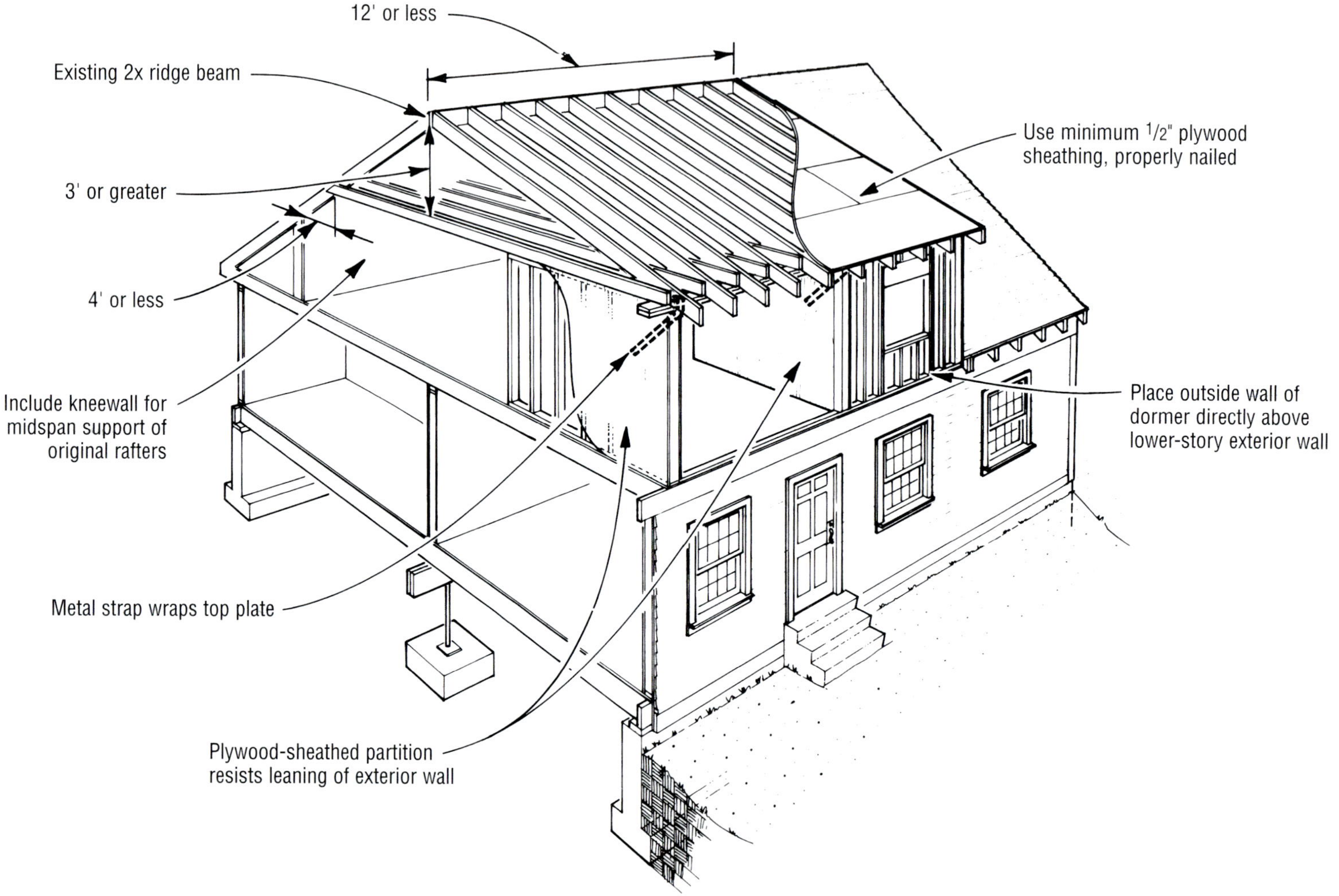

It is often possible to safely add a small shed dormer where a structural ridge or loadbearing wall is not feasible. The precautions shown here offer general guidelines that can work together to support the dormer and to prevent the ridge from sagging and the dormer wall from leaning.

I-BEAMS ARE BEING USED MORE OFTEN IN ROOF FRAMING, particularly on cathedral-type ceilings with long spans.

STRUCTURAL RIDGE REQUIRED

All I-beam roofs require a center loadbearing wall or a structural ridge beam, since there is no practical way to make a strong shear connection at the rafter ends to resist horizontal thrust.

Structural Ridge With Hangers

The most common connection at the ridge is a face-mounted hanger with a sloped seat. These hangers can be adjusted in the field to match the I-beam slope, and can be skewed side-to-side up to 45 degrees for hip and valley jack rafters.

Install beveled web-stiffeners on each side of an I-beam where it sits in the hanger. Leave a 1/8- to 1/4-in. gap at the top flange and secure the stiffeners by driving three or more 8d nails from each side, staggered and with the points clinched. Larger I-beams may require 2x stiffeners with 16d nails. In all cases, follow manufacturers' nailing schedules.

Structural Ridge Below Rafters

I-beams can sit on top of a structural ridge on a beveled plate (**Figure 2-101**). Never cut a birdsmouth in the I-beam at the ridge — this would require cutting though the bottom flange, thereby damaging the joist.

To prevent rafters from rolling, install blocking between the I-beams on either side of the ridge. The blocking can be I-beam material, dimensional lumber, or metal cross bracing. If you need to leave room for a ventilation channel, use dimensional lumber or engineered rim joist material notched to allow airflow, or use metal cross-bracing.

Figure 2–101. Structural Ridge Connections

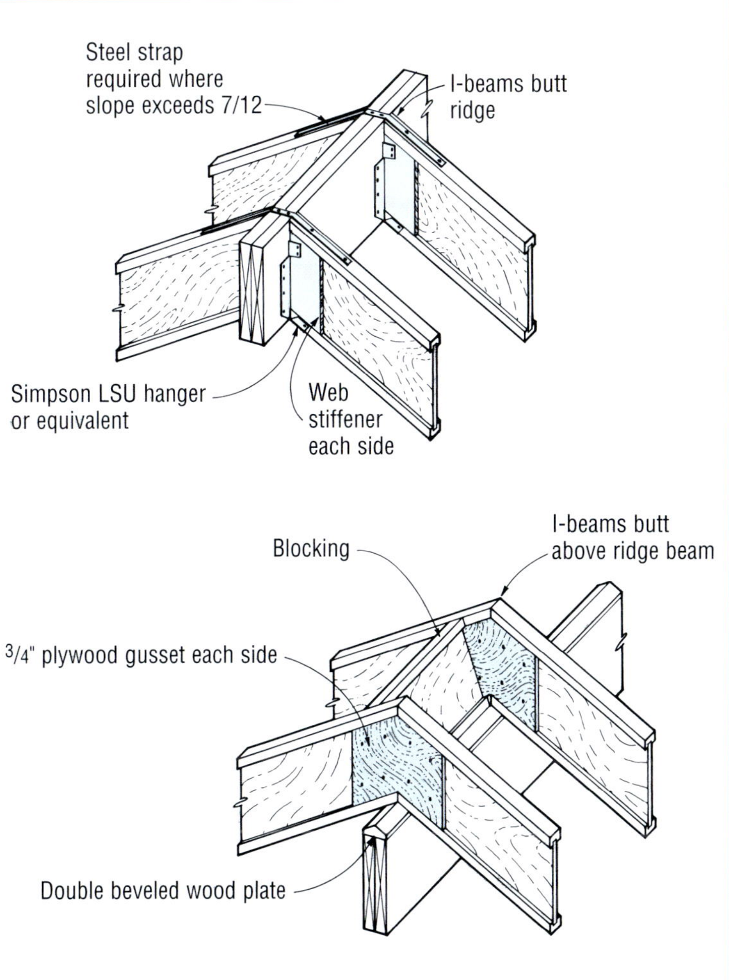

The most common ridge connection to I-beam rafters uses a sloped seat hanger (top). Steel strapping over the tops of the rafters is required with roof slopes greater than 7/12. Where I-beam rafters rest on top of the ridge (above), blocking is required on both sides of the ridge.

Structural Ridge for I-Beam Roofs

EAVES CONNECTIONS

A number of details are possible at the exterior wall plate, depending on the roof profile desired.

Birdsmouth for I-Beam Rafters

The most common approach to joining I-beam rafters to walls is to cut a birdsmouth at the plate. The seat must never overhang the inside face of the bearing wall, otherwise the I-beam will be compromised. Web stiffeners are required on both sides of the birdsmouth cut (**Figure 2-102**).

Beveled Wall Plates for I-Beam Rafters

I-joist rafters can sit on a beveled top plate (**Figure 2-103**). This option can save time since no birdsmouth or web stiffeners are required.

Blocking between rafters is required when using beveled wall plates. For blocking, use either I-beams, dimensional lumber, or metal cross-bracing. Some I-beam manufacturers may require additional connectors on roofs with steep slopes.

Figure 2–102. Birdsmouth Cut for I-Beam Rafters

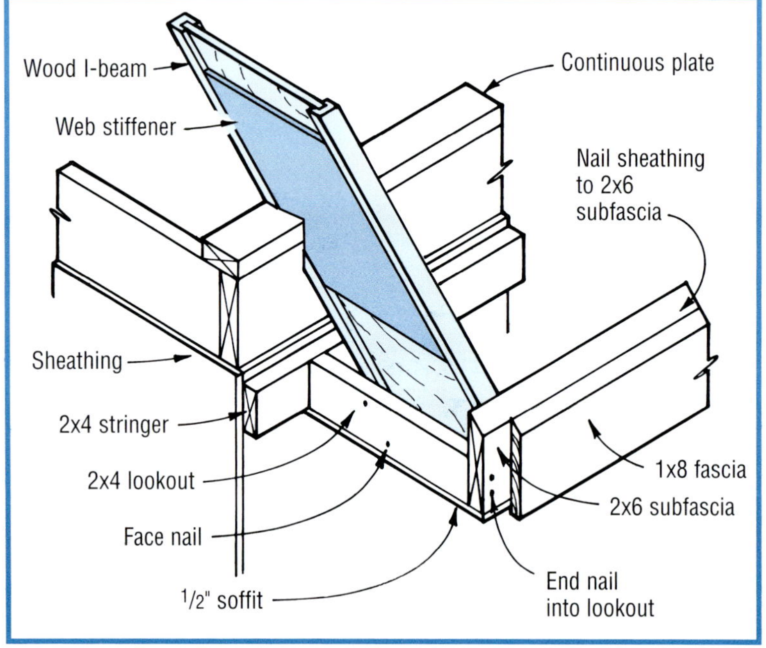

With a birdsmouth cut, it is critical that the entire bottom chord bear directly on the top plate. Provide web stiffeners and lateral bracing — metal cross bracing, dimensional lumber, engineered rim joist material, or I-beam material — on both sides of rafter.

Figure 2–103. Beveled Wall Plates for I-Beam Rafters

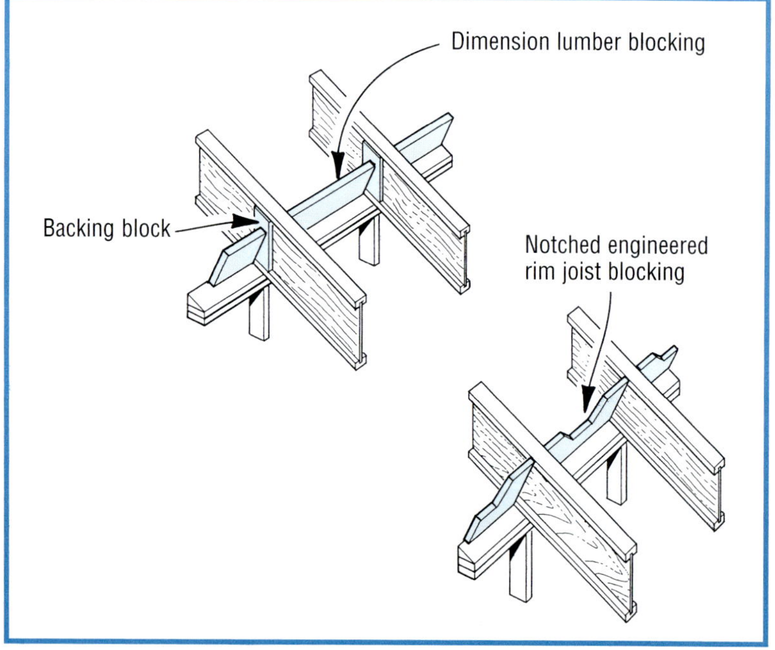

Joining rafters to the walls with a beveled top plate typically does not require web stiffeners, but does require blocking on both sides of the rafter.

For more information: "Roof Framing with Wood I-Joists," JLC 1/97

Sloped Seat Connectors

A third option to connect I-beam rafters to walls is to use sloped-seat metal connectors. As with the beveled-plate approach, these do not require a birdsmouth or web stiffeners, but do require blocking between rafters.

I-BEAM OVERHANGS

Because cutting through the bottom flange compromises the strength of an I-beam rafter, rafter tails with birdsmouths are limited to a 2-ft. horizontal projection (**Figure 2-104**).

Soffits

There are several ways to build traditional soffits. One common approach is to attach 2x4 blocking to the I-beam rafter tails as shown in **Figure 2-102**. Another option is to sister on rafter tails of dimensional lumber.

Gable-End Overhangs

For an overhang at the gable end, cantilever dimensional lumber outriggers across the rake-wall top plate (**Figure 2-105**). These are similar to the outriggers used in stick framing (**Figure 2-95, page 149**).

TEMPORARY BRACING FOR I-BEAMS

I-beam rafters (like floor I-joists) are unstable until fully braced and sheathed. It's crucial to install all blocking and provide lateral restraint before applying any weight to the rafters. As with floor I-joists, stabilize an unbraced end by installing the first 4 ft. of sheathing, then brace the rest of the rafters with 1x4 braces nailed flat to the top of the I-beam rafters every 6 to 8 ft. and lapped at the ends by at least two joists.

Figure 2–104. I-Beam Rafter Tails

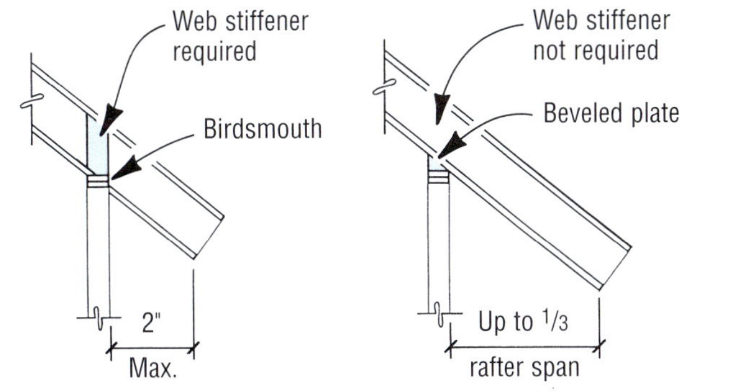

With a birdsmouth (left), the overhang for an I-beam rafter is limited to 2 ft. of run. With a beveled top plate or sloped hanger, the overhang may be as long as one-third the length of the rafter (right).

Figure 2–105. Rake Wall Overhangs

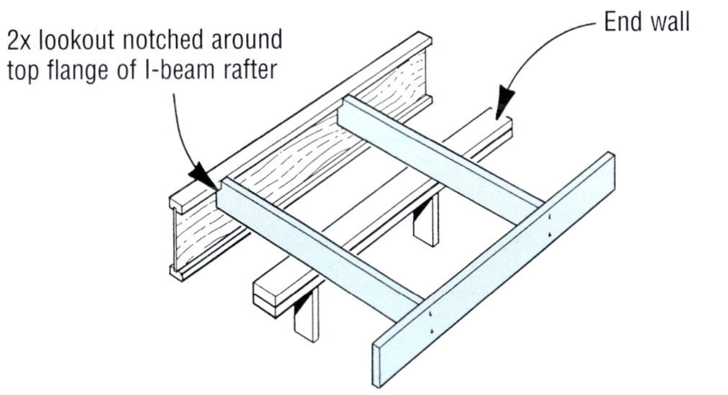

Frame rake-wall overhangs with dimensional lumber outriggers notched around the I-beam top flange. If the overhang exceeds the I-beam spacing, check with the manufacturer to see if a double I-beam is required.

THE MAIN COMPONENTS OF A TRUSS ARE SHOWN IN **Figure 2-106**. Roofs framed with trusses are engineered systems. They must be installed and braced properly to work as planned. Any changes from the original plan in bearing or loading or any field modifications to the trusses must be engineered.

TRUSS TYPES

Many types of standard trusses are now produced and almost any roof design can be custom fabricated. The most popular types of trusses are shown in **Figure 2-107**.

Truss Overhangs

When ordering trusses, specify the span, slope, width of the bearing walls (2x4 or 2x6), and the length and type of overhang (**Figure 2-108**).

The truss length is typically the length of the bottom chord; it does not include the overhang, which is most often a projection of the top chord.

Hip and Valley Trusses

Hip roofs and valleys between intersecting roofs may be built around an assembly of jack trusses attached to a girder truss.

Figure 2-106. Truss Components

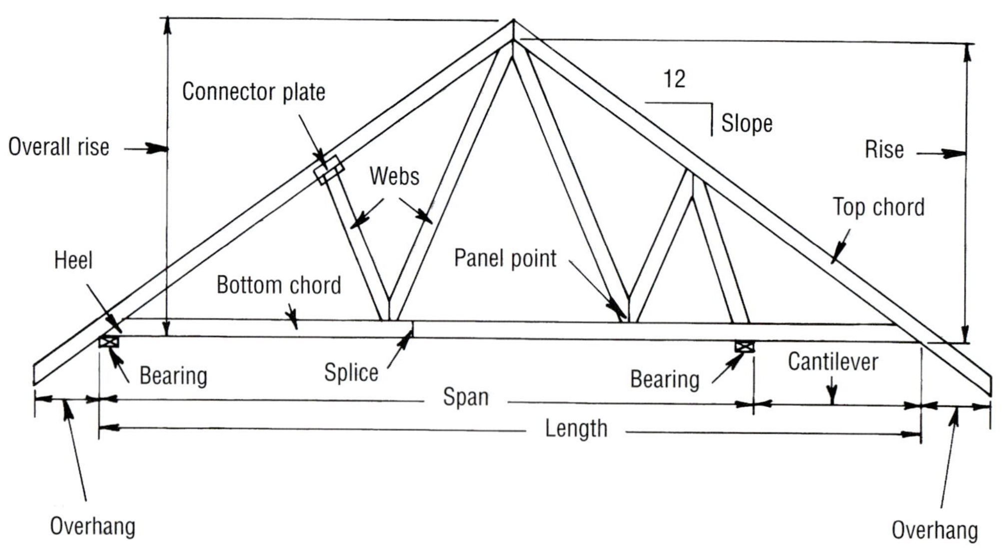

Most trusses are designed to clear-span between the exterior walls without requiring any interior support.

Figure 2–107. Truss Types

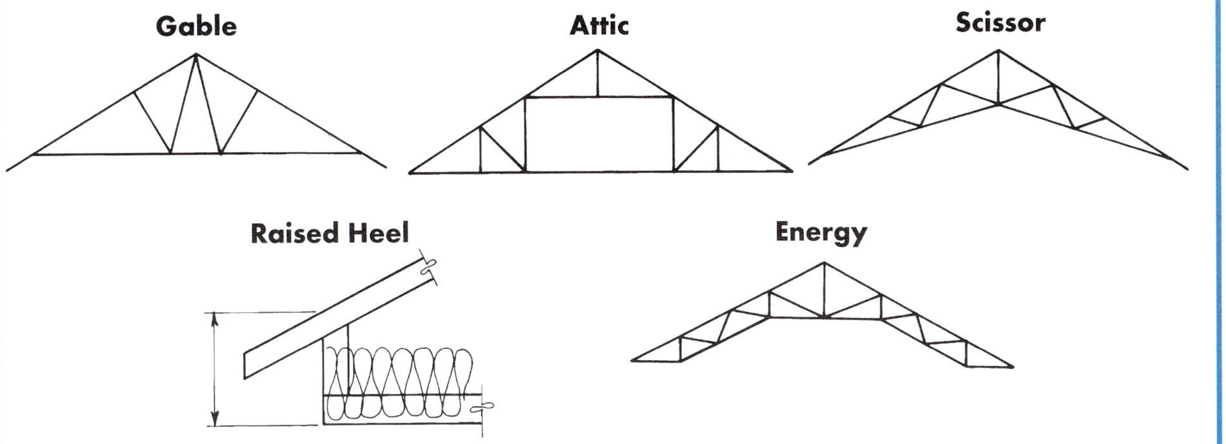

Truss Types

The common truss types provide a variety of ceiling profiles, including the full-height living space provided by the attic truss. Raised-heel trusses and energy trusses provide extra room for ceiling insulation.

Figure 2–108. Truss Overhangs

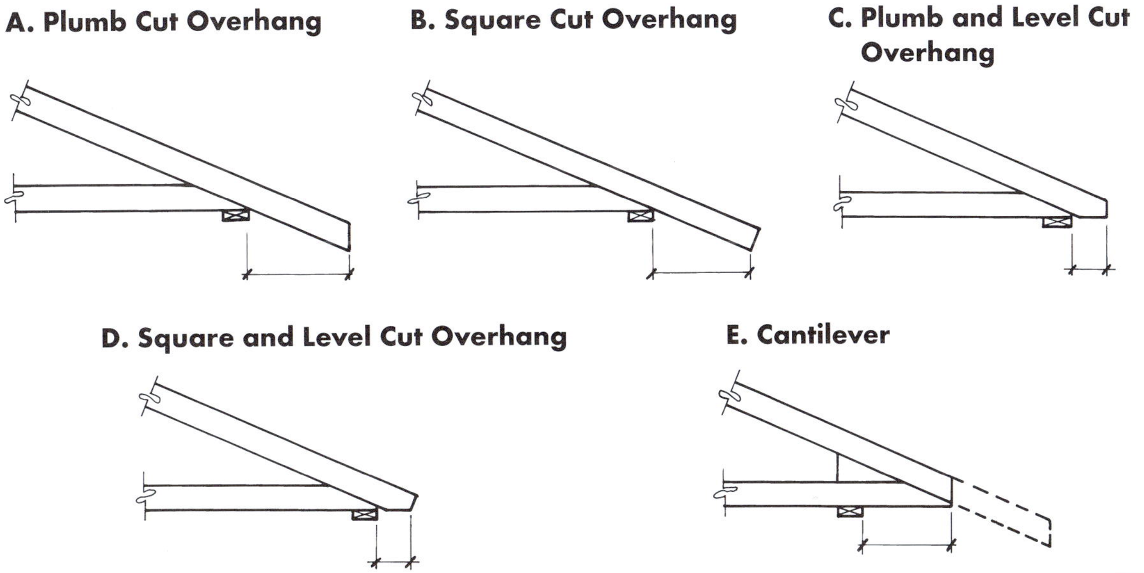

When ordering trusses, specify the length of the overhang, measured horizontally. The overhang usually is not considered part of the truss length; only cantilever overhangs (E) are included in the truss length. In addition, be sure to call out how the top chord is cut at the end (A, B, C, or D).

Hips. Step-down trusses complete the upper end of the hip series and smaller trusses fill in the lower corners (**Figure 2-109**). As with other trusses, it's critical that the dimensions of the building match the drawing and that walls are square and level.

Valleys on a trussed roof are usually framed conventionally. However, when a project includes more than one valley of the same size, a valley set may be more economical (**Figure 2-110**).

HANDLING AND STORING TRUSSES

Trusses are vulnerable to damage until they are installed, sheathed, and braced. To prevent damage, use the following precautions:

- Inspect trusses when receiving a shipment. Look carefully for loose plates and reject any damaged trusses.
- Banded trusses should be lifted off the truck from two points on the top chords. Individual trusses should be lifted the same way (**Figure 2-111**).
- Store trusses horizontally on relatively level, dry ground. To prevent any bending or twisting that could loosen truss plates, support the trusses with blocking placed every 8- to 10-ft.
- If trusses must be stored vertically, make sure they are braced and blocked in a stable manner. Gable trusses should be stored with their peaks up, scissors trusses with their peaks down.

Figure 2–109. Truss Hip Anatomy

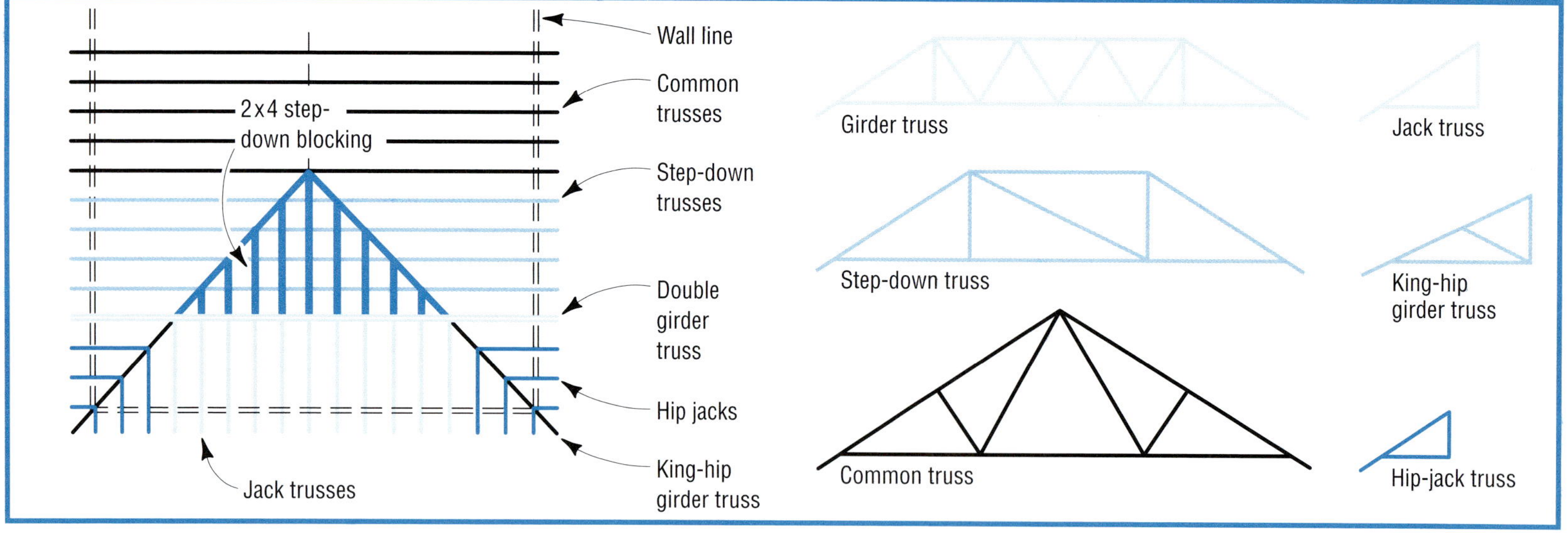

The most common hip system uses common trusses, which step down to a doubled girder truss. The hip is then completed with a pair of king hip girder and jack trusses.

For more information: "Installing Hip Roof Trusses," JLC 6/97

"Building Truss Roofs the Safe Way," JLC 12/01

BRACING ROOF TRUSSES

Trusses are inherently unstable until sheathed and require two types of bracing: temporary and permanent.

Use temporary bracing to safely erect the trusses; permanent bracing is required to stabilize the trusses throughout the life of the structure. The truss fabricator or "designer of record" should provide a permanent bracing plan.

Usually some of the elements of the temporary bracing stay in place as part of the permanent bracing. In general, all bracing should be placed near panel points (where truss members meet).

Figure 2–110. Valley Sets

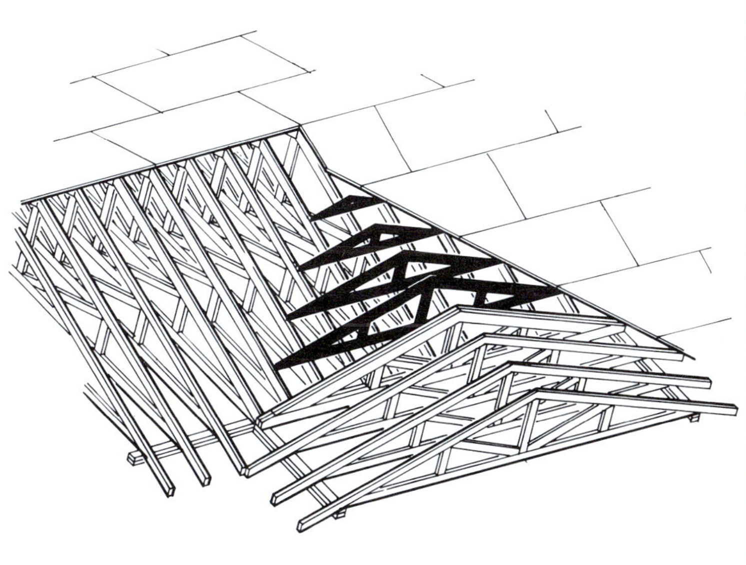

Although valleys are usually framed conventionally, valley truss systems may be more economical if the same size valley is used more than once in a project.

Temporary Bracing: The First Truss

It is important to accurately plumb the first truss and to brace it substantially, since additional trusses rely on the first one for stability.

Handling Trusses

All bracing must be of 2x material.

Start with the gable-end truss. Most builders start by sheathing the gable-end truss on the ground and lifting

Bracing Trusses

Figure 2–111. Lifting Trusses

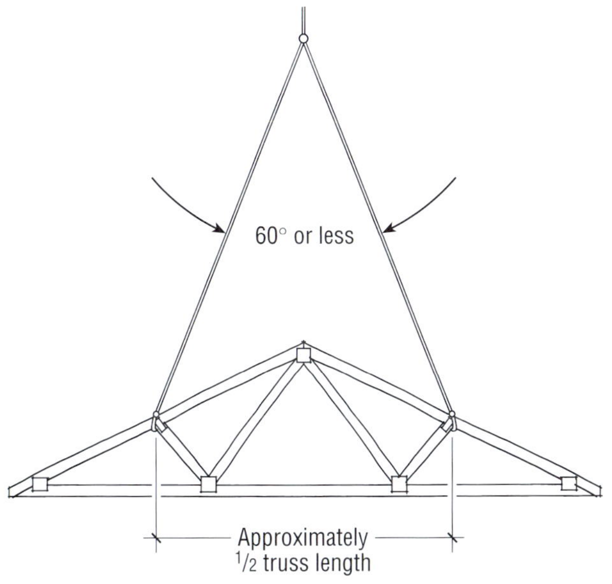

To prevent trusses from twisting and damaging plate connections, lift trusses less than 30 ft. long at two balance points. Longer trusses require reinforcement with strongbacks before lifting.

Figure 2–112. Bracing to the Deck

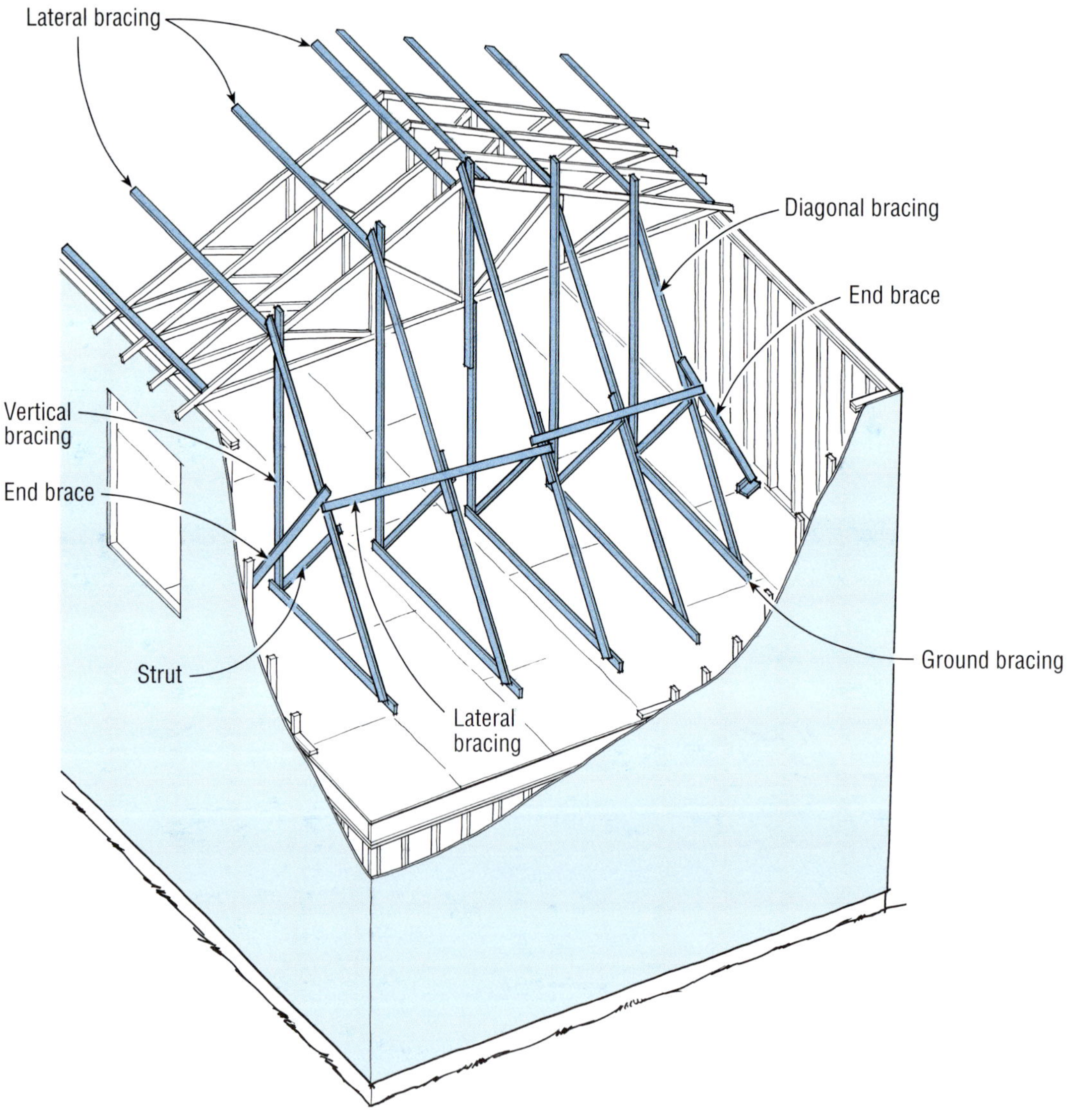

For more information: "Truss Bracing Tips," JLC 3/94

"Truss Roofs Production Style," JLC 5/99

The first truss must be straight and plumb and securely braced. This temporary bracing configuration, which starts several trusses in from the end and braces against the floor deck, is recommended by the Truss Plate Institute.

that first. Unfortunately, the gable-end truss is difficult to brace back to the deck without the braces interfering with the second truss. On a one-story building, the gable-end truss can be braced to stakes driven in the ground.

Start several trusses in. Alternatively, start the first truss 8- to 12-ft. in from the gable end and brace it to the deck (**Figure 2-112**). Then work away from the braced end, filling in the open 8- to 12-ft. section last, after permanently bracing the rest of the trusses.

Temporary Bracing

Or brace the trusses to an end wall (**Figure 2-113**). It may first be necessary to solidly brace the end wall to the deck to keep the top plate rigid.

Figure 2–113. Bracing to the End Wall

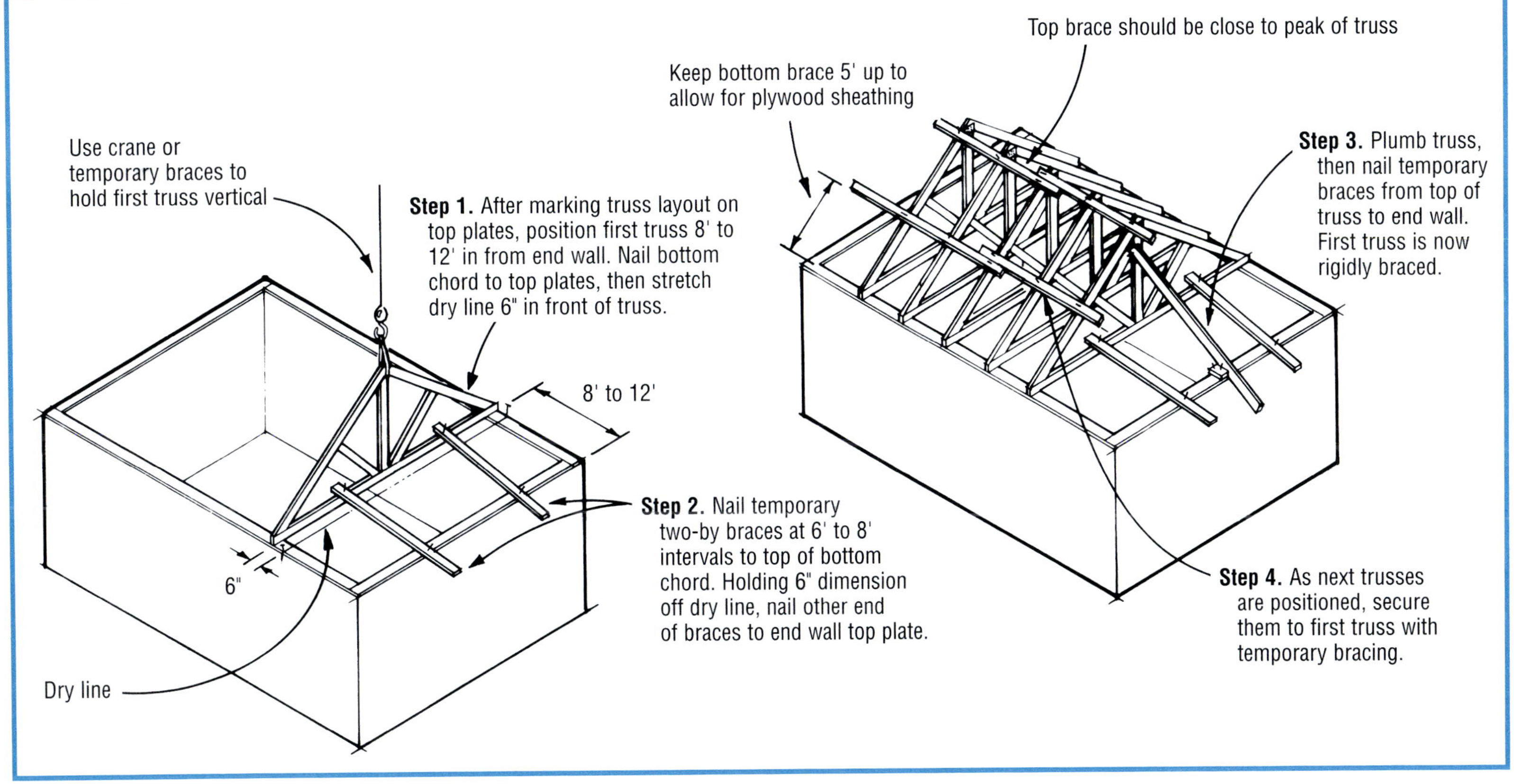

Builders have developed a variety of field techniques to brace the first truss. This approach requires that the end wall first be braced to keep the top plate rigid.

Temporary Bracing: Subsequent Trusses

Spacers. As trusses are added to the roof, they can be spaced from the preceding truss with short 1x4 or 2x4 spacers pre-cut to length, with light-gauge metal spacers or with heavy-duty fold-out metal braces. (The heavy metal braces can also serve as temporary lateral bracing for the top chord.)

Top-chord bracing. As soon as the fourth truss is up, start fastening lateral bracing across the top chords — one brace near the peak and one near the midpoint of each top chord. If required, place diagonal bracing at each end of the roof and at 20-ft. intervals in between (**Figure 2-114**).

All lateral bracing should be minimum construction-grade 2x4s, at least 10 ft. long and nailed with two 16d nails at each intersection. Individual pieces should lap at least one bay.

Roof sheathing. To quickly stabilize the top chords, some builders start installing sheathing after about a dozen trusses have been installed. The sheathing acts as both temporary and permanent bracing for the top chords.

Figure 2–114. Diagonal Top-Chord Bracing

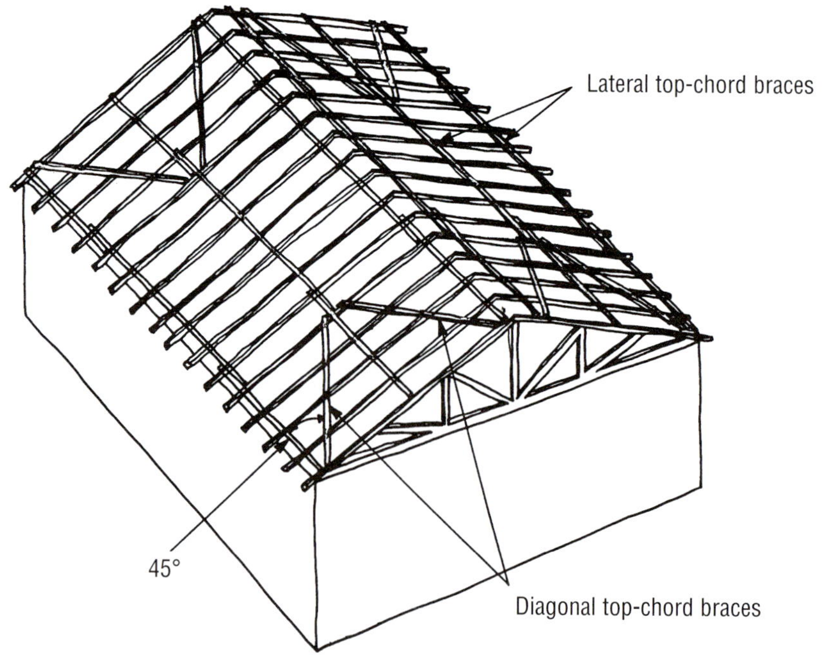

Diagonals are recommended to prevent racking until the sheathing is applied. Some builders place the lowest lateral bracing at least 4 ft. up to allow room for the first layer of plywood.

To simplify installation of the sheathing, some builders keep the lower top-chord braces 5 ft. up from the eaves line to allow room for the first course of sheathing. Unless the sheathing is installed shortly after erecting the trusses, diagonal top-chord braces may be needed to prevent racking in the roof or truss collapse.

Bottom-chord bracing. After the fourth truss is up, also start installing lateral bracing on the bottom chord (along its top edge if the bottom chords are receiving drywall). Install bottom-chord bracing according to the permanent bracing plan and leave in place. On large trusses, diagonal bottom-chord bracing may be required at each end and at 20-ft. intervals in between.

Web cross-bracing. After the fourth truss is up, also start fastening diagonal cross-bracing to the vertical webs. Cross-bracing of the vertical webs goes in an X pattern at each end of the building and every 20 ft. in between (**Figure 2-115**). Install web cross-bracing according to the permanent bracing plan and leave in place.

Temporary Bracing

Figure 2–115. Bottom-Chord Bracing and Web Cross-Bracing

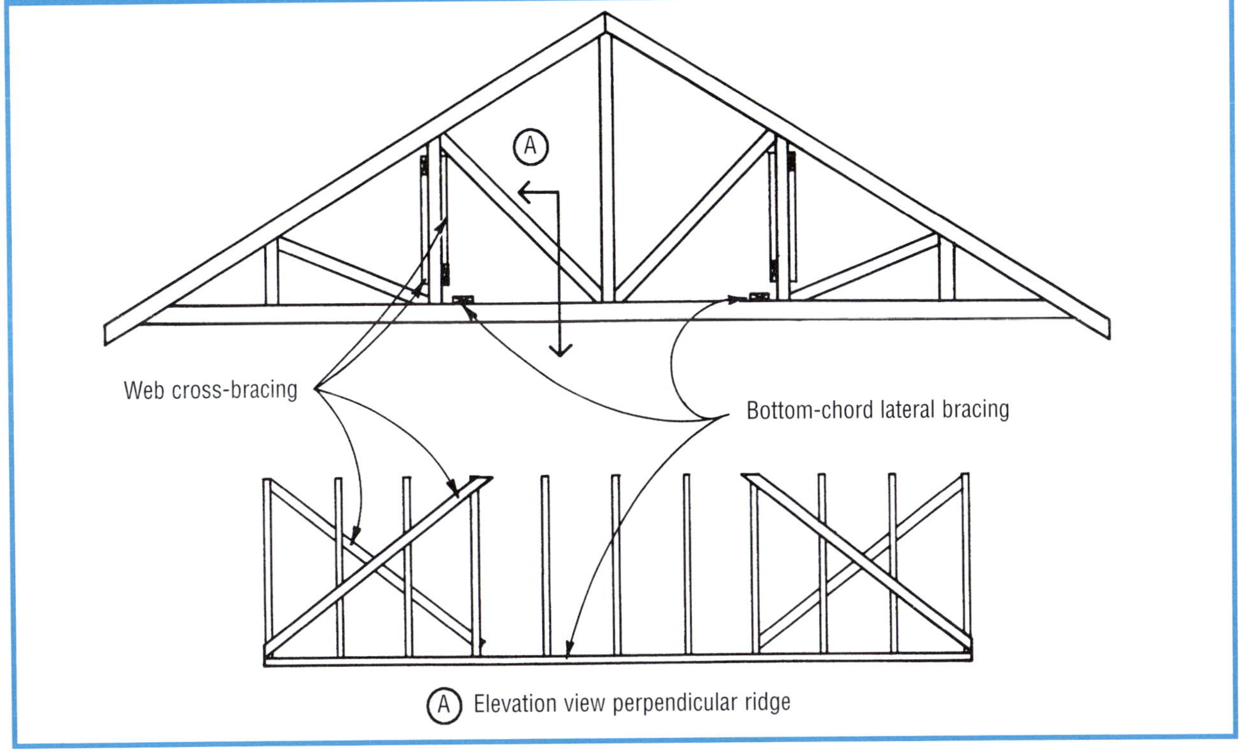

Bottom-chord bracing should be no more than 15 ft. o.c. Pairs of web cross-braces triangulate the trusses on each end of the building and every 20 ft. in between. These braces should be installed according to the permanent bracing plan and left in place.

Permanent Truss Bracing

To perform as designed, roof trusses require permanent bracing after assembly, as specified by the "designer of record" on the job. In the absence of an architect or engineer, this responsibility may fall to the general contractor.

Permanent bracing typically consists of many of the same elements as the temporary bracing (see above), most of which are left in place after construction. One exception is any top-chord bracing, which must be removed to allow installation of the roof sheathing. The roof sheathing then serves as permanent top-chord bracing.

PIGGYBACK TRUSSES

When trusses exceed 12- to 14-ft. in height, they are typically shipped in two sections to comply with highway transportation limits. These so-called "piggyback trusses" have special bracing requirements. Ignoring these requirements can lead to catastrophic failure under heavy loads such as snow.

In particular, it is critical to stabilize the top chords of the lower truss sections with either plywood or diagonal bracing (**Figure 2-116**). Other permanent bracing may also be required.

Figure 2–116. Bracing Piggyback Trusses

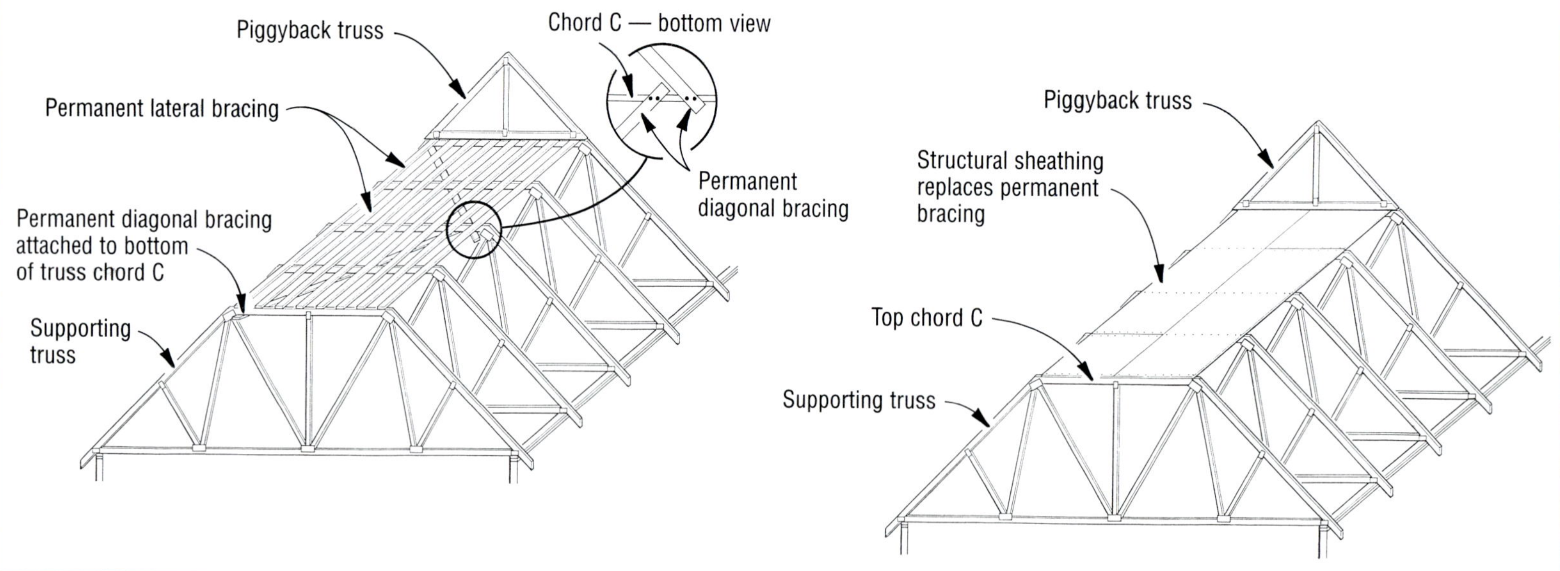

Proper bracing of chord C is critical in piggyback trusses. Use lateral bracing above and diagonal bracing below chord C (left), or use structural sheathing (right). With proper nailing, the sheathing acts as a diaphragm providing both lateral and diagonal bracing.

For more information: "Practical Engineering: Permanent Bracing for Piggyback Trusses," JLC 3/98

HIGH WINDS EXERT STRONG LATERAL FORCES ON WALLS AND uplift forces on roofs. Earthquakes also exert strong lateral forces, along with vertical shaking and other damaging forces. To help buildings in affected areas resist these forces, codes require that they have sufficient shear strength and have a continuous "hold down" path from roof to foundation (**Figure 2-117**). Metal connectors are typically required at the weak points between foundation and first floor, at level changes, and where upper walls join roofs. The details and hardware needed are similar in both seismic and high-wind areas.

Figure 2–117. Hold-Down Paths

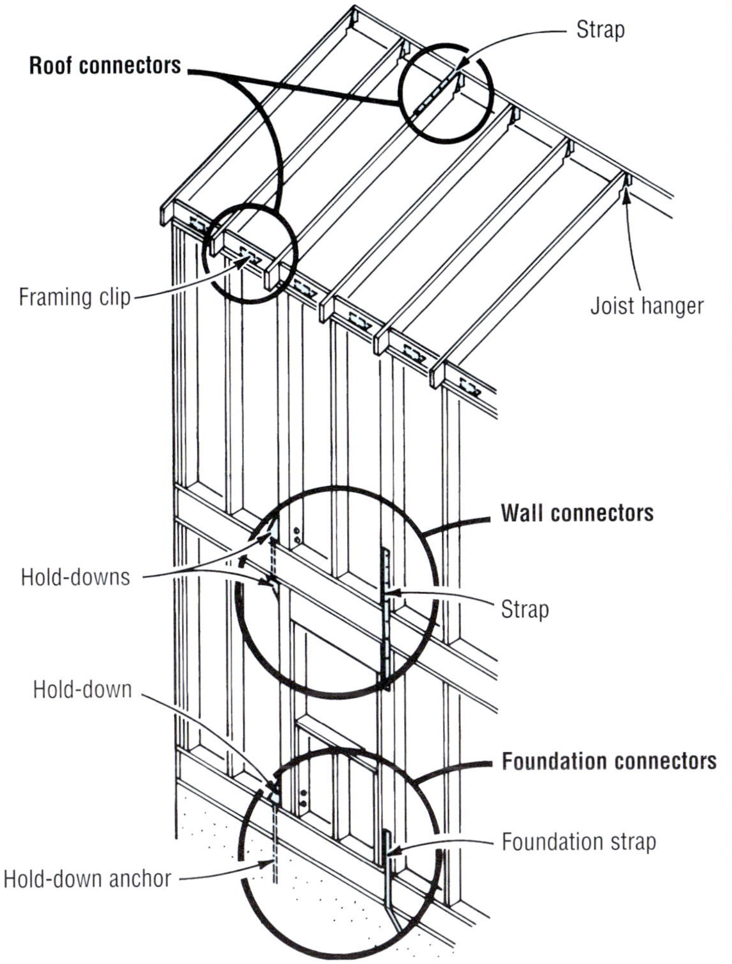

A variety of metal connectors are used to tie a house structure together and create a continuous "hold-down" path from roof to foundation.

SHEAR WALLS

Continuous plywood across wall planes and tighter nailing schedules are crucial elements of structural design for seismic and high-wind areas.

Shear Wall Nailing

Shear walls are designed to resist racking. Both interior and exterior shear walls have rigorous nailing schedules, with nails spaced only 2- to 3-in. apart on panel edges (**Figure 2-118**, next page). Nails must be held back $3/8$ in. from the edges of sheathing panels and $3/8$ in. in from the edges of framing members. Also, they must be staggered vertically (**Figure 2-119, page 167**).

Nail types and sizes are also specified. While galvanized nails can be used, they have lower shear values and require larger nails or closer spacing.

Nail heads must be large enough and must not be overdriven (**Figure 2-121, page 167**).

Figure 2–118. Typical Shear Wall Details

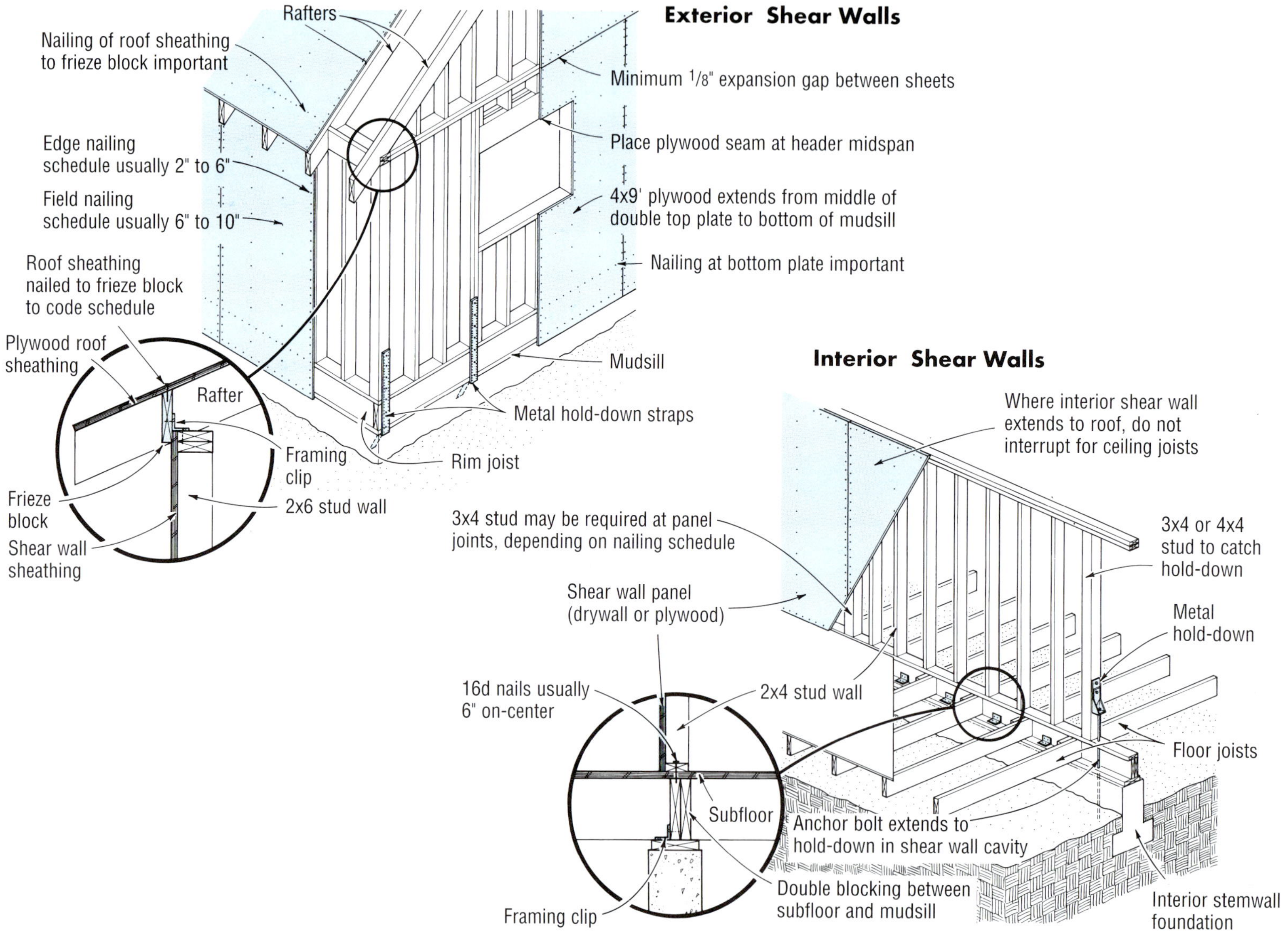

For more information: "Shear Wall Basics," JLC 1/93

"Tying Down the House," JLC 9/97

Figure 2–119. Proper Nailing at Shear Panel Joints

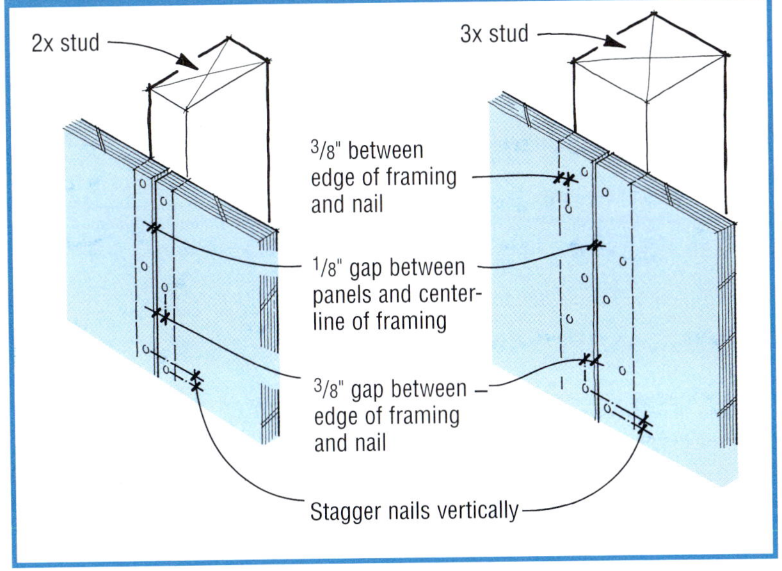

Where shear panels break on a stud (left), hold nails back 3/8 in. from the panel edges and 3/8 in. from the edge of the framing. Where nails are spaced closer than 6 in. o.c., code now requires minimum 3-by framing.

Figure 2–120. Shear Wall Boundaries

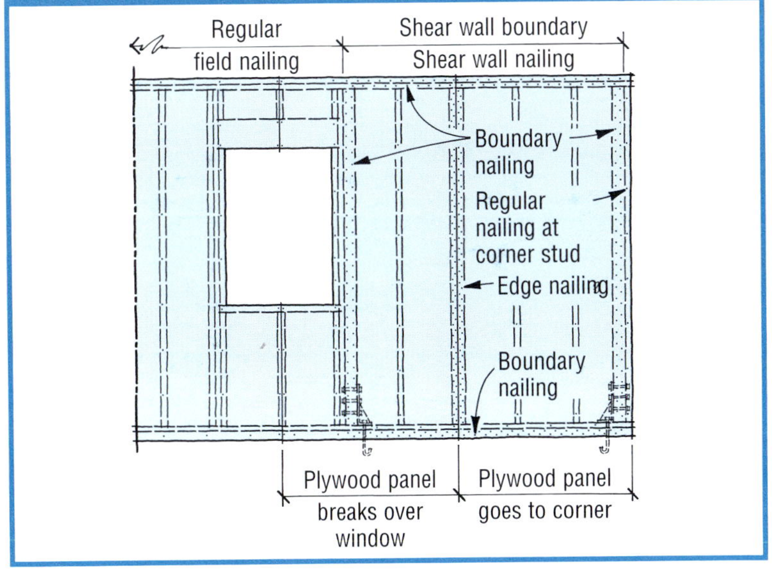

Where plywood edges fall within the shear wall boundary, 3-by framing is required to prevent splitting from the close nailing pattern. Where a plywood edge falls outside the shear wall boundary, however, as at the window opening, standard sheathing nailing is permitted.

Shear Wall Nailing

Figure 2–121. Setting Nails

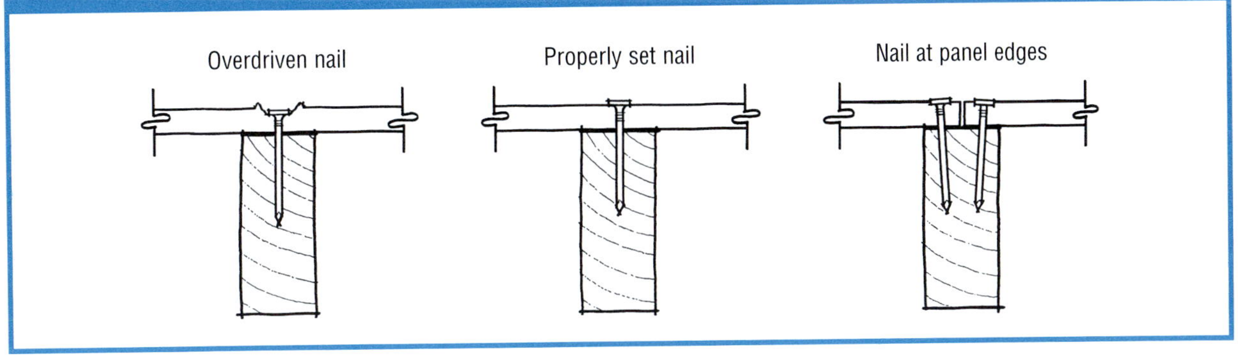

A nail that ruptures the outer plies of the panel (far left) has less shear strength. A properly set nail should either sit snuggly on top of the plywood (center) or slightly dimple the outer ply. When panels break on a stud, set the nails at a slight angle to ensure proper embedment in the framing (near left).

Figure 2–122. Hold-Down Installation

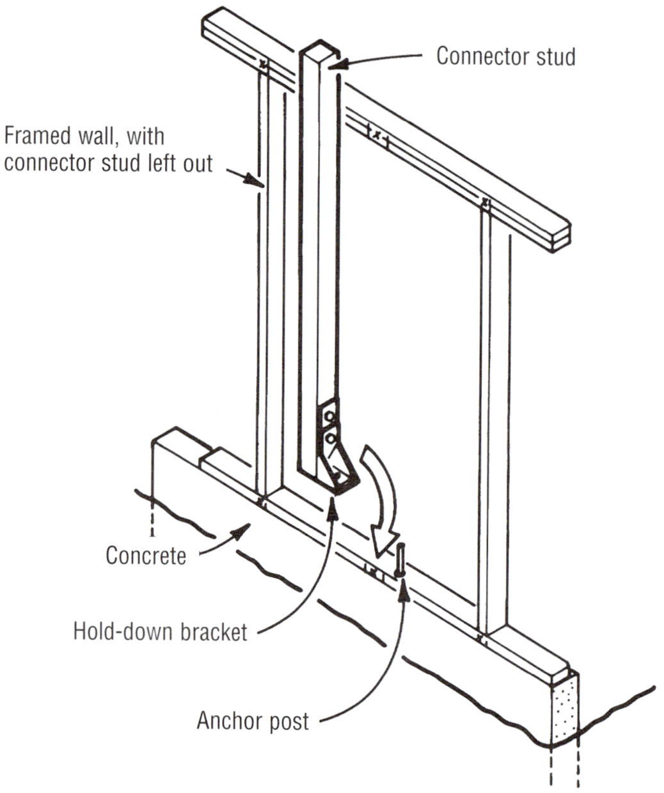

Where possible, an anchor-bolt layout can be simplified by leaving out the studs that connect to hold-downs until after the walls have been raised. (Note: This is not suitable where hold-downs attach to corners and king posts at openings.)

Figure 2–123. Out-of-Plumb Anchor Bolts

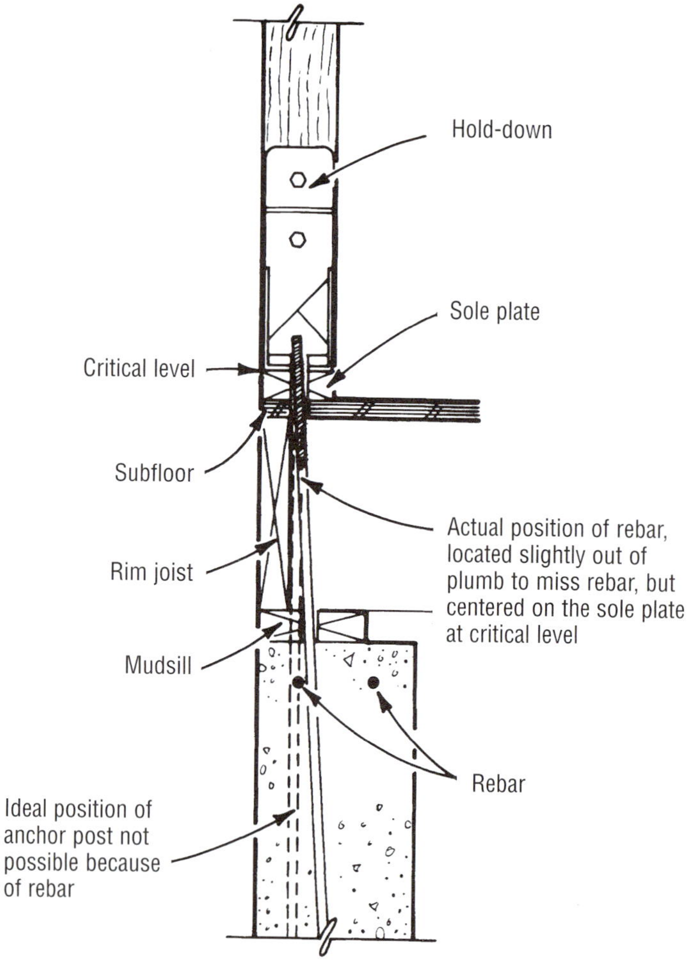

An anchor bolt may be slightly out of plumb if rebar is in the way. This will not affect the strength of the hold-down, but be sure the top end of the bolt is still centered where it comes through the sole plate.

For more information: "Practical Engineering: Shear Nonsense," JLC 7/98

Shear Wall Blocking

Because shear forces are transferred at panel edges, all edges of plywood and other sheathing panels must be fastened to solid blocking. Many builders use 4x9 or 4x10 plywood panels installed vertically (**Figure 2-118, page 166**).

Where sheathing panels meet, each edge must have at least 1/2 in. of bearing on the stud, plate, or other member it is nailed to. Where the design shear exceeds 350 lbs. per foot (typically, where nails are spaced closer than 6 in. o.c.), the Uniform Building Code requires minimum 3x lumber at the joint. Most builders use 4x4s (**Figure 2-120, page 167**).

ANCHORING TO FOUNDATIONS

Anchoring to Foundations

All shear walls must be mechanically fastened to the foundation with metal hold-downs, metal straps, closed-space anchor bolts, or some combination of these (see "Anchors," **page 48**).

Hold-Downs

These L-shaped brackets attach to anchor posts (anchor bolts) embedded in the foundation (**Figure 1-51, page 49**). For studs in a wall without openings, it's often best to install the studs that connect to the hold-downs after the wall is raised (**Figure 2-122**).

Where anchor posts attach to a corner stud or at one side of a wall opening, they must be located precisely. If possible, wire the anchor posts directly to the rebar. If the rebar is in the way, it may be necessary to install the anchor posts slightly out of plumb (**Figure 2-123**).

Strap Anchors

These concrete-embedded metal straps nail or bolt to the face of the stud. Because they usually require no drilling, they are easier to install than hold-downs (**Figure 1-51, page 49**). However, they have significantly lower design

Figure 2–124. Floor-to-Floor Connections

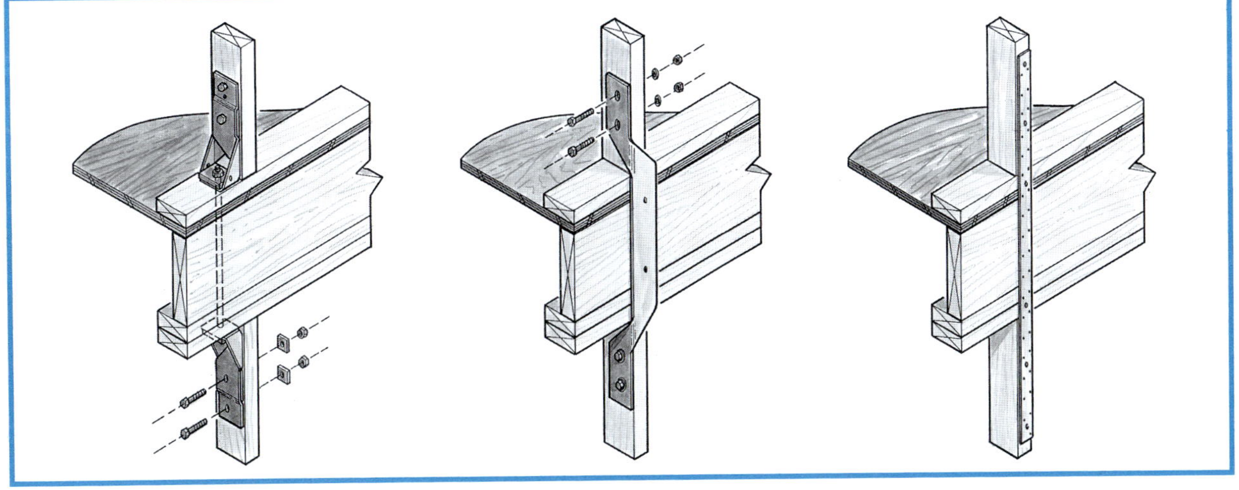

To provide a secure load path between stories, engineers specify bolted hold-downs, with the lower one turned upside down (far left), or straps that bolt (center) or nail to the studs (near left).

loads. Also, the studs they nail to may split under the heavy nailing required. If so, replace the stud with a double stud or 4x member.

STACKED WALL CONNECTIONS

There are several ways to reinforce the connection between floors (**Figure 2-124**). The most common approach for seismic work uses a pair of hold-downs with a large threaded rod between them. Other approaches use straps nailed directly to the studs or through the sheathing. In all cases, the studs must be aligned between floors. With metal straps, some builders use 4x4s because they are less prone to splitting with the large number of nails.

Figure 2–125. Eaves Connections

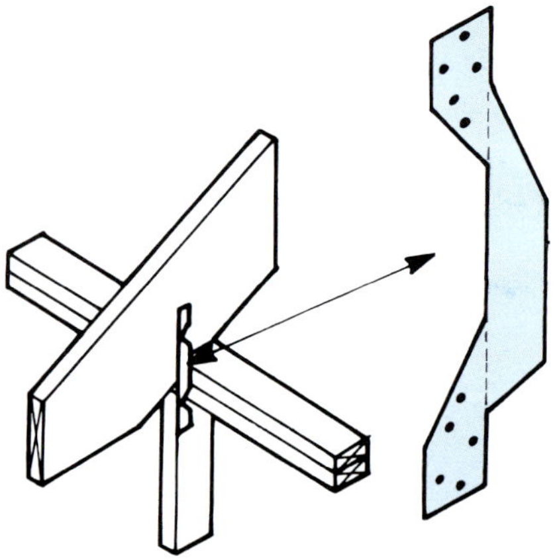

Nailing rafters to plates and plates to studs is not always enough to resist high winds. Some codes call for adequately rated hurricane anchors or rafter hold-downs at 4-ft. intervals in areas of high winds.

For more information: "Practical Engineering: Resisting Wind Uplift," JLC 1/97

ROOF CONNECTIONS

Winds can exert strong uplift forces. Roofs with low slopes or large overhangs are more susceptible to these forces. Minimum code nailing schedules are rarely adequate to resist these forces. Adding wind ties, hurricane anchors, or other steel connectors at eaves connections is often required, and is good insurance in high-wind areas.

Framing Clips

A common detail in seismic zones is to connect the frieze block to the double top plate with a metal framing clip (**Figure 2-118, exterior wall inset, page 166**). These must be installed before the roof sheathing.

Hurricane Anchors

In high-wind areas, hurricane clips or rafter ties may be required to connect rafters or trusses to top plates (**Figure 2-125**). As with all rated connectors, the manufacturer's nailing schedule must be followed.

Ridge Connections

Steel connectors also may be required to reinforce the rafter connection at the ridge (**Figure 2-117, page 165**). These may consist of joist hangers installed at the ridge beam or rated steel straps placed across the tops of the opposing rafters. As with other strap connectors, nailing into 2x material may be a problem due to the number of fasteners.

ENERGY CODES FOCUS ON AIR-SEALING DETAILS, VAPOR barriers, and standard insulation values throughout the house. Good framing practice helps to ensure there is an adequate amount of insulation in all corners of the exterior shell. In addition, slight alterations in practice can make it easier to install a continuous air/vapor barrier on the inside of the house, as required by most energy codes in the U.S.

SEALING THE RIM JOIST

The rim joist is one of the greatest sources of air leaks in a wood-frame house. However, it is also one of the most difficult places to seal effectively.

Figure 2-126. Insulating the Rim Joist

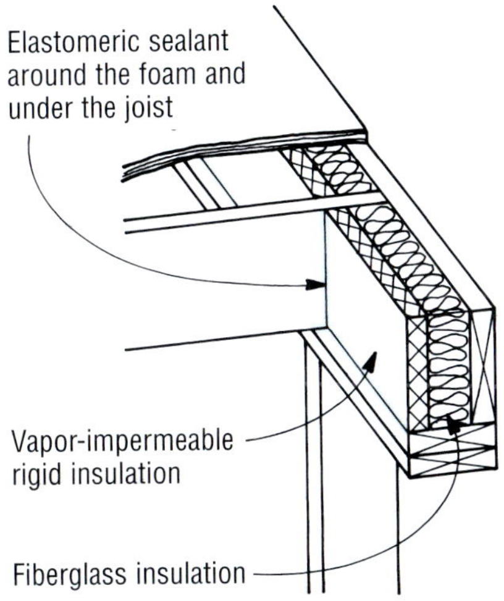

Fiberglass insulation alone is not enough to stop air from leaking through the rim joist. Use pieces of foam insulation board over a fiberglass batt, fitting the foam tightly and sealing around the edges. Seal below each joist where it sits on the plate, too.

Sealing the Rim Joist

The most common way to address this difficult area is to seal and insulate it after framing (**Figures 2-126**). Do not just stuff fiberglass into each joist bay. Foam insulation board, sealed tightly on all four sides, must be used to stop the flow of air and vapor.

When done thoroughly, fitting foam into each joist bay is time-consuming, so it's actually easier to address this area at the framing stage by offsetting the wall plates (**Figure 2-127**). This provides a 1$^1/_2$- to 2-in. cavity for strips of foam insulation board. Seal the corner between the rim joist and the bottom plate with urethane caulk

Figure 2-127. Offset Plates

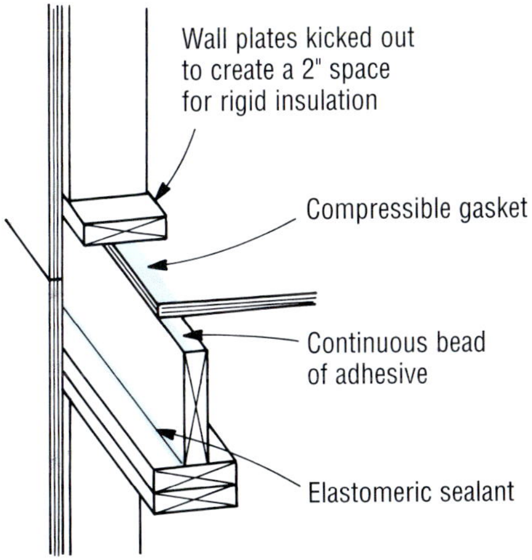

Keeping the rim joist warm is critical in cold climates where condensation can accumulate on cold framing. This is best done by offsetting the wall plates to provide a channel for insulation board around the perimeter.

Figure 2-128. Interior Wall Channels

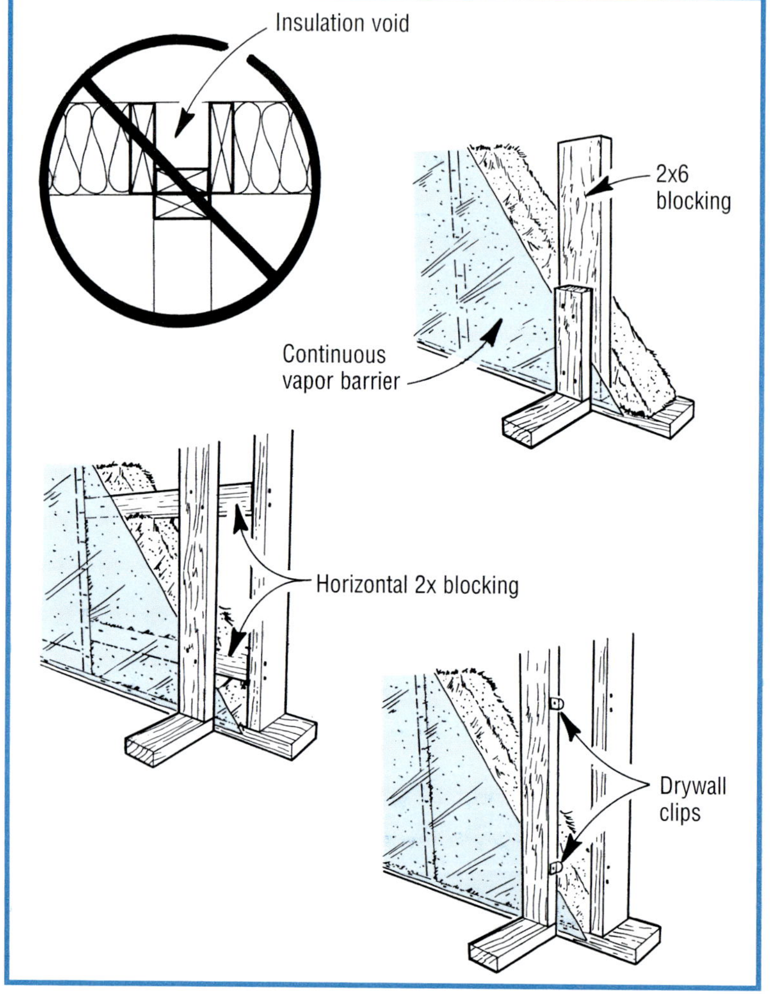

Conventional partition channels are impossible to insulate if the walls have been sheathed (top left). The other channel options shown here not only allow the area to be insulated, but make it easier to install a continuous air/vapor barrier as required by most building codes.

Figure 2-129. Knee Walls

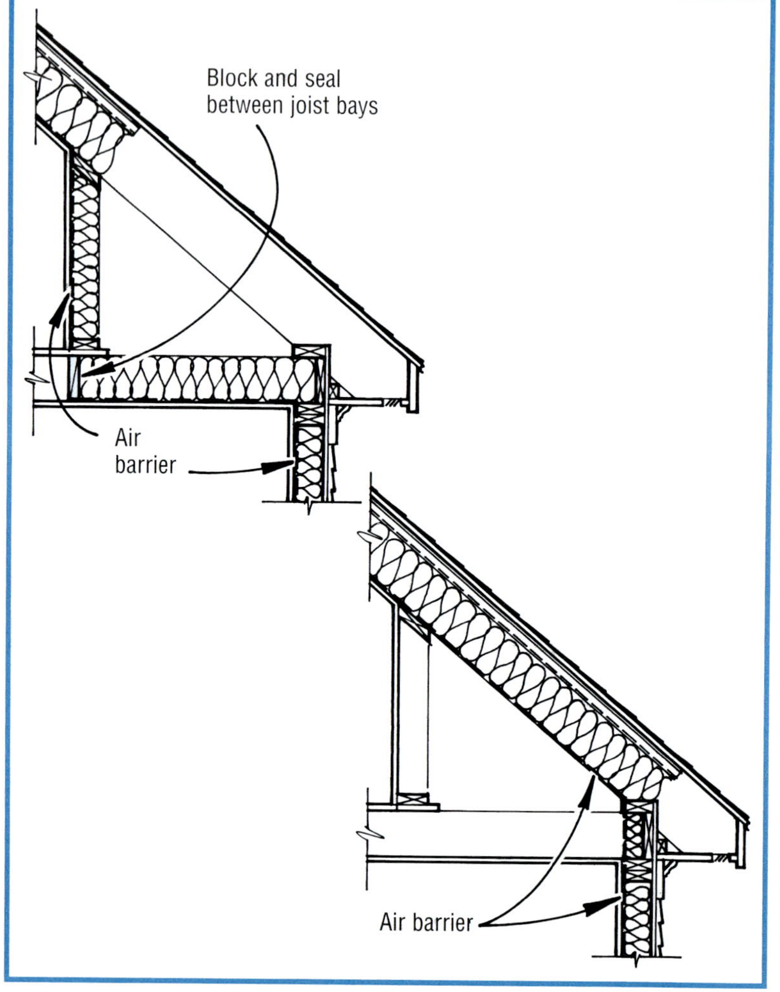

When building knee walls, make sure there is a continuous air barrier on the warm side of the insulation. If addressed before the knee wall is framed, it's easiest to insulate along the roof and exterior wall (above), rather than along the ceiling (top).

(see "Exterior Caulks and Sealants," **page 371**) and apply continuous beads of construction adhesive when installing floor sheathing and exterior walls.

PARTITION CHANNELS

Interior partition channels in framed walls often create voids, which are inaccessible once the exterior walls are sheathed. **Figure 2-128** shows several ways wall channel can be detailed to provide an easily accessible insulation bay. To provide a continuous air/vapor barrier on the exterior walls, insulate and install a poly vapor barrier before standing the interior walls.

SEALING KNEE WALLS

The knee walls in a story-and-a-half Cape often create gaps in the insulation and/or in the air/vapor barrier. There are two common ways to insulate the knee wall area (both shown in **Figure 2-129**):

Follow the wall and floor. The more difficult way is to insulate the knee wall and the floor of the knee-wall area. Install a poly air/vapor over the insulation of the knee wall, as well as on the ceiling area below, taping it to an air block (foil-faced foam works well) between the joists just below the knee wall.

Follow the roof. An easier way to handle knee walls is to insulate the roof all the way down to the exterior wall — including a continuous vapor barrier on the inside face of the rafters — and then build the knee walls later. This works especially well if the knee-wall cavity will be used for storage. In this case, however, it's critical that the rim joist be sealed using foam board, as well.

SEALING CANTILEVERS

Many Colonial-style houses have a section of second floor cantilevered over the first-story wall. Without proper attention, this area can lose a lot of heat.

Keep track of the barrier. During construction, treat cantilevered floors like the floor-knee-wall transition, but install the vapor barrier on the subfloor of the second floor (**Figure 2-130**). Where the poly air barrier on the wall meets the floor, tape the plastic to the plywood. Then seal all the seams in the plywood subfloor with tape or caulk — that way, the plywood will function as an effective air barrier on the warm side of the insulation.

Sealing Channels, Kneewalls, and Cantilevers

Figure 2-130. Cantilevers

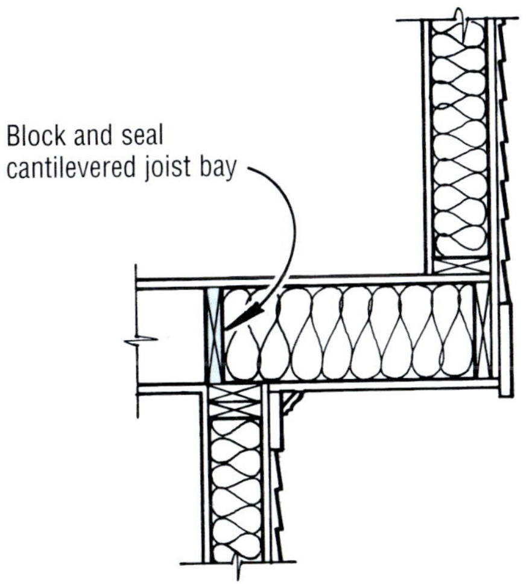

Cantilevered floor sections can lose a lot of heat unless an air block (2x blocking or rigid foam blocks) is added in the joist bay above the first-story top plate.

Block airflow at the floor. Between the cantilevered floor joists, fill the voids completely with fiberglass or cellulose insulation. Then install rigid foil-faced foam insulation in the bays where the joists bear on the first-floor wall and seal the edges where the foam meets the framing with caulk or foam. Tape the joint where the poly on the first-floor wall face meets the foam between the joists.

SEALING BALLOON-FRAMED WALLS

Gable-end walls and shed dormer front walls are often balloon-framed. In these situations, heated air enters the wall through a penetration — such as an electrical outlet or even a crack around the baseboard and drywall at the bottom of the wall. The warm air rises in the stud cavity like smoke in a chimney, escaping into the cold attic and exiting through the roof vents. But even where there is no penetration in the wall, cold air from the attic will drop into the stud cavity where it is warmed by the wall and rises back into the attic. This kind of *convection loop* carries heat out of the house.

To prevent this type of heat loss, install wall blocking in line with the ceiling joists (**Figure 2-131**).

Figure 2-131. Balloon-Framed Walls

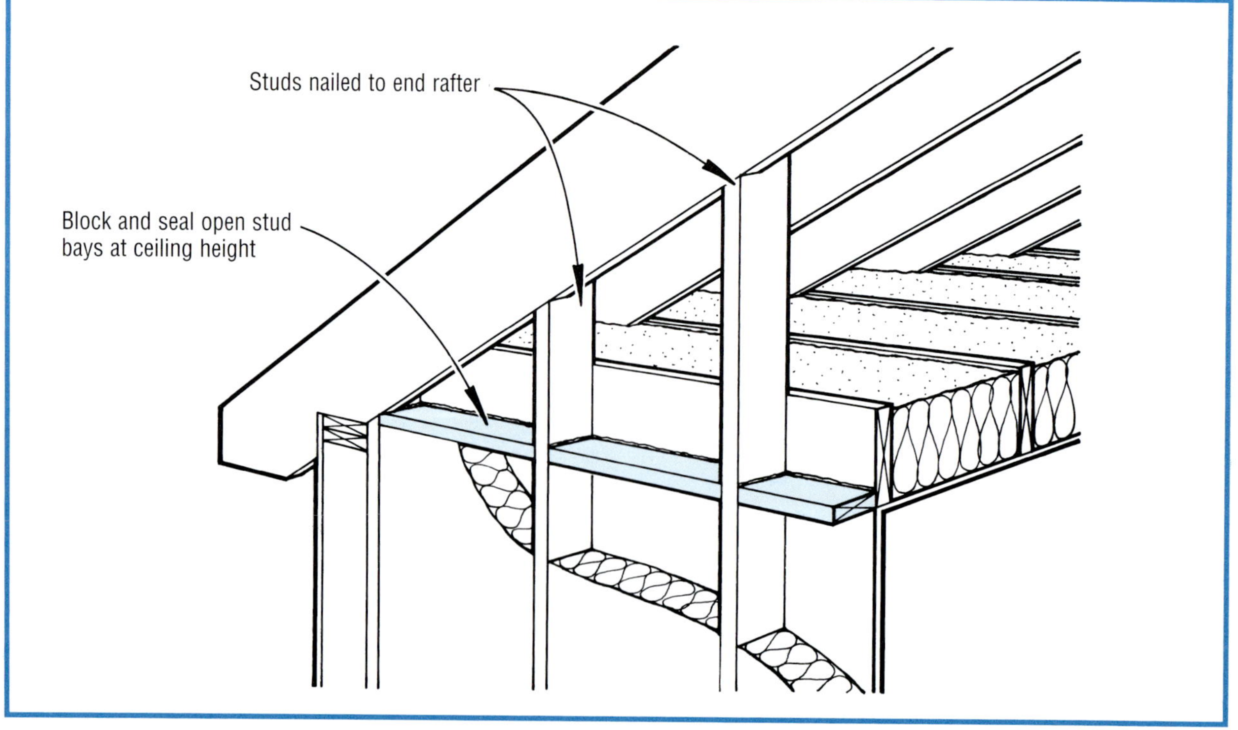

Balloon-framed exterior walls should always be blocked to prevent convective air movement into the attic.

For more information: "Air Sealing the Story-and-a-Half," JLC 8/95

SEALING FRAMING PROJECTIONS

Eyebrow roofs, bay window roofs, crickets, and other projections of the wall or roof lines are sometimes built before the building is sheathed. This allows air to penetrate the exterior walls and leak out the building.

To avoid these heat losses, always sheathe the exterior walls and then attach additional framing to a ledger (**Figure 2-132**).

CATHEDRAL CEILING OPTIONS

Cathedral ceilings must have adequate insulation (R-38 to R-40 in cold climates) and ventilation, but it can be a challenge to provide both in the space provided by a typical 12-in. rafter.

Hot vs. Cold Roofs

Most codes require new roofs to have an eaves-to-ridge flow of air (see "Roof Ventilation," **pages 285-288**). These are called *cold roofs* because they keep the underside of the sheathing cold in winter. In a heating climate, the flow of cold air under the sheathing discourages interior moisture buildup and ice damming. In a cooling climate, ventilation relieves heat buildup.

Cathedral ceilings that aren't ventilated are referred to as *hot roofs*, and typically rely on a well-sealed air and vapor barrier to keep moisture out of the roof. Solid-foam roof systems, using stress-skin panels or spray-in foam, are hot roof alternatives. However, many code officials do not permit these roofs. The cathedral ceiling options shown in this manual are all cold-roof options.

Sealing Gable-End Walls

Cathedral Ceilings

Figure 2-132. Eyebrow Roofs

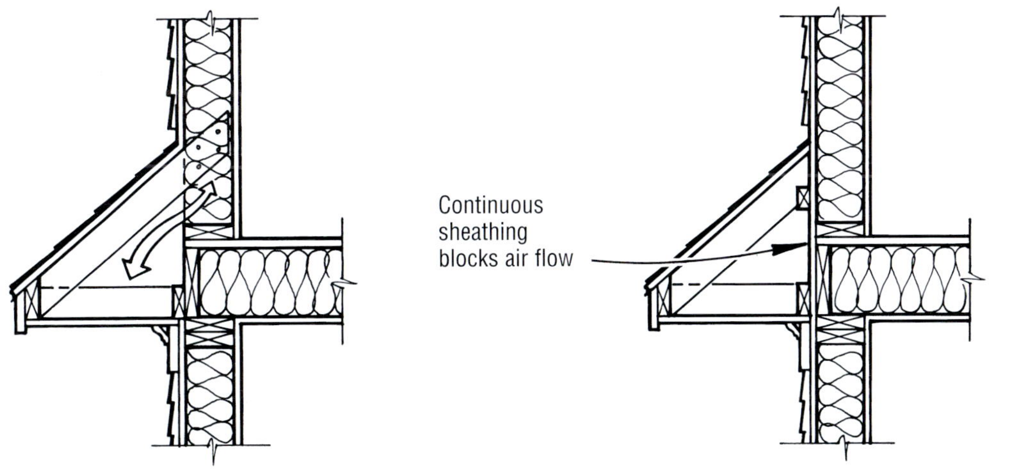

When building an eyebrow roof on a shed dormer, sheathe the exterior walls first, and then attach the rafter tails to a ledger (near left).

A. Foil-faced foam sheets applied to the bottom of 2x rafters boost the ceiling's R-value and provide a thermal break. Tape the seams between the foam sheets to reduce air and vapor movement into the ceiling cavity.
B. Hang 2x3s below rafters using plywood gussets to create as wide a space as needed for blown cellulose insulation.
C. Insulation space can be added by installing 2x3s toenailed across the bottom edge of the rafters, which also provides a thermal break.

Figure 2-133. Cathedral Ceiling Options

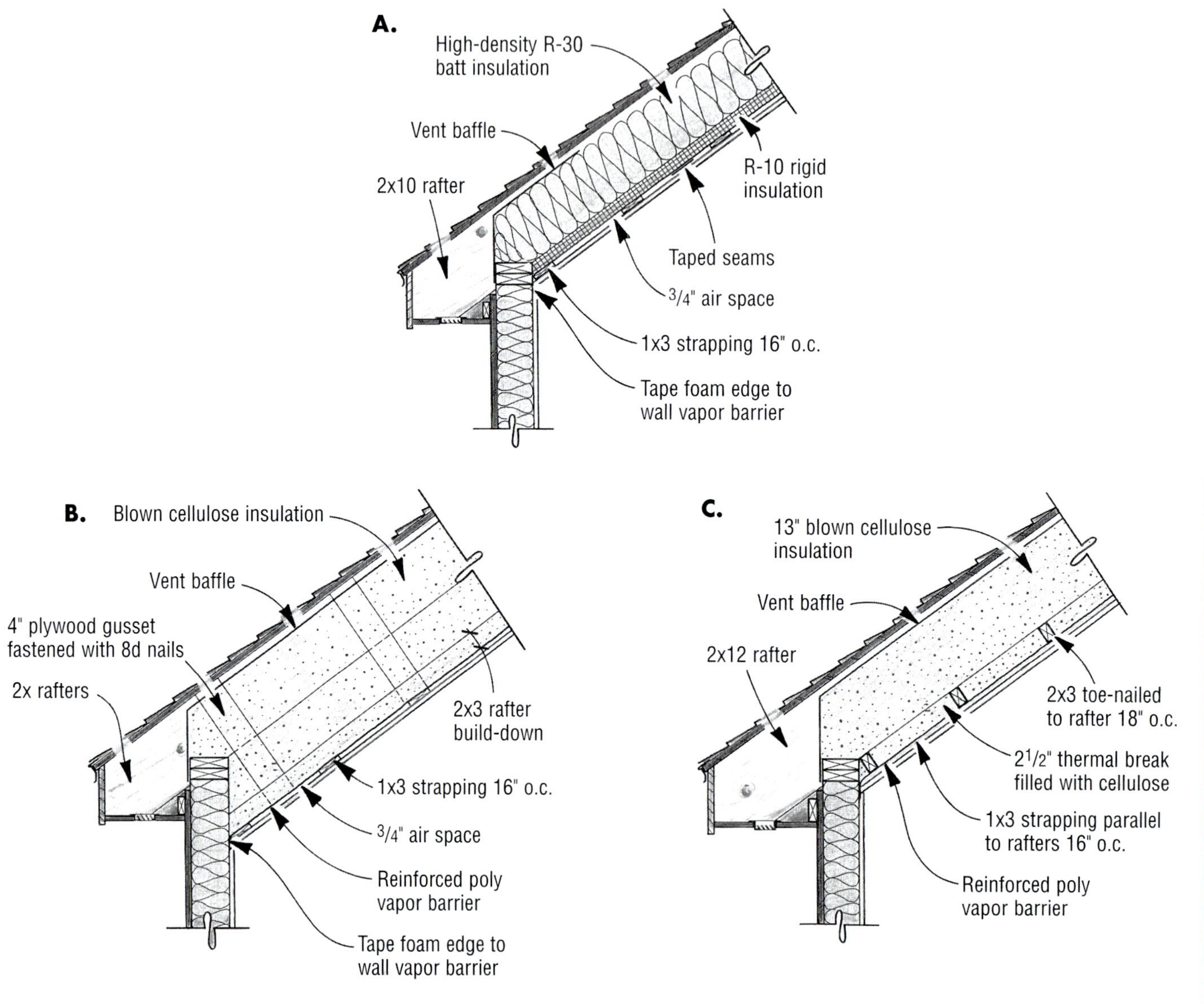

Figure 2-133. Cathedral Ceiling Options

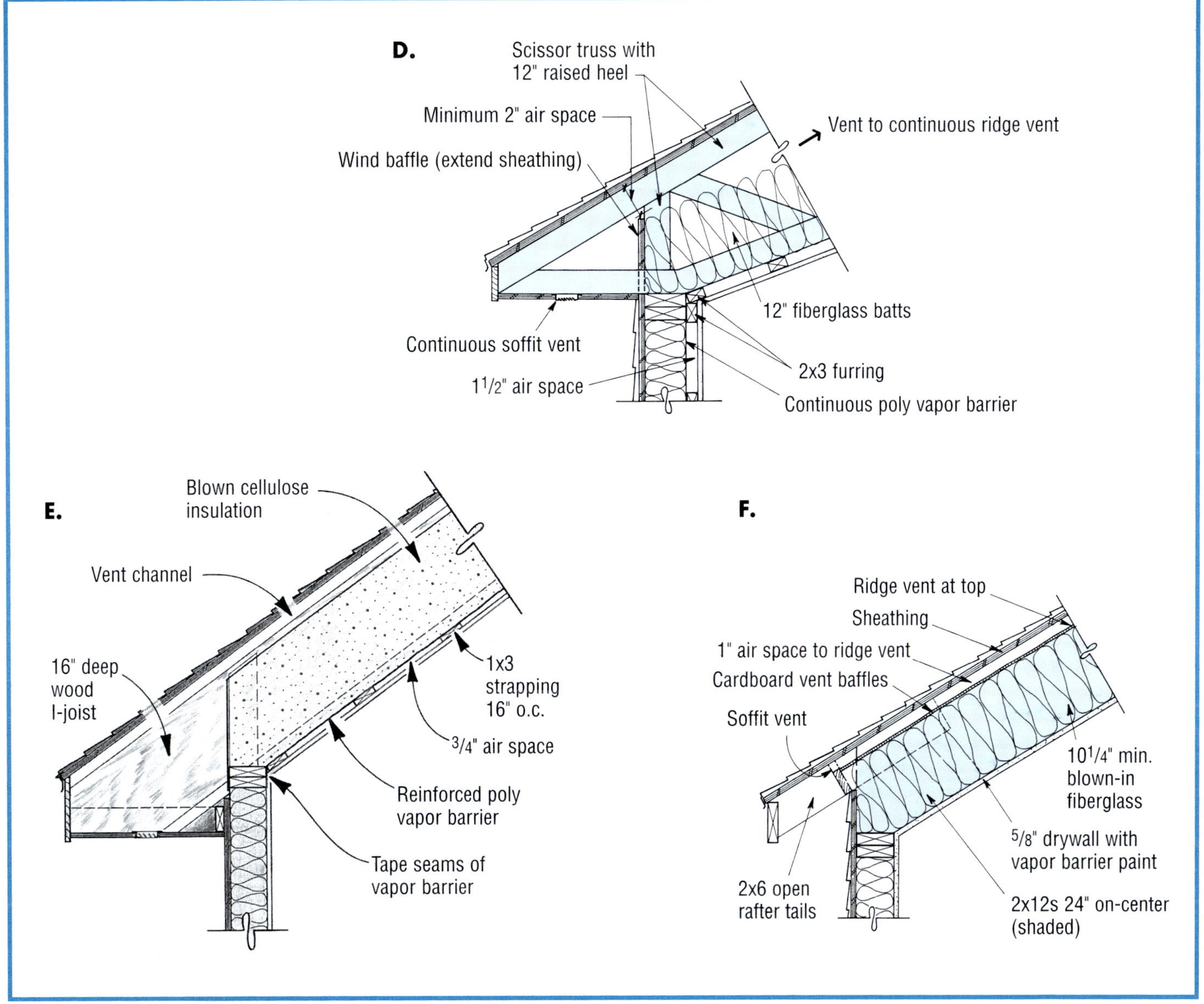

Cathedral Ceiling Options

D. Scissors trusses with raised heels provide adequate space for insulation. Batts must be carefully fitted to minimize short-circuiting at truss chords.
E. 16-in.-deep I-joists insulated with dense-blown cellulose.
F. High R-value insulation (blown-in fiberglass) can be substituted for standard fiberglass batts or blown-in cellulose.

Cathedral Framing Options

Cathedral roofs framed with conventional 2x10 or 2x12 rafters typically do not provide adequate space for insulation in cold climates. The rafters need to be packed out to provide more space or foam insulation should be added to the underside (**Figure 2-133**, previous two pages).

Thermal break. Since wood is not very good insulation, the rafters themselves conduct cold through the roof. Most of the framing options shown in **Figure 2-133** provide a *thermal break*: an interruption of the conducting material (in this case, wood) intended to reduce continuous heat flow through the rafter.

Insulation options. The R-values of insulation materials vary widely, as shown in **Figure 2-134**, so installing insulation with a higher R-value may be the easiest alternative for cathedral ceilings. For example, high-performance batts or blown-in fiberglass (BIBS) can be used to bring the R-value of a 2x12 rafter space to an acceptable R-38 to R-40 (**Figure 2-133**, previous page, option F).

Vent baffle. In every cold-roof option shown in **Figure 2-133** (previous page), a vent baffle (using cardboard or vent channel) must be installed to provide uninterrupted airflow from eaves vents to a ridge vent.

Alternative Framing Options

An alternative to conventional rafters is to use scissors trusses or deep I-joists — these provide enough space for the extra insulation needed to achieve a higher R-value (**Figure 2-133**, previous page, options D and E).

Figure 2-134. Insulation R-Values

Insulation	Approx. R-Value per in.
Fiberglass batt	3.2
High-performance fiberglass batt	3.8
Loose-fill fiberglass	2.5
Blown-in fiberglass (BIBS)	4
Loose-fill cellulose	3.5
Dense-pack cellulose	3.5
Perlite or vermiculite	2.7
Expanded polysyrene board	3.8
Extruded polystyrene board	4.8
Foil-faced polyisocyanurate board	7
Spray-urethane foam	5.9

For more information: "Energy-Efficient Cathedral Ceilings," JLC 2/97

"Cathedral Ceiling Solutions," JLC 8/91

BECAUSE OF THE HIGH BEARING CAPACITY OF STEEL, IT IS A useful alternative for carrying heavy loads, particularly where space is limited. **Figure 2-135** shows the types of steel that can be useful for columns, beams, and brackets in wood-framed construction.

Steel Types

STRUCTURAL STEEL HEADERS

Where height is limited, but a header must be installed to carry a heavy point load (a second-floor girder resting on a sliding-glass door header, for example), steel is often the only solution. **Figure 2-136** shows examples of standard

Steel Headers

Figure 2-135. Steel Types and Sizes

	Shape & Type	Symbol	Stock Size Specs	Lengths	Finish	Uses
I-Beams	Junior[1]	M	Height in inches x pounds/LF (Ex: W8x24)	20, 40, 50, and 60 feet	Hot-rolled, mill finish	Carrying beams, headers, ridge beams, cantilevered beams
	Standard	S				
	Wide Flange	W or WF				
Channels	Stringer[1]	MC	Height in inches x pounds/LF (Ex: C5x9)	20 and 40 feet	Hot-rolled, mill finish	Carrying beams, flitchplates, headers, ridge beams, columns
	Standard	C				
	Ship & Car[2]	MC				
Tee	Tee[3]	T	Height in inches x width in inches (Ex: T6x8)	20 feet	Hot-rolled, mill finish	Lintels, ledgers, light-load columns
Angles	Equal Legs Unequal Legs	Angle in degrees or L	Leg x leg x thickness (Ex: 3x6x1/4")	20 and 40 feet	Hot-rolled, mill finish	Lintels, ledgers, web and flange reinforcements, joint clips
Bar Stock	Flats Rounds Squares	N/A	Thickness x width (Ex: 1/2x8) Diameter (Ex: 2") Width of one side (Ex: 1")	20 feet, 12 feet and random	Hot-rolled, mill finish Cold-rolled, pickled and oiled	Column plates, splice plates, machinist parts, tools
Pipe	Sch. 10[1] Sch. 40 Sch. 80[2]	BPPE (black pipe plain end) or BPTC (black pipe threaded coupling)	Inside diameter x schedule weight (3" Sch. 40 BPPE)	21 to 24 feet	Hot-rolled, mill finish, painted, hot galv.	Columns
Tubing	Round Square Rectangular	ERW (elec.-res. welded) DOM (drawn over mandrel)	Outside dim. or diameter x wall thickness (Ex: 2x1/8" round; 2x4x1/4" rectangle)	20 and 40 feet	Hot- or cold-rolled, pickled and oiled	Handrails, balusters, specialties

(1) Also called "Lightweight" (2) Also called "Heavyweight" (3) Made by splitting I-beams in half

Each grade of structural steel has a specific quality as described by American Society of Testing and Materials (ASTM) standards. ASTM A-326 is the predominant grade in the structural steel market. It has a carbon content of .26%, which gives it relatively high strength (60,000 psi tensile), yet it is relatively easy to weld and fabricate.

steel shapes and sizes that will work as headers to carry such heavy point loads. Each of these pieces of steel can be worked into a 2x4 wall, although the W beams (commonly called I-beams) are a full 4 in. wide and require careful work to fit.

Jack Support

Headers should be supported by at least double 2x4 jack studs for up to a 12,000-lb. load and triple jacks for loads above 12,000 lb. up to 18,000 lb. The header should sit directly on the end grain of the jack studs. Don't use wood

Figure 2-136. Structural Steel Headers

	Span			
Point Load (lb.)	**3'**	**4'**	**5'**	**6'**
5,000	T3x3x5/16 (1.88)	T4x4x3/16 (2.50)	T4x3x1/4 (3.13)	C6x8.2 (3.75)
10,000	C6x8.2 (3.75)	C6x10.5 (5.00)	W6x12 (6.25)	T6x3x3/8 (7.50)
15,000	C6x13 (5.63)	T6x3x3/8 (7.50)	T7x3x3/8 (9.38)	W8x15 (11.25)
20,000	T6x3x3/8 (7.50)	W6x16 (10.00)	W10x15 (13.00)	W10x17 (15.00)

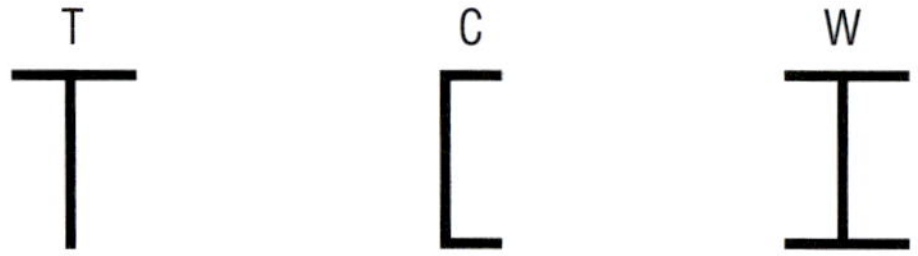

The number in parenthesis is the section modulus required for the given load and span; the steel section above it is the shallowest and lightest piece of steel that will work. Substitutions may have to be made because the steel yard may not carry every size and shape.

For more information: "Practical Engineering: Solve It with Steel," JLC 1/95

"Practical Engineering: Attaching Steel Beams and Columns," JLC 2/97

Figure 2-137. Splicing I-Beams

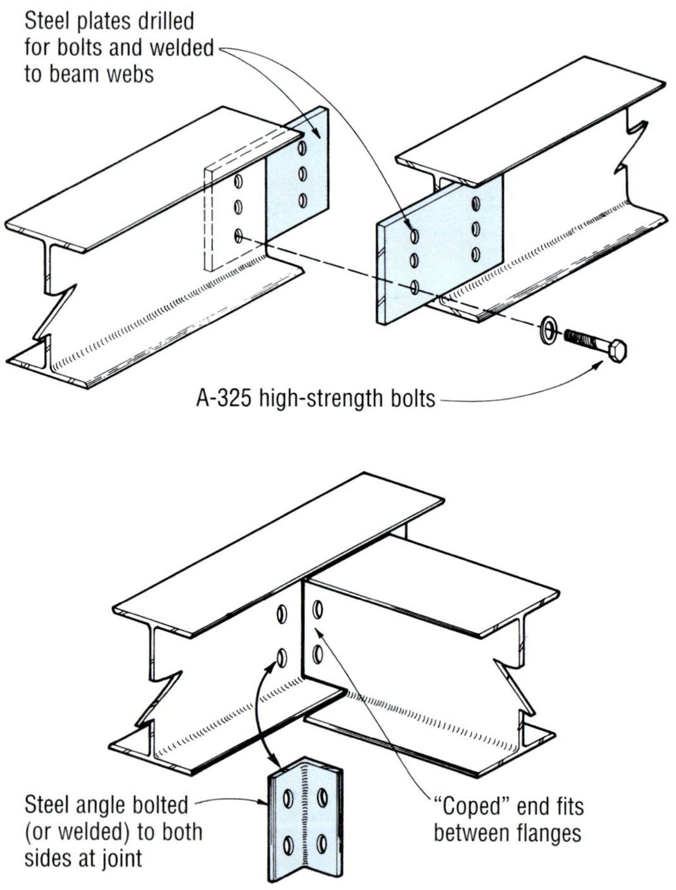

In an inline splice (top) the ends of the I-beams butt together with plates welded on each side. The plates are then through-bolted with high-strength (A-325) bolts placed in predrilled holes. Steel angles, or "clips," are used to bolt beams together after a steel fabricator has "coped" the ends to fit together (bottom).

Figure 2-138. Column Connections

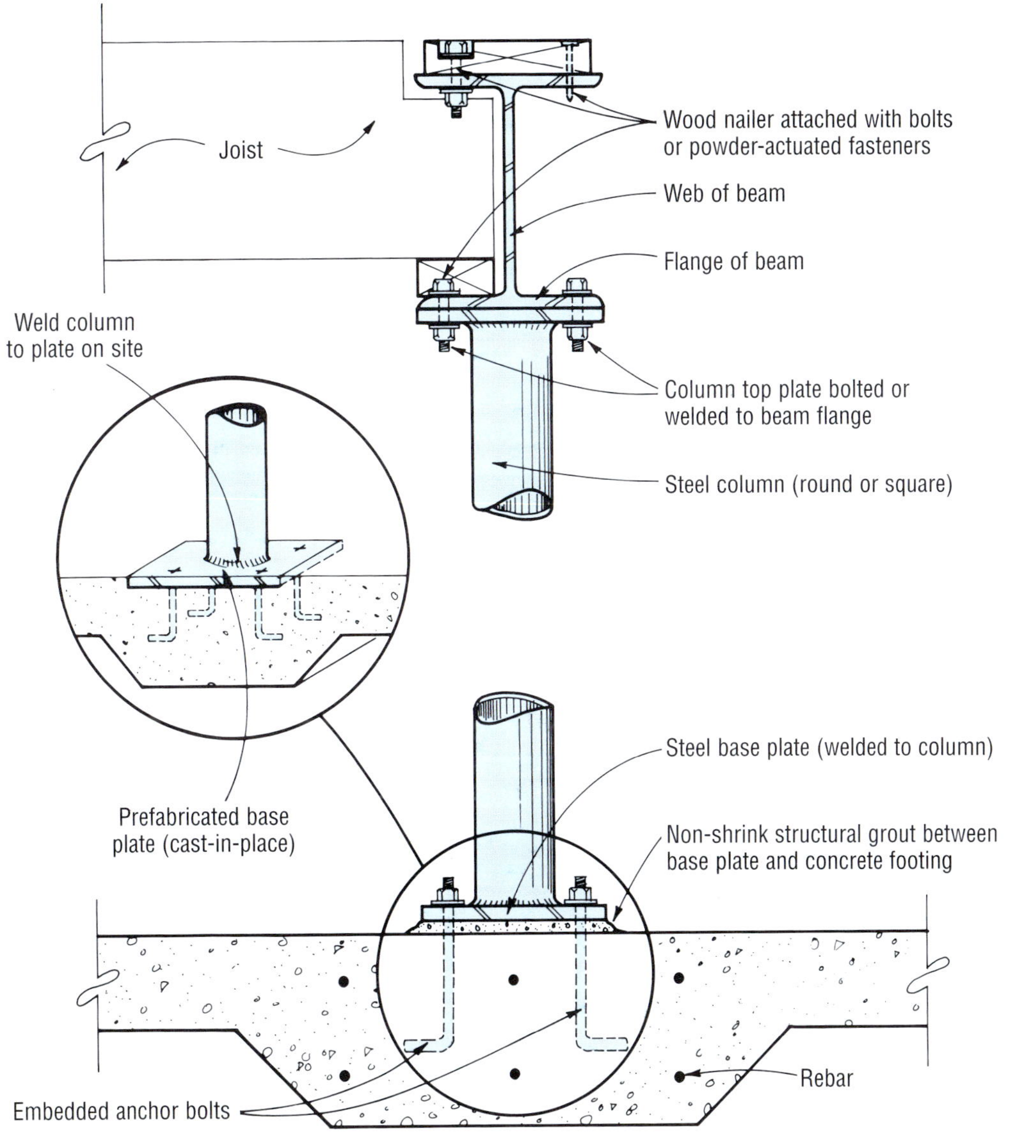

Strength of Steel Headers

Beam and Column Connection

Steel columns (not wood posts) should be used when freestanding support of steel beams is required. The column plates must be bolted to beam flanges (top) and to the thickened slab footing (bottom). An alternative is to weld column plates to beam flanges, and weld column bases to steel base plates cast into the slab footings (inset).

shims under the header — the load is perpendicular to the grain and tends to crush the wood.

STEEL-TO-STEEL CONNECTIONS

Most steel connections are either bolted or welded together. Welding is useful when the surface area to be joined is small, such as the edge of a steel column.

Bolted connections, on the other hand, don't require special skills. The most common bolted connections are in-line splices and beam-to-beam connections at right angles (**Figure 2-137, page 180**), and where column plates join footings and beam flanges (**Figure 2-138**, previous page).

Right-angle connections (**Figure 2-137, page 180**) require the flange of one beam to be cut back ("coped") at one end so that the webs of each beam fit together. It doesn't have to be perfect, since the bolts will carry the load.

For any bolted structural connection, an engineer should specify the diameter and length of the bolts needed for the type of connection. These will probably be high-strength, structural steel (A-325) bolts, which are available from steel supply houses or steelwork shops.

STEEL-TO-CONCRETE CONNECTIONS

Column bases must be bolted into concrete. Don't count on concrete alone to hold embedded steel in place. The concrete will shrink away over time, so you should attach the steel to anchor bolts. When embedding steel base plate for column supports, attach hooks to a portion of the steel that will be surrounded by concrete (**inset, Figure 2-138**, previous page). These will "grab" and prevent movement of the steel once the concrete has shrunk.

Figure 2-139. Steel-to-Wood Connections

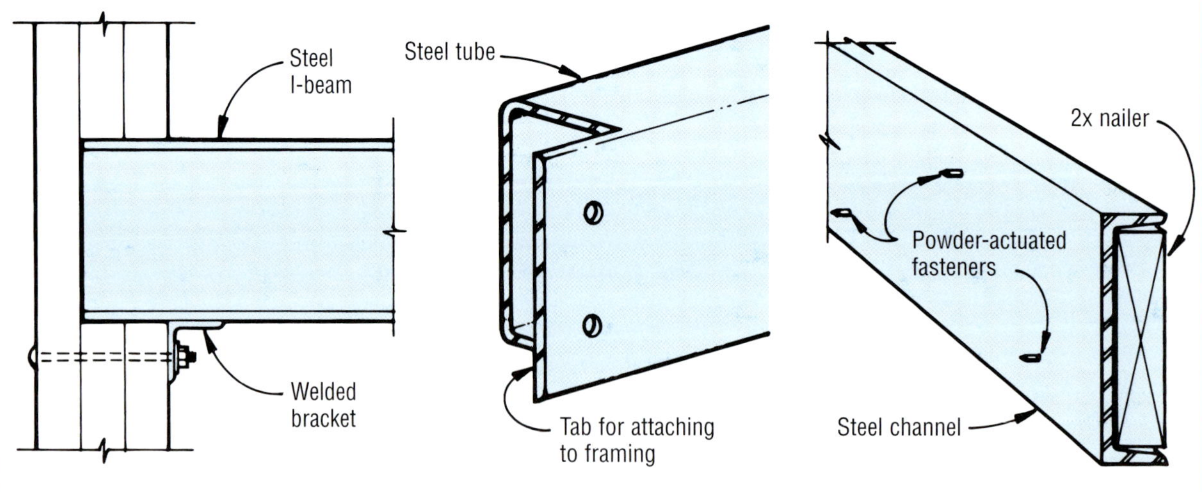

Different steel shapes require different connections. On site, the quickest method is to fasten 2x material to steel channel using powder-actuated fasteners (near right). The other methods require the services of a steel fabricator.

For more information: "Steelwork in Wood Frames," JLC 4/92

Steel-to-Steel

Steel-to-Wood

STEEL-TO-WOOD CONNECTIONS

There are several ways to attach steel to wood, depending on the circumstances. **Figure 2-139** shows some common steel-to-wood connections using different steel shapes.

Nailers

Wood nailers are easiest to attach with powder-actuated nails, spaced 24- to 36-in. o.c. and staggered side-to-side.

Header Connections

A common way to attach an I-beam header to wood framing is to have the steel yard weld on an angle bracket, as shown in the left-hand illustration of **Figure 2-139**. An alternative is to secure the connection using a top plate, which overlaps the steel and wood-framed walls (**Figure 2-140**).

For slim-profile tube headers where there is not enough room to "nail on" lumber (using a powder-actuated tool), ask the steel fabricator to leave a tab sticking out at each end, as shown in the center illustration of **Figure 2-139**. Then notch the framing to fit, and nail or lag the steel to the framing.

Figure 2-140. Steel Header Connection

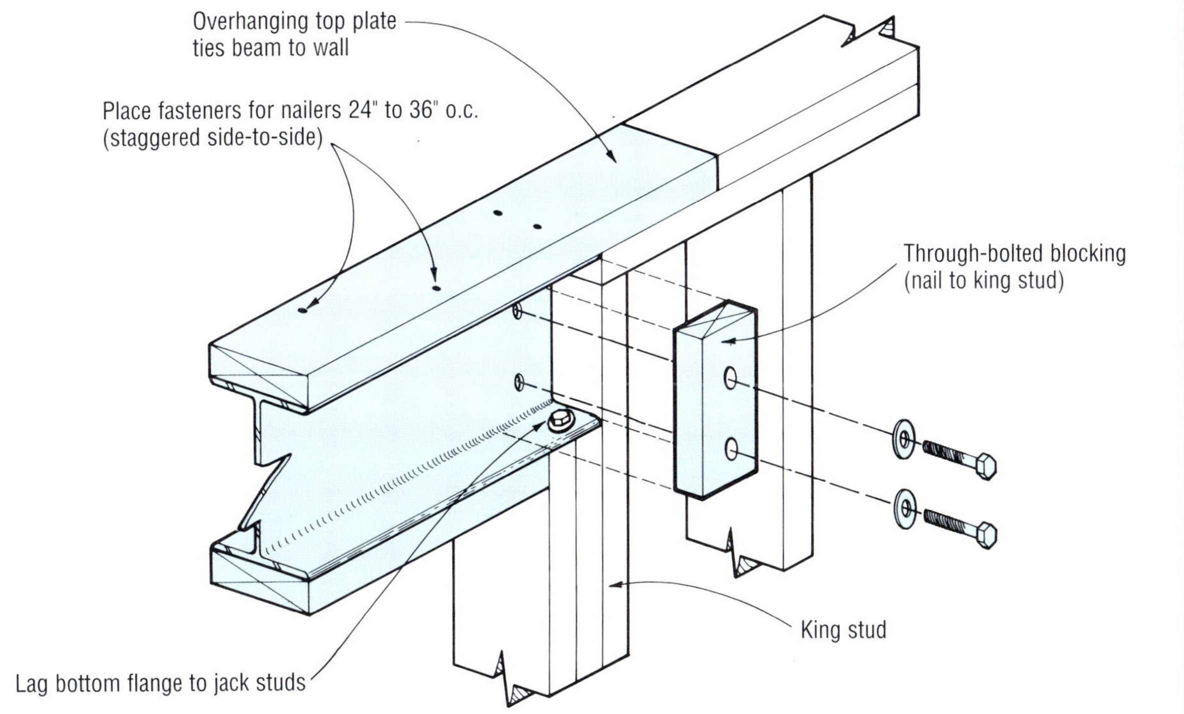

An I-beam header can be secured by the top plate that spans the connection between the beam and the wall. This connection should be reinforced with lag bolts into the bottom flange to the jack studs, and with bolt-on blocking that is toe-nailed to the king stud.

FLITCHPLATES

Flitchplates are used primarily in header assemblies when common framing lumber is not strong enough or stiff enough to carry the loads (**Figure 2-141**).

The steel is put in the middle for two reasons. First, this increases fire resistance. Wood can resist fire longer than steel, which softens very quickly in a fire. Second, the steel is not likely to buckle if placed between two wood members. A thin plate of steel will wrinkle and buckle under relatively small loads unless stabilized.

End Connections

An engineer should specify the size and number of bolts required to secure a flitch plate to the wood framing. Insufficient bolting and poorly sized bolt-holes will cause the wood around the bolts to crush and the plates to settle. In extreme cases, the steel can knife into the top of the post.

As a rule of thumb, bolt-holes should be placed 2 in. from all edges of a beam. Use the flitchplate as a template to precisely align the holes in the wood.

Figure 2-141. Equivalent Strengths: Flitch Beams vs. Built-Up Wood Beams & Steel I-Beams

Single Steel Flitch Beam: Wood	Steel	Built-Up Lumber Equivalents*	Steel I-Beam Equivalents
2x6	5" x 1/4"	2.3 - 2x6s	-
	5 x 3/8	3.5 - 2x6s	-
	5 x 1/2	4.7 - 2x6s	-
2x8	7 x 1/4	2.8 - 2x8s	-
	7 x 3/8	4.2 - 2x8s	-
	7 x 1/2	5.6 - 2x8s	W6x9
2x10	9 x 1/4	2.9 - 2x10s	-
	9 x 3/8	4.3 - 2x10s	-
	9 x 1/2	5.8 - 2x10s	W8x10
2x12	11 x 1/4	2.9 - 2x12s	-
	11 x 3/8	4.4 - 2x12s	-
	11 x 1/2	5.8 - 2x12s	W10x12

Double Steel Flitch Beam: Wood	Steel	Built-Up Lumber Equivalents*	Steel I-Beam Equivalents
2x6	5" x 1/4"	4.7 - 2x6s	-
	5 x 3/8	7 - 2x6s	-
	5 x 1/2	9.4 - 2x6s	-
2x8	7 x 1/4	5.6 - 2x8s	W6x9
	7 x 3/8	8.4 - 2x8s	W8x10 or W6x12
	7 x 1/2	11.3 - 2x8s	W8x13
2x10	9 x 1/4	5.6 - 2x10s	W8x10
	9 x 3/8	8.4 - 2x10s	W8x15
	9 x 1/2	11.5 - 2x10s	W8x18
2x12	11 x 1/4	5.8 - 2x12s	W10x12
	11 x 3/8	8.8 - 2x12s	W10x17
	11 x 1/2	11.7 - 2x12s	W10x22

* The strength equivalent shown for built-up lumber beams assumes Doug-fir-larch (E=1.6, Fb=1,200).

For more information: "Flitchplates," JLC 9/92

The strength equivalent shown for built-up lumber beams assumes Doug-fir-larch (E = 1.6, Fb = 1,200).

Figure 2–24. Maximum Allowable Spans (ft.-in.) for Floor Joists

Residential Living Areas: 40 PSF Live, 10 PSF Dead (L/360)

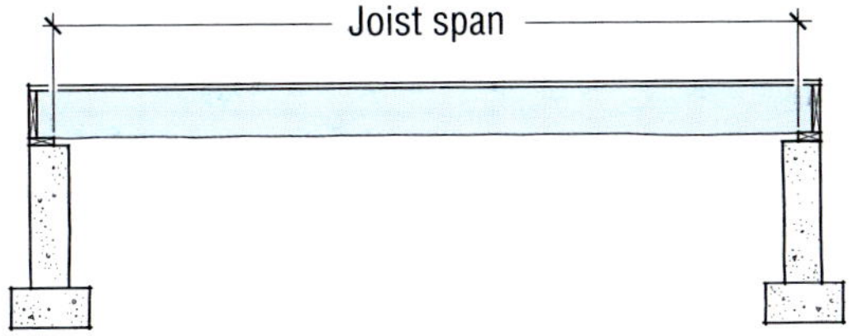

		2x6			2x8			2x10			2x12		
Species Group	**Spacing (in.)**	**No. 1**	**No. 2**	**No. 1-2 (Can.)**	**No. 1**	**No. 2**	**No. 1-2 (Can.)**	**No. 1**	**No. 2**	**No. 1-2 (Can.)**	**No. 1**	**No. 2**	**No. 1-2 (Can.)**
D-Fir-L	12	10-11	10-9	10-9	14-5	14-2	14-2	18-5	18-0	17-6	22-0	20-11	20-4
	16	9-11	9-9	9-9	13-1	12-9	12-5	16-5	15-7	15-2	19-1	18-1	17-7
	24	8-8	8-3	8-0	11-0	10-5	10-2	13-5	12-9	12-5	15-7	14-9	14-4
SPF	12	–	–	10-3	–	–	13-6	–	–	17-3	–	–	20-7
	16	–	–	9-4	–	–	12-3	–	–	15-5	–	–	17-10
	24	–	–	8-1	–	–	10-3	–	–	12-7	–	–	14-7
Hem-Fir	12	10-6	10-0	10-9	13-10	13-2	14-2	17-8	16-10	18-0	21-6	20-4	21-11
	16	9-6	9-1	9-9	12-7	12-0	12-10	16-0	15-2	16-5	18-10	17-7	19-1
	24	8-4	7-11	8-6	10-10	10-2	11-0	13-3	12-5	13-5	15-5	14-4	15-7
SYP	12	10-11	10-9	–	14-5	14-2	–	18-5	18-0	–	22-5	21-9	–
	16	9-11	9-9	–	13-1	12-10	–	16-9	16-1	–	20-4	18-10	–
	24	8-8	8-6	–	11-5	11-0	–	14-7	13-2	–	17-5	15-4	–

For information on framing joists, see **pages 114-122**.

2 - FRAMING: Joist Spans

Figure 2–25. Maximum Allowable Spans (ft.-in.) for Floor Joists

Residential Living Areas: 1 1/2-in. Concrete Floor (Tile Bed); 40 PSF Live, 20 PSF Dead (L/360)

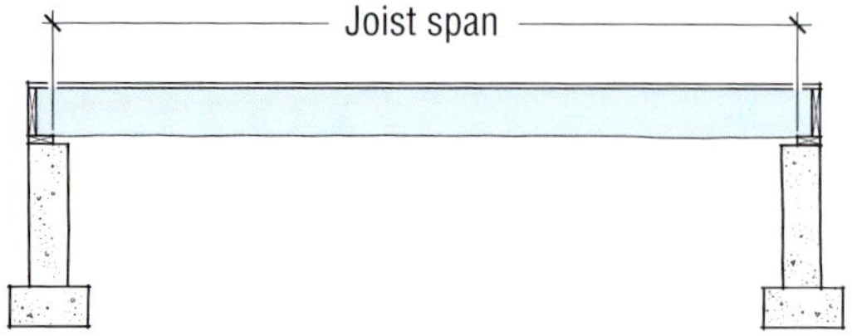

		2x6			2x8			2x10			2x12		
Species Group	**Spacing (in.)**	**No. 1**	**No. 2**	**No. 1-2 (Can.)**	**No. 1**	**No. 2**	**No. 1-2 (Can.)**	**No. 1**	**No. 2**	**No. 1-2 (Can.)**	**No. 1**	**No. 2**	**No. 1-2 (Can.)**
D-Fir-L	12	10-11	10-8	10-4	14-2	13-6	13-1	17-4	16-5	16-0	20-1	19-1	18-6
	16	9-8	9-3	8-11	12-4	11-8	11-4	15-0	14-3	13-10	17-5	16-6	16-1
	24	7-11	7-6	7-4	10-0	9-6	9-3	12-3	11-8	11-4	14-3	13-6	13-1
SPF	12	–	–	10-3	–	–	13-3	–	–	16-3	–	–	18-10
	16	–	–	9-1	–	–	11-6	–	–	14-1	–	–	16-3
	24	–	–	7-5	–	–	9-5	–	–	11-6	–	–	13-4
Hem-Fir	12	10-6	10-0	10-9	13-10	13-1	14-2	17-1	16-0	17-4	19-10	18-6	20-1
	16	9-6	8-11	9-8	12-2	11-4	12-4	14-10	13-10	15-0	17-2	16-1	17-5
	24	7-10	7-4	7-11	9-11	9-3	10-0	12-1	11-4	12-3	14-0	13-1	14-3
SYP	12	10-11	10-9	–	14-5	14-2	–	18-5	16-11	–	22-5	19-10	–
	16	9-11	9-6	–	13-1	12-4	–	16-4	14-8	–	19-6	17-2	–
	24	8-8	7-9	–	11-3	10-0	–	13-4	12-0	–	15-11	14-0	–

For information on framing joists, see **pages 114-122**.

For more information: "How to Use Span Tables," JLC 4/94

Figure 2–26. Maximum Allowable Spans (ft.-in.) for Ceiling Joists

Drywall Attached – No Future Rooms and No Attic Storage; 10 PSF Live 5 PSF Dead (L/240)

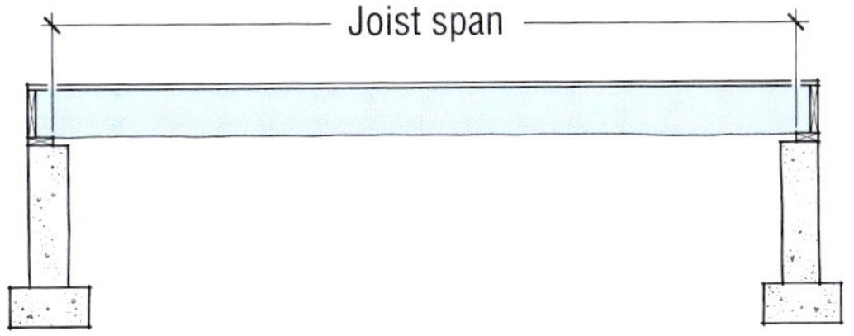

		2x4			2x6			2x8			2x10		
Species Group	**Spacing (in.)**	**No. 1**	**No. 2**	**No. 1-2 (Can.)**	**No. 1**	**No. 2**	**No. 1-2 (Can.)**	**No. 1**	**No. 2**	**No. 1-2 (Can.)**	**No. 1**	**No. 2**	**No. 1-2 (Can.)**
D-Fir-L	12	12-8	12-5	12-5	19-11	19-6	19-6	26-2	25-8	25-8	33-5	32-9	32-0
	16	11-6	11-3	11-3	18-1	17-8	17-8	23-10	23-4	22-8	30-0	28-6	27-8
	24	10-0	9-10	9-10	15-9	15-0	14-7	20-1	19-1	18-6	24-6	23-3	22-7
SPF	12	–	–	11-10	–	–	18-8	–	–	24-7	–	–	31-4
	16	–	–	10-9	–	–	16-11	–	–	22-4	–	–	28-1
	24	–	–	9-5	–	–	14-9	–	–	18-9	–	–	22-11
Hem-Fir	12	12-2	11-7	12-5	19-1	18-2	19-6	25-2	24-0	25-8	32-1	30-7	32-9
	16	11-0	10-6	11-3	17-4	16-6	17-8	22-10	21-9	23-4	29-2	27-8	29-9
	24	9-8	9-2	9-10	15-2	14-5	15-6	19-10	18-6	20-1	24-3	22-7	24-6
SYP	12	12-8	12-5	–	19-11	19-6	–	26-2	25-8	–	33-5	32-9	–
	16	11-6	11-3	–	18-1	17-8	–	23-10	23-4	–	30-5	29-4	–
	24	10-0	9-10	–	15-9	15-6	–	20-10	20-1	–	26-6	24-0	–

For information on framing joists, see **pages 114-122**.

2 - FRAMING: Joist Spans

Figure 2–27. Maximum Allowable Spans (ft.-in.) for Ceiling Joists

Drywall Attached – No Future Rooms, Limited Attic Storage; 20 PSF Live 10 PSF Dead (L/240)

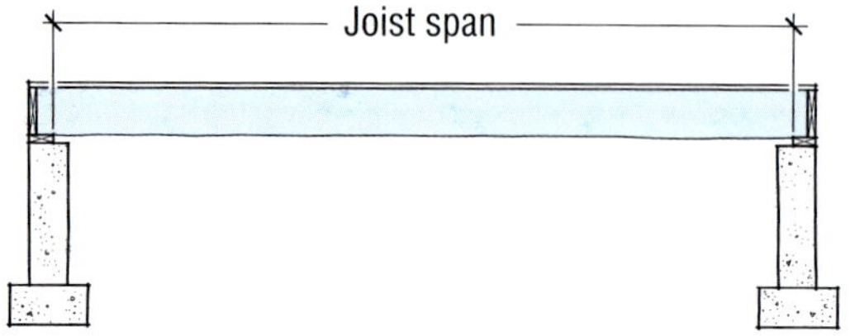

		2x4			2x6			2x8			2x10		
Species Group	**Spacing (in.)**	**No. 1**	**No. 2**	**No. 1-2 (Can.)**	**No. 1**	**No. 2**	**No. 1-2 (Can.)**	**No. 1**	**No. 2**	**No. 1-2 (Can.)**	**No. 1**	**No. 2**	**No. 1-2 (Can.)**
D-Fir-L	12	10-0	9-10	9-10	15-9	15-0	14-7	20-1	19-1	18-6	24-6	23-3	22-7
	16	9-1	8-11	8-8	13-9	13-0	12-8	17-5	16-6	16-0	21-3	20-2	19-7
	24	7-8	7-3	7-1	11-2	10-8	10-4	14-2	13-6	13-1	17-4	16-5	16-0
SPF	12	–	–	9-5	–	–	14-9	–	–	18-9	–	–	22-11
	16	–	–	8-7	–	–	12-10	–	–	16-3	–	–	19-10
	24	–	–	7-2	–	–	10-6	–	–	13-3	–	–	16-3
Hem-Fir	12	9-8	9-2	9-10	15-2	14-5	15-6	19-10	18-6	20-1	24-3	22-7	24-6
	16	8-9	8-4	8-11	13-7	12-8	13-9	17-2	16-0	17-5	21-0	19-7	21-3
	24	7-7	7-1	7-8	11-1	10-4	11-2	14-0	13-1	14-2	17-1	16-0	17-4
SYP	12	10-0	9-10	–	15-9	15-6	–	20-10	20-1	–	26-6	24-0	–
	16	9-1	8-11	–	14-4	13-6	–	18-11	17-5	–	23-2	20-9	–
	24	8-0	7-8	–	12-6	11-0	–	15-11	14-2	–	18-11	17-0	–

For information on framing joists, see **pages 114-122**.

For more information: "How to Use Span Tables," JLC 4/94

Figure 2–60. Header and Girder Spans (ft.-in.) for Exterior Bearing Walls

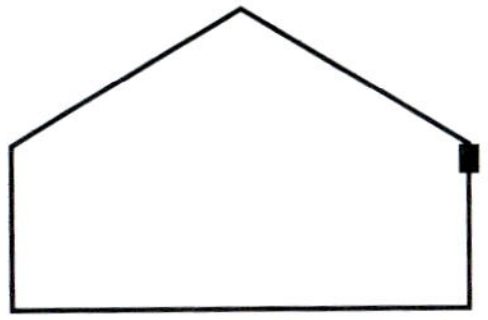

20 lb. Snow Load

Maximum Header Size	Building Width (ft.) 20 Span	20 No. of Jacks	36 Span	36 No. of Jacks
2-2x4	3-6	1	2-10	1
2-2x6	5-5	1	4-2	1
2-2x8	6-10	1	5-4	2
2-2x10	8-5	2	6-6	2
2-2x12	9-9	2	7-6	2
3-2x8	8-4	1	6-8	1
3-2x10	10-6	1	8-2	2
3-2x12	12-2	2	9-5	2
4-2x8	9-2	1	7-8	2
4-2x10	11-8	1	9-5	2
4-2x12	14-1	1	10-11	2

50 lb. Snow Load

Maximum Header Size	Building Width (ft.) 20 Span	20 No. of Jacks	36 Span	36 No. of Jacks
2-2x4	2-10	1	2-2	1
2-2x6	4-1	1	3-2	2
2-2x8	5-2	2	4-0	2
2-2x10	6-4	2	4-11	2
2-2x12	7-4	2	5-8	3
3-2x8	6-6	2	5-0	2
3-2x10	7-11	2	6-2	2
3-2x12	9-2	2	7-2	2
4-2x8	7-6	2	5-10	2
4-2x10	9-2	2	7-1	2
4-2x12	10-8	2	8-3	2

Note: Calculated spans assume #2 grade Doug.-Fir, Hem-Fir and SYP lumber.

2 - FRAMING: Header Spans

Figure 2–61. Header and Girder Spans (ft.-in.) for Exterior Bearing Walls

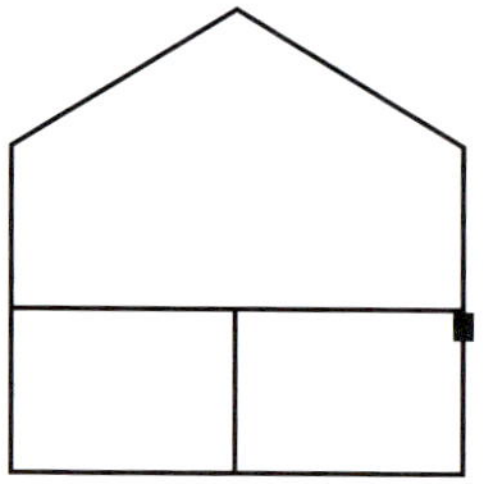

Supporting truss roof and ceiling (no attic) + one center-bearing floor and ceiling

20 lb. Snow Load

Maximum Header Size	Building Width (ft.) 20 Span	20 No. of Jacks	36 Span	36 No. of Jacks
2-2x4	3-1	1	2-5	1
2-2x6	4-6	1	3-7	2
2-2x8	5-9	2	4-6	2
2-2x10	7-0	2	5-6	2
2-2x12	8-1	2	6-5	2
3-2x8	7-2	1	5-8	2
3-2x10	8-9	2	6-11	2
3-2x12	10-2	2	8-0	2
4-2x8	8-1	2	6-7	2
4-2x10	10-1	1	8-0	2
4-2x12	11-9	2	9-3	2

50 lb. Snow Load

Maximum Header Size	Building Width (ft.) 20 Span	20 No. of Jacks	36 Span	36 No. of Jacks
2-2x4	2-6	1	2-0	1
2-2x6	3-8	1	2-11	2
2-2x8	4-8	2	3-8	2
2-2x10	5-8	2	4-5	2
2-2x12	6-7	2	5-2	2
3-2x8	5-10	1	4-7	2
3-2x10	7-1	2	5-7	2
3-2x12	8-3	2	6-6	2
4-2x8	6-8	2	5-3	2
4-2x10	8-2	2	6-5	2
4-2x12	9-6	2	7-6	2

Note: Calculated spans assume #2 grade Doug.-Fir, Hem-Fir and SYP lumber.

Figure 2–62. Header and Girder Spans (ft.-in.) for Exterior Bearing Walls

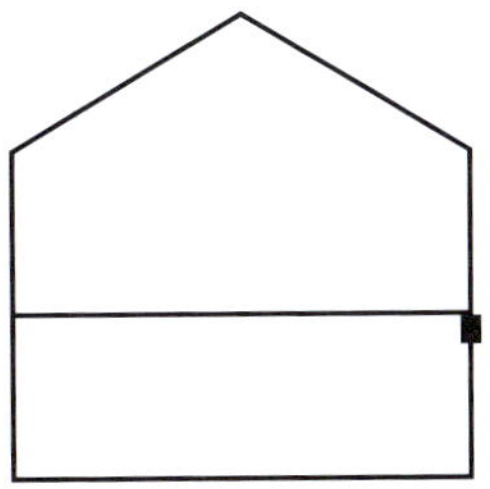

20 lb. Snow Load

Maximum Header Size	Building Width (ft.) 20 Span	20 No. of Jacks	36 Span	36 No. of Jacks
2-2x4	2-8	1	2-1	1
2-2x6	3-11	1	3-0	2
2-2x8	5-0	2	3-10	2
2-2x10	6-1	2	4-8	2
2-2x12	7-1	2	5-5	3
3-2x8	6-3	2	4-10	2
3-2x10	7-7	2	5-11	2
3-2x12	8-10	2	6-10	2
4-2x8	7-2	2	5-7	2
4-2x10	8-9	2	6-10	2
4-2x12	10-2	2	7-11	2

50 lb. Snow Load

Maximum Header Size	Building Width (ft.) 20 Span	20 No. of Jacks	36 Span	36 No. of Jacks
2-2x4	2-5	1	1-10	1
2-2x6	3-6	2	2-9	2
2-2x8	4-5	2	3-5	2
2-2x10	5-5	2	4-3	3
2-2x12	6-3	2	4-11	3
3-2x8	5-6	2	4-4	2
3-2x10	6-9	2	5-3	2
3-2x12	7-10	2	6-1	2
4-2x8	6-4	2	5-0	2
4-2x10	7-9	2	6-1	2
4-2x12	9-0	2	7-1	2

Note: Calculated spans assume #2 grade Doug.-Fir, Hem-Fir and SYP lumber.

Figure 2–63. Header and Girder Spans (ft.-in.) for Exterior Bearing Walls

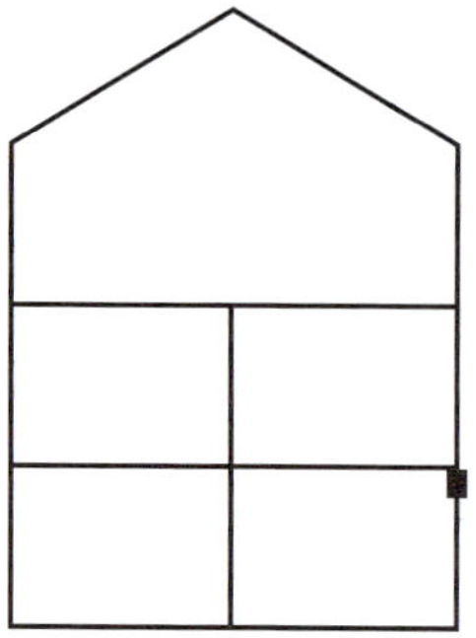

20 lb. Snow Load

Maximum Header Size	Building Width (ft.) 20 Span	20 No. of Jacks	36 Span	36 No. of Jacks
2-2x4	2-7	1	2-0	1
2-2x6	3-9	2	2-11	2
2-2x8	4-9	2	3-9	2
2-2x10	5-9	2	4-7	3
2-2x12	6-8	2	5-3	3
3-2x8	5-11	2	4-8	2
3-2x10	7-3	2	5-9	2
3-2x12	8-5	2	6-7	2
4-2x8	6-10	2	5-5	2
4-2x10	8-4	2	6-7	2
4-2x12	9-8	2	7-8	2

50 lb. Snow Load

Maximum Header Size	Building Width (ft.) 20 Span	20 No. of Jacks	36 Span	36 No. of Jacks
2-2x4	2-3	1	1-10	1
2-2x6	3-4	2	2-8	2
2-2x8	4-3	2	3-4	2
2-2x10	5-2	2	4-1	3
2-2x12	6-0	2	4-9	3
3-2x8	5-4	2	4-2	2
3-2x10	6-6	2	5-2	2
3-2x12	7-6	2	6-0	3
4-2x8	6-1	2	4-10	2
4-2x10	7-6	2	5-11	2
4-2x12	8-8	2	6-11	2

Note: Calculated spans assume #2 grade Doug.-Fir, Hem-Fir and SYP lumber.

Figure 2–64. Header and Girder Spans (ft.-in.) for Exterior Bearing Walls

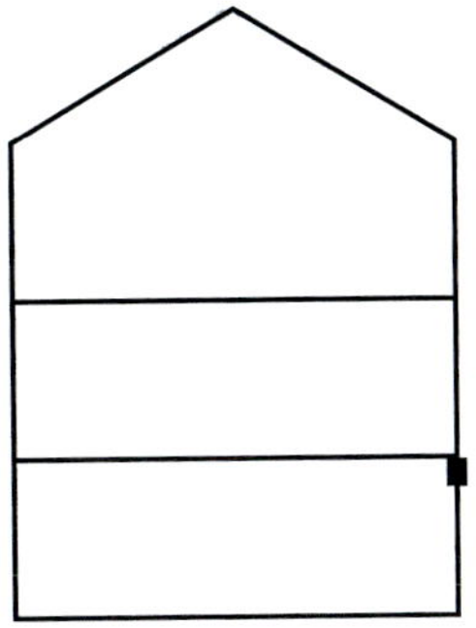

20 lb. Snow Load

Maximum Header Size	Building Width (ft.) 20 Span	20 No. of Jacks	36 Span	36 No. of Jacks
2-2x4	2-1	1	1-7	2
2-2x6	3-1	2	2-4	2
2-2x8	3-10	2	3-0	3
2-2x10	4-9	2	3-8	3
2-2x12	5-6	3	4-3	3
3-2x8	4-10	2	3-9	2
3-2x10	5-11	2	4-7	3
3-2x12	6-10	2	5-4	3
4-2x8	5-7	2	4-4	3
4-2x10	6-10	2	5-3	2
4-2x12	7-11	2	6-2	3

50 lb. Snow Load

Maximum Header Size	Building Width (ft.) 20 Span	20 No. of Jacks	36 Span	36 No. of Jacks
2-2x4	2-0	1	1-7	2
2-2x6	2-11	2	2-3	2
2-2x8	3-9	2	2-11	3
2-2x10	4-7	2	3-6	3
2-2x12	5-3	3	4-1	3
3-2x8	4-8	2	3-7	2
3-2x10	5-8	2	4-5	3
3-2x12	6-7	2	5-1	3
4-2x8	5-5	2	4-2	3
4-2x10	6-7	2	5-1	2
4-2x12	7-8	2	5-11	3

Note: Calculated spans assume #2 grade Doug.-Fir, Hem-Fir and SYP lumber.

2 - FRAMING: Header Spans

Figure 2-65. Header and Girder Spans (ft.-in.) for Interior Bearing Walls

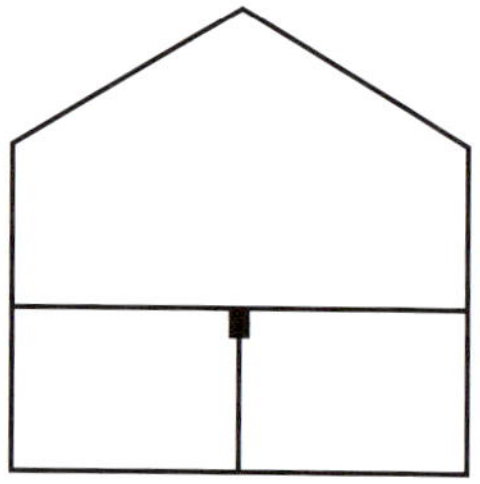

Supporting one center-bearing floor and ceiling only

Maximum Header Size	Building Width (ft.) 20 Span	20 No. of Jacks	36 Span	36 No. of Jacks
2-2x4	3-5	1	2-6	1
2-2x6	4-11	1	3-8	1
2-2x8	6-3	1	4-8	2
2-2x10	7-8	2	5-9	2
2-2x12	8-11	2	6-7	2
3-2x8	7-10	1	5-10	2
3-2x10	9-7	1	7-2	2
3-2x12	11-1	2	8-3	2
4-2x8	9-0	1	6-9	2
4-2x10	11-1	1	8-3	2
4-2x12	12-10	1	9-7	2

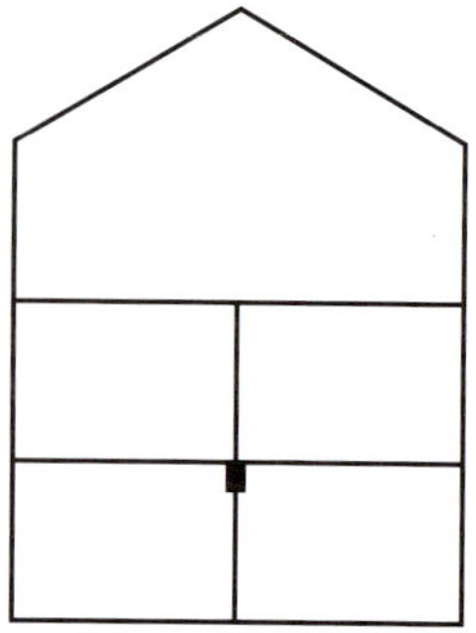

Supporting two center-bearing floors and ceiling only

Maximum Header Size	Building Width (ft.) 20 Span	20 No. of Jacks	36 Span	36 No. of Jacks
2-2x4	2-3	1	1-9	1
2-2x6	3-4	2	2-6	2
2-2x8	4-3	2	3-3	2
2-2x10	5-2	2	3-11	3
2-2x12	6-0	2	4-7	3
3-2x8	5-4	2	4-0	2
3-2x10	6-6	2	4-11	2
3-2x12	7-6	2	5-8	3
4-2x8	6-1	2	4-8	2
4-2x10	7-6	2	5-8	2
4-2x12	8-8	2	6-7	2

Note: Calculated spans assume #2 grade Doug.-Fir, Hem-Fir and SYP lumber.

Figure 2–69. Maximum Allowable Horizontal Spans (ft.-in.)

Rafters: 20 psf live load; 10 psf dead load. Non-snow region. Lightweight roof covering. No ceiling drywall.

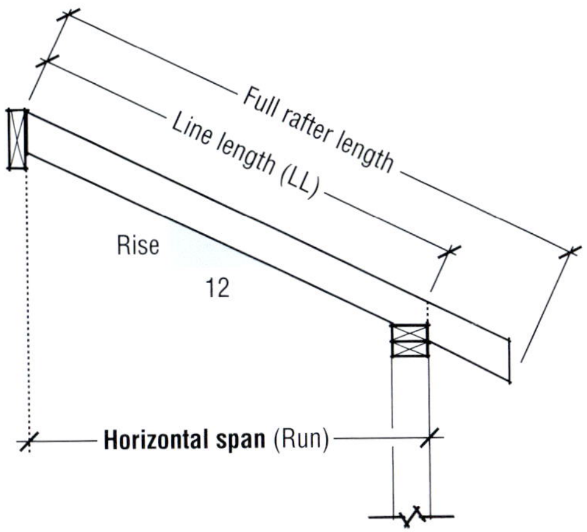

Species Group	Spacing (in.)	2x4 No. 1	2x4 No. 2	2x4 No. 1-2 (Can.)	2x6 No. 1	2x6 No. 2	2x6 No. 1-2 (Can.)	2x8 No. 1	2x8 No. 2	2x8 No. 1-2 (Can.)	2x10 No. 1	2x10 No. 2	2x10 No. 1-2 (Can.)
D-Fir-L	12	11-1	10-10	10-10	17-4	16-10	16-4	22-5	21-4	20-8	27-5	26-0	25-3
	16	10-0	9-10	9-8	15-4	14-7	14-2	19-5	18-5	17-11	23-9	22-6	21-11
	24	8-7	8-2	7-11	12-6	11-11	11-7	15-10	15-1	14-8	19-5	18-5	17-10
SPF	12	–	–	10-4	–	–	16-3	–	–	21-0	–	–	25-8
	16	–	–	9-5	–	–	14-4	–	–	18-2	–	–	22-3
	24	–	–	8-0	–	–	11-9	–	–	14-10	–	–	18-2
Hem-Fir	12	10-7	10-1	10-10	16-8	15-11	17-0	21-11	20-8	22-5	27-1	25-3	27-5
	16	9-8	9-2	9-10	15-2	14-2	15-4	19-2	17-11	19-5	23-5	21-11	23-9
	24	8-5	7-11	8-7	12-4	11-7	12-6	15-8	14-8	15-10	19-2	17-10	19-5
SYP	12	11-1	10-10	–	17-4	17-0	–	22-11	22-5	–	29-2	26-9	–
	16	10-0	9-10	–	15-9	15-1	–	20-10	19-6	–	25-10	23-2	–
	24	8-9	8-7	–	13-9	12-4	–	17-9	15-11	–	21-1	18-11	–

2 - FRAMING: Rafter Spans

Figure 2–70. Maximum Allowable Horizontal Spans (ft.-in.)

Rafters: 20 psf live load; 10 psf dead load. Non-snow region. Light-weight roof covering. Drywall ceiling. No Attic.

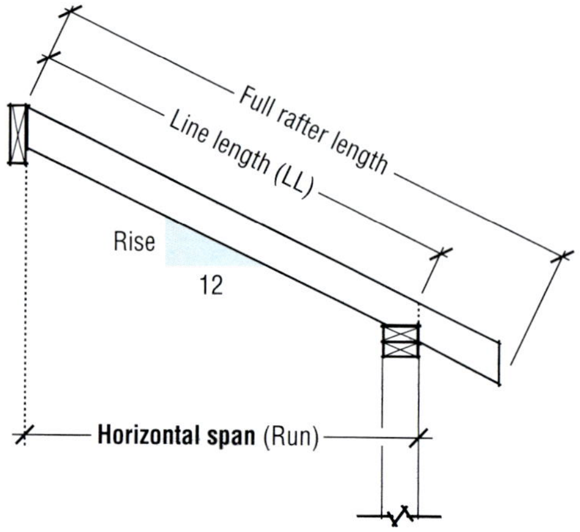

		2x6			2x8			2x10			2x12		
Species Group	**Spacing (in.)**	**No. 1**	**No. 2**	**No. 1-2 (Can.)**	**No. 1**	**No. 2**	**No. 1-2 (Can.)**	**No. 1**	**No. 2**	**No. 1-2 (Can.)**	**No. 1**	**No. 2**	**No. 1-2 (Can.)**
D-Fir-L	12	15-9	15-6	15-6	20-10	20-5	20-5	26-6	26-0	25-3	31-10	30-2	29-4
	16	14-4	14-1	14-1	18-11	18-5	17-11	23-9	22-6	21-11	27-6	26-1	25-5
	24	12-6	11-11	11-7	15-10	15-1	14-8	19-5	18-5	17-10	22-6	21-4	20-9
SPF	12	–	–	14-9	–	–	19-6	–	–	24-10	–	–	29-9
	16	–	–	13-5	–	–	17-9	–	–	22-3	–	–	25-9
	24	–	–	11-9	–	–	14-10	–	–	18-2	–	–	21-0
Hem-Fir	12	15-2	14-5	15-6	19-11	19-0	20-5	25-5	24-3	26-0	30-11	29-4	31-8
	16	13-9	13-1	14-1	18-2	17-3	18-6	23-2	21-11	23-8	27-2	25-5	27-6
	24	12-0	11-5	12-3	15-8	14-8	15-10	19-2	17-10	19-5	22-2	20-9	22-6
SYP	12	15-9	15-6	–	20-10	20-5	–	26-6	26-0	–	32-3	31-4	–
	16	14-4	14-1	–	18-11	18-6	–	24-1	23-2	–	29-4	27-2	–
	24	12-6	12-3	–	16-6	15-11	–	21-1	18-11	–	25-2	22-2	–

For more information: "How to Use Span Tables," JLC 4/94

Figure 2–71. Maximum Allowable Horizontal Spans (ft.-in.)

Rafters: 20 psf live load; 10 psf dead load. Snow region. Light-weight roof covering. No ceiling drywall.

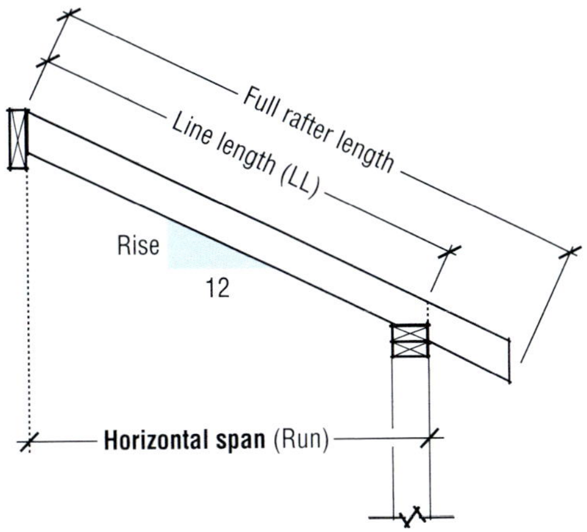

		2x4			2x6			2x8			2x10		
Species Group	**Spacing (in.)**	**No. 1**	**No. 2**	**No. 1-2 (Can.)**	**No. 1**	**No. 2**	**No. 1-2 (Can.)**	**No. 1**	**No. 2**	**No. 1-2 (Can.)**	**No. 1**	**No. 2**	**No. 1-2 (Can.)**
D-Fir-L	12	11-1	10-10	10-9	17-0	16-2	15-8	21-6	20-5	19-10	26-4	24-11	24-3
	16	10-0	9-7	9-3	14-9	14-0	13-7	18-8	17-8	17-2	22-9	21-7	21-0
	24	8-3	7-10	7-7	12-0	11-5	11-1	15-3	14-5	14-0	18-7	17-8	17-2
SPF	12	–	–	10-4	–	–	15-11	–	–	20-2	–	–	24-7
	16	–	–	9-5	–	–	13-9	–	–	17-5	–	–	21-4
	24	–	–	7-8	–	–	11-3	–	–	14-3	–	–	17-5
Hem-Fir	12	10-7	10-1	10-10	16-8	15-8	17-0	21-3	19-10	21-6	26-0	24-3	26-4
	16	9-8	9-2	9-10	14-6	13-7	14-9	18-5	17-2	18-8	22-6	21-0	22-9
	24	8-1	7-7	8-3	11-10	11-1	12-0	15-0	14-0	15-3	18-4	17-2	18-7
SYP	12	11-1	10-10	–	17-4	16-8	–	22-11	21-7	–	28-7	25-8	–
	16	10-0	9-10	–	15-9	14-5	–	20-10	18-8	–	24-9	22-3	–
	24	8-9	8-2	–	13-6	11-9	–	17-0	15-3	–	20-3	18-2	–

2 - FRAMING: Rafter Spans

Figure 2–72. Maximum Allowable Horizontal Spans (ft.-in.)

Rafters: 20 psf live load; 10 psf dead load. Snow region. Lightweight roof covering. Drywall ceiling. No Attic.

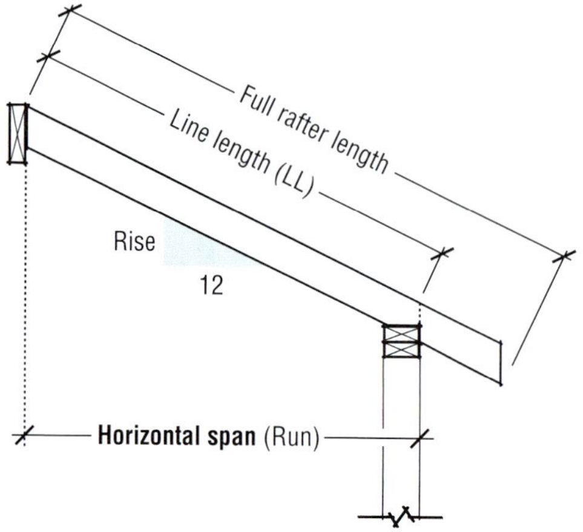

		2x6			2x8			2x10			2x12		
Species Group	**Spacing (in.)**	**No. 1**	**No. 2**	**No. 1-2 (Can.)**	**No. 1**	**No. 2**	**No. 1-2 (Can.)**	**No. 1**	**No. 2**	**No. 1-2 (Can.)**	**No. 1**	**No. 2**	**No. 1-2 (Can.)**
D-Fir-L	12	15-9	15-6	15-6	20-10	20-5	19-10	26-4	24-11	24-3	30-6	28-11	28-1
	16	14-4	14-0	13-7	18-8	17-8	17-2	22-9	21-7	21-0	26-5	25-1	24-4
	24	12-0	11-5	11-1	15-3	14-5	14-0	18-7	17-8	17-2	21-7	20-5	19-11
SPF	12	–	–	14-9	–	–	19-6	–	–	24-7	–	–	28-6
	16	–	–	13-5	–	–	17-5	–	–	21-4	–	–	24-8
	24	–	–	11-3	–	–	14-3	–	–	17-5	–	–	20-2
Hem-Fir	12	15-2	14-5	15-6	19-11	19-0	20-5	25-5	24-3	26-0	30-1	28-1	30-6
	16	13-9	13-1	14-1	18-2	17-2	18-6	22-6	21-0	22-9	26-1	24-4	26-5
	24	11-10	11-1	12-0	15-0	14-0	15-3	18-4	17-2	18-7	21-3	19-11	21-7
SYP	12	15-9	15-6	–	20-10	20-5	–	26-6	25-8	–	32-3	30-1	–
	16	14-4	14-1	–	18-11	18-6	–	24-1	22-3	–	29-4	26-1	–
	24	12-6	11-9	–	16-6	15-3	–	20-3	18-2	–	24-1	21-4	–

For more information: "How to Use Span Tables," JLC 4/94

Figure 2–73. Maximum Allowable Horizontal Spans (ft.-in.)

Rafters: 30 psf live load; 10 psf dead load. Snow region. Lightweight roof covering. No ceiling drywall.

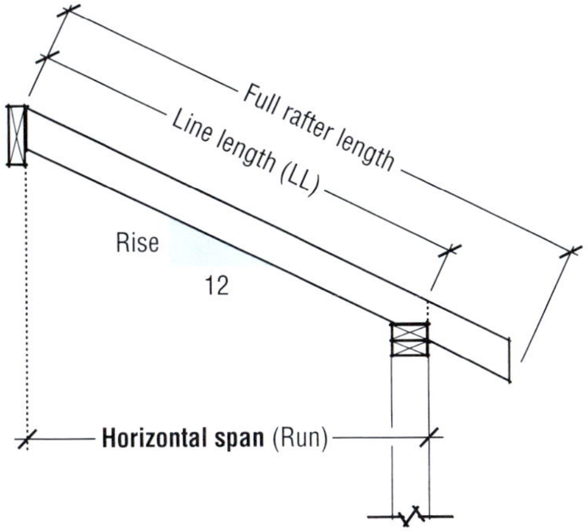

Species Group	Spacing (in.)	2x4 No. 1	2x4 No. 2	2x4 No. 1-2 (Can.)	2x6 No. 1	2x6 No. 2	2x6 No. 1-2 (Can.)	2x8 No. 1	2x8 No. 2	2x8 No. 1-2 (Can.)	2x10 No. 1	2x10 No. 2	2x10 No. 1-2 (Can.)
D-Fir-L	12	9-8	9-6	9-3	14-9	14-0	13-7	18-8	17-8	17-2	22-9	21-7	21-0
	16	8-9	8-3	8-0	12-9	12-1	11-9	16-2	15-4	14-11	19-9	18-9	18-2
	24	7-1	6-9	6-7	10-5	9-10	9-7	13-2	12-6	12-2	16-1	15-3	14-10
SPF	12	–	–	9-1	–	–	13-9	–	–	17-5	–	–	21-4
	16	–	–	8-2	–	–	11-11	–	–	15-1	–	–	18-5
	24	–	–	6-8	–	–	9-9	–	–	12-4	–	–	15-1
Hem-Fir	12	9-3	8-10	9-6	14-6	13-7	14-9	18-5	17-2	18-8	22-6	21-0	22-9
	16	8-5	8-0	8-7	12-7	11-9	12-9	15-11	14-11	16-2	19-6	18-2	19-9
	24	7-0	6-7	7-1	10-3	9-7	10-5	13-0	12-2	13-2	15-11	14-10	16-1
SYP	12	9-8	9-6	–	15-2	14-5	–	20-0	18-8	–	24-9	22-3	–
	16	8-9	8-7	–	13-9	12-6	–	18-0	16-2	–	21-5	19-3	–
	24	7-8	7-1	–	11-9	10-2	–	14-9	13-2	–	17-6	15-9	–

2 - FRAMING: Rafter Spans

Figure 2–74. Maximum Allowable Horizontal Spans (ft.-in.)

Rafters: 30 psf live load; 10 psf dead load. Snow region. Lightweight roof covering. Drywall ceiling. No Attic.

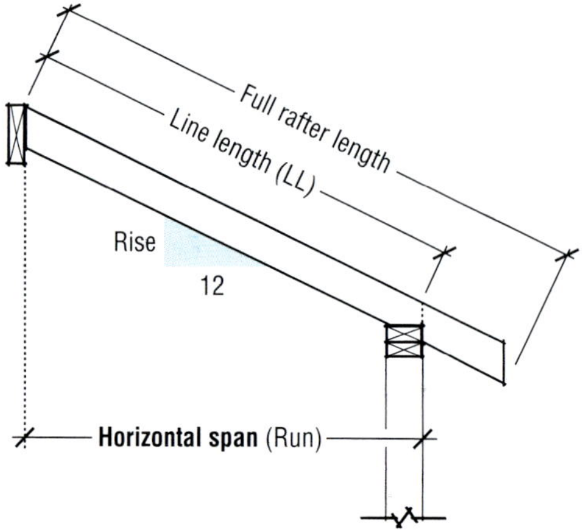

Species Group	Spacing (in.)	2x6 No. 1	2x6 No. 2	2x6 No. 1-2 (Can.)	2x8 No. 1	2x8 No. 2	2x8 No. 1-2 (Can.)	2x10 No. 1	2x10 No. 2	2x10 No. 1-2 (Can.)	2x12 No. 1	2x12 No. 2	2x12 No. 1-2 (Can.)
D-Fir-L	12	13-9	13-6	13-6	18-2	17-8	17-2	22-9	21-7	21-0	26-5	25-1	24-4
	16	12-6	12-1	11-9	16-2	15-4	14-11	19-9	18-9	18-2	22-10	21-8	21-1
	24	10-5	9-10	9-7	13-2	12-6	12-2	16-1	15-3	14-10	18-8	17-9	17-3
SPF	12	–	–	12-11	–	–	17-0	–	–	21-4	–	–	24-8
	16	–	–	11-9	–	–	15-1	–	–	18-5	–	–	21-5
	24	–	–	9-9	–	–	12-4	–	–	15-1	–	–	17-6
Hem-Fir	12	13-3	12-7	13-6	17-5	16-7	17-10	22-3	21-0	22-9	26-1	24-4	26-5
	16	12-0	11-5	12-3	15-10	14-11	16-2	19-6	18-2	19-9	22-7	21-1	22-10
	24	10-3	9-7	10-5	13-0	12-2	13-2	15-11	14-10	16-1	18-5	17-3	18-8
SYP	12	13-9	13-6	–	18-2	17-10	–	23-2	22-3	–	28-2	26-1	–
	16	12-6	12-3	–	16-6	16-2	–	21-1	19-3	–	25-7	22-7	–
	24	10-11	10-2	–	14-5	13-2	–	17-6	15-9	–	20-10	18-5	–

For more information: "How to Use Span Tables," JLC 4/94

Figure 2–75. Maximum Allowable Horizontal Spans (ft.-in.)

Rafters: 30 psf live load; 20 psf dead load. Snow region. Heavy roof covering. No ceiling drywall.

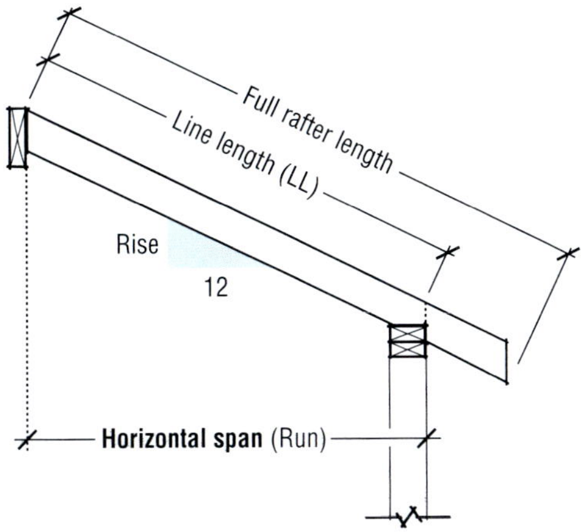

Species Group	Spacing (in.)	2x4 No. 1	2x4 No. 2	2x4 No. 1-2 (Can.)	2x6 No. 1	2x6 No. 2	2x6 No. 1-2 (Can.)	2x8 No. 1	2x8 No. 2	2x8 No. 1-2 (Can.)	2x10 No. 1	2x10 No. 2	2x10 No. 1-2 (Can.)
D-Fir-L	12	9-0	8-6	8-4	13-2	12-6	12-2	16-8	15-10	15-4	20-4	19-4	18-9
	16	7-10	7-5	7-2	11-5	10-10	10-6	14-5	13-8	13-4	17-8	16-9	16-3
	24	6-4	6-0	5-10	9-4	8-10	8-7	11-9	11-2	10-10	14-5	13-8	13-3
SPF	12	–	–	8-5	–	–	12-4	–	–	15-7	–	–	19-1
	16	–	–	7-3	–	–	10-8	–	–	13-6	–	–	16-6
	24	–	–	5-11	–	–	8-9	–	–	11-0	–	–	13-6
Hem-Fir	12	8-11	8-4	9-0	13-0	12-2	13-2	16-6	15-4	16-8	20-1	18-9	20-4
	16	7-8	7-2	7-10	11-3	10-6	11-5	14-3	13-4	14-5	17-5	16-3	17-8
	24	6-3	5-10	6-4	9-2	8-7	9-4	11-8	10-10	11-9	14-3	13-3	14-5
SYP	12	9-8	9-0	–	14-10	12-11	–	18-8	16-8	–	22-2	19-11	–
	16	8-8	7-9	–	12-10	11-2	–	16-2	14-5	–	19-2	17-3	–
	24	7-1	6-4	–	10-6	9-1	–	13-2	11-10	–	15-8	14-1	–

2 - FRAMING: Rafter Spans

Figure 2–76. Maximum Allowable Horizontal Spans (ft.-in.)

Rafters: 30 psf live load; 20 psf dead load. Snow region. Heavy roof covering. Drywall ceiling. No Attic.

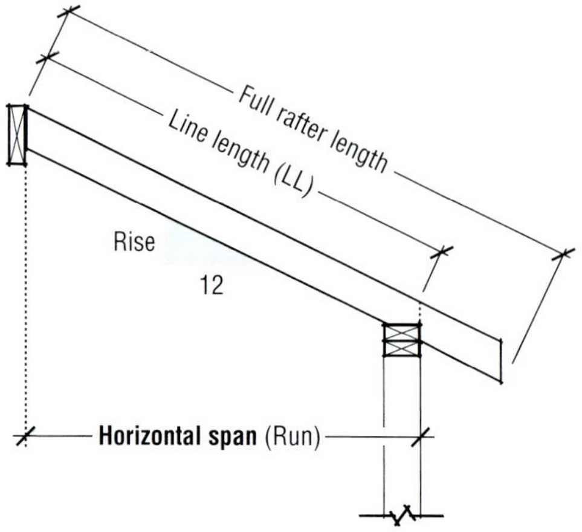

		2x6			2x8			2x10			2x12		
Species Group	**Spacing (in.)**	**No. 1**	**No. 2**	**No. 1-2 (Can.)**	**No. 1**	**No. 2**	**No. 1-2 (Can.)**	**No. 1**	**No. 2**	**No. 1-2 (Can.)**	**No. 1**	**No. 2**	**No. 1-2 (Can.)**
D-Fir-L	12	13-2	12-6	12-2	16-8	15-10	15-4	20-4	19-4	18-9	23-7	22-5	21-9
	16	11-5	10-10	10-6	14-5	13-8	13-4	17-8	16-9	16-3	20-5	19-5	18-10
	24	9-4	8-10	8-7	11-9	11-2	10-10	14-5	13-8	13-3	16-8	15-10	15-5
SPF	12	–	–	12-4	–	–	15-7	–	–	19-1	–	–	22-1
	16	–	–	10-8	–	–	13-6	–	–	16-6	–	–	19-2
	24	–	–	8-9	–	–	11-0	–	–	13-6	–	–	15-7
Hem-Fir	12	13-0	12-2	13-2	16-6	15-4	16-8	20-1	18-9	20-4	23-4	21-9	23-7
	16	11-3	10-6	11-5	14-3	13-4	14-5	17-5	16-3	17-8	20-2	18-10	20-5
	24	9-2	8-7	9-4	11-8	10-10	11-9	14-3	13-3	14-5	16-6	15-5	16-8
SYP	12	13-9	12-11	–	18-2	16-8	–	22-2	19-11	–	26-5	23-4	–
	16	12-6	11-2	–	16-2	14-5	–	19-2	17-3	–	22-10	20-2	–
	24	10-6	9-1	–	13-2	11-10	–	15-8	14-1	–	18-8	16-6	–

For more information: "How to Use Span Tables," JLC 4/94

Figure 2–77. Maximum Allowable Horizontal Spans (ft.-in.)

Rafters: 40 psf live load; 10 psf dead load. Snow-region. Lightweight roof covering. No ceiling drywall.

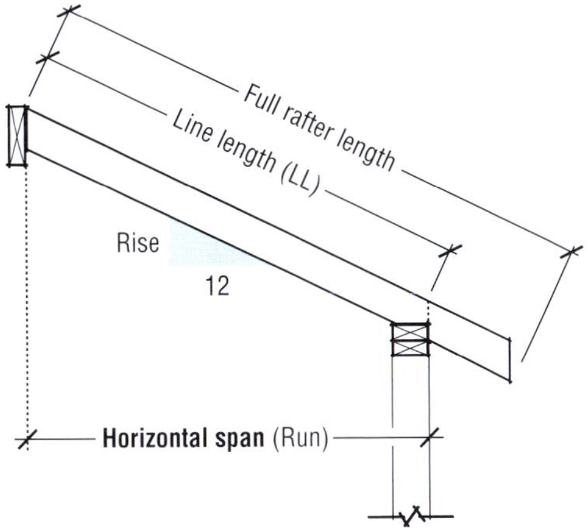

		2x4			2x6			2x8			2x10		
Species Group	**Spacing (in.)**	**No. 1**	**No. 2**	**No. 1-2 (Can.)**	**No. 1**	**No. 2**	**No. 1-2 (Can.)**	**No. 1**	**No. 2**	**No. 1-2 (Can.)**	**No. 1**	**No. 2**	**No. 1-2 (Can.)**
D-Fir-L	12	8-9	8-6	8-4	13-2	12-6	12-2	16-8	15-10	15-4	20-4	19-4	18-9
	16	7-10	7-5	7-2	11-5	10-10	10-6	14-5	13-8	13-4	17-8	16-9	16-3
	24	6-4	6-0	5-10	9-4	8-10	8-7	11-9	11-2	10-10	14-5	13-8	13-3
SPF	12	–	–	8-3	–	–	12-4	–	–	15-7	–	–	19-1
	16	–	–	7-3	–	–	10-8	–	–	13-6	–	–	16-6
	24	–	–	5-11	–	–	8-9	–	–	11-0	–	–	13-6
Hem-Fir	12	8-5	8-0	8-7	13-0	12-2	13-2	16-6	15-4	16-8	20-1	18-9	20-4
	16	7-8	7-2	7-10	11-3	10-6	11-5	14-3	13-4	14-5	17-5	16-3	17-8
	24	6-3	5-10	6-4	9-2	8-7	9-4	11-8	10-10	11-9	14-3	13-3	14-5
SYP	12	8-9	8-7	–	13-9	12-11	–	18-2	16-8	–	22-2	19-11	–
	16	8-0	7-9	–	12-6	11-2	–	16-2	14-5	–	19-2	17-3	–
	24	7-0	6-4	–	10-6	9-1	–	13-2	11-10	–	15-8	14-1	–

2 - FRAMING: Rafter Spans

Figure 2–78. Maximum Allowable Horizontal Spans (ft.-in.)

Rafters: 40 psf live load; 10 psf dead load. Snow region. Lightweight roof covering. Drywall ceiling. No Attic.

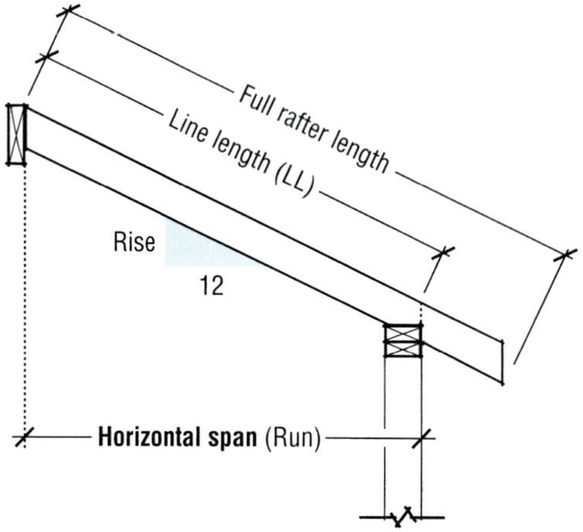

		2x6			2x8			2x10			2x12		
Species Group	**Spacing (in.)**	**No. 1**	**No. 2**	**No. 1-2 (Can.)**	**No. 1**	**No. 2**	**No. 1-2 (Can.)**	**No. 1**	**No. 2**	**No. 1-2 (Can.)**	**No. 1**	**No. 2**	**No. 1-2 (Can.)**
D-Fir-L	12	12-6	12-3	12-2	16-6	15-10	15-4	20-4	19-4	18-9	23-7	22-5	21-9
	16	11-5	10-10	10-6	14-5	13-8	13-4	17-8	16-9	16-3	20-5	19-5	18-10
	24	9-4	8-10	8-7	11-9	11-2	10-10	14-5	13-8	13-3	16-8	15-10	15-5
SPF	12	–	–	11-9	–	–	15-6	–	–	19-1	–	–	22-1
	16	–	–	10-8	–	–	13-6	–	–	16-6	–	–	19-2
	24	–	–	8-9	–	–	11-0	–	–	13-6	–	–	15-7
Hem-Fir	12	12-0	11-5	12-3	15-10	15-1	16-2	20-1	18-9	20-4	23-4	21-9	23-7
	16	10-11	10-5	11-2	14-3	13-4	14-5	17-5	16-3	17-8	20-2	18-10	20-5
	24	9-2	8-7	9-4	11-8	10-10	11-9	14-3	13-3	14-5	16-6	15-5	16-8
SYP	12	12-6	12-3	–	16-6	16-2	–	21-1	19-11	–	25-7	23-4	–
	16	11-5	11-2	–	15-0	14-5	–	19-2	17-3	–	22-10	20-2	–
	24	9-11	9-1	–	13-1	11-10	–	15-8	14-1	–	18-8	16-6	–

For more information: "How to Use Span Tables," JLC 4/94

Figure 2–79. Maximum Allowable Horizontal Spans (ft.-in.)

Rafters: 40 psf live load; 15 psf dead load. Snow region. Medium-weight roof covering. No ceiling drywall.

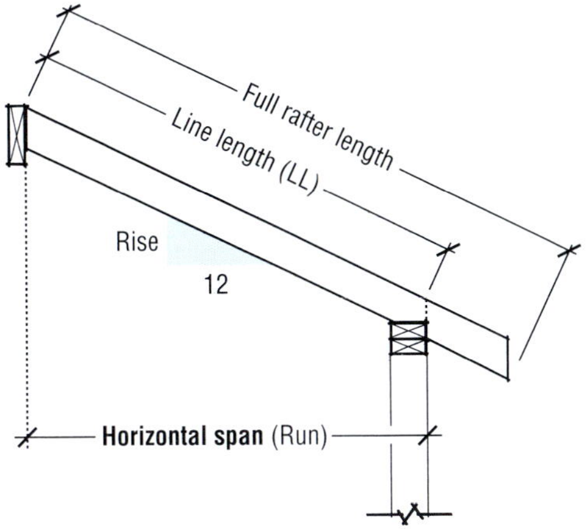

Species Group	Spacing (in.)	2x4 No. 1	2x4 No. 2	2x4 No. 1-2 (Can.)	2x6 No. 1	2x6 No. 2	2x6 No. 1-2 (Can.)	2x8 No. 1	2x8 No. 2	2x8 No. 1-2 (Can.)	2x10 No. 1	2x10 No. 2	2x10 No. 1-2 (Can.)
D-Fir-L	12	8-7	8-2	7-11	12-7	11-11	11-7	15-11	15-1	14-8	19-5	18-5	17-11
	16	7-5	7-1	6-10	10-10	10-4	10-0	13-9	13-1	12-8	16-10	15-11	15-6
	24	6-1	5-9	5-7	8-11	8-5	8-2	11-3	10-8	10-4	13-9	13-0	12-8
SPF	12	–	–	8-0	–	–	11-9	–	–	14-10	–	–	18-2
	16	–	–	6-11	–	–	10-2	–	–	12-11	–	–	15-9
	24	–	–	5-8	–	–	8-4	–	–	10-6	–	–	12-10
Hem-Fir	12	8-5	7-11	8-7	12-5	11-7	12-7	15-8	14-8	15-11	19-2	17-11	19-5
	16	7-4	6-10	7-5	10-9	10-0	10-10	13-7	12-8	13-9	16-7	15-6	16-10
	24	6-0	5-7	6-1	8-9	8-2	8-11	11-1	10-4	11-3	13-7	12-8	13-9
SYP	12	8-9	8-7	–	13-9	12-4	–	17-9	15-11	–	21-1	19-0	–
	16	8-0	7-5	–	12-3	10-8	–	15-5	13-9	–	18-3	16-5	–
	24	6-9	6-1	–	10-0	8-8	–	12-7	11-3	–	14-11	13-5	–

2 - FRAMING: Rafter Spans

Figure 2–80. Maximum Allowable Horizontal Spans (ft.-in.)

Rafters: 40 psf live load; 15 psf dead load. Snow region. Medium-weight roof covering. Drywall ceiling. No attic.

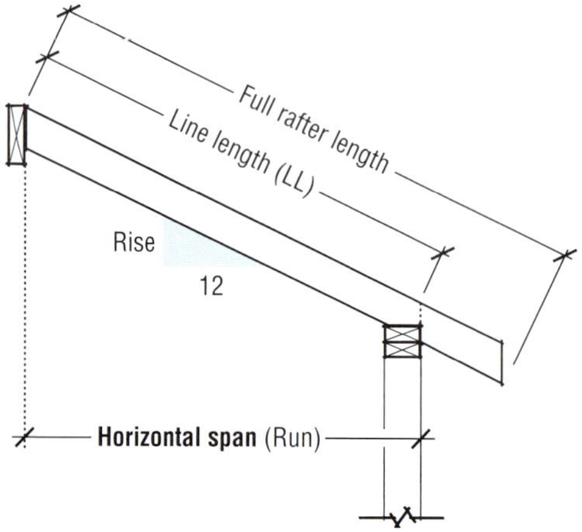

		2x6			2x8			2x10			2x12		
Species Group	**Spacing (in.)**	**No. 1**	**No. 2**	**No. 1-2 (Can.)**	**No. 1**	**No. 2**	**No. 1-2 (Can.)**	**No. 1**	**No. 2**	**No. 1-2 (Can.)**	**No. 1**	**No. 2**	**No. 1-2 (Can.)**
D-Fir-L	12	12-6	11-11	11-7	15-11	15-1	14-8	19-5	18-5	17-11	22-6	21-4	20-9
	16	10-10	10-4	10-0	13-9	13-1	12-8	16-10	15-11	15-6	19-6	18-6	18-0
	24	8-11	8-5	8-2	11-3	10-8	10-4	13-9	13-0	12-8	15-11	15-1	14-8
SPF	12	–	–	11-9	–	–	14-10	–	–	18-2	–	–	21-1
	16	–	–	10-2	–	–	12-11	–	–	15-9	–	–	18-3
	24	–	–	8-4	–	–	10-6	–	–	12-10	–	–	14-11
Hem-Fir	12	12-0	11-5	12-3	15-8	14-8	15-11	19-2	17-11	19-5	22-3	20-9	22-6
	16	10-9	10-0	10-10	13-7	12-8	13-9	16-7	15-6	16-10	19-3	18-0	19-6
	24	8-9	8-2	8-11	11-1	10-4	11-3	13-7	12-8	13-9	15-9	14-8	15-11
SYP	12	12-6	12-3	–	16-6	15-11	–	21-1	19-0	–	25-2	22-3	–
	16	11-5	10-8	–	15-0	13-9	–	18-3	16-5	–	21-9	19-3	–
	24	9-11	8-8	–	12-7	11-3	–	14-11	13-5	–	17-9	15-9	–

For more information: "How to Use Span Tables," JLC 4/94

Figure 2–81. Maximum Allowable Horizontal Spans (ft.-in.)

Rafters: 50 psf live load; 10 psf dead load. Snow region. Lightweight roof covering. Drywall ceiling. No Attic.

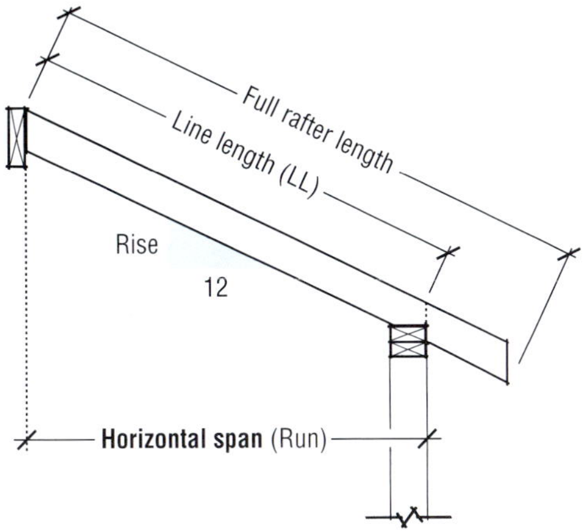

Species Group	Spacing (in.)	2x6 No. 1	2x6 No. 2	2x6 No. 1-2 (Can.)	2x8 No. 1	2x8 No. 2	2x8 No. 1-2 (Can.)	2x10 No. 1	2x10 No. 2	2x10 No. 1-2 (Can.)	2x12 No. 1	2x12 No. 2	2x12 No. 1-2 (Can.)
D-Fir-L	12	11-8	11-5	11-1	15-3	14-5	14-0	18-7	17-8	17-2	21-7	20-5	19-11
	16	10-5	9-10	9-7	13-2	12-6	12-2	16-1	15-3	14-10	18-8	17-9	17-3
	24	8-6	8-1	7-10	10-9	10-3	9-11	13-2	12-6	12-1	15-3	14-6	14-1
SPF	12	–	–	10-11	–	–	14-3	–	–	17-5	–	–	20-2
	16	–	–	9-9	–	–	12-4	–	–	15-1	–	–	17-6
	24	–	–	7-11	–	–	10-1	–	–	12-4	–	–	14-3
Hem-Fir	12	11-2	10-8	11-5	14-8	14-0	15-0	18-4	17-2	18-7	21-3	19-11	21-7
	16	10-2	9-7	10-4	13-0	12-2	13-2	15-11	14-10	16-1	18-5	17-3	18-8
	24	8-5	7-10	8-6	10-8	9-11	10-9	13-0	12-1	13-2	15-1	14-1	15-3
SYP	12	11-8	11-5	–	15-4	15-0	–	19-7	18-2	–	23-9	21-4	–
	16	10-7	10-2	–	13-11	13-2	–	17-6	15-9	–	20-10	18-5	–
	24	9-3	8-4	–	12-0	10-9	–	14-4	12-10	–	17-0	15-1	–

Figure 2–82. Maximum Allowable Horizontal Spans (ft.-in.)

Rafters: 50 psf live load; 15 psf dead load. Snow region. Medium-weight roof covering. Drywall ceiling. No Attic.

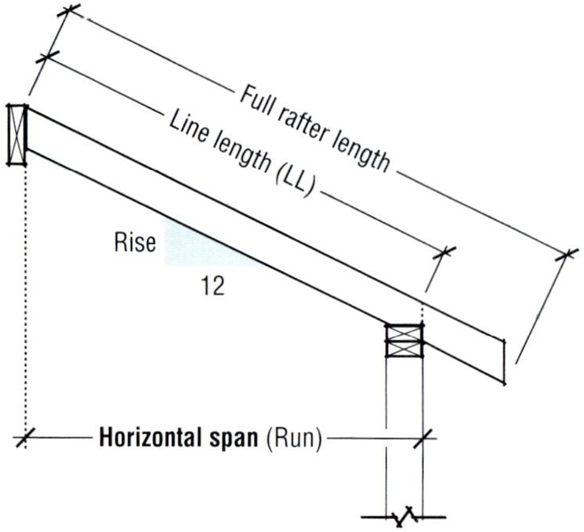

Species Group	Spacing (in.)	2x6 No. 1	2x6 No. 2	2x6 No. 1-2 (Can.)	2x8 No. 1	2x8 No. 2	2x8 No. 1-2 (Can.)	2x10 No. 1	2x10 No. 2	2x10 No. 1-2 (Can.)	2x12 No. 1	2x12 No. 2	2x12 No. 1-2 (Can.)
D-Fir-L	12	11-7	10-11	10-8	14-7	13-10	13-6	17-10	16-11	16-6	20-9	19-8	19-1
	16	10-0	9-6	9-3	12-8	12-0	11-8	15-6	14-8	14-3	17-11	17-0	16-6
	24	8-2	7-9	7-6	10-4	9-10	9-6	12-8	12-0	11-8	14-8	13-11	13-6
SPF	12	–	–	10-10	–	–	13-8	–	–	16-9	–	–	19-5
	16	–	–	9-4	–	–	11-10	–	–	14-6	–	–	16-9
	24	–	–	7-8	–	–	9-8	–	–	11-10	–	–	13-8
Hem-Fir	12	11-2	10-8	11-5	14-5	13-6	14-7	17-8	16-6	17-10	20-5	19-1	20-9
	16	9-10	9-3	10-0	12-6	11-8	12-8	15-3	14-3	15-6	17-9	16-6	17-11
	24	8-1	7-6	8-2	10-3	9-6	10-4	12-6	11-8	12-8	14-6	13-6	14-8
SYP	12	11-8	11-4	–	15-4	14-8	–	19-5	17-6	–	23-2	20-6	–
	16	10-7	9-10	–	13-11	12-8	–	16-10	15-1	–	20-0	17-9	–
	24	9-2	8-0	–	11-7	10-4	–	13-9	12-4	–	16-4	14-6	–

For more information: "How to Use Span Tables," JLC 4/94

Slope and Pitch

Calculating Roof Area

SLOPE AND PITCH

Slope is defined as a roof's vertical rise (in inches) per foot of horizontal run (**Figure 3-1**). For example, a 6:12, or "six-in-twelve," roof rises 6 in. for every ft. of run.

SLOPE = RISE/RUN

Pitch is different from slope; it's the ratio of rise to span. For example, a roof sloped at 6:12 has a pitch of 1/4.

PITCH = RISE/SPAN

Flat Roofs

Even "flat" roofs should have some slope. To prevent pools of standing water, they should slope at least 1/4 in. per ft.

Figure 3–1. Slope vs. Pitch

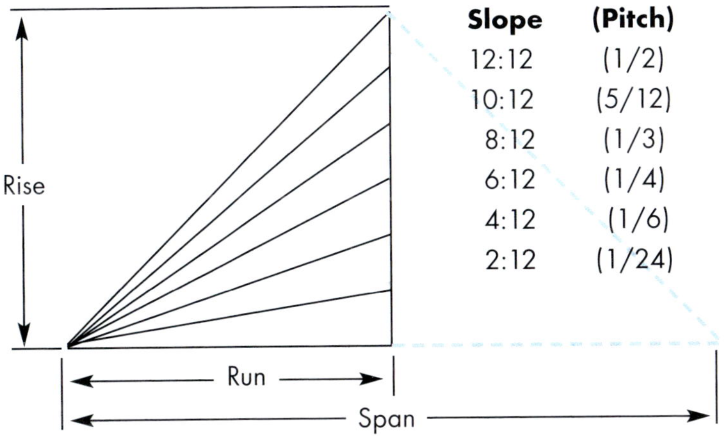

Slope	(Pitch)
12:12	(1/2)
10:12	(5/12)
8:12	(1/3)
6:12	(1/4)
4:12	(1/6)
2:12	(1/24)

Don't confuse "slope" and "pitch." Slope is in. of rise per ft. of run. Pitch is the ratio of rise to span.

CALCULATING ROOF AREA

Gable Roof Area

To find the length of a rake on a gable roof, multiply its horizontal distance (the roof's run) by the slope conversion factor in **Figure 3-2**, then multiply by the length of the ridge to find the area.

RAKE LENGTH = RUN × SLOPE FACTOR

GABLE ROOF AREA = RAKE LENGTH × RIDGE LENGTH

Hip/Valley Length

To find the length of a hip or a valley, multiply its horizontal distance (run) by the hip/valley factor in **Figure 3-2**.

HIP OR VALLEY LENGTH = RUN × HIP/VALLEY FACTOR

Figure 3–2. Slope, Hip/Valley Conversion

Slope (in. per ft.)	Conversion Factor Slope	Hip/Valley
4:12	1.054	1.452
5:12	1.083	1.474
6:12	1.118	1.500
7:12	1.157	1.524
8:12	1.202	1.564
9:12	1.250	1.600
10:12	1.302	1.642
11:12	1.356	1.684
12:12	1.414	1.732

Multiply a roof's run by the correct factor to get the length of its rake. To find the length of hips and valleys, multiply the correct factor by the horizontal distance of a hip or valley.

MATERIAL QUANTITY

Most roofing materials are sold by the square.

1 SQUARE = 100 SQ. FT. ROOF AREA

Order extra for hips, ridges, valleys, starter courses, and cutting waste, as described for each material.

Estimating Underlayment

- **No. 15 asphalt felt:** One roll (144 ft. x 36 in.) covers 4 squares using standard coverage (**Figure 3-7, page 213**) or 2 squares for half-lap applications (**Figure 3-8, page 214**).
- **No. 30 asphalt felt:** One roll (72 ft. x 36 in.) covers 2 squares using standard-coverage, or 1 square in half-lap applications.

Estimating Asphalt Shingles

Most asphalt shingles are sold three bundles per square (assuming a 5-in. weather exposure). After calculating the appropriate number of squares for the roof field, add the following:

- Add 10% to the total roof area for waste;
- Add one square of shingles for every 100 lin. ft. of hips, valleys, ridges, and starter courses;
- Add up the lengths of all hips and valleys and divide the sum by 20. Then add this many sq. ft. of shingles for additional waste.

Estimating Wood Shakes and Shingles

A nominal square is four bundles, and a nominal square of shakes is five bundles, but this depends on the course exposure (**Figure 3-4** and **Figure 3-5**, facing page). The steeper the slope, the more shingle that can be exposed.

- For starter courses, add one square for every 240 lin. ft. of eaves.
- For valleys, order one extra square per 100 lin. ft. of valley to account for waste.

Estimating Nails

To estimate nail quantities, see **Figure 3-3**.

Estimating Clay and Concrete Tile

- Field tile is usually sold in full pallets, and accessories in full boxes. Consult the manufacturer's information for the number of field tiles per square, the number of squares per pallet, and the number of tile accessories packaged per box.
- It takes about 90 concrete field tiles to cover a square of roof area when using a 3-in. top lap.
- Add 3% to field tile quantities for waste and breakage.

Figure 3–3. Fasteners per Square

Roofing Type	Nail Quantity Nail Type	(per square)
Asphalt Shingles		
New roof	$1\frac{1}{4}$" galvanized roofing	2 lbs.
Reroofing	$1\frac{1}{2}$"- $1\frac{3}{4}$" galvanized roofing	3 lbs.
Wood Shakes or Shingles*		
24"	4d shingle	2 lbs.
16"-18"	3d shingle	2 lbs.
Hand-split shakes	6d shingle	2 lbs.

* Note: Use only hot-dipped, stainless-steel, or aluminum nails for shakes and shingles. Do not use electrogalvanized or copper.

For more information: "Estimating Roofing Without Leaving the Ground," JLC 5/89

Figure 3–4. Wood Shingle Coverage

Length & Thickness*	Approximate Coverage of One Square (4 bundles) Weather Exposure (in.)								
	3 1/2	4	4 1/2	5	5 1/2	6	6 1/2	7	7 1/2
16" x 5/2"*	70	80	90	100**	–	–	–	–	–
18" x 5/2 1/4"	–	72 1/2	81 1/2	90 1/2	100**	–	–	–	–
24" x 4/2"	–	–	–	–	73 1/2	80	86 1/2	93	100**

* Sum of the shingles stacked; e.g., 5/2" means 5 butts = 2" thick.
** Maximum exposure recommended for roofs.

Adapted from Cedar Shake and Shingle Bureau

Wood Shingle Coverage

Figure 3–5. Wood Shake Coverage

Approximate Coverage (sq. ft.) of One Square (shakes applied with 1/2-in. spacing)

Shake type, length, and thickness	No. of bundles	Weather Exposure (in.)			
		5 1/2	7 1/2	8 1/2	10
18" x 1/2" Hand split-and-resawn mediums	4	55*	75**	–	–
18" x 3/4" Hand split-and-resawn heavies	5	55*	75**	–	–
24" x 1/2" Hand split-and-resawn mediums	4	–	75*	85	100**
24" x 3/4" Hand split-and-resawn heavies	5	–	75*	85	100**
24" x 1/2" Tapersplit	4	–	75*	85	100**
18" x 3/8" Straight-split	5	65*	90**	–	–
24" x 3/8" Straight-split	5	–	75*	85	100**

* Maximum recommended exposure for 3-ply roof.
** Maximum recommended exposure for 2-ply roof.

Adapted from Cedar Shake and Shingle Bureau

Wood Shake Coverage

THE TERM ASPHALT SHINGLES REFERS TO BOTH ORGANIC-asphalt and fiberglass shingles. The main difference is in the mat, or base sheet. Unless otherwise specified, all references to asphalt shingles here include both types. For general information on asphalt shingles, refer to **Figure 3-6**.

SELECTING ASPHALT SHINGLES

When choosing asphalt shingles look for:

Organic vs. Fiberglass Mat

Asphalt shingles may have either an organic felt or fiberglass mat. Organic-mat shingles are generally thicker, more tear resistant, and better able to withstand expansion and contraction. With a fiberglass shingle, a heavy mat or double-layer mat is desirable for added strength.

Figure 3–6. Typical Asphalt Shingles

	Tabbed	Multi-Tabbed	Laminated	No Cutout
Configuration	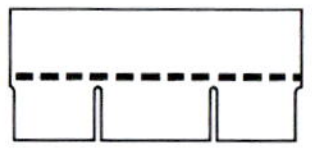	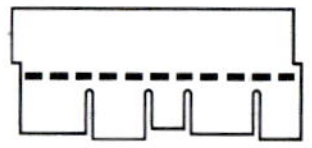	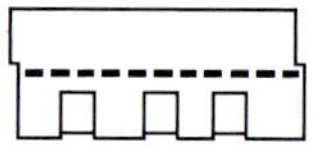	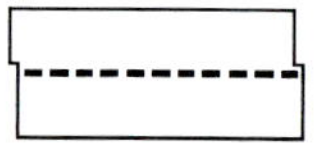
Width	12 to 13 1/4"	12 to 17"	11 1/2 to 14 1/4"	12 to 13 1/4"
Length	36 to 40"	36 to 40"	36 to 40"	36 to 40"
Low Slope(1)	2:12 to 4:12	2:12 to 4:12	2:12 to 4:12	2:12 to 4:12
Normal Slope	4:12 to 21:12	4:12 to 21:12	4:12 to 21:12	4:12 to 21:12
Steep Slope(2)	21:12 & beyond	21:12 & beyond	21:12 & beyond	21:12 & beyond
Exposure	5 to 5 5/8"	4 to 7 1/2"	4 to 6 1/8"	5 to 5 5/8"
Approx. ship. wt. per square	200 to 300 lbs.	240 to 300 lbs.	240 to 360 lbs.	200 to 300 lbs.
No. of shingles per square	65 to 80	65 to 80	64 to 90	65 to 81
Bundles per square	3 or 4	3 or 4	3 or 4	3 or 4
Fire/wind ratings	A or C fire rating; wind resistant	A or C fire rating; wind resistant	A or C fire rating; wind resistant	A or C fire rating; wind resistant
Life expectancy(3)	15-25 yrs.	15-25 yrs.	20-30 yrs.	15-25 yrs.

(1) See **Figure 3-8**, page 214.
(2) Use six fasteners per shingle and place two spots of quick-setting asphalt cement under each tab.
(3) Life expectancy depends on weight and tear strength of shingle.

Adapted from Asphalt Roofing Manufacturers Assoc.

For more information: "Choosing an Asphalt Shingle: Organic vs. Fiberglass," JLC 5/93

"Have Asphalt Shingles Improved?" JLC 7/01

Shingle Specs

Shingle Weight

In an **organic-asphalt shingle,** weight is a good measure of shingle quality. A 210-lb. (per square) shingle should last about 15 years, a 230- to 240-lb. shingle should last 20 years and a 320-lb. shingle should last 25 years.

In a **fiberglass** or high-profile **"architectural"** shingle, however, weight is not a good measure of quality because the weight may be due to heavy mineral fillers that do not add strength.

Fire Ratings

Use only Class-A rated shingles for residential construction.

Tear Resistance

Shingle strength is largely a function of the quality of the mat and the quality of the asphalt. Use only ASTM D3462 rated shingles. Fiberglass shingles should also have a UL listing.

Reroofing

Over existing shingles, use traditional organic-asphalts, because they're thicker and much less likely than fiberglass to let the old roof "telegraph" through.

Figure 3–7. Standard Underlayment (4:12 and steeper)

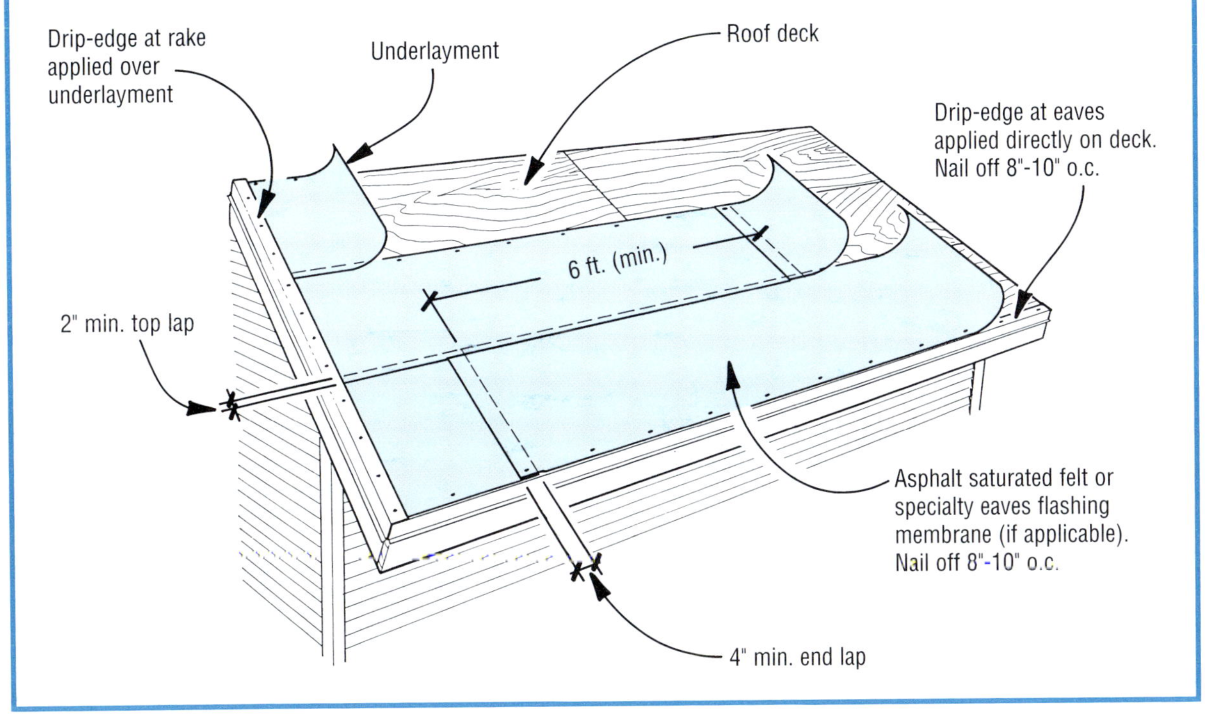

Adapted from Asphalt Roofing Manufacturers Assoc.

Use single coverage underlayment on normal slopes (over 4:12). In regions where ice damming is a problem, use eaves flashing membrane (see Figure 3-66, page 270).

UNDERLAYMENT FOR ASPHALT SHINGLES

Codes and warranties require underlayment. It keeps the house dry until the roof is finished, provides an extra barrier against wind-driven rain and ice dams, and separates shingles from resins in wood sheathing.

Underlayment Weight

Roofing felt is classified by its weight per square. Use 15-lb. in the field and 30-lb. to line valleys. Thirty-pound felt is stiff and may telegraph wrinkles if used in the field.

Underlayment for Steep Roofs

On roofs 4:12 and steeper, use one layer of 15-lb. roofing felt, lapped at least 2 in. at overlaps and 4 in. at end laps (**Figure 3-7**).

Underlayment for Low-Slope Roofs

On roofs from 2:12 to 4:12, start with a 19-in. starter course, then lap succeeding courses by 19 in. (**Figure 3-8**). With a standard 36-in. roll, this will provide double coverage. For cold-weather treatment, see "Eaves Flashing," **Figure 3-66, page 270**.

FASTENERS FOR ASPHALT SHINGLES

Most blow-offs are caused by poorly applied, or badly chosen, fasteners.

Nails

Nails should penetrate thick plank decking by 3/4 in. minimum; they should go all the way through thin sheathings. See **Figure 3-9** for proper nailing depth.

For more information: "Leakproof Details for Shallow Roofs," JLC 5/94

"Asphalt Shingle Q&A," JLC 5/02

*For low-slope roofs (from 2:12 to 4:12), use two layers of roofing felt. In snowy regions, use peel-and-stick eaves flashing, as shown in **Figure 3-12, page 218**.*

Figure 3–8. Low-Slope Underlayment

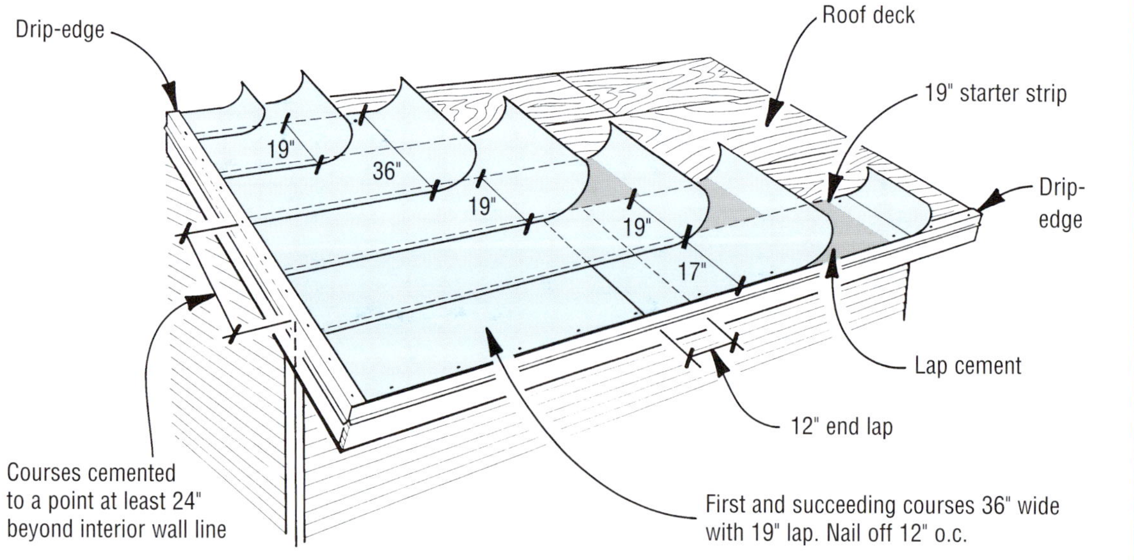

Adapted from Asphalt Roofing Manufacturers Assoc.

Staples

- Avoid using staples on asphalt roofs, particularly in high-wind areas; hand-driven or air-driven roofing nails are stronger.
- If you insist on using staples, adjust the staple gun so that the staple's crown rests evenly against the shingle's surface without breaking it (**Figure 3-9**). Any more or less, and it won't hold the shingle reliably in place.

Nailing patterns for low and normal slopes. The nailing for asphalt shingles depends on roof slope. Fasten with four nails, directly above each cutout and 1/2 in. in from each end (**Figure 3-10**).

Nailing pattern for steep slopes. Fasten with six nails on steep roofs, such as mansards, that slope more than 21 in. per ft.

Seal each tab with a 1-in.-diameter dab of asphalt adhesive cement placed 2 in. above the butt at the center of each tab.

High winds. Use six nails per shingle in hurricane-prone areas, plus a quarter-size dab of asphalt adhesive under each tab corner (a total of six dabs).

Install shingles in warm weather so the self-sealing strip is soft enough to stick.

SHINGLE INSTALLATION

Starter Course

The starter course along the eaves sheds water that may migrate through the cutouts and joints of the first full shingle course.

- Use self-adhering asphalt shingles for starter courses.

Figure 3-9. Proper Nailing and Stapling

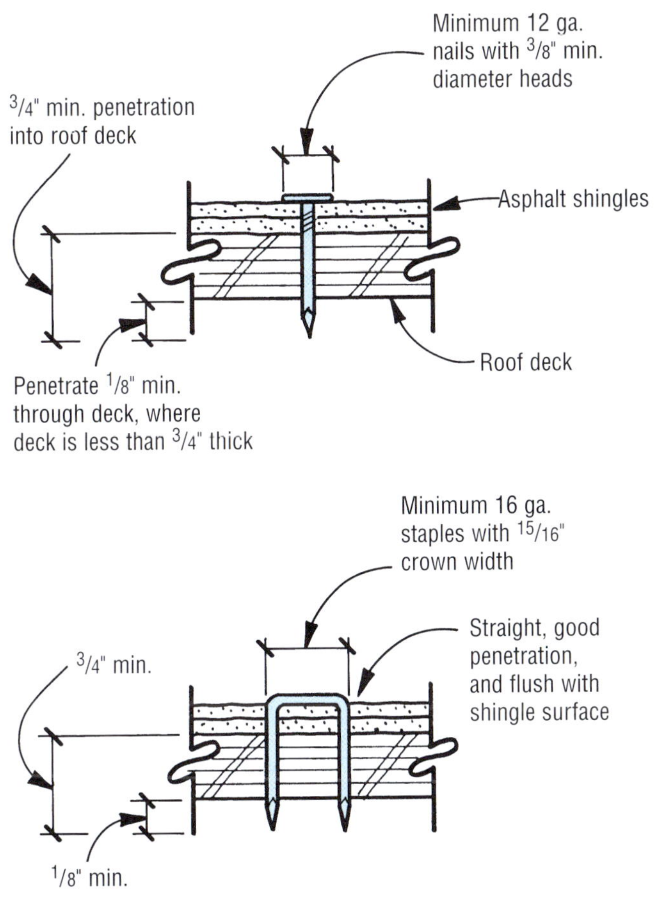

Adapted from Asphalt Roofing Manufacturers Assoc.

In order for fasteners to be effective, the nail head (or the crown of the staple) must bear tightly against the shingle without penetrating it. Any more or less and it will loosen over time. When reroofing over shingles, fasteners must penetrate the roof deck 3/4 in. minimum. Use 1 1/2- to 1 3/4-in.-long roofing nails.

Underlayment for Asphalt Shingles

Fasteners for Asphalt Shingles

- Cut off the tabs and use the remaining part of the shingle, installing it with the seal-strip along the bottom edge.
- Starter shingles should overlap the eaves and rake edges 1/4- to 3/4-in. depending on the shingle manufacturer's recommendations.

Figure 3–10. Nailing Patterns

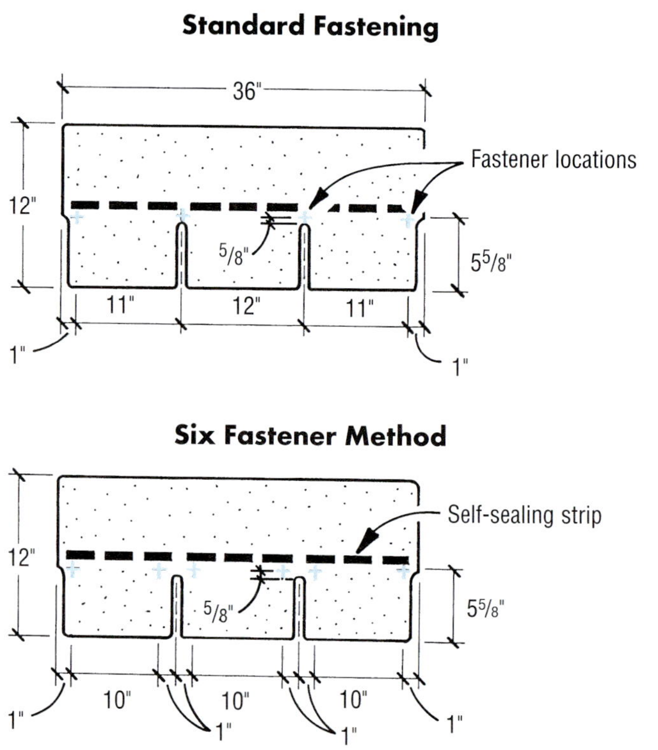

Do not nail onto or above sealing strip, or shingles may not seal properly. Use six nails on steep slopes or in hurricane-prone areas.

- Nail the starter strip 2 1/2- to 4-in. above the overhanging edge, and 6- to 9-in. o.c.

First Full Course

Start the first course with a full shingle, and succeeding courses with varying cut lengths to produce the familiar "stair-stepped" pattern.

Stacking Bundles

Lay bundles flat on the roof; don't drape over the ridge or edges. Also distribute bundles over roof area to avoid creating concentrated loads.

Alignment Guides

Most shingles include 1/2-in. alignment slits across the top edge and half-slots along the edges. Use these to line up adjacent shingles and successive courses. In addition, shadow marks may run the length of each shingle near the top of the slots. These are meant as an aid to horizontal alignment.

Shingle Offset

How well three-tab shingles hold up over time depends to some degree on the patterns used to lay them. Although many customers prefer to see the cutouts aligned on the roof, this directs runoff into channels that erode the shingle surface.

There are four basic patterns to laying out three-tab shingles: straight-up method, half-pattern, 4-in. offset, and 5-in. offset (**Figure 3-11**). The 5-in.-offset pattern provides the best all-around protection.

Installation Temperature

The proper temperature range for applying asphalt shingles is 40°F to 80°F. Above or below that range they're either

Installing Asphalt Shingles

Offset Options

Figure 3–11. Shingle Offset

A. Straight-Up Method

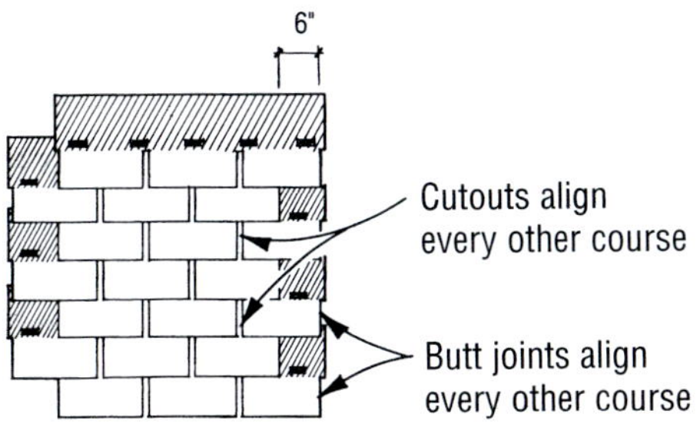

The straight-up method of laying shingles offsets alternating courses 6 in. from a vertical line up the roof (half a tab width for metric shingles). The alignment of butt joints and cutouts in every other course, however, can cause premature weathering of surface granules.

B. Half-Pattern

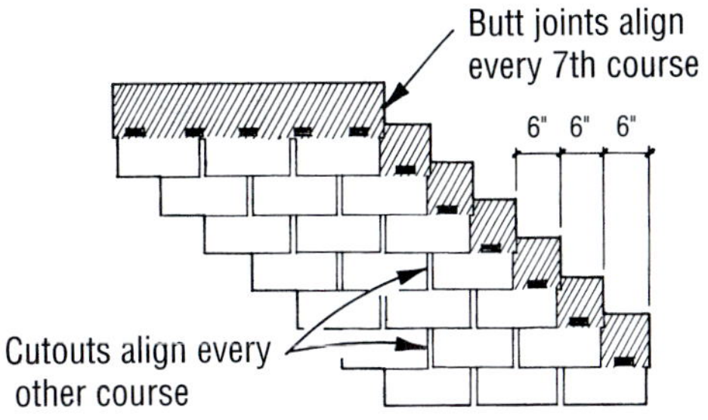

A half pattern (also called half-tab or sixes) staggers each course with a 6-in. offset (half a tab width for metric shingles). Butt joints are better protected from water migrating horizontally, but cutouts still align every other course.

C. 4-In. Offset

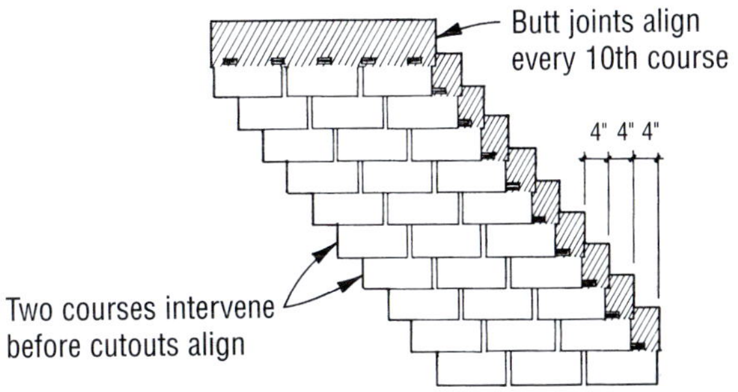

A 4-in. offset keeps cutouts separated by two courses, and butt joints align every ten courses. But the short lap doesn't provide enough protection in wet climates, or in areas with severe freeze-thaw cycles.

D. 5-In. Offset

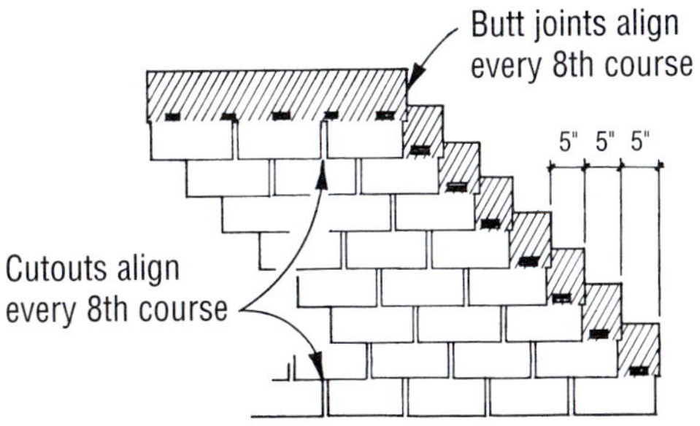

A 5-in. offset provides the best all-around protection. Both the cutouts and the butt joints align only every eight courses, so runoff is less likely to cut channels into the shingle granules. It also hides shingle irregularities as well.

too brittle or too soft, and the sealing strips (or dabs of adhesive) do not seal well and can lead to blow-offs in windy areas. If installed in cool weather, however, shingles will seal when warmer air and sunlight hit them.

Low-Slope Asphalt-Shingle Roofs

You can use asphalt shingles on roofs as shallow as 2:12. However, for slopes below 4:12, most codes insist that the shingles be installed over double coverage roofing felt (**Figure 3-8, page 214**) with a bituminous eaves flashing.

- For low-slope roofs, use shingles that are labeled "wind-resistant," or use standard asphalt shingles and put a dab of asphalt adhesive cement beneath each tab.
- For low-slope roofs (shed dormers, for example) on sites subject to heavy wind-driven rain or heavy snow loads, cover the entire roof deck with a bituminous eaves membrane before applying shingles (**Figure 3-12**).

Figure 3–12. Waterproof Deck

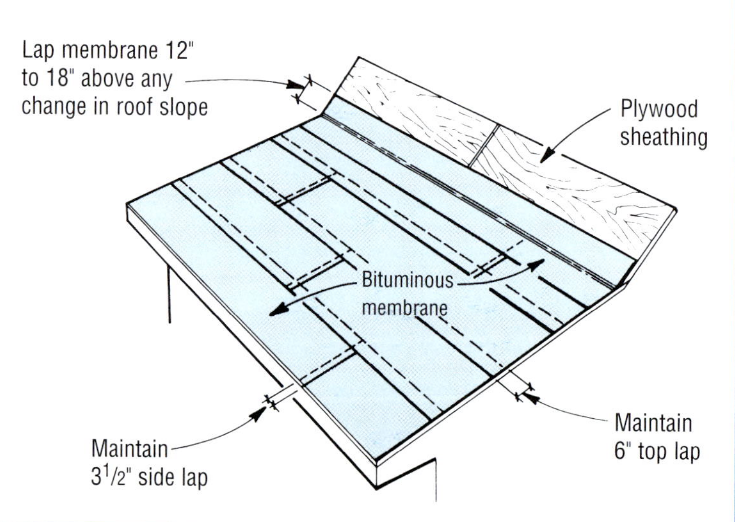

In snowy regions, shallow slope roofs (below 2:12) and low-slope roofs (4:12 to 2:12) should be waterproofed with a peel-and-stick bituminous membrane. Laps between membrane layers should be at least 6 in. along the top edge and at least $3^1/2$ in. on the sides.

For more information: "Reroofing Over Asphalt Shingles," JLC 5/99

Windy Locations

In very windy locations, you can special-order wind-resistant shingles or put a dab of asphalt adhesive cement beneath each shingle tab. In hurricane-prone areas, use six nails per shingle.

REROOFING

When to Strip

If more than 10% of an old roof needs repair, the best solution is usually a new roof.

Code Requirements

Most codes allow either two to three layers of asphalt or just one layer of asphalt over a layer of wood shingles. Some limit you to one layer of roofing, period. Check local code to be sure.

Incompatible Materials

The following materials just won't work together:

- **Asphalt over slate.** (Yes, it's been done.) Because of its high thermal mass, the slate acts as a heat sink. The asphalt shingles are cooked from the back. This dries them out and cuts their service life in half.
- **Asphalt over metal.** The results are the same as putting asphalt over slate, but the damage is done by radiant heat that's reflected back from the surface of the metal.

Reroofing

Layering Asphalt Shingles

Figure 3–13. Maximum Rafter Span (inches) for Multiple Layers of Asphalt Shingles

		Two Layers of Shingles								Three Layers of Shingles							
Rafter Spacing		16″ o.c.				24″ o.c.				16″ o.c.				24″ o.c.			
Lumber Grade		#1	#2	#3	U	#1	#2	#3	U	#1	#2	#3	U	#1	#2	#3	U
Size	**Live Load**																
2x4	10 lb.	[126]	[123]	99	85	[110]	[107]	82	70	[126]	[121]	92	79	104	100	76	65
	20 lb.	[100]	[97]	80	68	[87]	[85]	65	56	[100]	[97]	76	65	86	82	62	53
	30 lb.	[87]	[85]	69	59	[76]	74	56	48	[87]	[85]	66	57	75	71	54	46
	40 lb.	[79]	[77]	61	52	69	66	50	43	[79]	[77]	59	51	67	64	49	42
	50 lb.	[74]	[72]	56	48	63	60	46	39	[74]	[71]	54	47	61	59	44	38
2x6	10 lb.	197	188	143	133	163	156	118	101	183	175	133	114	151	144	110	94
	20 lb.	[157]	152	116	99	131	125	95	82	151	145	110	94	125	119	91	78
	30 lb.	137	131	100	86	113	108	82	70	132	127	96	82	109	104	79	68
	40 lb.	122	117	89	76	100	96	73	63	119	114	86	74	97	93	71	61
	50 lb.	111	107	81	70	91	88	66	57	109	104	79	68	89	85	65	56
2x8	10 lb.	246	236	179	153	204	196	148	127	229	219	166	143	190	182	138	118
	20 lb.	200	191	145	125	165	158	120	103	190	182	138	119	157	150	114	98
	30 lb.	173	165	125	108	142	136	103	89	166	159	121	104	137	131	99	85
	40 lb.	154	148	112	96	127	121	92	79	150	143	109	93	123	118	89	77
	50 lb.	141	135	102	88	115	111	84	72	137	131	100	86	113	108	82	70
2x10	10 lb.	296	284	215	185	247	237	180	154	276	265	201	172	230	220	167	143
	20 lb.	242	232	176	151	200	192	146	125	231	221	168	144	191	183	139	119
	30 lb.	209	201	152	131	173	165	126	108	202	193	147	126	167	159	121	104
	40 lb.	187	179	136	117	154	148	112	96	182	174	132	113	150	143	109	93
	50 lb.	171	164	124	107	141	135	102	88	167	160	121	104	137	131	100	86
2x12	10 lb.	339	325	246	212	284	272	206	177	317	303	230	198	264	253	192	165
	20 lb.	278	266	202	173	231	221	168	144	265	254	193	165	220	211	160	137
	30 lb.	241	231	175	150	200	191	145	124	233	223	169	145	192	184	140	120
	40 lb.	216	207	157	135	178	171	129	111	210	201	153	131	173	166	126	108
	50 lb.	197	189	143	123	163	156	118	101	193	185	140	120	159	152	115	99

Notes: Chart assumes asphalt shingles @ 3 psf per layer, plywood @ 1.5 psf, L/240 deflection. Rafter spans in brackets [ ] are controlled by deflection.

To use the chart, go down the left-hand column to the correct row for the rafter size and snow load for your area. Move across the chart to the correct rafter spacing and lumber grade to find the maximum length that the rafters can span. The chart is based on Douglas fir grades 1, 2, and 3. In cases where no grade stamp is available, use the "U" column for unknown species. Span is measured horizontally, not on the slope. This chart is not applicable to cathedral ceilings, which have ceiling materials attached to rafters.

Preparing To Reroof Checklist

- Repair all leaks, damaged framing members, and any damaged or rusted flashing.
- Even out the substrate by filling in any missing areas of old roofing, nail down all loose or curled shingles, and remove or drive home all loose nails.
- Install drip edge along the rakes and eaves, and eaves flashing if required.
- Over 4:12, reroofing usually doesn't require a new underlayment; the existing roof should serve the same purpose. However, some codes may require a new underlayment on shallower slopes.

ASPHALT OVER ASPHALT

Some codes may permit three layers of asphalt, but it's usually better to have no more than two.

Do Not Use Too Many Layers

There are several reasons not to install too many layers of asphalt shingles:

- The layers may not provide adequate lateral support for nails, increasing the chance of a blow-off;
- The roof retains more heat, reducing the new shingles' life expectancy;
- The total weight might exceed the roof's design load. See **Figure 3-13, previous page**, to evaluate the roof structure.

Use **organic-mat shingles** for second layer. They're thicker than fiberglass shingles, so they won't telegraph the old roof contours as readily.

Use **nails, not staples,** to fasten the second layer.

Do not reroof over **high-profile shingles** because of the added weight and the difficulty of producing an even surface.

NESTING PROCEDURE FOR REROOFING

The nesting procedure described in **Figure 3-14** minimizes any unevenness that might result from installing new shingles. It also ensures that the new horizontal fastening pattern is 2 in. below the old one.

Starter Strip

- Cut the starter strip shingles wide enough to overhang the eaves and carry water into the gutter. Usually this means removing the tabs plus 2 in. from the top, so that the remaining portion is equal in width to the exposure of the old shingles (normally 5 in.).
- Apply the starter strip so that it is even with the existing roof at the eaves. Do not overlap the existing course above.
- Remove 3 in. from the rake end of the first starter strip shingle to ensure that joints between adjacent starter strip shingles will be covered when the first course is applied.

First Course

- Cut 2 in. or more from the butts of the first course of shingles so that the shingles fit between the butts of the existing third course and the eaves edge of the new starter strip.
- Start at the rake with a full-length shingle.
- Use four fasteners per shingle, locating them in the same positions as in new construction — 5/8 in. over the cut-outs and 1 in. in from the sides (**Figure 3-10, page 216**). Do not fasten into or above the adhesive sealing strip.

Installing Asphalt Over Asphalt

Second and Succeeding Courses

- Use full-width shingles.
- Remove 6 in. from the rake end of the first shingle in each succeeding course, through the sixth.
- Place the top edge of the new shingles against the butt edge of the old shingles in the course above.
- Repeat the cycle by starting the seventh course with a full-length shingle.

Nesting Procedure for Reroofing

Figure 3–14. Reroofing Using the Nesting Procedure

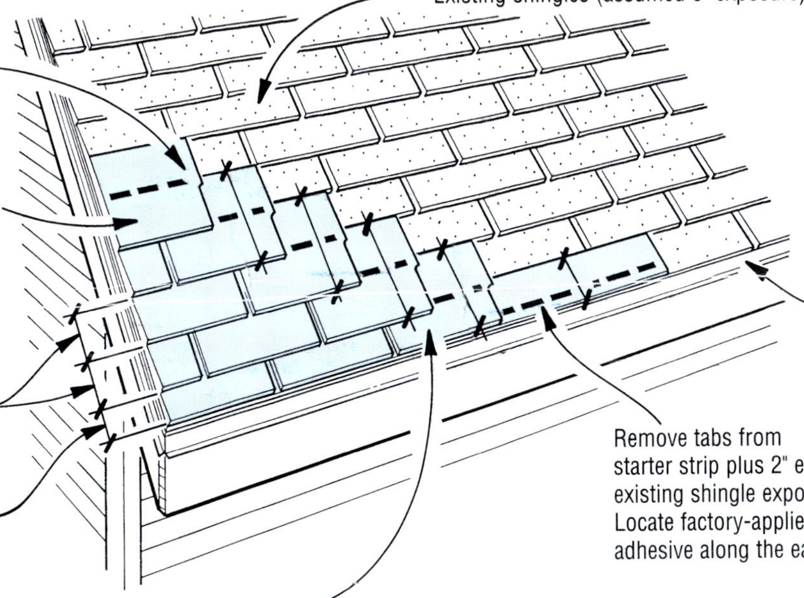

Adapted from Asphalt Roofing Manufacturers Assoc.

When applying new asphalt shingles over existing asphalt shingles using the nesting procedure, first provide a 5-in. starter strip. Second, cut the first shingle course width down to 10 in. Then apply second and succeeding courses using the half-pattern method.

GRADES AND TYPES

Wood shingles are sawn on both sides to produce a smooth, uniform surface. Shakes, on the other hand, are split from logs, leaving both sides rough (see **Figure 3-15** for both shake and shingle types).

Lengths and Weights

Shingles come in three lengths: 16-in. (sometimes called 5X), 18-in. (perfections), and 24-in. (royals). See **Figure 3-16** for specification information.

Shakes come in 18- and 24-in. lengths, as well as in a 15-in.-long starter/finish shake. They're available in two weights: mediums have 1/2-in. butts; heavies have 3/4-in. butts on average.

EXPOSURES AND SLOPE

Normal-exposure shingles and shakes can be used on slopes as low as 4:12.

Low Slopes

- For roofs from 3:12 to 4:12 you can use shingles if you reduce the exposure as noted in **Figure 3-16.**
- For slopes lower than 3:12 (4:12 for shakes), you need to first create a waterproof deck, then lay vertical strapping up the roof with spaced sheathing over it (**Figure 3-20, page 227**). The wood roof on top, in this case, is purely decorative.

For more information: "Long-Lasting Wood Roofs," JLC 5/97

"Wood Shakes and Shingles," JLC 10/92

Figure 3–15. Types of Wood Shingles and Shakes

Adapted from Cedar Shake and Shingle Bureau

No. 1 Blue Label®
The premium grade of shingles for roofs and sidewalls. These top-grade shingles are 100% heartwood, 100% clear, and 100% edge-grain.

No. 2 Red Label
A good grade for many applications. Not less than 10" clear on 16" shingles, 11" clear on 18" shingles, and 16" clear on 24" shingles. Flat grain and limited sapwood are permitted in this grade.

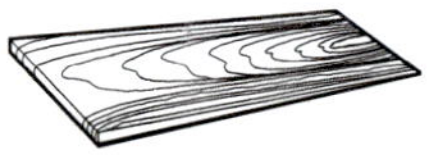

No. 3 Black Label
A utility grade for economy applications and secondary buildings. Not less than 6" clear on 16" and 18" shingles, 10" clear on 24" shingles.

No. 4 Undercoursing
A utility grade for starter course undercoursing.

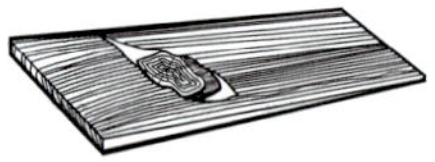

No. 1 Hand-split & Resawn
These shakes have split faces and sawn backs. Cedar logs are first cut into desired lengths. Blanks or boards of proper thickness are split and then run diagonally through a bandsaw to produce two tapered shakes from each blank.

No. 1 Certi-Sawn®
These shakes are sawn both sides. No. 2 and 3 are also available.

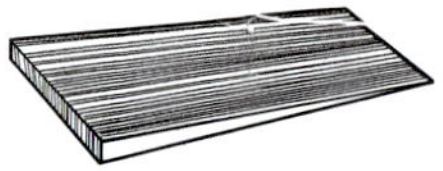

No. 1 Tapersplit
Produced largely by hand, using a sharp-bladed steel froe and a mallet. The natural shingle-like taper is achieved by reversing the block, end-for-end, with each split.

No. 1 Straightsplit
Produced by machine or in the same manner as tapersplit shakes except they are split from the same end of the block, so the shakes have the same thickness throughout.

Snowy Regions

Use a double layer of felt (preferably cemented together) at the eaves, or a bituminous eaves flashing (see "Eaves Flashing," page 270).

FIRE RETARDANTS

Fire-Rated Materials

The only shingles with permanent fire-resistance are pressure-treated at the factory and carry the Cedar Shake and Shingle Bureau's Certi-Guard label.

Shingle/Shake Grades

Shingle/Shake Exposure

Figure 3–16. Typical Wood Shingles and Shakes

Shingles

Grade	Length	Exposure 3:12 to 4:12	Exposure 4:12 and beyond	Courses per Bundle	Bundles per Square
No. 1	16"(1)	3 3/4"	5"	20/20	4 bundles
Blue Label	18"(2)	4 1/4"	5 1/2"	18/18	4 bundles
No. 2	24"(3)	5 3/4"	7 1/2"	13/14	4 bundles
Red Label	16"(1)	3 1/2"	4"	20/20	4 bundles
	18"(2)	4"	4 1/2"	18/18	4 bundles
	24"(3)	5 1/2"	6 1/2"	13/14	4 bundles
No. 3	16"(1)	3"	3 1/2"	20/20	4 bundles
Black Label	18"(2)	3 1/2"	4"	18/18	4 bundles
	24"(3)	5"	5 1/2"	13/14	4 bundles
No. 4 Undercoursing	16"	N/A	N/A	14/14 or 20/20	2 bundles
	18"	N/A	N/A	14/14 or 18/18	2 bundles
No. 1 or No. 2	16"(1)	3 3/4"	5"	33/33	1 carton
Rebutted/	18"(2)	4 1/4"	5 1/2"	28/28	1 carton
Rejointed	24"(3)	5 3/4"	7 1/2"	13/14	4 bundles

(1) 5X; (2) perfections; (3) royals

Slope – 4:12 to 12:12 is normal. 3:12 is considered minimum. For 3:12 to 4:12, reduce exposure (see exposure column). Below 3:12, create a waterproof deck (see **Figure 3-12, page 218**).

Life Expectancy – 8 to 25 years (See note at right regarding life expectancy.)

Shakes

Grade	Length & Thickness	Maximum Exposure (4:12 & steeper)	Courses per Bundle	Bundles per Square
			18" pack*	
No. 1	15" starter-finish	7 1/2" (for 18")	9/9	5
Hand-split	18" x 1/2" to 3/4"		9/9	5
& Resawn	18" x 3/4" to 1 1/4"		9/9	5
	24" x 3/8"*	10" (for 24")	9/9	5
	24" x 1/2" to 3/4"		9/9	5
	24" x 3/4" to 1 1/4"		9/9	5
No. 1 Tapersplit	24" x 1/2" to 5/8"	10"	9/9	5
			20" pack	
No. 1	18" x 3/8" *	7 1/2"	14 straight	4
Straightsplit	18" x 3/8"	7 1/2"	19 straight	5
	24" x 3/8"	10"	16 straight	5

*True-edge (24 x 3/8" hand-split shakes limited to 5" maximum weather exposure, per U.B.C.)

Slope – 4:12 to 12:12 is normal. Below 4:12, create a waterproof deck (see **Figure 3-12, page 218**).

Life Expectancy – 10 to 30 years

Note: Wood can be broken down by excessive moisture as well as by moss and other fungi, both of which are more common on north-facing surfaces. In the absence of moisture problems, however, the south side of a roof tends to degrade quicker because of its higher temperature and increased UV exposure. You can get a 30-year warranty if shingles or shakes have been pressure-treated at the factory.

Adapted from Cedar Shake and Shingle Bureau

Certi-Guard shingles are available with a Class-B or Class-C fire rating at about twice the price of untreated shingles.

Field-applied fire retardants give protection for six months to one year at most. You can get comparable results from treating with a high-phosphate fertilizer, which is inexpensive and non-corrosive to flashings. Phosphates give a Class-C fire rating for about six months (they'll take up to 30 in. of rainfall before losing their fire resistance). Re-apply every spring and fall.

UNDERLAYMENT

A permeable 15# or 30# felt is typically installed only over the solid sheathing area at eaves (**Figure 3-17**).

Figure 3-17. Sheathing and Underlayment for Wood Roofs

	Sheathing	Interlayment	Underlayment(1)
Wood Shingles	Spaced(3)	none	15# or 30# felt over section of solid sheathing at eaves
	Solid(2 & 4)	none	15# or 30# felt at eaves. Optional over field of roof
Wood Shakes	Spaced(3)	30# asphalt-saturated felt	30# felt over section of solid sheathing at eaves
	Solid(2 & 4)	30# asphalt-saturated felt	30# felt at eaves

(1) Underlayment at eaves should extend to a point 12–24 in. inside building. Bituminous eaves flashing recommended in cold, snowy climates.
(2) Use to protect against windblown snow or where required for structural reasons. May shorten life of shingles and shakes.
(3) Recommended in hot, humid climates.
(4) Cedar Breather® is recommended to provide ventilation between underlayment and shingles.

For more information: "Protecting Wood Roofs from Fire," JLC 8/94

SHEATHING OPTIONS FOR SHINGLES

It is best to install wood roofing over spaced sheathing — evenly spaced 1x4s or 1x6s (**Figure 3-18**). Shingles installed on solid sheathing can't dry out as well. Because their back sides stay wet while the sun beats on their face, they decay faster and are more likely to cup, curl, and split.

Spaced Sheathing

- One option is to space 1x4s to coincide with the weather exposure of the shingles as shown in **Figure 3-19**.
- The other option is to use 1x6s such that two courses of shingles are nailed to each 1x6, as shown in **Figure 3-18**.

Solid Sheathing

Solid sheathing is required for wood shingles only for structural reasons — such as in seismic zones — or for protection against wind-blown snow. If the job demands solid sheathing, strap the roof or use a ventilating underlayment, as follows:

Strapping Over Plywood

- Sheathe the roof deck with plywood;
- Cover the plywood with 15- or 30-lb. asphalt-impregnated felt paper;
- Install horizontal 1x3 strapping over the plywood, cutting 1/2-in. notches at 4 ft. o.c. on the bottom of the strapping for drainage;
- Install the shingles over the strapping, following the installation details for spaced sheathing. An alternative to notching is to lay vertical strapping first, and then horizontal strapping.

- The second strip starts two exposure widths up from the bottom edge of the roof.
- Install the remaining strips on centers that equal the weather exposure. Make sure the top edge of each layer of felt rests on a section of sheathing, or it won't provide a good baffle.

Underlayment on Shallow Roofs

CABO sets a minimum slope of 3:12 for wood shingles and 4:12 for shakes. For lower slopes, NRCA and the Cedar Shake and Shingle Bureau recommend installing wood roofing on a lattice-like framework of pressure-treated strapping over a watertight membrane (**Figure 3-20**).

FASTENERS FOR SHINGLES AND SHAKES

Shingles and shakes can be fastened with nails or staples, as follows:

Nails

- Use only corrosion-resistant, hot-dipped galvanized, stainless-steel, or aluminum nails.
- For hip and ridge caps, use nails two sizes larger than in the field (**Figure 3-21**).
- Drive nails so they rest on the surface of the wood. Nails that crush the surface will encourage the shingle to split (**Figure 3-22**).

Interlayment for Shake Roofs

Fasteners for Shingles and Shakes

Figure 3–20. Low-Slope Shingle and Shake Installation

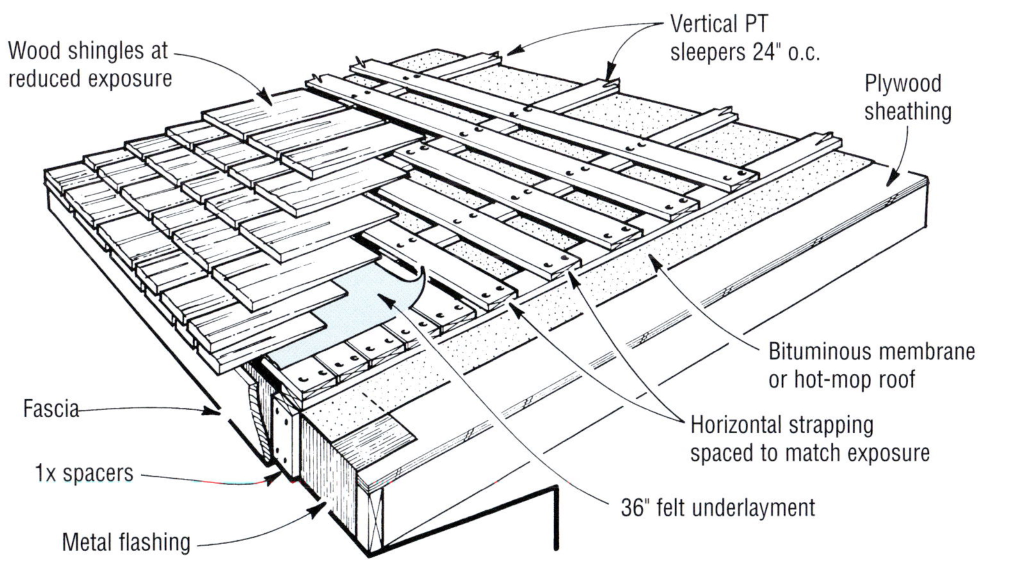

On shallow slopes, install wood shingles and shakes over a watertight subroof and a framework of pressure-treated strapping. The fascia is spaced out from the metal flashing at the eaves to allow any water that gets under the roofing to drain from the subroof.

- Wide shakes and shingles will eventually split. To avoid problems, make a vertical score down its center with a utility knife and drive a nail on either side of the score. Treat the score as a joint, offsetting it at least 1½ in. from any joints directly above or below.
- Refer to **Figure 3-23** for nailing patterns.

Staples

- Use only aluminum or Types 304 or 316 stainless-steel fasteners. Galvanized staples on wood roofs will rust.
- When using a pneumatic nailer, it's important to adjust it so that the staples don't tear the wood fibers, as this will reduce their holding power.

INSTALLING SHINGLES AND SHAKES

Keyway

Like all wooden materials, shingles and shakes swell when wet. To compensate, make a "keyway" by separating shingles by ¼ in. to ⅜ in. (about the width of a pencil), or by separating shakes ⅜ in. to ⅝ in. This keyway can be omitted when working with wet shingles, since they'll shrink rather than expand.

Rake Treatment

To eliminate dripping water at gable end walls, nail a clapboard along each rake under the shingles or shakes. The flair will direct water toward the center of the roof.

Ridges and Hips

- Cap ridges and hips to ensure a weathertight joint.
- Use either premanufactured ridge and hip caps (sold in bundles) or assemble them on site. Either type must have alternate overlaps and concealed nailing.
- Weather exposures should be the same as for shingles or shakes in the field of the roof.
- Nails must penetrate ¾ in. into the sheathing or completely pass through it (**Figure 3-24**).

Vented Ridges

To vent wood shingle roofs, hold the plywood sheathing back at the ridge, leaving a 1-in. gap (**Figure 3-25**).

Figure 3-21. Nails for Shingles and Shakes

Shingles	Nail Type	Minimum Length
16" and 18"	3d box	1¼"
24"	4d box	1½"
Shakes		
18" Straight-Split	5d box	1¾"
18" and 24" Hand-split-and-Resawn	6d box	2"
24" Tapersplit	5d box	1¾"
18" and 24" Taper-sawn	6d box	2"

Adapted from Cedar Shake and Shingle Bureau

Figure 3-22. Nailing Depths

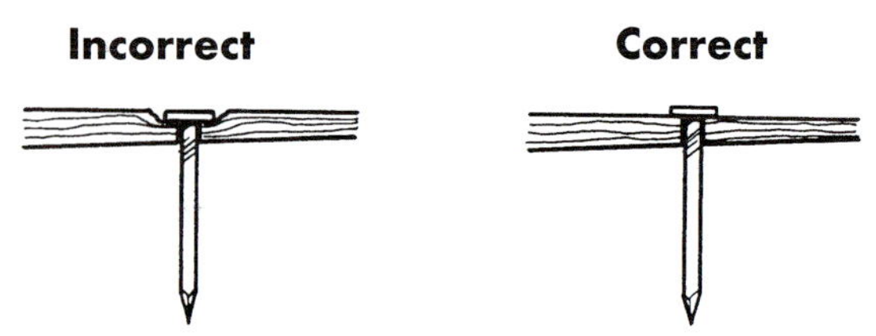

Nails should rest on the surface of the wood. If the nailheads crush the wood fibers, movement will be limited and the shingle will be more apt to split.

- After the roof shingles are installed, cover the ridge with a synthetic matrix material such as Roll Vent® (Benjamin Obdyke, Horsham, Pa., www.obdyke.com).
- Center matrix over the ridge, tack it in place, and then cover it with a 7-in.-wide strip of 30-lb. felt.
- When the Roll Vent® is in place, cap the ridge with manufactured cedar ridge caps.
- Attach the ridge caps using 2-in.-long 1/2-in.-crown galvanized staples. Use a staple gun, as hand-nailing the ridge caps through the mushy Roll Vent® material will result in excessive splitting of the cap shingles.

Installing Shingles and Shakes

Figure 3-23. Shingle and Shake Layout

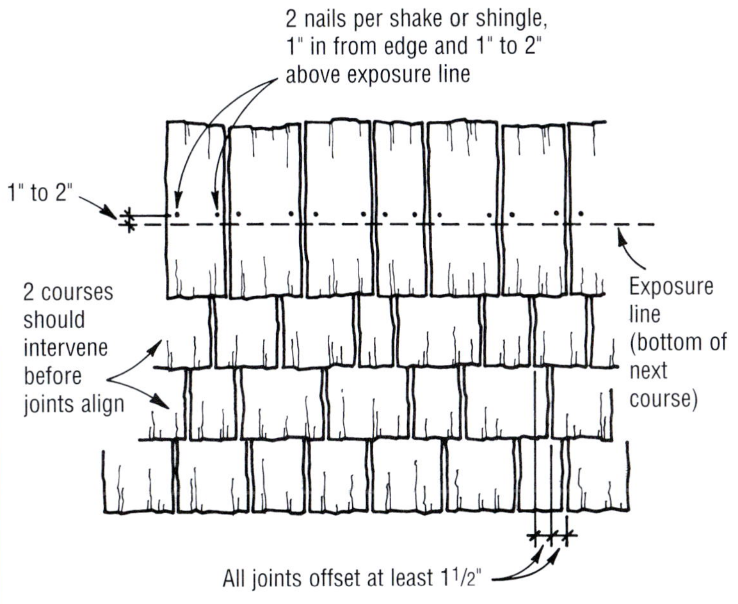

All joints in wood shake and shingle roofs should be offset by at least 1 1/2 in. in adjacent courses. Treat a knot or similar defect as if it were a joint.

Figure 3-24. Hip and Ridge Details

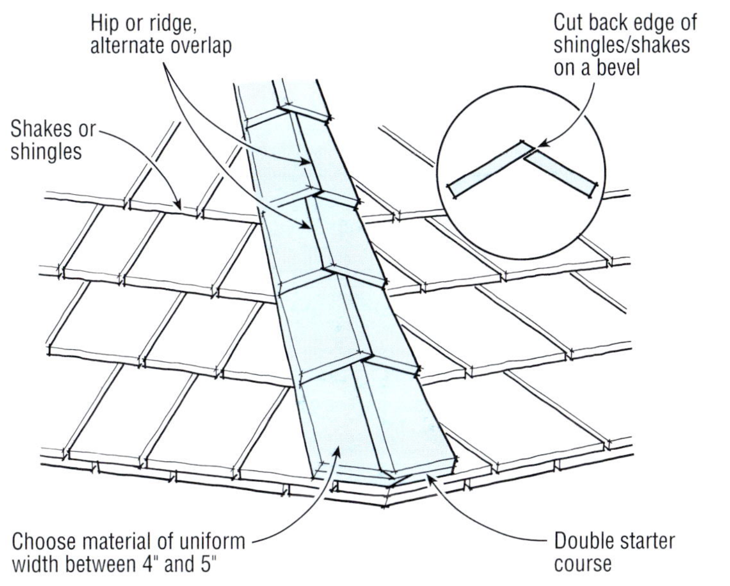

Alternate overlaps and conceal nailing at ridges and hips to provide a weathertight joint. Use nails long enough to penetrate the sheathing 3/4 in. minimum. Factory-assembled ridge and hip units are available. They come with shingles ganged together for ease of installation.

MAINTENANCE AND REPAIR

Shingles and shakes need more care than other types of roofs. Fungus and rot are very much at home on wood roofs, particularly low-pitched roofs in humid, wooded areas. Parts of the roof beneath overhanging trees are especially vulnerable since wet debris such as needles and leaves can trap moisture, leading to fungus growth, premature rot, and leaks.

Cleaning a Wood Roof

Cleaning with a power washer or garden sprayer can double a wood roof's service life. Clean roofs yearly, paying particular attention to the keyways between adjacent shingles.

Figure 3–25. Vented Ridge

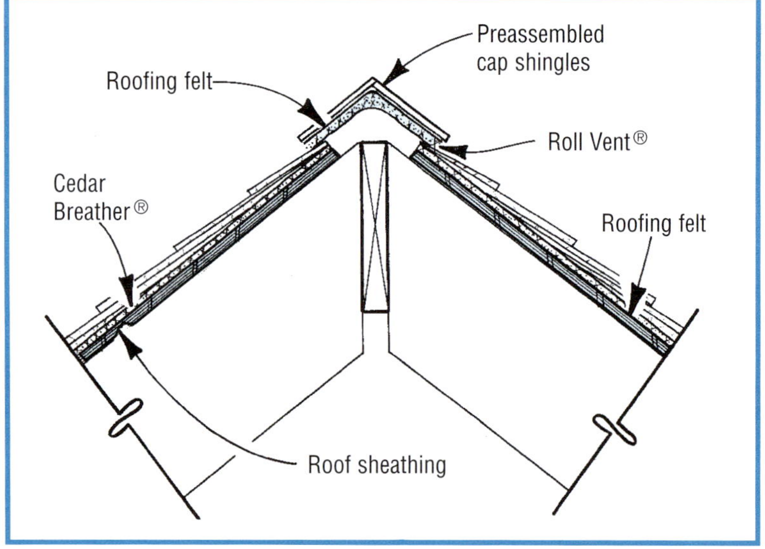

A vented ridge can be created using Roll Vent®, 30-lb. felt paper, and preassembled ridge cap shingles. Install with a pneumatic staple gun to prevent splitting.

- A high-pressure water sprayer works best. Use a power washer with a flow rate of 4 to 6 gallons per minute.
- Or use a 50/50 solution of water and household bleach (for an even stronger solution, replace the bleach with 2 to 4 ounces of swimming-pool chlorine) in a garden pump. Spray 1 to 1 1/2 gallons of this solution over each square of roof area. Let it stand for 15 to 30 minutes, and then rinse it off with a power washer or garden hose.
- Protect the surrounding shrubs and grass: Hose them down with clean water before, during, and after application, or cover them with tarps (see "Landscape Protection," **page** 377).

MINOR REPAIRS

Cracked Shingles

- If both sides of a cracked shingle are still present, nail them to the roof deck.
- Drill pilot holes first to prevent splitting the shingle and cover both the joint and the nail heads with roofing cement.

Temporary Repair

- A damaged shingle can be temporarily repaired by inserting a piece of galvanized steel or aluminum under it and nailing through the shingle and metal with two nails.
- The metal should extend 2 in. beyond both edges of the shingle and 1 in. under the butt line of the overlying shingle in the next course.

Shingle Replacement

- Individual shingles can be removed by first splitting them and removing the pieces. Cut the roofing nails with a shingle ripper or hacksaw blade. Make sure that the nails are cut off flush with the sheathing; use caution when cutting the nails to avoid damage to the roof deck sheathing or underlayment.
- Next, trim the replacement shingle to the required width and slide it into place. Tap it gently into place with a hammer and wood block. Nail it to the roof deck and cover the nail heads with roofing cement.

Bowed Shingles

- Split a bowed shingle down the center and remove about 1/4 in. of wood from the inside edge of one section to form a joint for roofing cement.
- Nail the two sections on either side of the joint and cover the joint and nail heads with roofing cement.

REROOFING

When to Strip

As with asphalt shingles, if more than 10% of an old roof needs repair, the best solution is usually a new roof. Avoid roofing over wet or water-damaged wood shingles and wood shakes; they're too irregular to make a good base.

Repairing Wood Roofs

When to Replace Shingles/Shakes

ROOFING TILE TYPICALLY MEANS CLAY OR CONCRETE TILE (**Figure 3-26**). A good tile roof should include a 50-year warranty and can last over a century. When properly installed, most come with a Class-A fire rating.

Weight of Roofing Tile

A tile roof can weigh three to five times as much as an asphalt one (900 to 1,000 pounds per square is typical, although you can get lightweight Portland cement and mineral-fiber tiles that weigh only 400 to 800 pounds per square). This means tile roofs impose three to five times the design dead loads on the roof structure. Be sure to evaluate the framing before applying tile (see "Rafter Loads," **page 140**).

When installing tile, make sure to stack the tile properly to avoid overloading the roof (see "Loading the Roof with Tile," **page 240**).

Slope Limitations of Roofing Tile

With tile, as with slate, the steeper the roof the better.

- The minimum recommended slope for a concrete one-piece barrel tile or interlocking flat-ribbed tile is 4:12.
- For flat tile, the minimum slope should be 5:12.

Figure 3–26. Typical Clay & Concrete Tile

		Types	Min. Slope	Overall Length x Width	Exposure Length x Width	Approx. Weight/Square	Approx. Pieces/Square
Clay	Pan & Cover	Tapered or straight mission, barrel or barrel mission: overlapping	4:12	16" x 8"	13 1/2" x 5 3/4"	1,115 lbs.	192
				19" x 10"	16" x 13 1/2"	1,005 lbs.	134
				18" x 12"	14" x 9 1/2"	1,040 lbs.	108
		Spanish or "S"	4:12	13 1/4" x 9 3/4"	10 1/4" x 8 1/4"	900 lbs.	171
				12" x 12"	10 1/2" x 9 1/2"	900 lbs.	144
				17" x 13 1/2"	14" x 12"	950 lbs.	92
				20" x 13 1/4"	17" x 11 3/8"	1,100 lbs.	75
	Flat Tile	English shingle, closed shingle, or French: interlocking	3:12	11" x 8 3/4"	8" x 8"	900 lbs.	225
				14" x 9"	11" x 8 1/4"	800 lbs.	158
				16 1/2" x 13"	13 1/2" x 11 3/4"	900 lbs.	90
		Plain tile, English flat or slab shingle: overlapping	4:12	12" x 7"	5" x 7"	1,840 lbs.	412
				15" x 7"	6 1/2" x 7"	1,600 lbs.	31
				18" x 8"	7 1/2" x 8"	1,200 lbs.	240
Concrete		High-Profile: interlocking	4:12	17" x 12 3/8"	14" x 10 3/4"	950 lbs.	94
		Low-Profile: interlocking	4:12	16 1/2" x 13"	13 1/2" x 11 3/4"	900 lbs.	90
		Flat Tile: interlocking	4:12	11" x 8 3/4"	8" x 8"	1,000 lbs.	225

For more information: "Roofing with Concrete Tile," JLC 5/01

- Code prohibits the installation of any concrete tile on slopes less than 3:12, except with permission of the building official.

CLAY TILE

Clay is porous, but when tiles are fired in a kiln at about 2,000°F, the individual particles fuse together and eventually "vitrify," forming a waterproof glaze at the surface. Tile with substantially damaged edges or surfaces should be discarded.

Clay Tile Styles

There are two categories of clay tile (**Figure 3-27**):

- **Pan and cover tile** includes two-piece (Barrel or Mission, Roman and Greek) or one-piece (Spanish or "S").
- **Flat tile,** such as English flat, or "scab," or French "flat ribbed" tile, typically have interlocking edges.

Most manufacturers make ridge, hip, and rake tile to match both styles of clay tile. In either style, tiles may interlock with one another or overlap.

Clay Tile Color

The color of the tile is determined by the color of the clay. Most are natural earth tones: tan, reddish brown, and orange. A range of additional colors made with metallic glazes and paints may also be available.

Freeze-Thaw Protection

In regions with frequent freeze-thaw cycles, use tiles that have been fired to full vitrification, giving them a glass-like surface. To resist damage from ice, these tiles need an absorption rate of less than 3% and a compressive strength of at least 8,000 psi.

CONCRETE TILE

Selecting Tile Roofing

Concrete tiles are made from sand and Portland cement in a 3:1 or 4:1 ratio that's extruded under high pressure, making them impermeable to water. Each has an interlocking channel on its left edge that's lapped by the next tile. On the underside, a head lug fits snugly over the edge of a 1x batten. A series of ridges called weather checks rest on the tile below and act as a barrier against windblown rain.

Concrete Tile Styles

Concrete tiles fall into three main categories (**Figure 3-28**):

- **High-profile** styles imitate traditional clay tiles (Mission, Spanish, Barrel, or S tiles).
- **Low-profile** tiles have less pronounced curves, but still have water channels running down the face of each tile.
- **Flat tiles** imitate shakes or slate, and are laid with a staggered bond (each joint is offset from the courses above and below). Flat tiles are a little easier to fudge when trying to keep courses straight than high- or low-profile tiles. And unlike profiled tiles, they don't leave big voids at rakes, hips, and ridges that have to be sealed up with mortar.

Manufacturers make a ridge/hip tile with a 120° spread, as well as a 90° rake tile to match most concrete tile types. Some (but not all) also offer apex tiles for hip and ridge intersections and hip starter courses.

Concrete Tile Color

Tile color can either be integral or applied to the surface as an acrylic spray or slurry. Integral tiles are known as color-through. They come in fewer colors, and the colors are darker.

Courtesy of NRCA Roofing and Waterproofing Manual

Figure 3–27. Types of Clay Tile

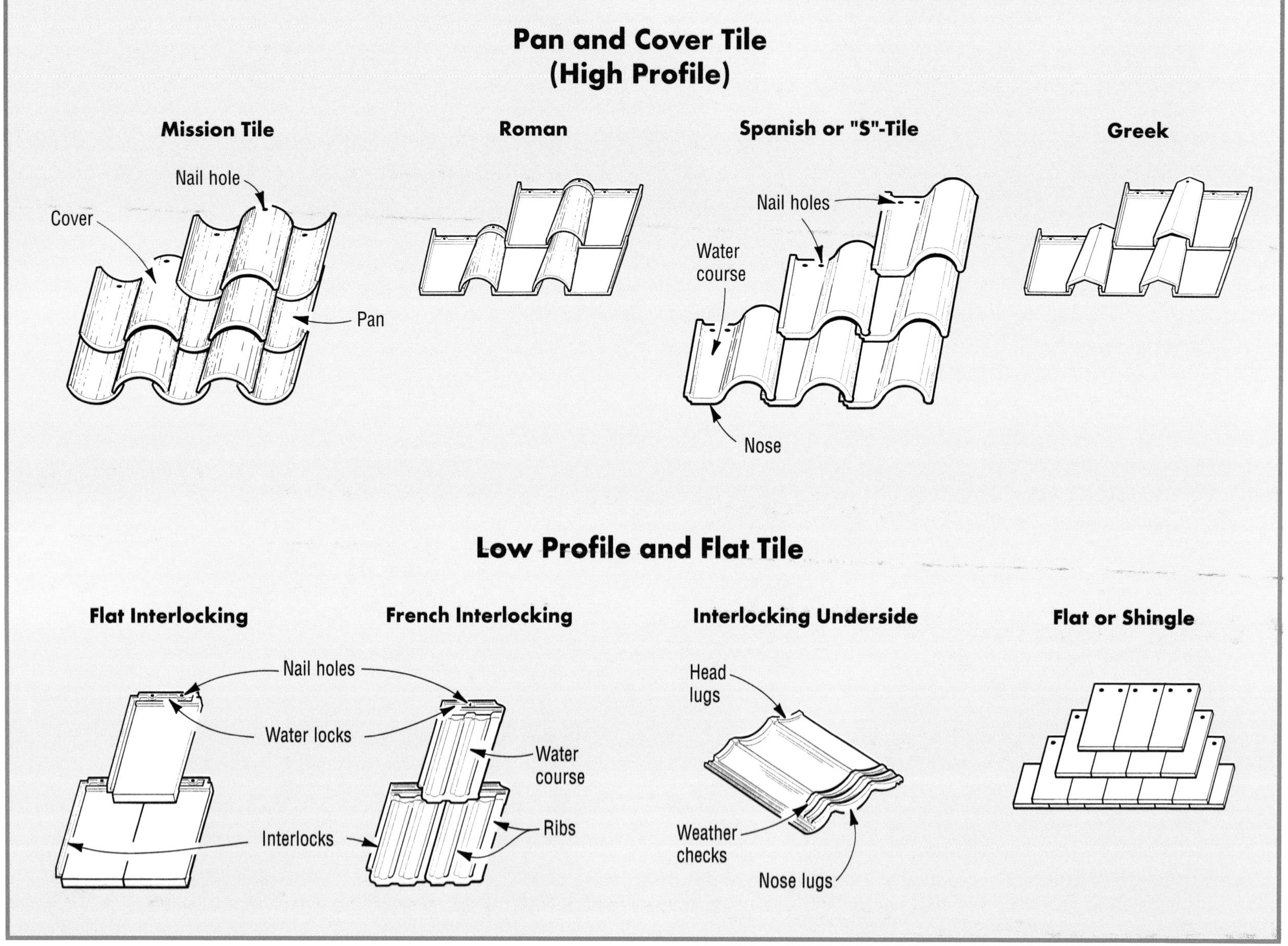

Freeze-Thaw Protection

Due to the impermeability of extruded concrete, freeze-thaw protection is generally much better with concrete tiles than with clay tiles.

SHEATHING AND UNDERLAYMENT

Sheathing Options for Tile

A tile roof can be installed either over a solid sheathing or spaced sheathing.

Underlayment Options for Tile

Tile Roof Sheathing and Underlayment Options

- For slopes 4:12 and over, solid sheathing and 30-lb. felt will do. See **Figure 3-29**.
- On solid sheathing, 1x2 battens are mounted to the roof deck over the underlayment and between the roof eaves and the ridge. Battens are required on roofs with slopes in excess of 7:12, and are generally recommended on all tile installations. See **Figure 3-30** for underlayment application over solid sheathing.

Figure 3–28. Types of Concrete Tile

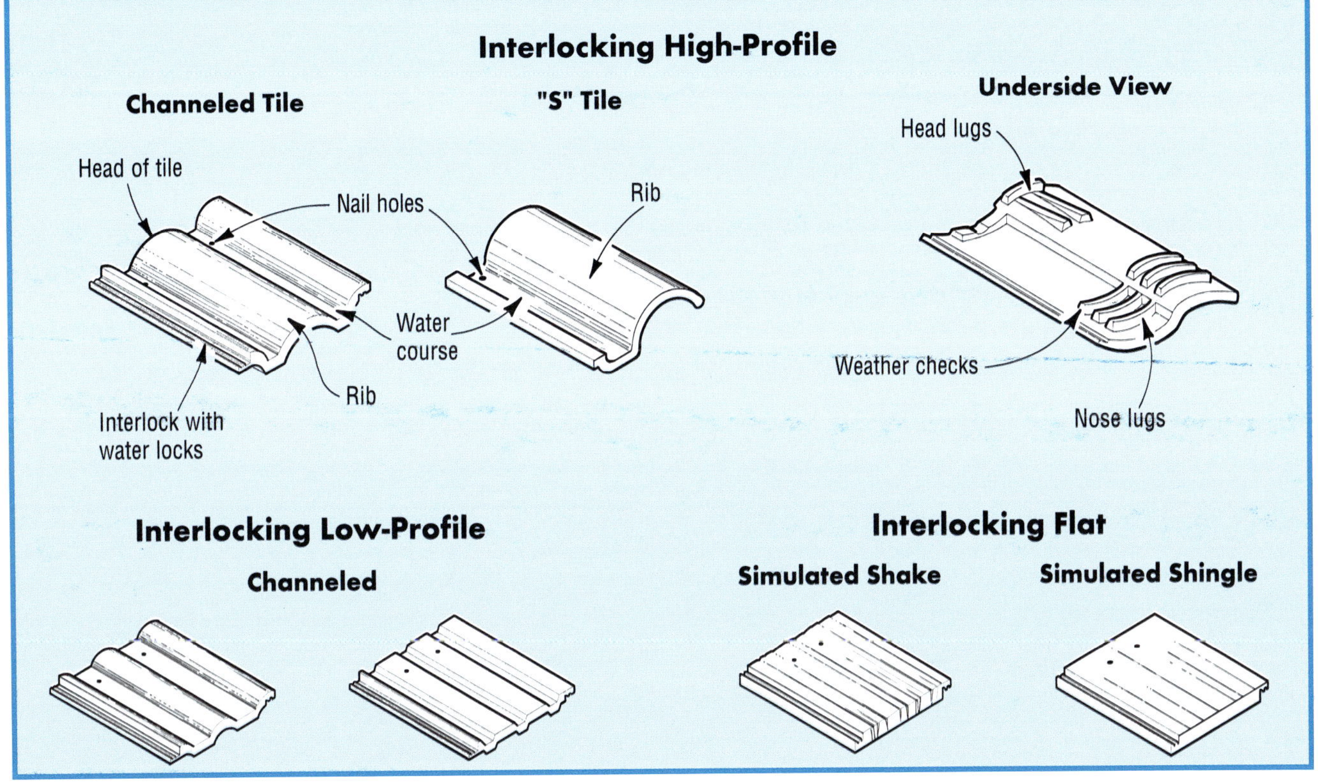

Because the manufacturing process for concrete tiles differs from that of clay tile, the selection of textures and finishes is more varied.

- Check manufacturers' suggested layout procedures. Each manufacturer should supply information regarding batten spacing and kick-strip installation, as well as a layout calculation chart for its product.
- Tiles on shallow slopes must be installed over a waterproof low-slope roof, either built-up or single-ply (see "Low-Slope and Flat Roofs," **pages 262** to **269**). In this case, the tiles are mostly decorative, although they do shield the membrane against ultraviolet light, high winds, and mechanical damage.
- Flat concrete tiles should be installed over preservative-treated battens (**Figure 3-38, page 245**). In climates with snow accumulation, install the counter battens over battens that run from ridge to eaves (**Figure 3-30**). These enhance drainage and ventilation, and help to keep the roof uniformly cold, which lessens the chance of ice damming.

ANCHORING SYSTEMS

Before installing tile, determine the type of anchoring system appropriate to prevailing conditions.

Mortar-Set

In areas where wind velocities approach 80 mph, tile is often set in a mortar bed over a built-up membrane. Soak the tile before installation to reduce the chance of dry tile drawing moisture out of the mortar, and weakening the bond.

Figure 3–29. Underlayment for Tile Roofs

Roof Slope	Use Conditions	Underlayment (based on standard size tile with 3" min. overlap)
More than 20:12 slope	Normal conditions	One layer of 30# asphalt-saturated felt (min.)
	Heavier/Thicker tile application	One layer of 40# A-S, coated felt. Check with tile manufacturer if heavier felt is required.
	In areas with severe weather	Two layers of 30# A-S felt or one layer of 40# (A-S, coated felt) recommended
8:12 to 20:12 slope	Normal conditions	Two layers of 30# A-S felt recommended, but one layer of 40# (A-S, coated felt) recommended
	Heavier/Thicker tile application	One layer of 45# min. A-S, coated felt. Check with tile manufacturer if heavier felt is required.
	In areas with severe weather	Two layers of 30# A-S felt or one layer of 40# A-S coated felt acceptable
4:12 to 8:12 slope	Normal conditions	Two layers of 30# A-S felt or one layer of 40# A-S coated felt recommended
	Heavier/Thicker tile application	One layer of 45# min. A-S, coated felt. Check with tile manufacturer if heavier felt is required.
	In areas with severe weather	On lower slopes an upgraded underlayment system and/or waterproof membrane is recommended. Check with the manufacturer and local building officials.
Less than 4:12 slope	All conditions	At these slopes, tile is considered decorative. A more stringent underlayment, waterproof membrane system, and reduced tile exposures may be required. Check with tile manufacturer and local building officials.

Nail-On

With the "nail-on" method, each roof tile is nailed directly to the roof sheathing (for low-slope roofs) or to a wooden batten attached to the sheathing. Nailing down the tiles is much faster and less labor-intensive than cementing tiles. Also, unlike the mud-on system, nailed tiles will "give" a little in big winds.

Nails for attaching tile. See **Figure 3-31** for tips on fastening roof tile.

- Do not use smooth-shank nails; they will work their way out in heavy winds.
- Use either a ring-shank nail or a hot-dipped galvanized nail.
- Two nails are better than one for fastening roofing tile.

Twisted Wire Systems

Designed for 2:12 to 24:12 slope roofs, this system uses fewer nails, so it punctures the waterproof underlayment less. Twisted wire systems are typically specified in earthquake zones and areas with moderate winds. Because the roof is not rigidly attached, the wire yields without transferring the shifting load caused by wind or seismic activity to the building.

Tile Roof Anchoring Systems

Figure 3–30. Underlayment and Batten Layout

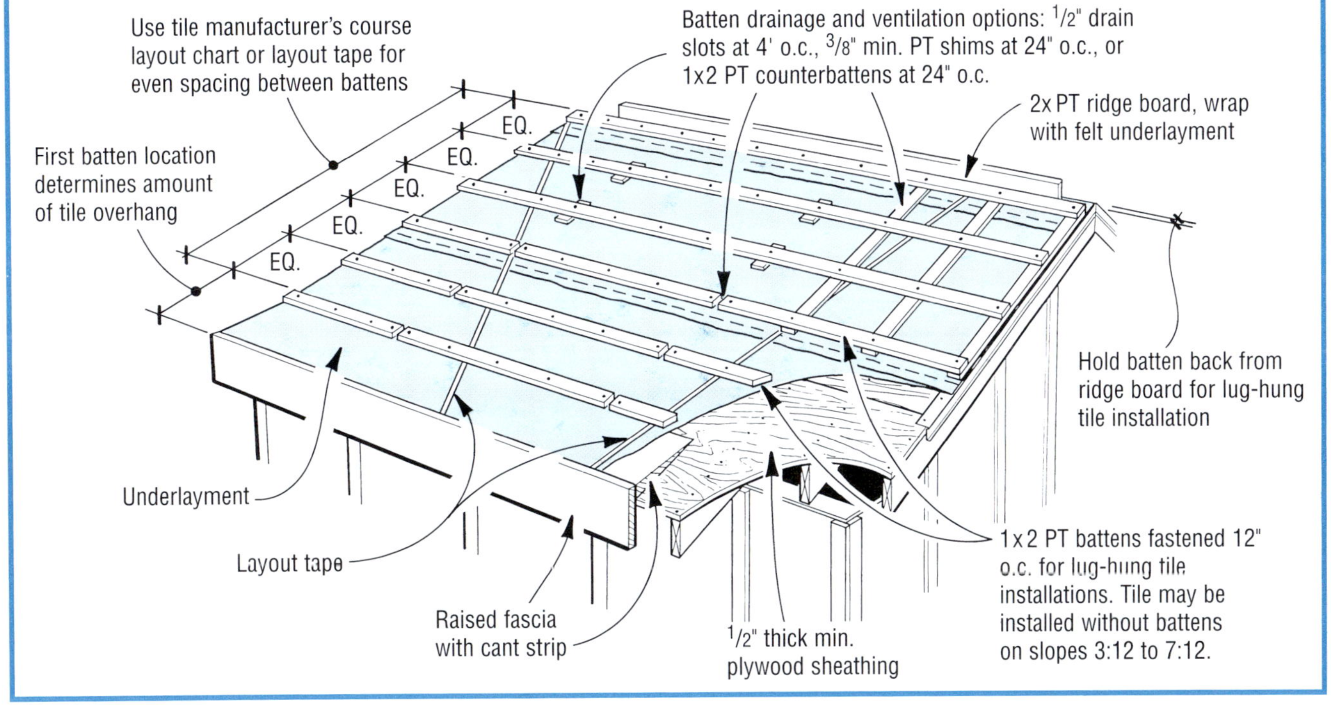

When installing clay and concrete tile over solid sheathing, cover the plywood with 15- or 30-lb. felt paper and install strapping over the plywood, cutting notches or shimming under battens for drainage. For optimum drainage and ventilation, install counterbattens.

- **Vertical wire runs.** The twisted wire system consists of two wires (12-gauge galvanized, with stainless or copper also available) wound together with an expanded loop or eyelet approximately every 6 in. The wires run from ridge to eaves between vertical courses of tile and are attached with approved anchors (**Figure 3-32**).
- **Anchor wires every 10 feet.** Nail or screw the twisted wires into the roof at the ridge, eaves, and every 10 ft. along the roof slope.
- **Tie-wires secure tiles.** Secure individual tiles to the twisted wire with separate tie-wires. The individual tie-wires are threaded through the loops on the wire runs, then twisted closed.
- **Replacing damaged tile.** Damaged tiles fastened with the tie-wire system are easy to replace by folding back the damaged tile, untwisting the wires, and adding the new tile.

For high-wind areas and tall buildings, the UBC currently requires one nail in the head of every tile, plus one nose clip for eaves tile and an extra nail for rake tiles. The National Tile Roofing Manufacturers Association recommends two nails or a nail and a clip for every tile in seismic and high-wind areas, and three fasteners for perimeter tiles. Check with your local officials for new code developments if you work in high-wind or seismic zones.

For more information: "Roof Tile Fasteners for High-Wind Regions," JLC 5/96

Figure 3–31. Fastening Tile in High-Wind Areas

		Field Tile			
	Roof Slope	Solid Sheathing with Battens	Solid Sheathing without Battens(1)	Eaves Course	Perimeter Tile and Tile on Cantilevered Area(2)
Fastening required on roofs 40 feet or less above ground (measured to the eaves) in areas with wind velocities up to 80 miles per hour	3:12 to 5:12	one 10d nail every tile	one 10d nail every tile	one 10d nail & one clip every tile (3)	one 10d nail every tile
	5:12 to less than 7:12	one 10d nail every tile	one 10d nail every tile	one 10d nail & one clip every tile (3)	one 10d nail every tile
	7:12 to less than 12:12	one 10d nail every tile	N/A	one 10d nail & one clip every tile (3)	one 10d nail every tile
	12:12 and over	one 10d nail and clip every tile	N/A	one 10d nail & one clip every tile (3)	one 10d nail and clip every tile
Fastening required on roofs 40 feet or less above ground (measured to the eaves) in areas with wind velocities up to 80 miles per hour	All slopes	one 12d nail or two 10d nails and one clip every tile		one 12d nail or two 10d nails and one clip every tile	one 12d nail or two 10d nails and one clip every tile
Fastening between 80 and 120 m.p.h.	All slopes	two 10d nails and one clip every tile		two 10d nails and one clip every tile	two 10d nails and one clip every tile

(1) For slopes exceeding 7:12, battens are required.
(2) Perimeter nailing areas include the distance equal to three tiles (but not less than 36") from the edges of hips, ridges, eaves, rakes, and major roof penetrations.
(3) You can use two nails per tile instead of clips.

Fasteners for High-Wind Regions

FASTENERS FOR HIGH-WIND REGIONS

Nose clips. Nail-on and twisted wire systems attach the tile only at the top — the butt end, or "nose," of the tile is free to rotate and break in a big wind. Special wire clips, known variously as nose clips, butt hooks, S hooks, and tile locks, provide a hold-down for the exposed edge of each cover tile (**Figure 3-33**).

- Use nose clips to secure the front of tiles to keep them from "chattering," or worse, from prying the nail out or snapping the tile at the attachment point.
- Nose clips can be used with any combination of fasteners and are available in stainless, brass, and galvanized.

Tile Nail. A new clip that functions as both a nail and a nose clip is the Tile Nail® (Newport Fasteners, Anaheim, Calif.). This is a 12-gauge galvanized wire (also available in brass or stainless) approximately 10 in. long with a 1 1/4-in. tail and a hook on the other end (**Figure 3-33**).

- Slip the end of the Tile Nail® through the hole in the tile, then nail it into the sheathing 6 in. above the tile — far enough away so there is no fear of breaking the tile.
- The next piece slips into the "nose" hook, positively restraining it as well as setting the head lap distance for the next course.
- "Butter" the nail end with roofing cement.

Figure 3–32. Twisted-Wire Fastening Method

The twisted wire method uses a double strand of 14-gauge wire fastened at the ridge and eaves with special anchors (left). Each tile is secured with an individual tie-wire (center) that is threaded through the open loops in the twisted wire (right).

Hurricane clips. A less effective clip is a hurricane clip, or side strap made of 18-gauge galvanized steel. This is a locking device for the side edge of concrete, clay, and clay S tiles.

- Bend the clip into an L shape with the top curved to fit snugly over the side edge of the tile.
- Keep the bottom of the clip flat; nail or screw it to the sheathing or batten.
- Hurricane clips should be used only as a backup fastener to help roofs survive high winds. It's hard to get these clips to fit snugly to the tile, so after several storms they tend to bend and work loose.

LOADING THE ROOF WITH TILE

Gable Roofs

On a gable roof, stack the tile to properly distribute the weight of the tile during installation. This avoids overloading the roof framing system at its weakest point, the midspan of the rafters.

- Pick tiles from different pallets to avoid color patterning.

Figure 3–33. Fasteners for High-Wind Areas

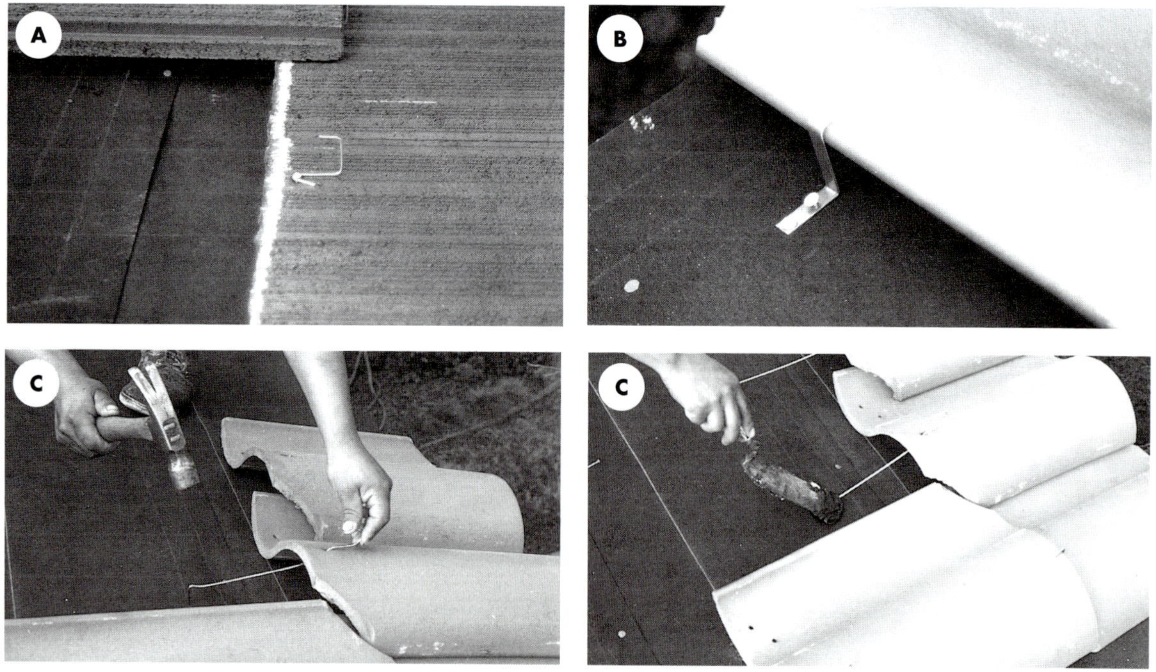

Nose clips (A and B) hold down the front end of the tile. The same nail secures both the nose clip and the tile underneath. ***The Tile Nail®*** *from Newport Fasteners (C) has several advantages over a conventional nose clip. Because it is driven in a few inches away from the tile, there is less risk of cracking the tile while nailing. Its length allows enough movement to keep the tile from breaking during high winds or an earthquake. Also, the holes can be patched with mastic to prevent leaks in the membrane.*

Loading the Roof with Tile

- Starting from the right, lay out the stacks in numbers as shown at left in **Figure 3-34**, beginning with the fourth course. Space stacks about 12 in. apart.
- Repeat the procedure every fourth course. Odd numbers of courses should have two additional tiles for each additional course on each stack along the ridge.

Hip Roofs

Loading a hip roof is the same as loading a gable roof except that the number of tiles in stacks at the far left may vary due to the size of the roof (**Figure 3-34**, at right).

CUTTING TILE

Cutting Tile

Straight cuts at rakes can be made by scoring a line on the tile, then breaking it with a roofing hatchet. Although this leaves a rough edge, it will be covered by the rake tiles.

Valley and hip cuts. Use a power saw — a cut-off saw or a standard tile saw — and a diamond blade. Tiles may be cut wet or dry. When dry cutting, workers should wear respirators because of the fine dust. Also sweep any accumulated dust off the roof so that it does not stain the tiles or cause a fall.

Figure 3–34. Loading the Roof

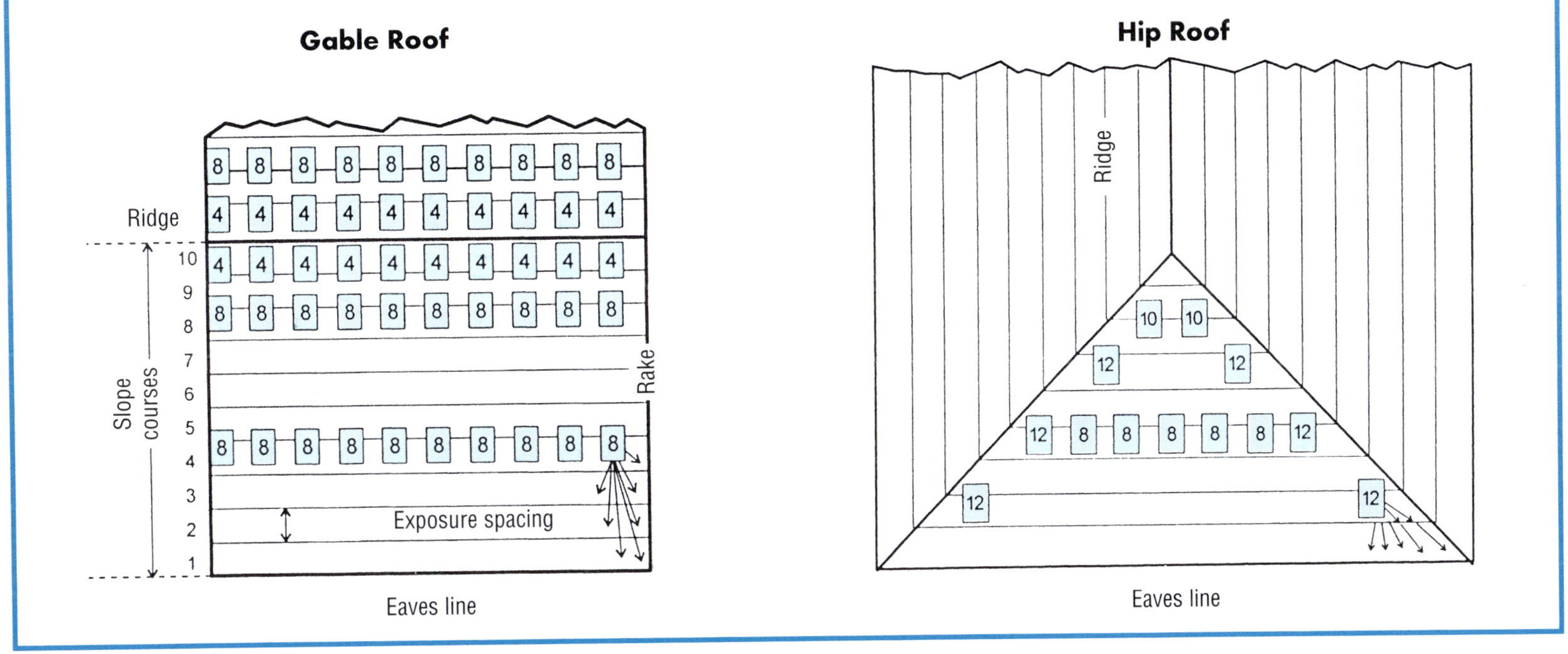

Properly loading the roofing tile avoids overloading the framing system at its weakest point, the midspan of the rafters.

Figure 3–35. Clay Tile Installation

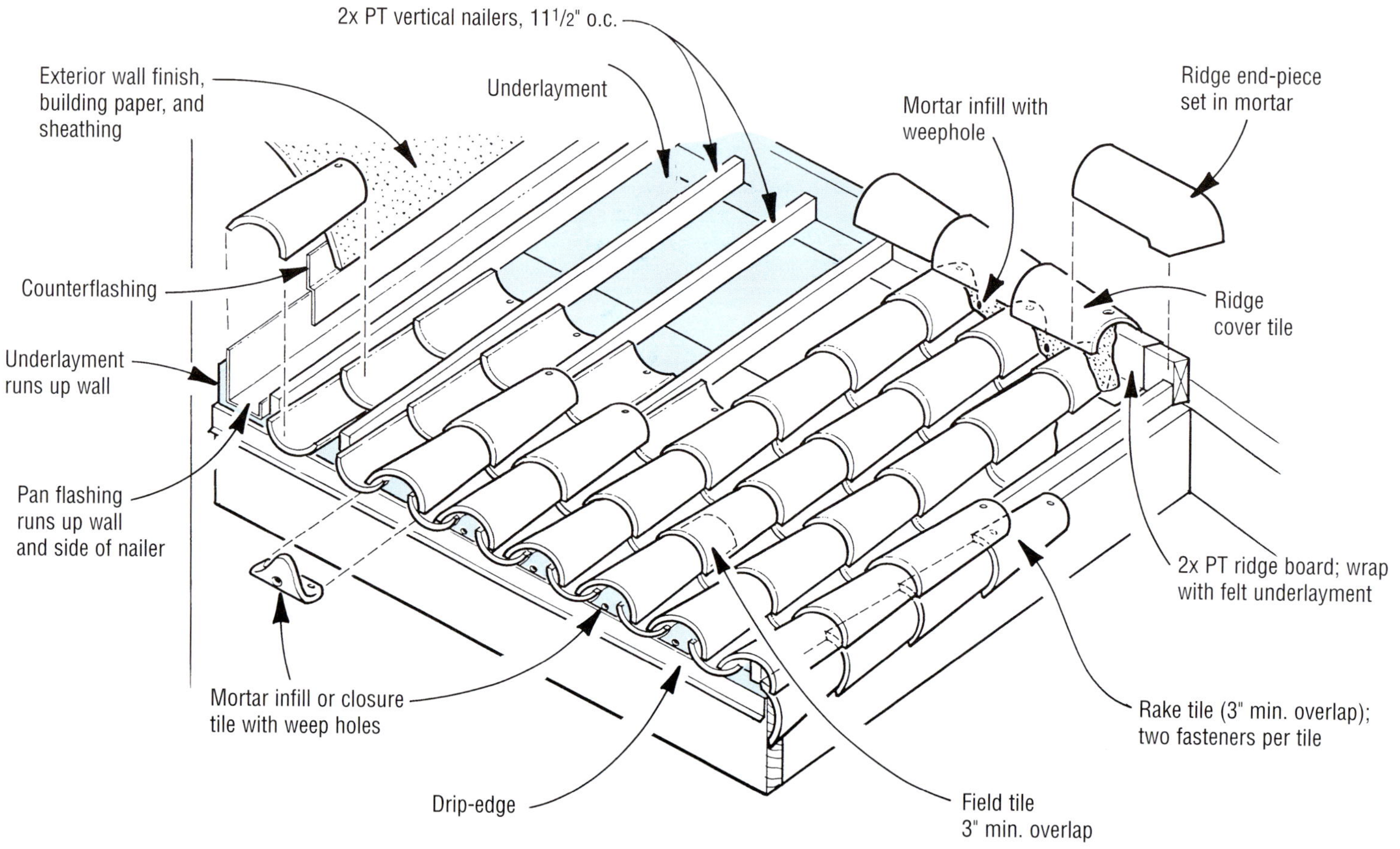

Typical installation of pan and cover clay tile: *Vertical nailers are optional but can provide good support and anchoring for the cover tile. In some parts of the country, tile is wire tied, straw nailed, or attached with other types of fasteners.*

Figure 3–36. Concrete Tile Installation

Installing Tile Roofs

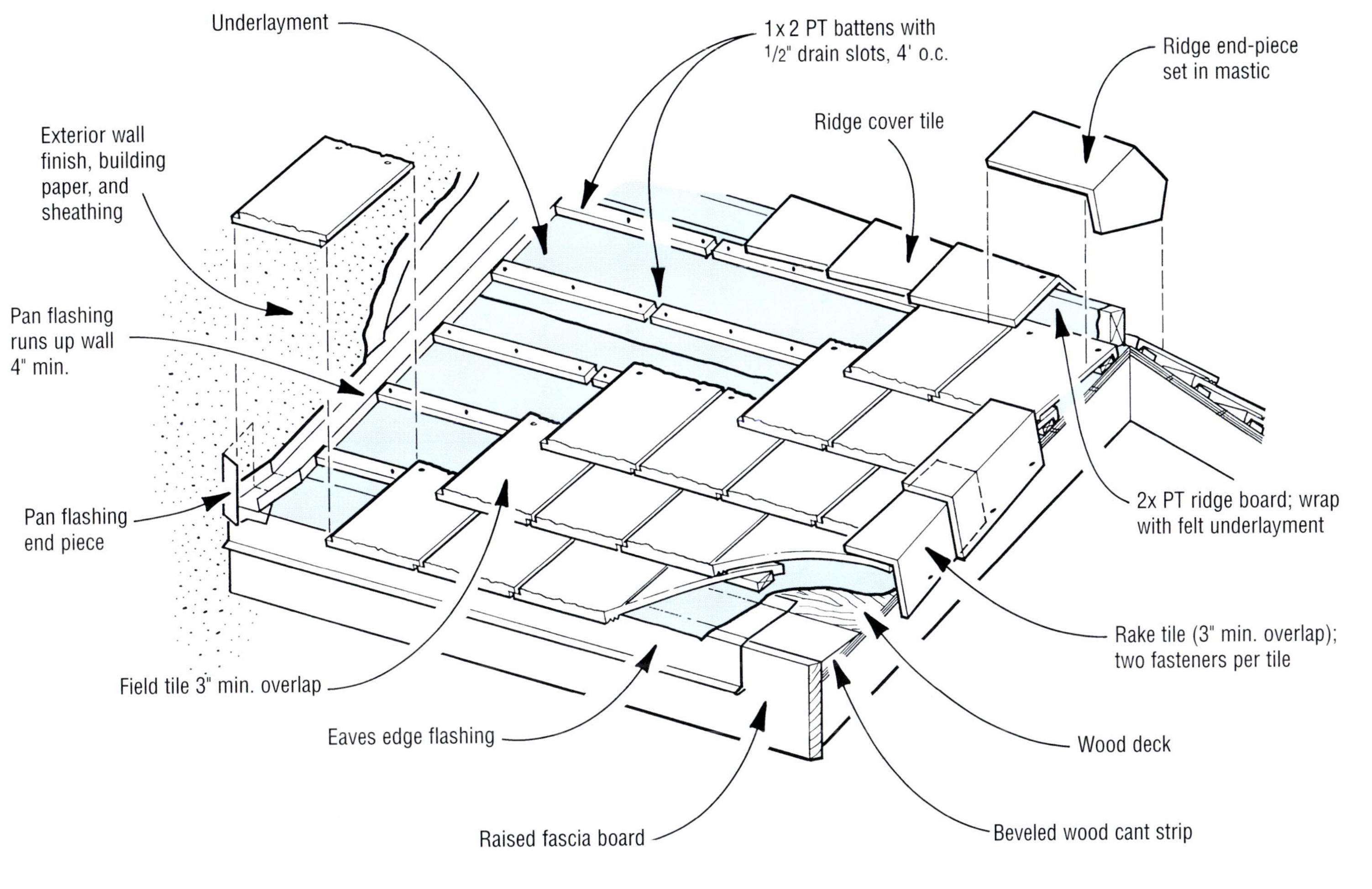

Typical installation of flat-interlocking concrete tile: *At the roof-to-wall connection, the underlayment is run up the wall 4 in. behind the pan flashing. The exterior wall finish acts as counterflashing and is held off the finish roofing. The pan flashing end-piece diverts water that runs down the pan flashing away from the wall finish (see* ***Figure 4-21, page*** *306 and* ***Figure 4-79, page*** *358).*

INSTALLATION DETAILS FOR TILE

Most tile companies publish thorough installation manuals, so it's worthwhile to compare these before deciding on a particular brand.

Figure 3-37. Flashing Roof Penetrations

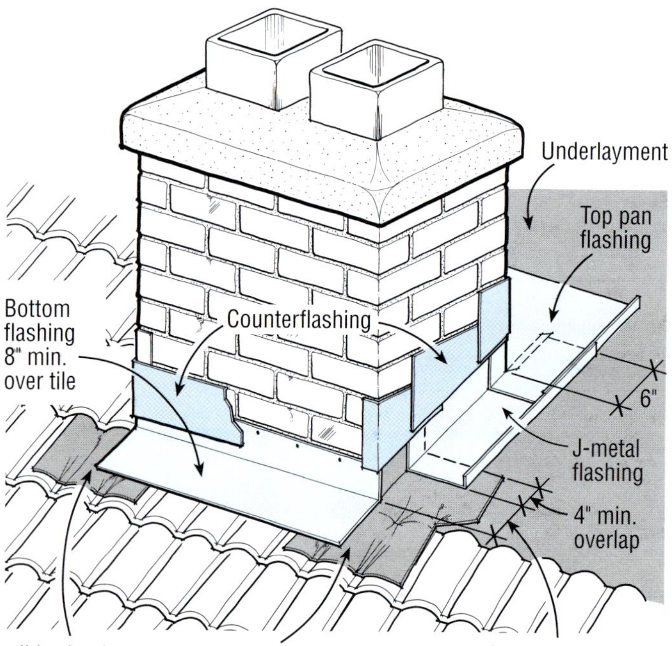

On a tile roof, the sides of dormers, skylights, and chimneys are flashed with galvanized pan flashing, which terminates on a piece of flexible aluminum. The aluminum flashing conducts water onto the top of tiles below the roof penetration.

Basic installation details for roofing tile are shown in **Figures 3-35** and **3-36, previous pages.**

Cold-climate eaves details. In cold climates with snow accumulation, install counterbattens over battens that run from eaves to ridge (**Figure 3-30, page 237**). These enhance drainage and ventilation, and help keep the roof uniformly cold to reduce the chance of ice damming.

Sidewall flashing requires a pan, as shown in **Figures 3-35** and **3-36, previous pages.**

Penetration flashing. The downhill corners of dormers, chimneys, and skylights should be flashed with both a flexible aluminum flashing, as well as a galvanized-steel pan flashing, as shown in **Figure 3-37**.

TILE MAINTENANCE AND REPAIRS

In general, problems with tile roofs are similar to those with slate roofs, and the maintenance procedures are the same (see "Slate," **page 259**).

Low-Slope Tile Repairs

On low slopes, a leaky, water-damaged subroof may be hidden by undamaged tiles. Unfortunately, the only way to fix a damaged subroof may be to tear off the tiles.

Working on Existing Tile

The most difficult part of repairing a few broken tiles is removing them without breaking the surrounding tiles. You can walk on flat shingle tiles if you're careful, but with high-profile tiles, lay down sheets of plywood, wide planks, or burlap bags filled with sand to spread your weight. In some cases, you can work off a ladder hooked to the ridge.

Removing Damaged Tile
Use a a slate ripper or hacksaw blade to cut the nails.

Attaching Replacement Tiles
Attach replacement tile with a copper nail, slate hook, or copper strap, as with slates (**Figure 3-56, page 260**).

TILE REROOFING

Tile Maintenance and Repairs

Always strip the old roofing material before applying a new clay or concrete tile roof. Most roof structures can't support the extra weight. If the old roofing is something other than slate or tile, then it may be necessary to reinforce the rafters.

Tile Reroofing

Figure 3–38. Eaves Details for Roofing Tile

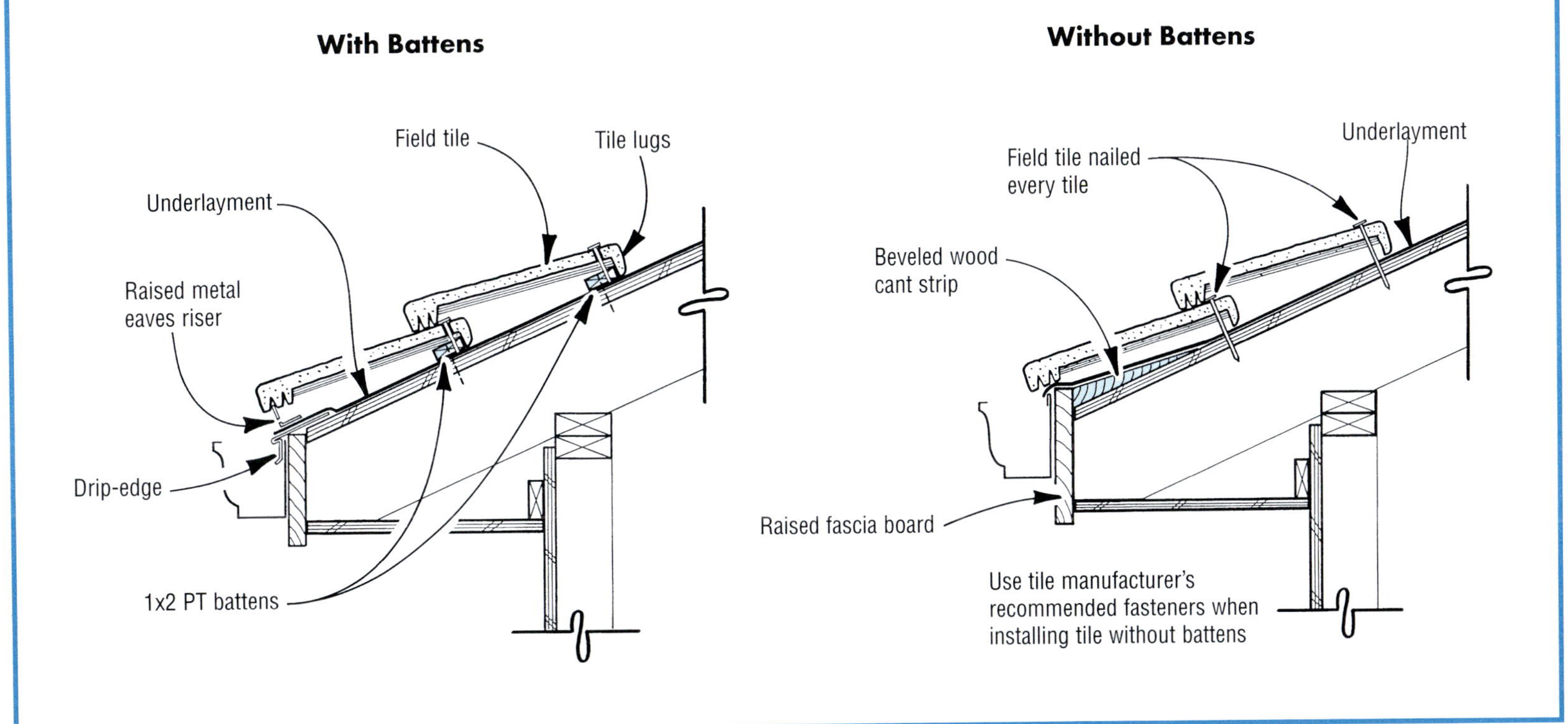

***Tile installed with battens and on raised metal eaves riser** (left): Underlayment should overhang drip-edge and be completely weathertight. **Tile installed without battens** (right): Check with the tile manufacturer for proper nail type and nailing depth. A continuous cant strip stops water from collecting along raised fascia board. Underlayment should overhang drip-edge and be completely weathertight.*

TRADITIONAL METAL ROOFS ARE MADE FROM FLAT-LOCK metal and site-formed standing seam panels. However, the increasing majority of metal roofs are made from several types of pre-fab panels, including screw-down, ribbed panels (similar to those used on agricultural buildings), and "modular" metal panels that mimic shingle, shake, and tile roofing.

Figure 3–39. Flat-Lock Panel Pattern

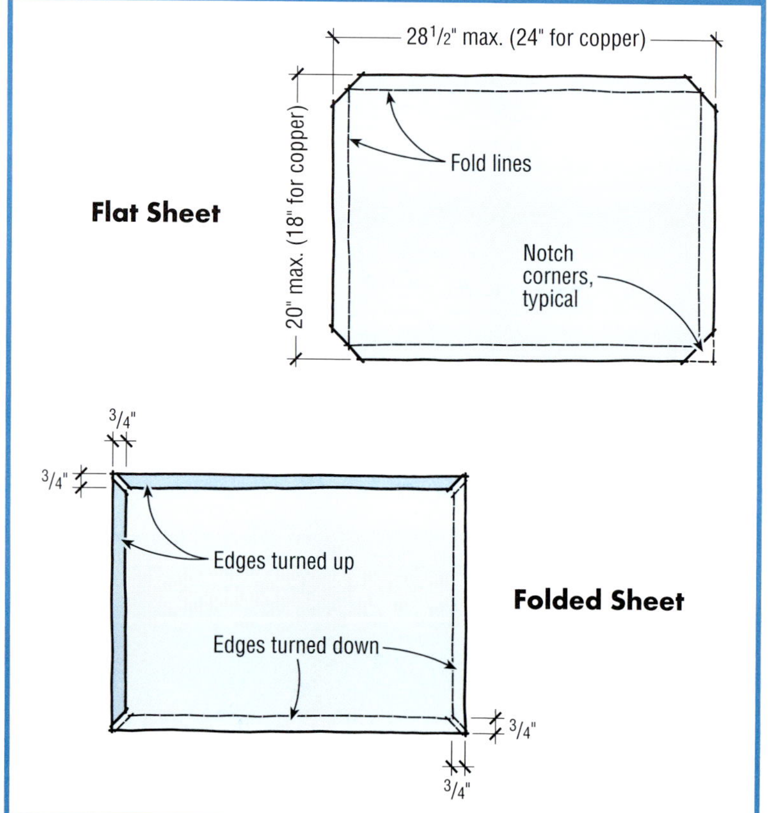

Cut flat-lock metal roof panels from 20x28-in. panels (for terneplate) or 18x24-in. (for copper), and then bend as shown.

For more information: "Restoration Primer: Flat-Lock Metal Roofing," JLC 9/87

FLAT-LOCK METAL ROOFS

What used to be called a "tin roof" is actually made from terne-coated steel. These days, it's used mostly on restoration projects. These roofs have a series of rectangular metal
panels, or pans, folded together along the edges. The seams must be beaten flat with a mallet, and then hand-soldered.

Slope

Flat-lock roofing is called for on any roof with a pitch less than 3:12. On steeper roofs, standing-seam metal should be used.

Flat-Lock Metals

Terneplate. This is the most common and inexpensive material used for flat-lock metal roofs. Originally, terne metal was a copper-bearing steel coated on both sides with an alloy of 80% lead and 20% tin. Since lead was outlawed, only Terne II coated steel, using a 50% zinc, 50% tin alloy coating, is available (Follansbee Steel, 800/624-6906, www.follansbeeroofing.com). This material comes in precut 20x28-in. panels or 28-in.-wide coil stock.

Unfinished terneplate must be painted on the underside before installation, and on the top side after installation. Use a rapid-dry water-based acrylic from Follansbee that is formulated specifically for Terne II. Typically, roofs should be repainted every 10 or 12 years.

Prefinished terneplate cannot be soldered; it is used for field-formed standing seam, not for flat-lock roofs.

TCS (terne-coated stainless steel) is a traditional material that doesn't need to be painted to prevent the base metal from corroding. However, it has the same silver color as unfinished terneplate and may be painted for appearance.

Copper is expensive but looks great. Be forewarned that the runoff stains the house. Copper also expands and contracts more than steel, so it should be used only on short roof runs. Lead-coated copper used to be a popular choice but is no longer widely available; existing roofs are often painted for appearance with a slow-drying linseed-oil-based paint.

Flat-Lock Installation Checklist

- Prepaint the underside of Terne II panels, including all cleats, flashing, and drip-edge stock using a rapid-dry acrylic paint. Never use an aluminized paint, which will promote galvanic corrosion (see "Galvanic Corrosion," **page 272**).
- Lay roofing panels over a wood deck, preferably plywood. Do not use treated wood; the salts will attack the metal coating.
- Install metal panels over a rosin-paper slip sheet. Do not use asphalt-impregnated felt; the asphalt will attack the metal coating.
- Form pans from 20x28-in. sheets, using a metal brake (see pattern, **Figure 3-39**).
- Panel edges along rakes and eaves lock over a drip-edge flashing (follow standing-seam detail, **Figure 3-44, page 250**). Lock panels to L-flashing bent from Terne II metal at vertical walls.
- Stagger panels, and secure to roof deck, as shown in **Figure 3-40**.
- Use 2x4-in. prepainted cleats and two 1-in.-long roofing nails to secure panels. Fold over flap on cleat to cover fastener heads.
- Fold seams closed, crimp tightly, and beat down with a mallet and wood block. Seams should be single locked a minimum of 1/2 inch.
- Solder seams with a 50/50 solder and rosin flux, using a 3-pound minimum soldering iron. Be sure to wipe off all excess flux.
- Paint soldered roof with a minimum two coats of the same rapid-dry acrylic paint used to prepaint the underside of the panels.

Flat-Lock Metal Roofs

Traditional Standing Seam

TRADITIONAL STANDING SEAM

Standing-seam roofs shed snow accumulations quickly. Standing seam is a good choice for steep gable and shed roofs in cold climates.

Figure 3-40. Flat-Lock Panel Layout

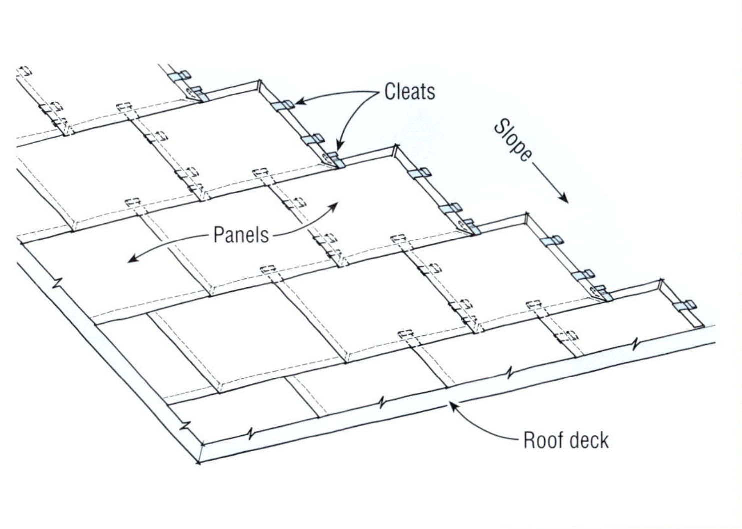

A typical layout shows how the staggered pans cover the roof. Uphill pans should always overlap the edges of lower pans.

Ordering Standing-Seam Metal

Traditional standing-seam roofs can be field-formed using the same materials used for flat-lock metal roofs. Pans are attached to the roof deck with cleats along one edge, while the seam along the other edge is covered by the next pan (**Figure 3-41**).

Preformed pans made of terneplate (24- or 20-in. wide) are available from Follansbee Steel (800/624-6906, www.follansbeeroofing.com).

Figure 3-41. Standing-Seam Detail

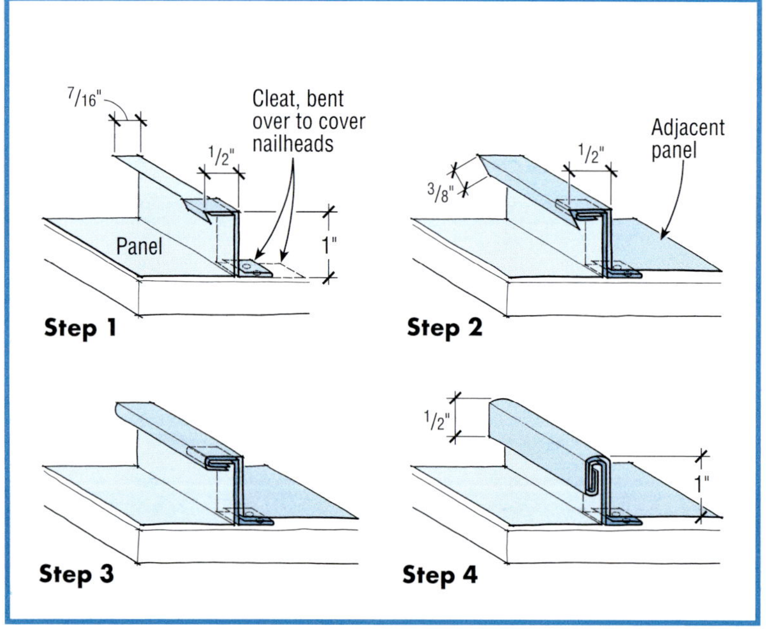

To secure a standing seam panel, a metal cleat (1) is bent over the panel stand (2), then nailed to the roof deck. The adjacent panel fits over the two pieces (3) and the entire seam is folded closed (4).

For more information: "Restoration Primer: Standing-Seam Roofs," JLC 5/86

Panel length. Standard lengths for preformed pan are 8-, 10-, and 12-ft. It's hard to handle a pan longer than 8 ft. without wrinkling the metal.

- When figuring pan lengths, allow extra for cross-seams and ridge seams (**Figure 3-42**).

Figure 3-42. Standing-Seam Panel Joints

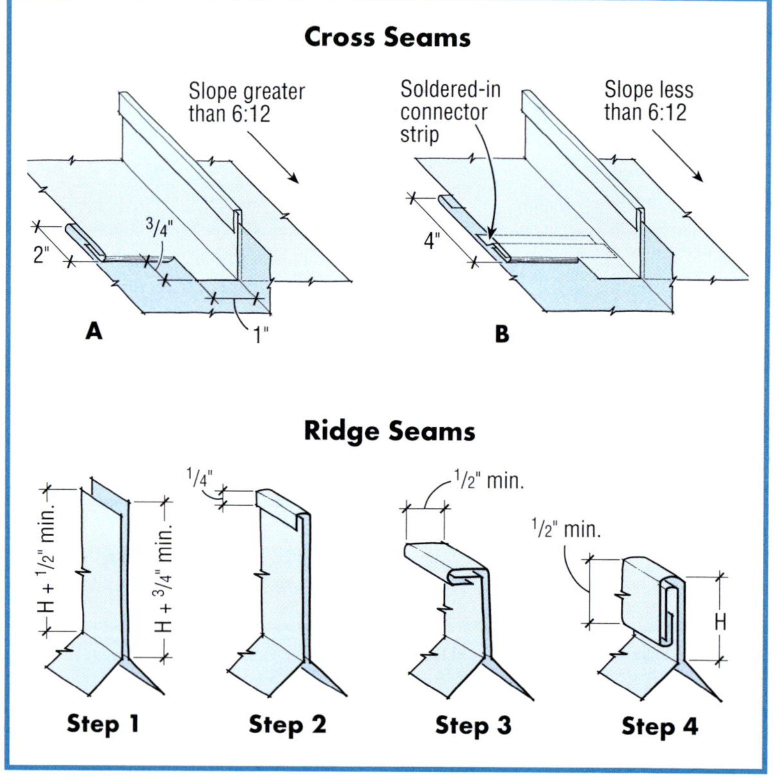

***Cross seams:** Use a standard cross seam (A) on roofs greater than 6:12. Roofs less than 6:12 require a more watertight cross seam (B), using a soldered-in cleat. **Ridge seams:** The stand at the ridge should be a minimum of 1 in. (H).*

- For shallow roofs (less than 6:12), allow a minimum 4-in. overlap, plus an additional couple of inches to form a deeper cross seam that accommodates a cross-seam connector strip (**Figure 3-42**).
- Do not use standing-seam panels below 3:12.

Number of pans. Measure the roof and plan for a standing seam at the centerline. Make a sketch showing the layout of the pans on the roof, and then calculate the number of pans you'll need (**Figure 3-43**).

Width of gable-end pans. At gable ends, cut off the upstand on the gable side of a preformed pan, and allow a 1/2-in. fold over a drip-edge. This will reduce the pan width on the roof by 1 1/2 in. (**Figure 3-44**).

Standing-Seam Installation Checklist

Installing Standing-Seam Metal Roofs

- Roll out rosin paper as a slip-sheet between the roof deck and the pans. Do not use roofing felt, as the asphalt will attack the metal.
- Snap chalklines at the position of each vertical seam to help keep seams straight.
- Start at the bottom left edge of the roof and hook the first pan into the drip-edges.
- Using a mallet and a block of hardwood, flatten the seams at the drip-edges. Then, using locking-plier crimpers or roofing tongs, crimp the pan/drip-edge seam tightly together.
- Install hold-down cleats at 12-in. spacing up the right side of the pan and across the top edge of the pan. Fasten the cleats to the deck with two 1-in. roofing nails. Use prepainted terne or TCS cleats fastened with galvanized nails. For copper pans, use copper cleats and copper nails. Fold the tail of the cleats over the nail heads.
- Install the second pan above the first by hooking it into the top of thc first pan and into the drip-edge. Close the seam at the drip-edge as before. Flatten the cross-seam between the two pans with a mallet and wood block.
- Anchor the second pan with cleats, and then install the remainder of the first row of pans all the way to the ridge of the roof.

Figure 3–43. Standing-Seam Layout

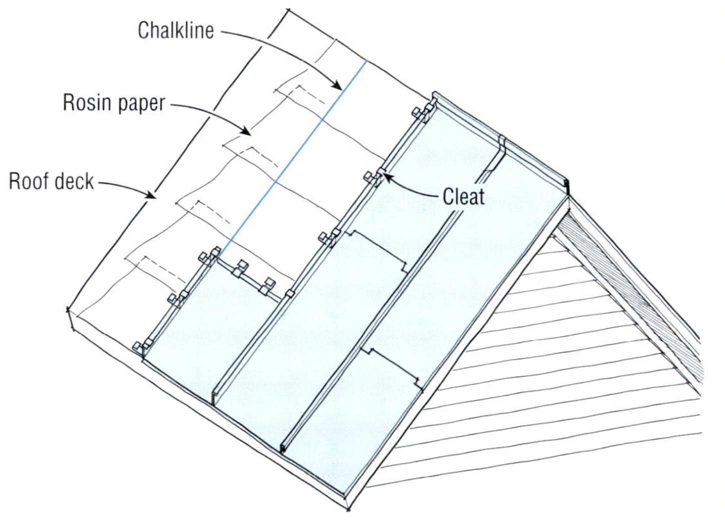

Installing a standing-seam roof is a lot like hanging wallpaper: Small errors at the beginning of the run create impossible problems at the end. Snap chalklines over a rosin-paper slip sheet to guide the installation.

Figure 3-44. Standing-Seam Edge Detail

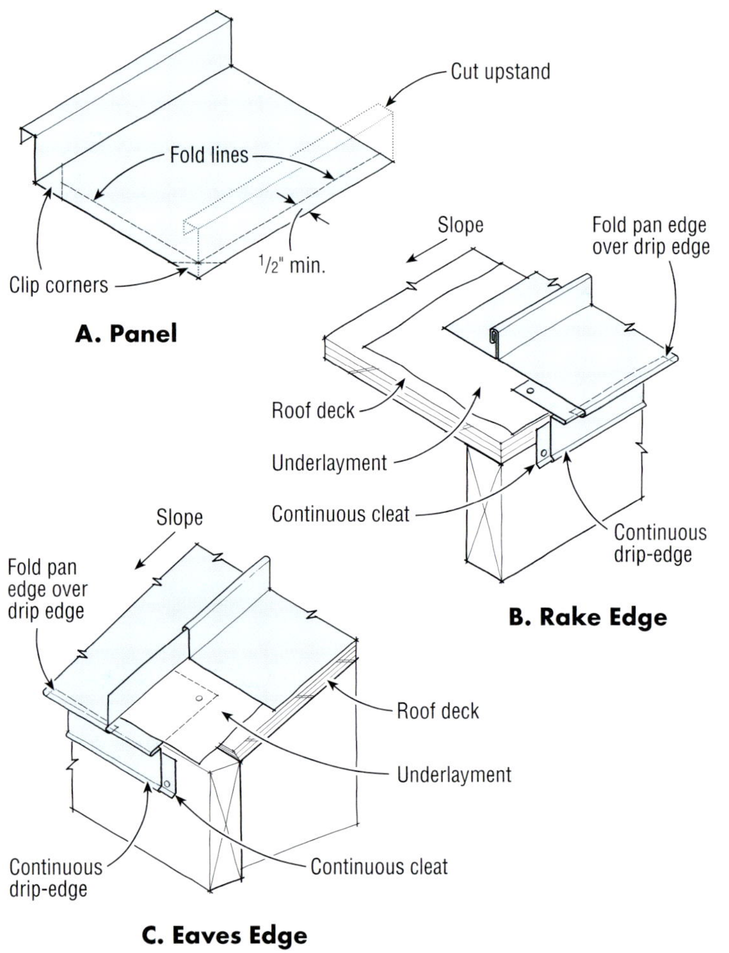

On a corner panel (along a gable end), cut off the upstand, then clip the pan corner at 45 degrees, as shown (A). Install a continuous drip-edge around the roof perimeter, then fold the edges of the pan 3/4 in. over this drip-edge, and lock in place with crimpers.

For more information: "Low-Cost Metal Roofing," JLC 6/92

- Install the first pan of the next row by hooking the triple bend over the double bend and sliding the pan up until it hooks on the drip-edge. Crimp the seam at the drip-edge, and install cleats along the right edge and top edge of the pan.
- Fold the first standing seam closed, as shown in **Figure 3-45**.
- Install the rest of the pans. Remember to stagger the cross-seams and keep the pans aligned with your chalklines.

Modern standing-seam panels are factory-formed from prefinished galvanized or Galvalume-coated steel (see "Metal Panel Coatings," **page 253**). These panels are made to custom dimensions and factory-formed with seams that snap in place. Some panel systems use cleats (or "clips"), but more commonly on residential products, the panels have a screw-flange along one edge of the panel that is concealed by the next panel.

Modern standing-seam panels are treated here as one variation of a pre-fab, vertical seam panel (see "Pre-Fab Metal Panels," below). A standing-seam profile will not leave voids in valleys like a ribbed panel; it is a good choice for complex roofs that must be made watertight (**Figure 3-41**).

PRE-FAB METAL PANELS

Similar to "ag panels" that have kept barns dry for years, pre-fab, vertical-seam panels typically use exposed fasteners — usually a screw with a rubber gasket. Some (particularly those with a standing-seam profile) use concealed clips or screws. Pre-fab panels are typically less expensive than field-formed standing-seam and are considerably easier for contractors to install without specialized equipment.

Selecting Metal Roof Panels

Steel. Most metal roofing panels are made of steel. All steel panels have a metallic coating, and may also have a paint finish. The most common thicknesses of steel roofing panels are 24-, 26-, and 29-gauge. (The higher the number, the thinner the metal.)

Aluminum. Some vertical-seam roofing panels are made with an aluminum base metal. Not to be confused with an aluminized coating (see "Metal Panel Coatings," **page 253**), aluminum panels are much more corrosion-resistant than steel panels. Residential "modular" panels may be as thin as .015 to .02 in. For larger vertical-seam panels, opt for a commercial-grade aluminum at .032-in.-thick.

Aluminum also has an increased potential for expansion and contraction with changes in temperature (see "Thermal Response," below). This is not a problem for a small modular panel, but long vertical-seam aluminum panels should be installed panels over Z-channel purlins with clips, per manufacturer's instructions (**Figure 3-46**).

Rib profiles. The rib pattern in vertical-seam metal roofing varies from manufacturer to manufacturer and region to region. In the Northeast and Southeast, a popular profile is called "5-V crimp," which has a 24-in.-wide exposure, with a double-vee profile along each edge and a single-vee profile down the center.

For cold, wet climates where ice dams and snow accumulations may occur, use a panel with the narrowest possible rib. The narrower the rib, the easier it is to seal.

THERMAL RESPONSE OF METAL

Metal expands and contracts with changes in temperature, which can be quite extreme on a roof. For example, in winter it's not unusual for a roof panel to go from 180°F on a sunny day to below 0° at night. Under these conditions, a 20-ft.-long steel panel would expand over $1/4$ in. An aluminum panel this size would expand more

Figure 3–45. Standing-Seam vs. Ribbed Panels

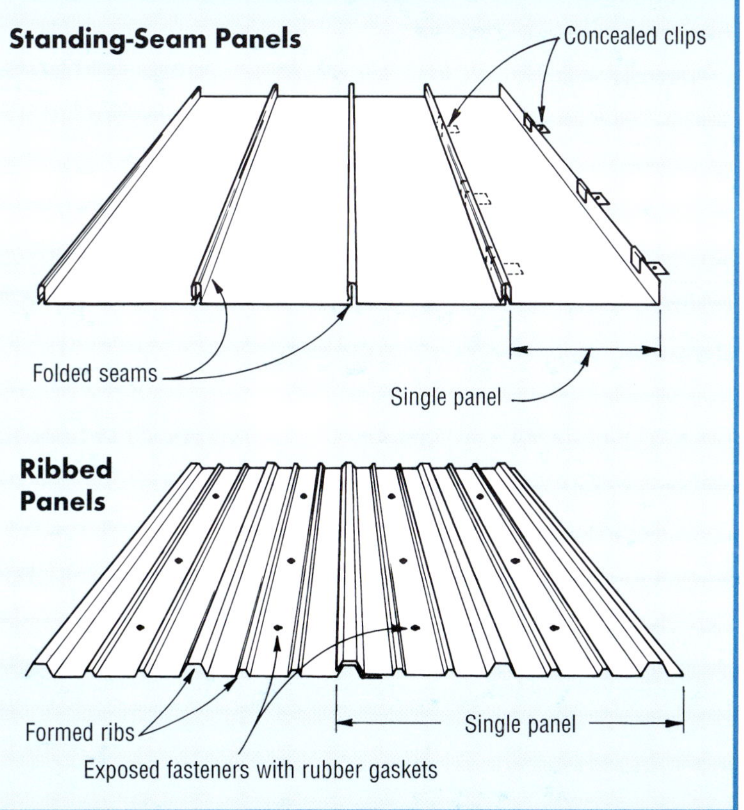

***Standing-seam** roof panels (at top) have folded seams and are fastened to the roof deck with concealed clips. **Ribbed panels** (at bottom) have formed ribs spaced by flat sections and are secured with exposed fasteners.*

than 1/2 in. If attached to plywood sheathing, the panel will "crown" between fasteners, and the force can cause screws to shear or the metal to tear around screws.

Evaluate the expected amount of change (**Figure 3-47**). If the change in the length of the panel exceeds .15 in., consider these strategies:

- Reduce panel length;
- Install panels over Z-channel to absorb movement (**Figure 3-46**);
- Choose a light (preferably white) panel in hot climates to reduce panel temperature.

Figure 3–46. Thermal Response Over Z-Channel

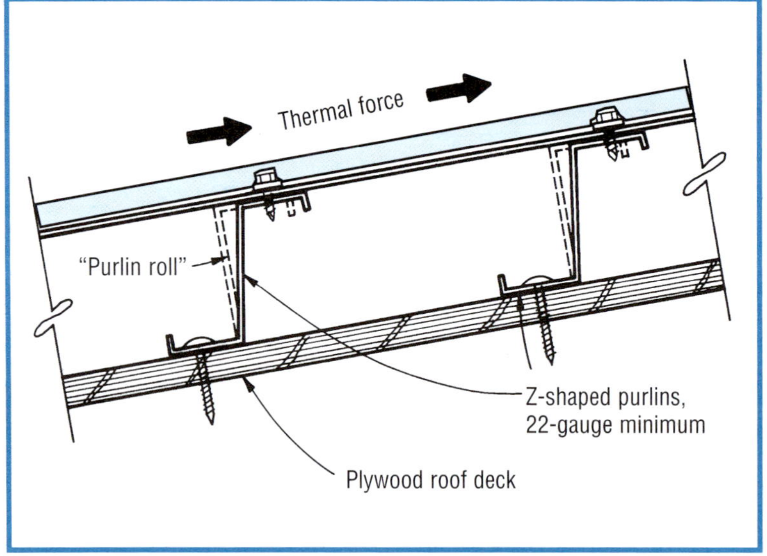

Metal roof panels installed over Z-shaped purlins will "roll" with thermal expansion of the panels. Wood strapping or hat channel will not absorb thermal expansion as well.

For more information: "Metal Roofing Options," JLC 1/03

"Installing Metal Panel Roofs," JLC 5/00

Galvanic Corrosion

Do not combine dissimilar metals. Always use fasteners, clips, purlins, and flashings of a similar or compatible metal. For a complete understanding of this, see description under "Flashing: Galvanic Corrosion," **page 272**.

- Don't use a pencil to mark cuts and fastener layouts. Graphite is listed on the galvanic scale and can corrode a roofing panel's metallic coating. Use a felt-tipped marker.

FASTENERS FOR METAL ROOFS

Pre-fab metal roof panels may have exposed or concealed fasteners.

Exposed fasteners for metal roofs are usually screws with rubber washers. Always install the screw in the flat portion of the panel, not on the raised rib. The rubber washer will seal around the screw. If installed through the rib, the screw will not get enough purchase to hold the panel securely, and because most of its length is unsupported, it may shear easily. Both increase the chances for blow-off.

Concealed fasteners for metal roofing may be a screw inserted through a concealed flange or through a panel clip designed to absorb expansion.

- Never use nails. The increased expansion and contraction of the panels will eventually work nails loose.
- Use screws that penetrate all the way through the sheathing 1/4- to 1/2-in.

ORDERING METAL ROOF PANELS

Agricultural panels may be sold in standard lengths, but most residential roofing panels are often ordered in any length (down to the fraction of an inch) from a lumberyard or roofing supplier. In general, the length of the panels will be the ridge-to-eaves measurement.

- In most cases, the 1 1/2 in. added for the eaves overhang is balanced by the 1 1/2 in. subtracted for ventilation at the ridge.
- The cap flashing installed at the ridge will allow room to fudge 1 or 2 in. at the top.

METAL PANEL COATINGS

All sheet steel is coated with a thin layer of corrosion-resistant metal.

Galvanized

The most common coating is pure zinc, known as a galvanized coating, which oxidizes, chemically sealing the base metal from corrosion.

- The thickness of a galvanized coating is designated as G-60, G-90, or G-100. The number here refers to minimum application rates equal to .60, .90, and 1.0 ounce of zinc per sq. ft. (total coating thickness both sides).
- Zinc is water soluble. As it dissolves, it tends to recoat scratches or cut ends and is considered "self healing." However, because the coating dissolves, it eventually washes away and the base metal begins to rust. The thicker the coating, the longer the panel will last.

Aluminized

Pure aluminum coatings over steel carry a Type II designation. (Type I is used in the automotive industry.) Owing to the truly inert nature of aluminum, these coatings don't dissolve or wear away, providing significantly better protection than galvanizing.

Galvalume

Also known as *Zincalume*, if the panel manufacturer originated in Australia or New Zealand, this coating is made from an aluminum-zinc alloy. The typical coating weight AZ-55 designates 55% aluminum and 45% zinc by weight. As one might expect, Galvalume falls between a pure zinc and a pure aluminum coating in its performance.

PAINTS FOR METAL ROOFING

Nearly all roofing panels have some type of factory-applied finish over the metallic coating.

Figure 3–47. Calculating Thermal Response of Metal

To get an accurate measurement of how much a metal panel will move in any climate, use the formula:

$$\Delta L = \Delta T \times L \times Ce$$

ΔL = the change in length of the panel

ΔT = the change in surface temperature throughout the year (in °F)

L = the panel length (in inches)

Ce = the coefficient of expansion for the panel material (see list below)

Coefficients of Expansion for Metal Roofs

Steel	.0000067
Copper	.0000094
Aluminum	.0000129
Stainless	.0000096

Fasteners for Metal Roofs

Metal Panel Coatings

Paints for Metal Roofing

Finish Warranty

Factory-applied paint rarely peels, but eventually will fade. It usually fades unevenly, so it looks unsightly. When a paint fades, it's a breakdown of the pigment. Depending on the quality of the paint resin, which creates a protective coating around the pigment, panel manufacturers will warrant the finish against fading. However, the warranty period on the finish is considerably shorter than the warranty against corrosion of the base metal.

Figure 3-48. Wall Flashing for Metal Roofs

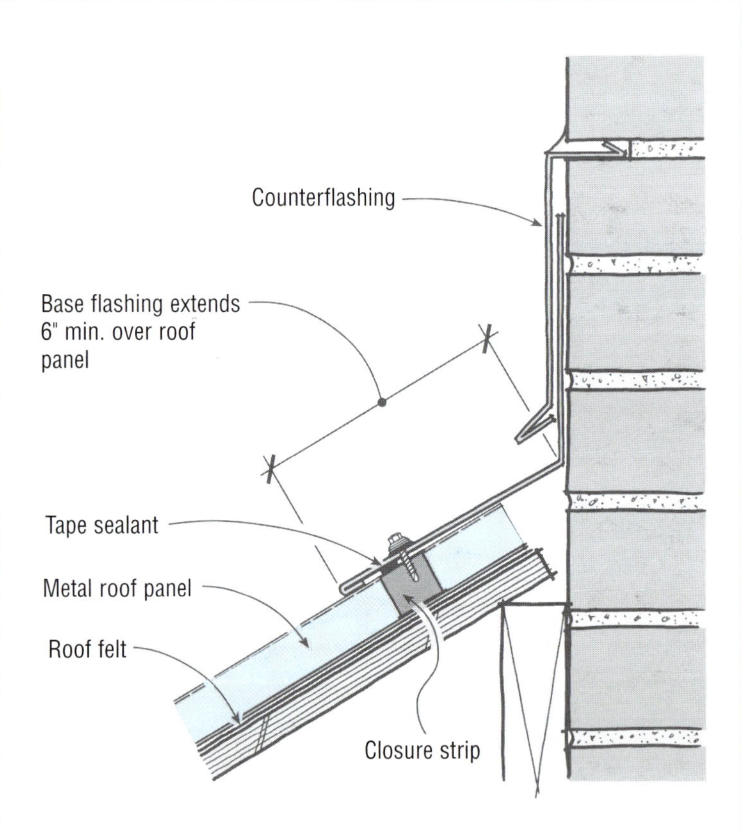

Use a two-piece metal flashing to accommodate the differential movement between walls (or chimneys, skylights, and roof hatches) and metal roof panels. Fasten the base flashing to the roof panel, but not to the counterflashing.

For more information: "Sealing and Flashing Metal Roofs," JLC 1/00

"Watertight Skylights in Metal Roofs," JLC 4/97

Color Choices

Red, bright blue, and black pigments tend to fade the fastest. Grey (the lighter the better), tan, and white fade the least.

Paint Resins

Better quality resins in paints bond tighter and provide better protection, but even the best ones eventually chalk. There are three types:

- **Polyester** resins are the least expensive and of the lowest quality.
- **Silicon-modified** resins will retain their gloss much longer than polyesters.
- **Fluoropolymers,** known by such brand names as Kynar® and Fluropon®, are the most expensive and longest lasting. The more resin, the longer the finish lasts. Top-quality panels typically use a 70% formulation, while less expensive products advertising this superior resin use thinner concentrations.

Field-applied paints. These will last only 5 to 7 years before peeling, even if applied under the best conditions.

METAL-PANEL SEALING DETAILS

House roofs, with dormers, hips, valleys, chimneys, skylights, and plumbing vent penetrations, put more demands on metal roofs. Metal roofs on houses must be built watertight by sealing all joints, side seams, and endlaps.

Sealants for Metal Panels

Thermal movement, freeze-thaw cycles, and exposure to ultraviolet light will degrade even the best quality sealants. For best results:

- Use butyl tape (1/2-in.-wide works well for panel seams). Tape sealants, which have a high solids content (about 97%), don't shrink as much as gun-grade sealants. Most tapes for metal panels are butyl polymers, which will last 30 years or more if protected from exposure to ultraviolet light. Gun-grade sealants are usually about 70% solids, which means that when the 30% solvent evaporates, the sealant shrinks.
- Do not use silicone. Many silicone sealants contain acids, which can damage metallic coatings. Also, silicone has poor adhesive strength on metal. If butyl tape is not available, use a urethane instead.
- Always seal between panels, so the sealant is hidden from the damages of ultraviolet light. Sealants smeared on the exterior of the panel will surely fail.
- At eaves, in valleys, and on the upslope side of chimneys and skylights, use rubber closure strips to seal the void created by the profile of the panel ribs. Most panel suppliers offer these closure strips in both male and female (inside and outside) configurations. Make sure a metal panel or flashing extends over the rubber closure to protect it from sunlight.

Wall Flashing For Metal Roofs

When flashing against walls — as well as skylights, chimneys, and roof hatches — use a two-piece flashing with closure strips (rubber blocks that mirror the panel profile) or preformed curb flashings (Figure 3-48).

- Connect the base flashing to the roof panel with washered screws.
- Apply tape sealant between the roof panel and the flashing, and be sure the screw pierces the sealant or is placed on the dry side of the tape, not on the weather side.

Flashing Metal Roofs

Figure 3–49. Sealing Valleys in Metal Roofs

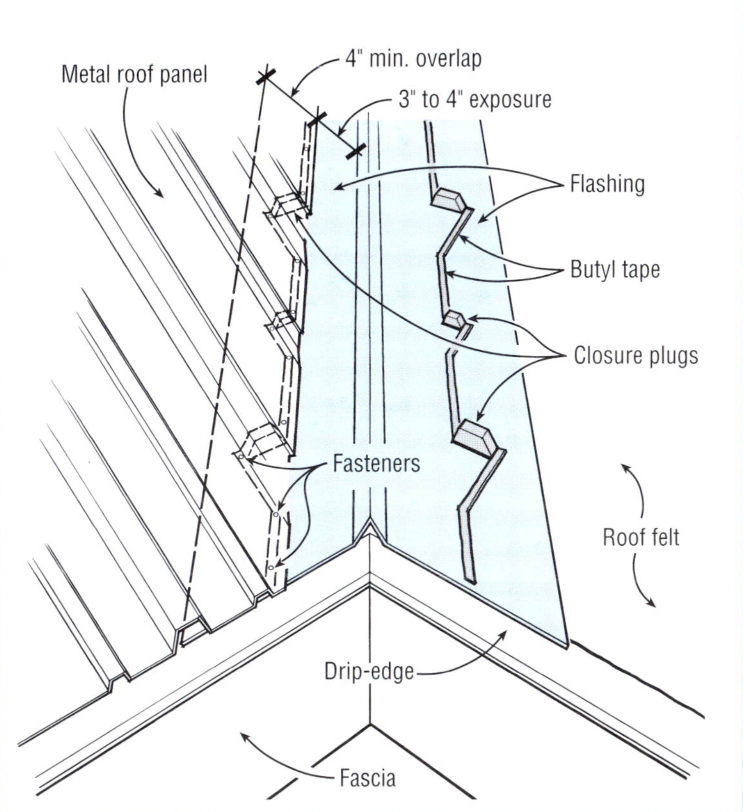

Seal the diagonal valley cuts on the roof panel with plugs cut from a closure strip. The plugs must be positioned square to the panel's length and sealed top and bottom with butyl tape.

- Seal the joint between the base flashing and the roof panel with tape sealant.
- On the upslope side of a chimney or skylight, use a cricket, or diverter, to direct the flow of water around the penetration (see "Flashing Chimneys," **page 280**). This cricket should be inserted beneath the upslope panel, sealed with butyl tape, and the voids at the cut edges of the roof panel sealed with rubber closure strips.

Sealing Valleys

Cutting a panel diagonally across a rib produces a large irregularly shaped void that is difficult to close.

- Use a valley flashing made from flat stock with the same coating and finish as the roof panels. Bend it to the roof slopes with an inverted "V" at the centerline (see "Open Valleys," **page 275**).
- Cut off the rib portion of a standard male closure so you are left with several "plugs." Position these plugs square to the panel's length.
- Install roof panels at least 3 to 4 in. upslope of the valley centerline, and make sure the flashing extends at least 4 in. past the panel edge so the closures seal to the flashing.
- When installing roof panels over the valley flashing, apply tape sealant on the top and bottom of the plugs, and use a close fastener spacing in the pattern shown in **Figure 3-49**. Screws must pierce the sealant or be placed on the dry side of the joint.
- On long valleys, use several valley sections to avoid problems with thermal expansion. Overlap the sections at least 3 in. and use a double row of tape sealant.

Figure 3–50. Sealing Vertical Seams

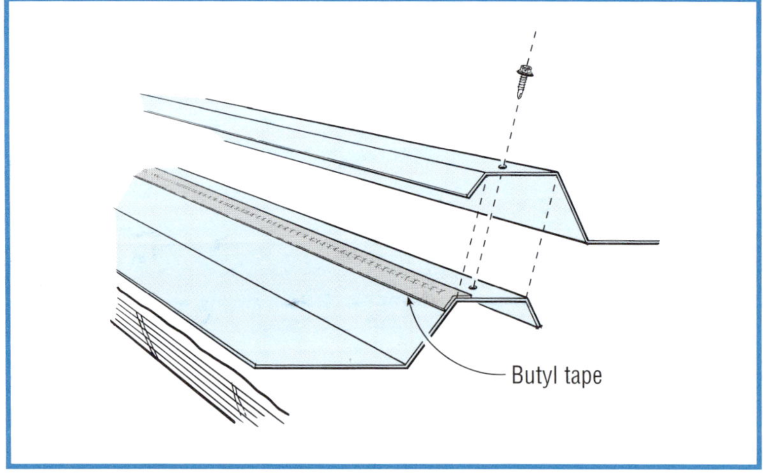

To seal the lap between adjacent panels, apply butyl tape to the weather side of the rib shoulder.

Figure 3–52. Sealing Eaves

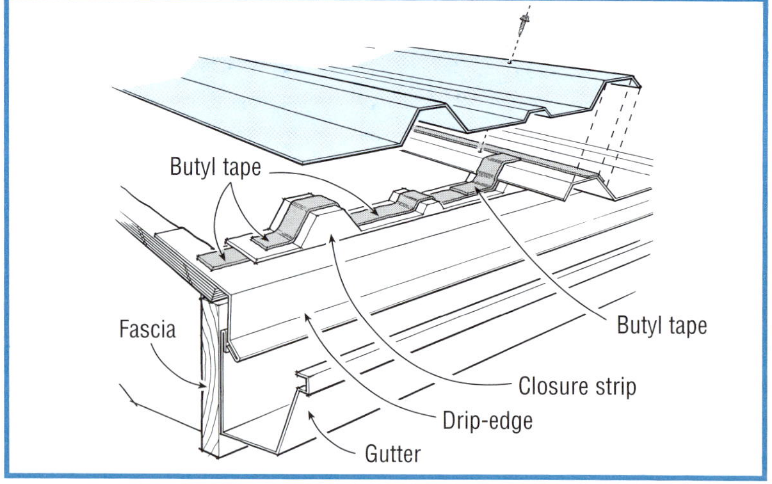

When gutters are used, *seal the eaves as if they will be submerged in water. Without gutters, roof panels should overhang the drip-edge at least 4 in.*

Sealing Round Penetrations

Round penetrations, such as plumbing vents and furnace flues, must be flashed with special rubber boot jacks that have a moldable aluminum compression ring (available from specialty contractor and commercial roof suppliers).

Sealing Vertical Seams

- Apply the tape to the top shoulder of the panel rib, making sure that the majority of the sealant winds up on the weather side of any fasteners (**Figure 3-50**).
- Do not peel away the release paper until you're ready to install the overlapping panel. Align the panel exactly before it contacts the sealant. Once a panel touches the tape, even a small adjustment is difficult.
- Screw panel sides together, piercing through the sealant on the "dry" side with 3/4-in.-long 1/4x14 hex head galvanized screws with a #1 drill point.

Sealing Endlaps

- Make sure the upslope panel always overlaps the downslope panel (**Figure 3-51**).
- Apply sealant on the weather side of the endlap fasteners.
- Place screws that secure the panel to the deck (or purlins) above the endlap sealant.

Sealing Eaves

- Drip flashing should be nailed down just enough to hold it in place prior to installing the roof panels.
- Screws to hold down roof panels should pierce through the panel, drip flashing, and deck.
- If gutters are used, take extra care to seal with rubber closures and butyl tape to prevent snow, rain, and ice from entering the panel voids (**Figure 3-52**).

Sealing Rakes

Prefabricated commercial rake flashings are available, but these are usually too large and "boxy" for use on a residence. Instead, ask the panel supplier to furnish flat sheet

Flashing Metal Roofs

Figure 3-51. Sealing Panel Endlaps

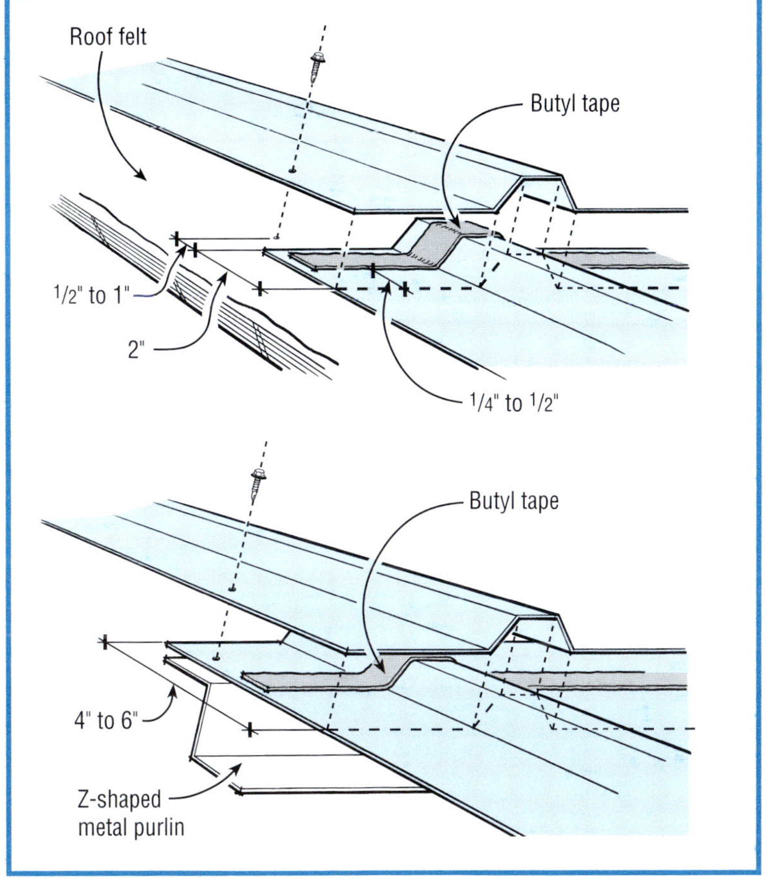

Use butyl tape to seal panel endlaps, making sure the screws that fasten the panel to the deck (at top) or to purlins and sleepers (at bottom) are placed above the sealant, as shown.

stock with a matching finish to be bent into a simple L-flashing, as shown in **Figure 3-53**.

- Attach L-flashing to the roof at a high rib with sealant tape. This might mean extending the horizontal leg a few extra inches.
- Fasten the L-flashing to the fascia using a metal cleat that matches the angle of the drip leg. (Alternatively, fasten through the face of the flashing with a washered screw.)

Hip and Ridge Caps

Most panel suppliers offer standard ridge components for use with their panels.

- Use a sealant tape to seal the joint between roof panel and ridge cap.

Figure 3-53. Rake Flashing

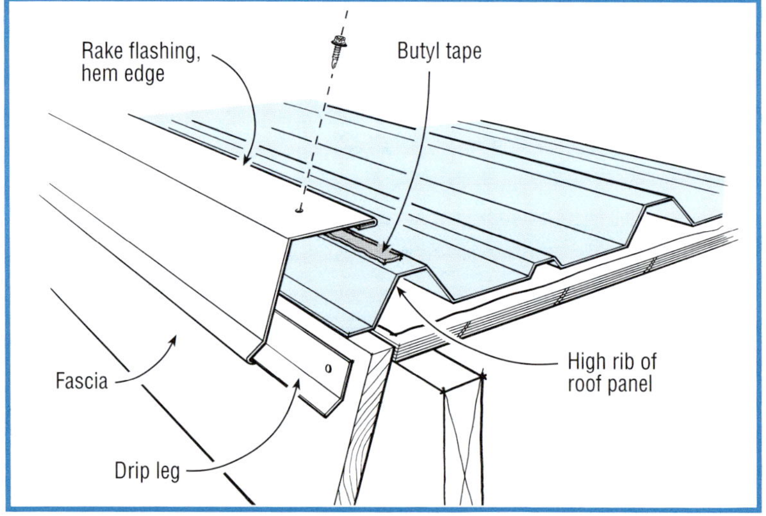

Fabricate rake flashing from flat sheet stock provided by the roof panel manufacturer. Extend the horizontal leg of the flashing so it attaches with butyl tape and screws to the roof at a high rib.

- When using long panels that are subject to substantial thermal movement, make sure the ridge cap and the panels are fastened separately to the deck.
- For hips, metal roof manufacturers typically provide metal J-trim, which is topped by a metal ridge cap (**Figure 3-54**).
- On complex roofs where snowmelt and ice damming patterns are difficult to predict, make the hip assembly watertight with plug closures and tape sealant.

REROOFING WITH METAL

- If there are two or more layers of shingles, the roof should be stripped to the sheathing.
- If there is a single layer of shingles, and the shingles aren't curled, metal panels can be installed over wood strapping or purlins.

Figure 3-54. Sealing Hips

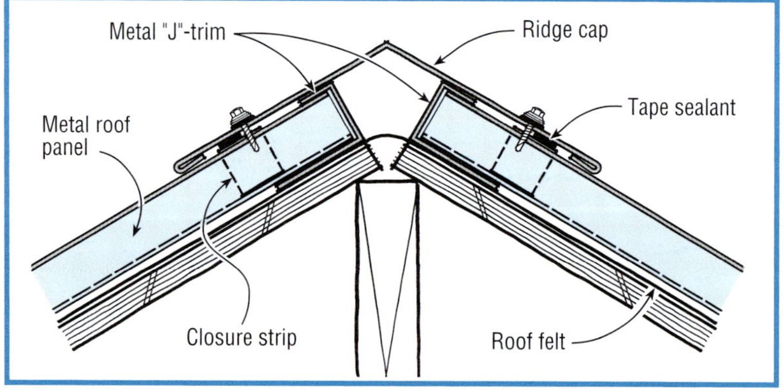

To seal hip ridges, use metal J-trim set square to the panel's length. On simple hips in steep roofs, this assembly can be installed without tape sealant; on complex or low-slope roofs, use with rubber closures and butyl tape.

New slate is not installed much these days, so this section focuses on inspection and repair.

WORKING ON SLATE

When working on existing slates, never walk directly on the roof surface. Dry slates are slippery; wet ones are as slick as ice. All slates are brittle and easily broken.

- If you must walk on the roof, spread your weight with foam insulation or plywood underfoot. Walk in the valleys, or hook a ladder over the ridge and rest it on some sort of cushion (a double layer of batt insulation will do).
- If the nails look very rusty, use extreme caution: when walked on, they could come loose.

INSPECTING SLATES

If possible, inspect the roof from the attic.

Evaluate Slate Conditions

The natural weathering of slates slowly chips and flakes off paper-thin laminations. Also, slates become soft and spongy as the inner layers begin to delaminate. Old, weathered slates tend to hold more moisture than new slates. This leads to further deterioration and even rot in the roof sheathing underneath.

Inspect Nails

The tips of nails can give a good indication of their condition. On the roof surface, signs of deteriorated nails include loose slates or a large number of repairs.

Inspect Transitions

When inspecting a slate roof, look closely at ridges, valleys, dormers, and other changes in roof direction. Cracked slates around a valley are a tipoff that something is trapping water. When the water freezes, it cracks the slate above it. Also look for roof cement — it's a sign of improper maintenance.

Moss and Mineral Deposits

Excessive moss and mildew are as harmful to slates as to other roof coverings. Moss traps moisture and its roots can penetrate and damage the slate. High moisture levels in the attic can also cause deterioration. Look for mineral deposits on the top or bottom surface of the slates, which indicate such underlying moisture problems and result in shortened shingle life.

Working on Slate

Inspecting Slates

Figure 3–55. Sizing Replacement Slate

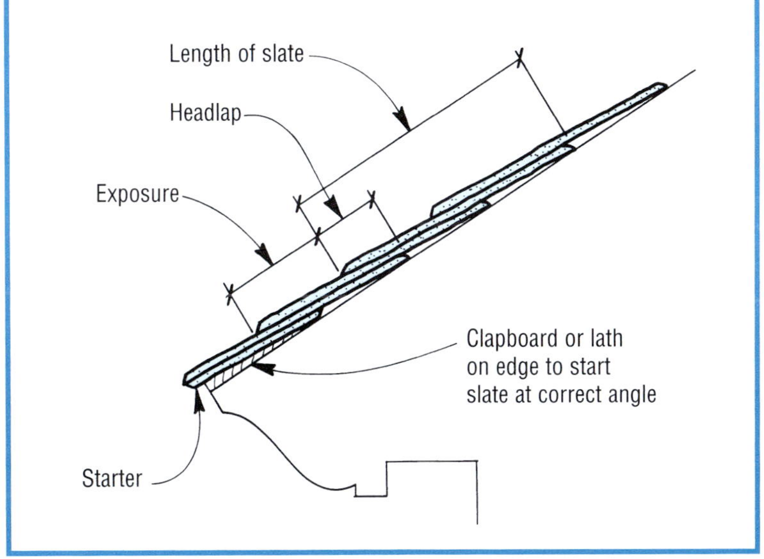

To find the size of the replacement slate, multiply the exposure by 2, and add 3 in. for the headlap. Round up to 12-, 14-, or 16-in.

Finding Leaks

In a severe windblown rain, even a good slate roof may get an occasional leak. Such leaks show up at inside corners, near valleys, and around chimneys. They're most common on shallow roofs.

Nail Holes

If an otherwise sound and well-designed roof leaks during a long rainstorm, look for nail holes that were punched too low or too close to the slates' centers. They may be letting water seep in from the end laps.

SLATE REPAIR CHECKLIST

Replacing a few broken slates is a simple procedure, but it must be done carefully or you can end up breaking more slates.

Measure Slates

To find the length required for a replacement slate, you can sometimes measure an exposed side at a gable end or valley. Another way is to multiply the exposure by 2 and add 3 in. as shown in **Figure 3-55**.

Figure 3–56. Replacing Damaged Slate

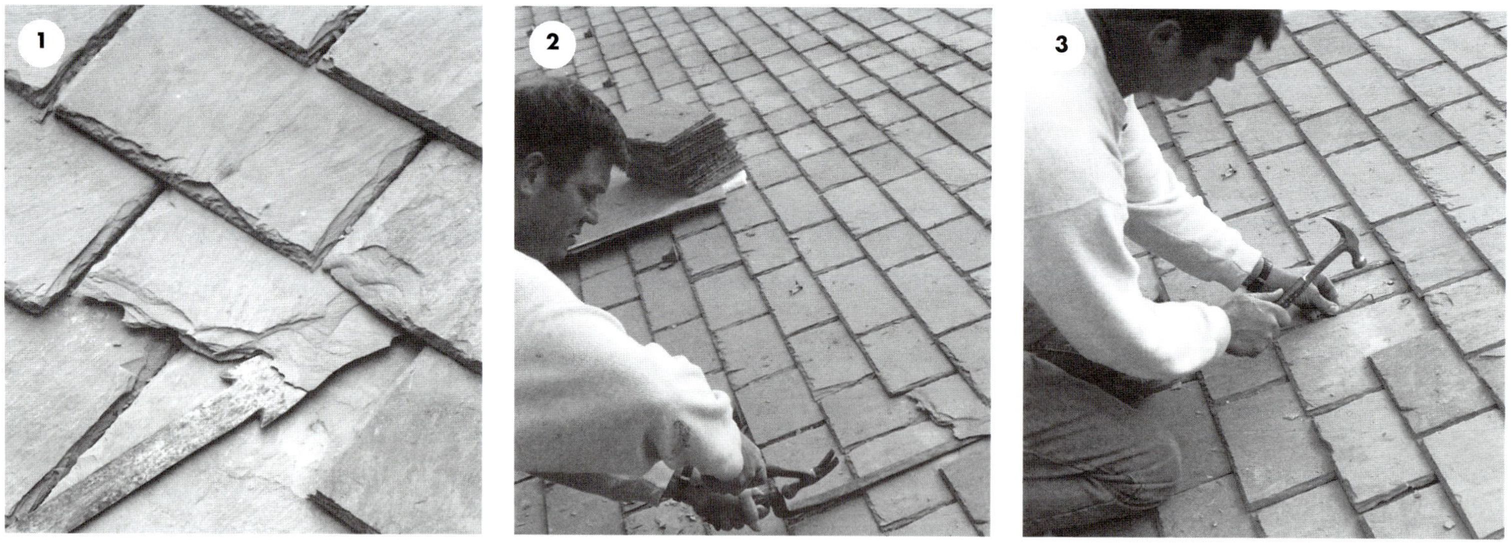

For more information: "Slate Roof Repairs," JLC 5/88

*(1) To remove the damaged slate, first hook the ripper on one of the two nails holding the slate. (2) Then hammer downward on the ripper to cut or pull the nail. (3) Next, nail a new slate into place with one nail and slide in a piece of copper flashing to cover the nail hole. Or hang the slate from a slate hook (**Figure** 3-57) or copper nail.*

Remove Damaged Slate
To remove a damaged slate, slide a slate ripper beneath it and hook onto one of the nails. Taking care not to put any upward pressure on the ripper (you don't want to crack any undamaged slates), hammer down on it until you either cut the nail or pull it out. Repeat for any other nails.

Install Replacement
The traditional technique is to nail punch or drill one hole in the replacement slate to line up with the joint between the two slates above. Nail in the joint with a copper nail, with the head trimmed if necessary. Then slide a 3x8-in. piece of copper flashing in place to cover the nail hole (**Figure 3-56**).

Nailing
Most slate roof failures are caused by improper nailing or by corroded nails or flashings. Galvanized nails, for example, may wear out long before the slates. For best results, use copper flashings and 14-gauge copper nails. They're expensive, but they'll last the life of the roof.

Nails should be set lightly so slates hang freely from the nails. Nails driven too far will crack the slate. Underdriven nails will crack the slate above.

Slate Hook
Another method is to fasten a copper or stainless-steel slate hook (available from most slate suppliers) in the joint beneath the slate you're replacing, as shown in **Figure 3-57**. Drive the hook's 3-in. shaft into the roof sheathing above the headlap of the slate below it. Then slide the new slate into place and pull it down until it rests in the bend of the hook.

Repairing Slate

In mild climates where snow and ice are not a problem, a strip of copper flashing bent to hold the slate can be used instead of a slate hook.

CUTTING SLATE

Cutting Slate

For a small number of cuts, a carborundum blade on a circular saw will do. When cutting a lot of slates, use a power tile saw or snap cutter.

Punching Nail Holes
To make nail holes, drill with carbide, or punch a hole (from the back of the slate) using the sharp point on a slater's hammer.

Figure 3–57. Installing a Slate Hook

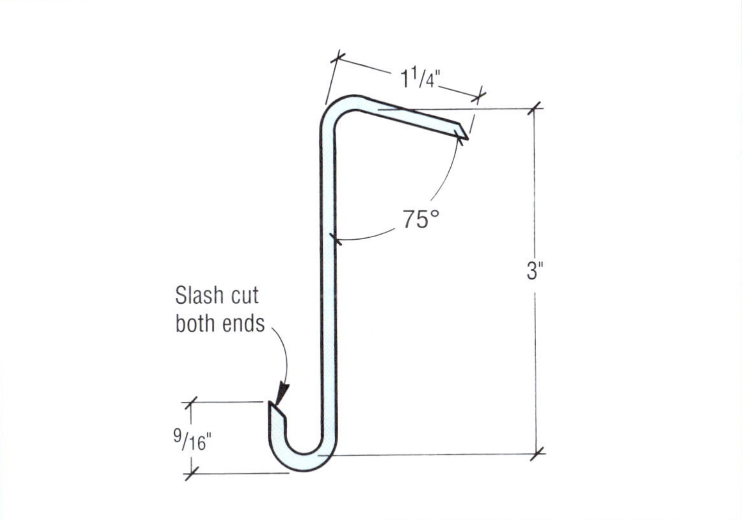

A slate hook is driven into the sheathing in the joint under the replacement slate. Use copper or stainless steel for a permanent replacement repair.

LOW-SLOPE ROOFS ARE THOSE WITH SLOPES BETWEEN 2:12 and 4:12. Flat roofs are those with slopes below 2:12. To prevent standing water, flat roofs should slope at least 1/4 in. per foot.

Roof Covering Choices

Roll roofing is a good, inexpensive choice for sheds or other outbuildings, but the best materials for residential flat and low-slope roofs are now the single-ply membranes. There are many single-ply membranes on the market, but the most proven are modified bitumen, EPDM rubber, and PVC. Some manufacturers sell only to certified installers and only for commercial installations. Refer to **Figure 3-58** for information regarding these different materials.

Figure 3–58. Roofing Materials For Low-Slope and Flat Roofs

Product	Advantages	Disadvantages
Roll Roofing	• Easy to install • Least expensive • Can have exposed nails on roofs with slopes 2:12 and above. Double coverage, with concealed nails, can be used on roofs with slopes of 1:12	• Short life expectancy (5 to 12 years depending on weather conditions) • Tendency to curl during installation • Should not use in temperatures below 45°F • Class "C" fire rating for both single & double coverage types
Built-Up Roofing	• Life expectancy 15 to 20 years (depending on the number of plies in the roof membrane and the method of application)	• Installed by specialists • Heavy equipment needed on site
Hot-Mopped Modified Bitumen	• Installation techniques based on BUR techniques • Slop in liquid bitumen can seal small holes	• Heavy equipment needed on site • Hot, dirty conditions during application may affect quality control
Self-Adhering Modified Bitumen	• Ease of installation • Minor adjustments of sheet placement are possible	• Needs careful substrate preparation; may be too much dust after tearoff to use self-adhering system • Rain-lapped application difficult on higher slopes
Torch-Applied Modified Bitumen	• Low-temperature applications (20°F) possible • Simple flashing of penetrations and terminations • Good adherence and good seams with experienced application	• Fire hazard from use of torches and flammable deck materials and cant strips • Client's and builder's insurance may not cover open-flame application • Worker's judgment required for proper flow of heated surface
EPDM	• Good resistance to ozone, UV light, weathering, and abrasion • Comes in larger sheets than PVC • Resistant to some acids, alkalis, ketones, esters and alcohols • Retains strength in extreme temperatures • Very stable across wide temperature range; less stress on flashings • Life expectancy between 15 to 20 years	• Membrane swells and distorts with exposure to various solvents and animal/vegetable oils • Lap seams are weak • Usually comes in black only; available in white at greater cost • Less puncture-resistant than asphaltic products
PVC	• Stronger seams than EPDM; lap seams are "welded" • Available in white, grey, and other light colors	• Life expectancy averages 8 1/2 years (less than EPDM) • Less resistant to chemicals than EPDM; Inferior puncture resistance • Membrane prone to shattering with cold temperatures • Easily damaged by contact with asphalt

For more information: "Leakproof Details for Shallow Roofs," JLC 5/94

ROLL ROOFING

Roll roofing is similar in composition to asphalt shingles. It comes in 3-ft.-wide rolls that include an asphalt-impregnated organic or fiberglass mat and mineral granule surfacing.

Installing Roll Roofing

Roll roofing may be single- or double-coverage, as in Figure 3-59.

Single-coverage roofing has only a 2-in. overlap and exposed nails.

Double-coverage. This two-ply system uses about twice as much material as single coverage, but results in a more durable and watertight roof. In this case, the nails are hidden by the overlying course and sealed with lap cement.

Repairing Roll Roofing

Repair roll roofing by cutting away the damaged area and cementing in an oversized patch.

EPDM RUBBER

EPDM (ethylene propylene diene monomer) is a single-ply, synthetic, rubber roofing membrane. It is dimensionally stable across a wide temperature range, so it puts less stress on flashings than other roof systems and retains its strength at extreme temperatures. Because the sheets are big, ranging in size from 50x50 ft. to 50x200 ft., lap splices can be kept to a minimum (lap splices are a weak link in most roofing membranes).

Selecting EPDM Roofing

EPDM sheets can be fully adhered or mechanically attached with special fittings. On residential projects, fully-adhered systems prove to be the most practical.

Availability

EPDM is available in two thicknesses: .045 in. and .060 in. The .060-in. or 60-mil EPDM lies flatter, wrinkles less, and is more durable than the 45-mil material.

Roll Roofing

EPDM membranes are usually black. If excessive heat buildup is a problem, however, they're also available in white (at a premium).

EPDM Rubber

Durability

EPDM has excellent UV resistance, so it requires no further covering for protection from sunlight. It deteriorates about 1 mil per year, but its life span is usually guaranteed by the manufacturer in the 15- to 20-year range.

EPDM will stretch without tearing, and it holds up well to the rigors of installation. It also performs well in extreme temperatures: Typical products will stay flexible down to 49°F and resist heat up to 300°F without cracking or deforming.

Avoid Grease and Solvents

EPDM should not be exposed to grease, solvents, oils, or petroleum products; fortunately, these are not usually a factor on residential roofs.

Decks over EPDM

EPDM can be covered with a wood deck on pressure-treated sleepers, with outdoor carpet or Astroturf, and even with concrete pavers.

Painting EPDM

EPDM can be painted with Hypalon paint, but the paint will have to be recoated every three or four years. (If you use EPDM on a rooftop deck, you should warn the homeowner about grease spills from a grill.)

Figure 3–59. Roll Roofing Installation: Single- and Double-Coverage

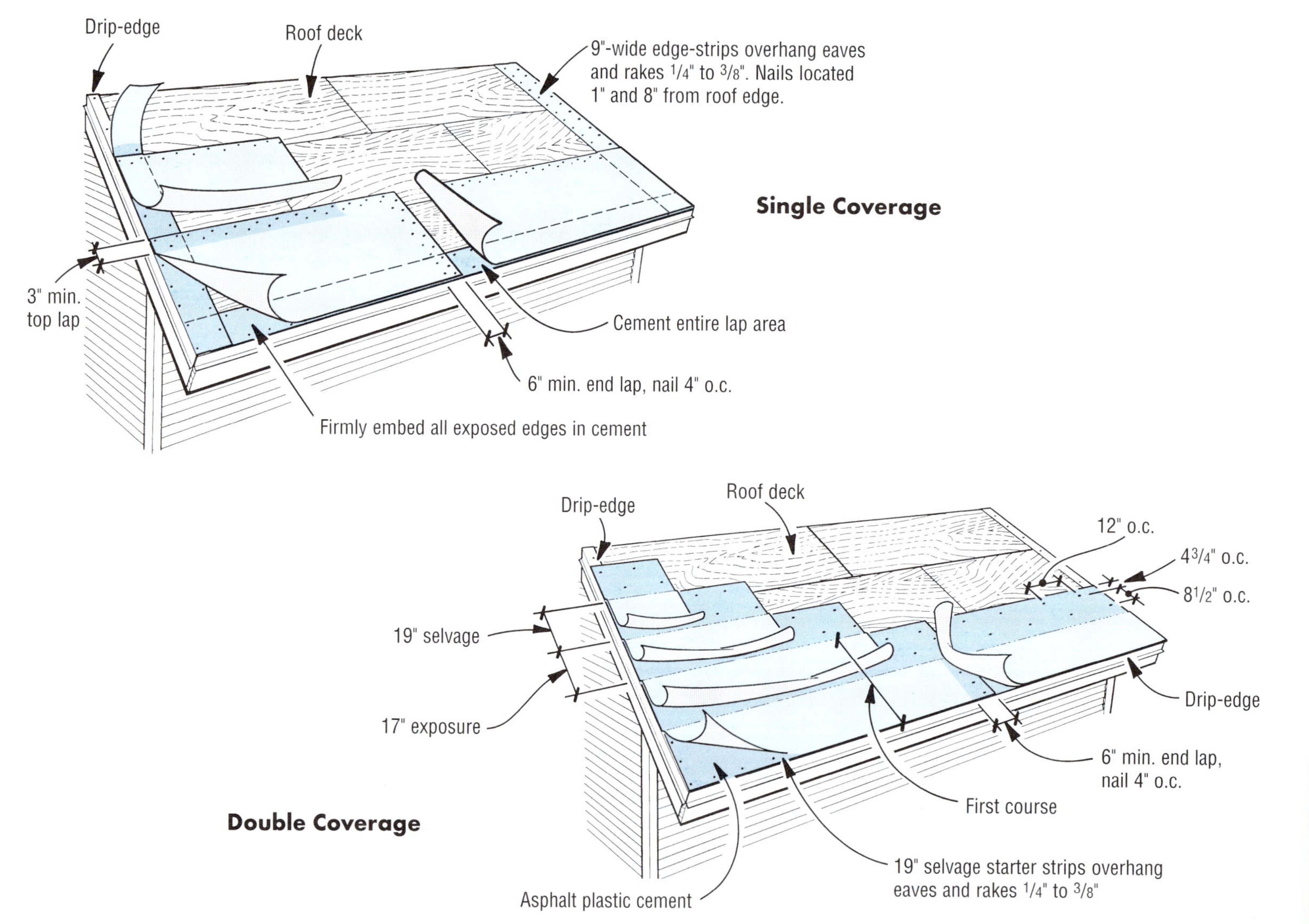

Nails are typically left exposed with single-coverage roll roofing. Double-coverage offers better protection and concealed nails.

For more information: "Installing EPDM Rubber Roofs," JLC 11/96

"Torch Down Roofing Basics," JLC 8/01

FULLY-ADHERED EPDM

The most practical method for installing EPDM on a residential project is to install a fully-adhered system where the membrane is installed over foam board made with a special glass-fiber-reinforced felt skin that is compatible with EPDM adhesive.

Foam-Board Substrate

- Attach the foam board to the roof deck with screws and large metal plates installed on 2-ft. centers in both dimensions (Figure 3-60). The fasteners sit slightly proud of the surface, so the plates show through the rubber with the manufacturer's distinct dimple pattern. That way, for warranty work, the manufacturer's inspector can identify and count the plates.

Fully-Adhered EPDM

Membrane

- Spread the rough-sized membrane sheet out on the deck, smooth out the wrinkles, and get it into final position. Then roll the sheet back onto itself halfway.
- Using rollers, spread the contact-type adhesive on both the rubber and deck.

Figure 3–60. EPDM Edge and Roof-to-Wall Details

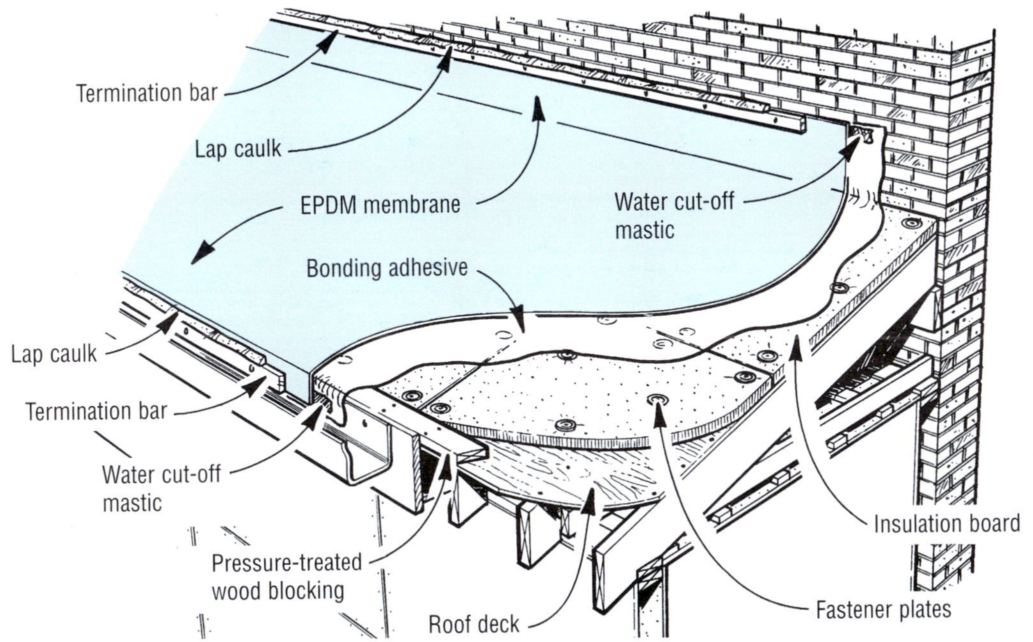

Use aluminum termination bars to secure the edges of an EPDM roof. A special lap caulk protects the top of the termination bar, while a water-block tape or mastic seals the bottom edge of the membrane.

- Once the adhesive is dry, carefully roll the membrane back into place and brush the surface with a stiff deck broom for good adhesion.
- Repeat the process for the other half of the sheet.

SELF-ADHERING EPDM

Self-adhering EPDM membrane has recently become available, eliminating the need to field-apply adhesive.

Seams

Splice seams together with either splicing adhesive or seam tape.

Option 1: Splicing Adhesive

- Allow for a 6-in. overlap.
- Adhere the bottom (downhill) sheet into the deck first.
- Carefully clean the overlapping surfaces with white gas or a cleaning solvent recommended by the EPDM manufacturer.
- Then brush the special splicing adhesive onto both mating surfaces. When gluing the top sheet, be careful not to get any adhesive on the overlap area.
- When adhesive is dry to the touch, carefully roll the top piece into place.

Figure 3–61. Turning Corners with EPDM

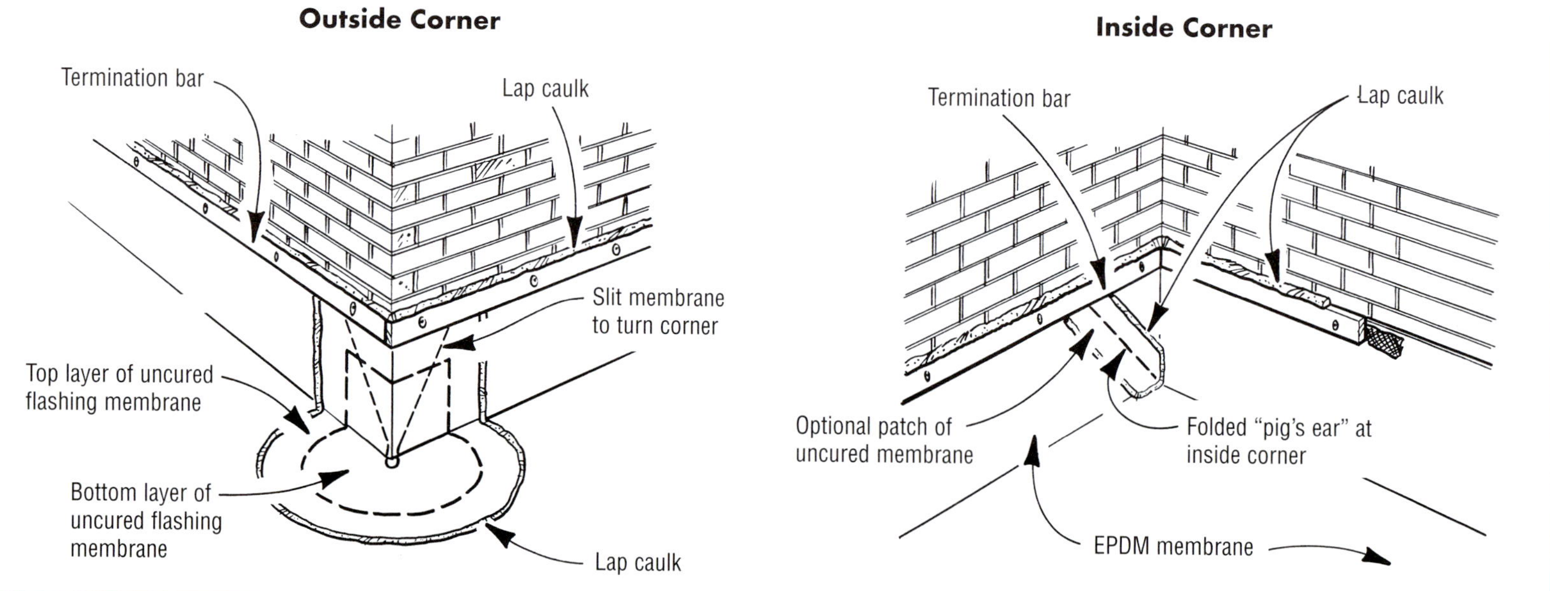

At outside corners, slit the membrane and apply a two-layer patch of uncured flashing membrane. Inside corners can be either slit and patched like an outside corner, or folded and secured with a termination bar.

For more information: "Low-Slope Re-roof with EPDM," JLC 11/00

"Building a Deck Over Living Space," JLC 5/02

- Make sure not to wrinkle or stretch the material, since this puts a built-in stress on the seam.
- Roll the seam with a hand roller to get 100% adhesion.
- Apply a bead of a special lap caulk to the edge of the seam to keep water out.

Option 2: Seam Tape

- Clean and prime the mating surfaces.
- Apply the tape (a 6-in.-wide, two-sided tape with a release paper on one side) to the bottom sheet, and hand-roll it with the release paper in place.
- Then fold the lapping piece over the seam and pull the release paper off the tape as you press the top piece into place.
- Carefully trim the top sheet before sticking it down. Its edge should be parallel to and 1/8 in. back from the edge of the seam tape.
- The edge can then be finished with lap caulk or left as is, since the exposed edge of the seam tape acts as a water stop.

EPDM EDGE DETAILS

There are a variety of ways to detail roof edges. Contact your EPDM supplier for installation specs.

Roof-to-brick. Where a roof meets a brick wall, glue the EPDM to the brick 8- to 12-in. up and secure it with an aluminum termination bar (**Figure 3-60**).

Roof-to-wood. For a wood-sided wall, run the EPDM behind the bottom course of siding.

Corners. Where the roof meets an inside corner of a sidewall or parapet, you can either fold the corner or slit the membrane and patch (**Figure 3-61**). At outside corners, a patch is always necessary.

Penetrations Through EPDM

For waterproofing roof penetrations, use uncured EPDM, which comes in narrow rolls designed for patching and flashing applications. Unlike cured EPDM, uncured EPDM has no memory. When you stretch and glue it around a vent pipe, for instance, it forms itself to the shape of the pipe (**Figure 3-62**).

Self-Adhering EPDM

EPDM Edge Details

Figure 3–62. Sealing around a Vent Pipe with EPDM

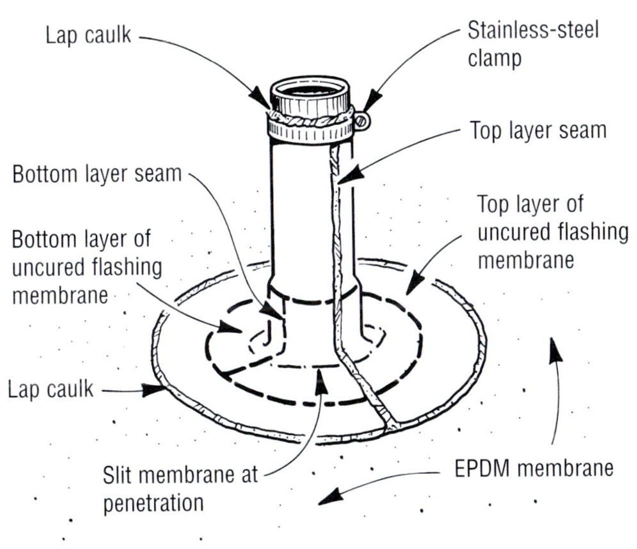

Uncured EPDM flashing membrane molds easily to roof penetrations like plumbing vent pipes. All flashing patches should be sealed with a fully-glued double layer of EPDM.

Patching EPDM

Patches require two layers of flashing membrane — a smaller first layer and a wider top layer that completely covers the bottom layer. Patches and flashing are glued with splicing adhesive.

BUILT-UP ROOFS

The most common encounter residential contractors have with built-up roofs (BUR) is when one is leaking. The most common problems are shown in **Figure 3-63**.

Leaks aren't always where they seem to be. With a membrane roof, there's a fair chance a leak will be right above the wet spots on the ceiling. With a BUR, however, water can travel through the plies and emerge almost anywhere.

Blistering usually indicates that a built-up roof wasn't installed right. Dirt or moisture may have gotten between the plies during installation, fasteners may not have been driven tightly enough, or the contractor may have used incompatible materials. Isolated blisters can be fixed following the patching procedure below. If the roof is covered with them, however, you may have to tear it off and start over.

Alligatoring means that the surface of the roof has dehydrated and begun to come apart. If the damage is minor, and if the roofing felts haven't yet hardened, you may be able to restore them with a mineral rubber resurfacer and sealant. If the felts have dried out, apply a primer first. Whatever the scope of the repair, remove all gravel first.

Lack of gravel is what causes alligatoring. One of the functions of a gravel layer is to protect the surface of a built-up roof from the sun's ultraviolet rays, which cause the roof to dry out. If the gravel wears away, the roof won't be far behind.

Figure 3–63. Troubleshooting Built-Up Roofing

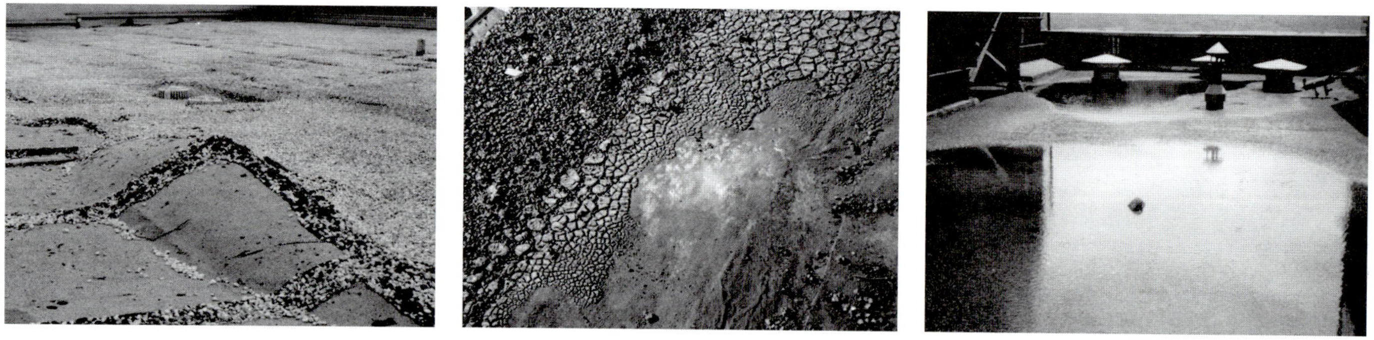

Blistering (left) indicates that the built-up roofing wasn't installed properly. ***Alligatoring*** *(center) is caused by lack of gravel in the flood coat. The surface of the roof has dried out and is starting to deteriorate.* ***Ponding*** *(right) is caused by the lack of positive drainage — due to either poor roof design and/or installation or to settlement of the roof structure.*

Ponding water may be caused by a deflection of the roof frame. Ponding can cut the life of a roof by half. Unfortunately, there's not much you can do about it short of a radical restructuring of the roof frame.

Slope transitions. Flat built-up roofs that meet a sloped roof must have a gravel stop to prevent the asphalt flood coat and gravel from migrating down the sloped roof (**Figure 3-64**).

Figure 3-64. Flat Roof Transition

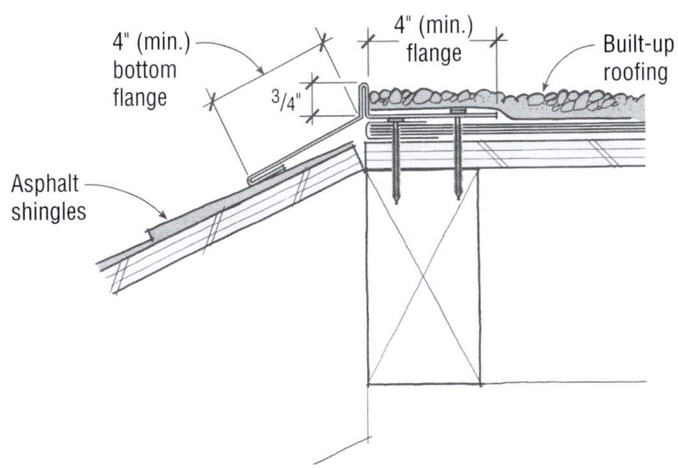

PATCHING A BUILT-UP ROOF

Small BUR repairs can be accomplished using plastic roofing cement and felt (**Figure 3-65**).

Surface Prep

Clear away the aggregate in the top or "flood" coat with the claw of a hammer or, for larger areas, use a chipping bar (be careful not to damage the existing BUR membrane below the coatings). Smaller stones can be scrubbed away with a wire brush.

Felt Patch

- Use a minimum of two layers of felt or glass fiber mesh to repair BUR. The bottom layer should be $1^{1}/2$ in. smaller around the edges than the top layer, so that each will bond separately to the existing membrane.
- Apply the patch with a thin coat of plastic roofing cement.

Modified Bitumen Patch

High-quality repairs also can be made with a single layer of modified bitumen.

- Heat the membrane on a scrap of plywood with a propane torch.
- Then apply it over the hole, being careful not to wrinkle it.

Figure 3-65. Patching a Built-up Roof

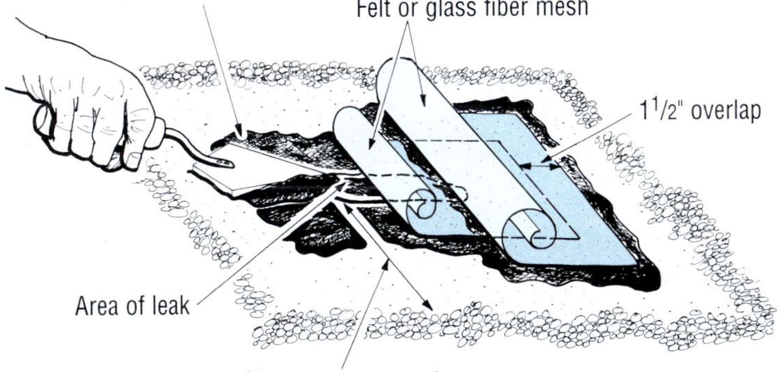

Apply plastic roofing cement carefully: Too much will weaken the bond between the existing BUR and the new patch.

FLASHING MATERIALS SHOULD BE AT LEAST AS DURABLE AS THE roof covering. For example, it would not make sense to flash a 100-year-old slate roof with thin aluminum. Once the aluminum begins to leak, the slates would have to be torn off to replace it. Copper flashing would cost more initially, but because it could last as long as the roof, its life cycle costs would be lower.

ASPHALT-BASED EAVES FLASHING

Most roofing systems can be installed over asphalt eaves flashings and roofing cement.

Eaves Flashing

An eaves flashing is a minimum 3-ft.-wide bituminous membrane that's installed beneath the roof covering at the lower edge of a sloped roof (**Figure 3-66**). It's used in regions with design temperatures of 0°F or cooler as a secondary defense against leakage from ice dams.

Not a cure for ice dams. Although some codes require eaves flashings, remember that they're merely a backup; the only proven cure for ice dams is proper insulation and ventilation (see "Ice Dams," **page** 287).

In cold climates, an eaves flashing is recommended as back-up protection against leaks from ice dams. (To prevent ice dams completely, address the ventilation and insulation issues in the attic. See "Roof Ventilation," ***pages 285 to 288.****) Self-adhering bituminous membranes are preferred since they seal around nail penetrations.*

Figure 3–66. Eaves Flashing

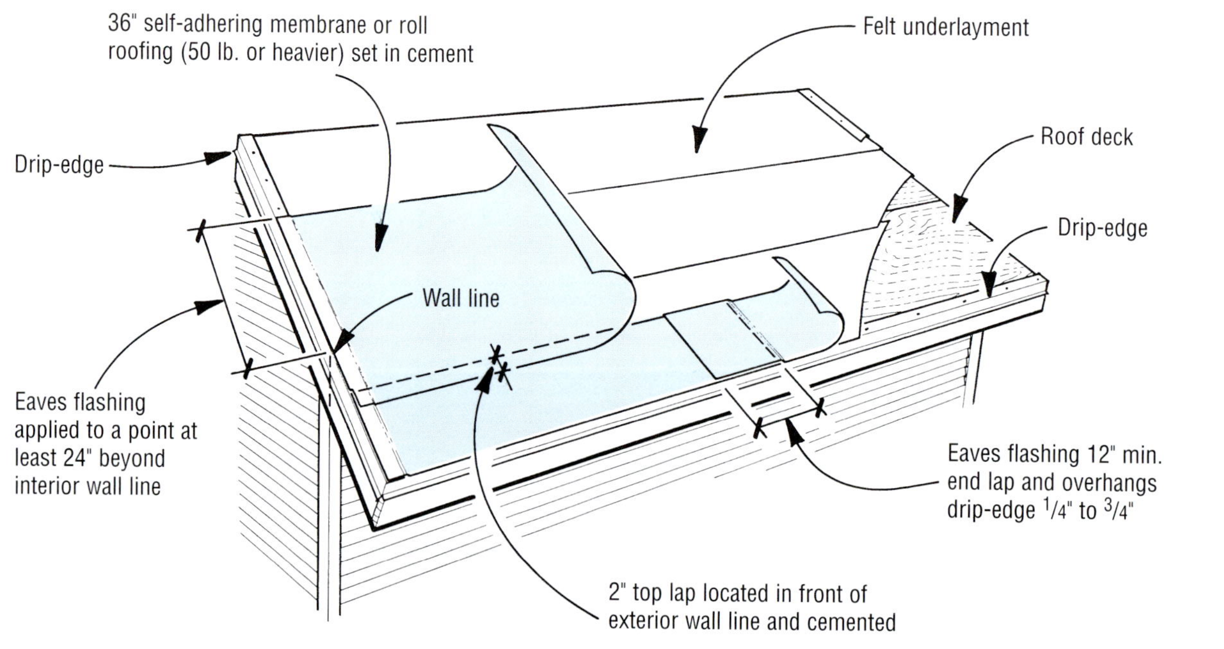

Roofing Cement

Although roofing cement has many uses, it's not a substitute for flashing. Also, not all roofing cement is alike. The three main types are *plastic asphalt cement*, *lap cement*, and *asphalt adhesive cement*. Use each according to Figure 3-67.

FLASHING METALS

The most familiar residential flashing metals are aluminum, galvanized steel, zinc, lead, and copper. Which one to use depends on the roofing material and the surrounding environment. Galvanized steel and aluminum are more than adequate for most asphalt roofs, for example, but they won't last as long as a good slate or tile roof. Copper and lead are more durable but also much more expensive.

Asphalt-Based Eaves Flashing

Gauge Thickness

Whatever the metal, be sure it's thick enough for the job. See Figure 3-68.

Durability of Metal Flashings

The longevity of any flashing depends on the environment.

Selecting Flashing Metals

Figure 3-67. Roofing Cement Types

Material	Characteristics	Uses	Cautions
Plastic asphalt cement (flashing cement)	Thick consistency; Won't flow when hot or become brittle when cold; Remains flexible enough when dry to expand and contract with the structure	Gluing down flashing assemblies, caulking joints, and minor repairs	Incompatible with roll roofing
Lap cement	Thinner than plastic cement; Brushable	Makes a waterproof bond between layers of roll roofing	Shouldn't be used on asphalt shingles
Quick-setting asphalt adhesive cement	More adhesive than plastic asphalt cement	Used to glue down strip shingles on shallow slopes	–

Figure 3-68. Minimum Thickness for Flashing Materials

Material	Base or Step and Counterflashing	Wall Openings Head & Sill	Roof Edges, Ridges, and Hips	Crickets, Valleys, or Gutters	Roof Penetrations
Aluminum	.019"	.019"	.019"	.019"	.040"
Galv. Steel	26 gauge	26 gauge	24 gauge	24 gauge	24 gauge
Copper	16 oz.	10 oz.	16 oz.	16 oz.	16 oz.
Lead	1 1/2 lb.	2 1/2 lb.	3 lb.	3 lb.	3 lb.
Painted Terne	20 lb.	40 lb.	20 lb.	40 lb.	40 lb.

For example, the alloy used in an aluminum or copper step flashing will corrode quickly when exposed to salt. Therefore, do not to use aluminum or copper flashing near the ocean. Use zinc instead — it's not harmed by salt. Lead and copper, on the other hand, stand up well to urban air pollutants, while zinc and galvanized steel don't.

Copper and Cedar Caution

There has been some speculation that the acids in cedar and redwood might corrode copper flashings in a damp environment. This hasn't been proven, but for the time being the cedar industry recommends that copper nails be avoided and that copper valleys be made from 16-ounce copper, minimum. To isolate wood shingles from a copper valley, use a cant strip.

Galvanic Corrosion

When dissimilar metals come into contact in the presence of moisture, a small electrical current flows between them, corroding one metal while protecting the other. This is known as a galvanic couple. The speed and direction of the current — and thus the rate of corrosion — is determined by each metal's place on the galvanic series (**Figure 3-69**).

The galvanic series ranks metals from active to passive. In a galvanic couple, the more active metal sacrifices itself to protect the more passive one. The farther apart the two metals in the series, the greater the potential for corrosion.

Figure 3–69. Galvanic Series

Active

1. Zinc
2. Aluminum
3. Steel
4. Iron
5. Nickel
6. Tin
7. Lead
8. Brass
9. Copper
10. Stainless Steel

Passive

Some metals can be used together — lead and galvanized steel, for example — with minimal problems, while other metals — such as copper and aluminum — should never be used together (**Figure 3-70**).

The Area Effect

Galvanic corrosion takes place when the anode gives up its electrons to the cathode — the faster the exchange, the faster the corrosion. Exactly how fast depends, in part, on the relative amounts of each metal. The larger the area of the cathode in relation to the anode, the more intense the coupling, and the faster the corrosion.

The area effect explains why an aluminum nail in a copper flashing corrodes much faster than an aluminum flashing fastened with a copper nail. In the latter case, the nail exerts a very small load on the flashing, while the flashing spreads the load over a relatively large area.

Figure 3–70. Galvanic Corrosion Between Construction Metals

	Zinc	Alum.	Iron/ Steel	Gal. Steel	Lead	Brass/ Bronze	Copper	Stainless Steel
Zinc	–	0	2	0	0	2	2	2
Aluminum	0	–	1	0	1	2	2	0
Galv. Steel	0	0	1	–	0	1	1	1
Lead	0	1	0	0	–	1	1	1
Copper	2	2	2	1	1	1	–	2

0 = Galvanic action minimal. Metals may come into contact under normal conditions.
1 = Galvanic action possible under some conditions or over long duration.
2 = Strong galvanic action, so direct contact not recommended.

For best results, never combine metal flashings with a 1 or 2 corrosion rating.

Galvanizing

Galvanic corrosion is the secret behind galvanizing. The zinc coating on a piece of galvanized steel is sacrificial, which means it gives way to protect the underlying steel. Because zinc is close to steel on the galvanic series, the reaction takes place slowly enough to give the steel long protection. Once the galvanized coating wears away, the underlying steel begins to rust.

Hot-Dipped vs. Electrogalvanized

Hot-dipping differs from electrogalvanizing in that the former diffuses the zinc part way into the steel's surface, while the latter just leaves a surface coating of zinc. Therefore, hot-dipped galvanized materials last longer than electrogalvanized before the base metal begins to corrode.

Galvanized Metals

Figure 3–71. Closed Valley Details

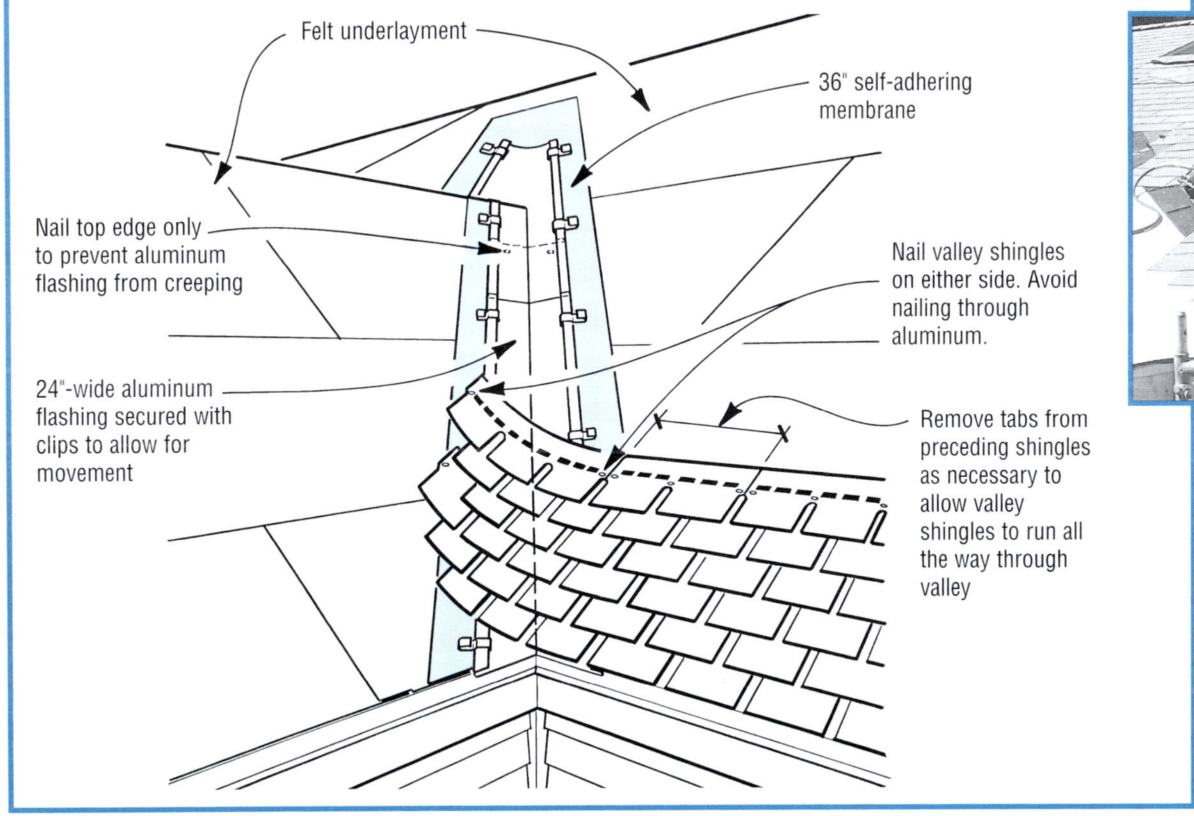

Use a built-up multi-layer flashing system for a more dependable valley. Avoid nailing through the aluminum valley, which would not allow the flashing to expand and contract with temperature changes. The shingles that run through the valley are laid out so that nails can be placed on either side of the aluminum.

VALLEY FLASHING DETAILS

Valleys can be open, closed, or woven (in the case of asphalt shingles), but it's almost always worthwhile to provide more than one level of protection at this juncture. **Figure 3-71** illustrates a "belt and suspenders" approach to valleys that is prudent in harsh climates.

Woven Valleys

Woven valleys can be quite durable if done right.

- Shingles must extend at least 12 in. beyond the centerline (**Figure 3-72**).
- Before shingling, line the valley with 55-lb. roll roofing or with self-adhering eaves membrane.

Figure 3-72. Asphalt Woven Valley

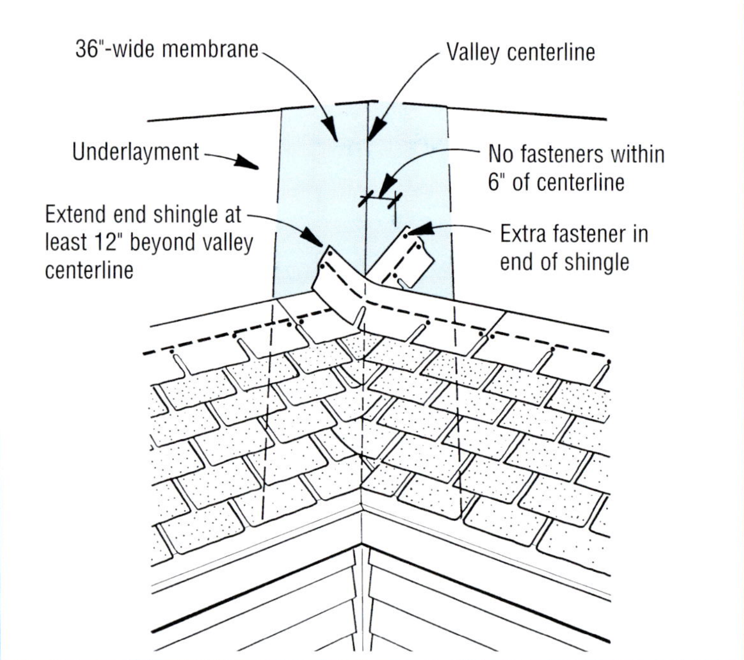

Woven valleys can be installed on roof slopes of 4:12 or greater. Caution: In some regions, moss accumulates between the shingle cutouts and impairs water run-off. Keep nails a minimum of 6 in. away from the center of the valley to prevent leakage (8 in. or more away on low-sloped roofs with valleys facing prevailing winds and in regions with annual freeze-thaw cycles).

For more information: "Bending V-Crimp Valley Flashing," JLC 8/99

Figure 3-73. Asphalt Closed-Cut Valley

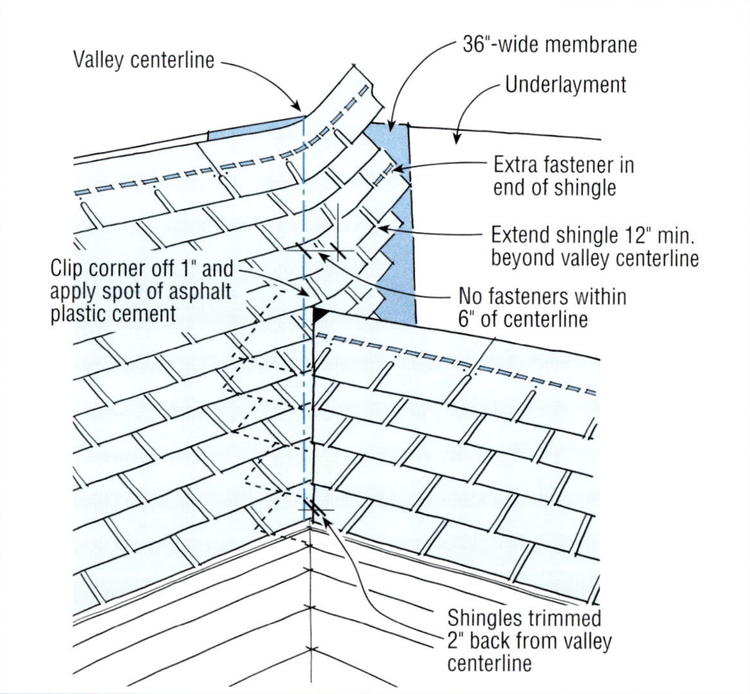

In a closed-cut valley installation, cut shingles should be on the side of the roof that has the greater runoff (the steeper or longer roof plane). You don't want the cut shingles to buck water from any "wash-over" that may occur due to unequal amounts of runoff.

Closed Valleys

Closed valleys have clean lines, but if done wrong they can trap ice and debris.

The covering course of a closed valley must be cut back a couple of inches from the centerline so that water flows away from the cut edge rather than into it (**Figure 3-73**).

Open Valleys

On an asphalt roof, an open valley may be lined with two layers of roll roofing. On all other roofs, metal valley flashing is the only long-term choice (**Figure 3-74**).

Some critical aspects of a metal valley flashing are its width, length, the configuration of its centerline and edges, and the way that it's fastened.

Valley Flashing

Width. A valley must channel runoff from two roofs, while having a shallower slope than either. It's critical that the valley be designed to handle the flow:

- At the peak, a valley should be 5 or 6 in. wide on either side of the centerline.
- As it descends toward the eaves, it should widen 1/8 in. for every foot of run.

Figure 3–74. Asphalt Open-Cut Valley

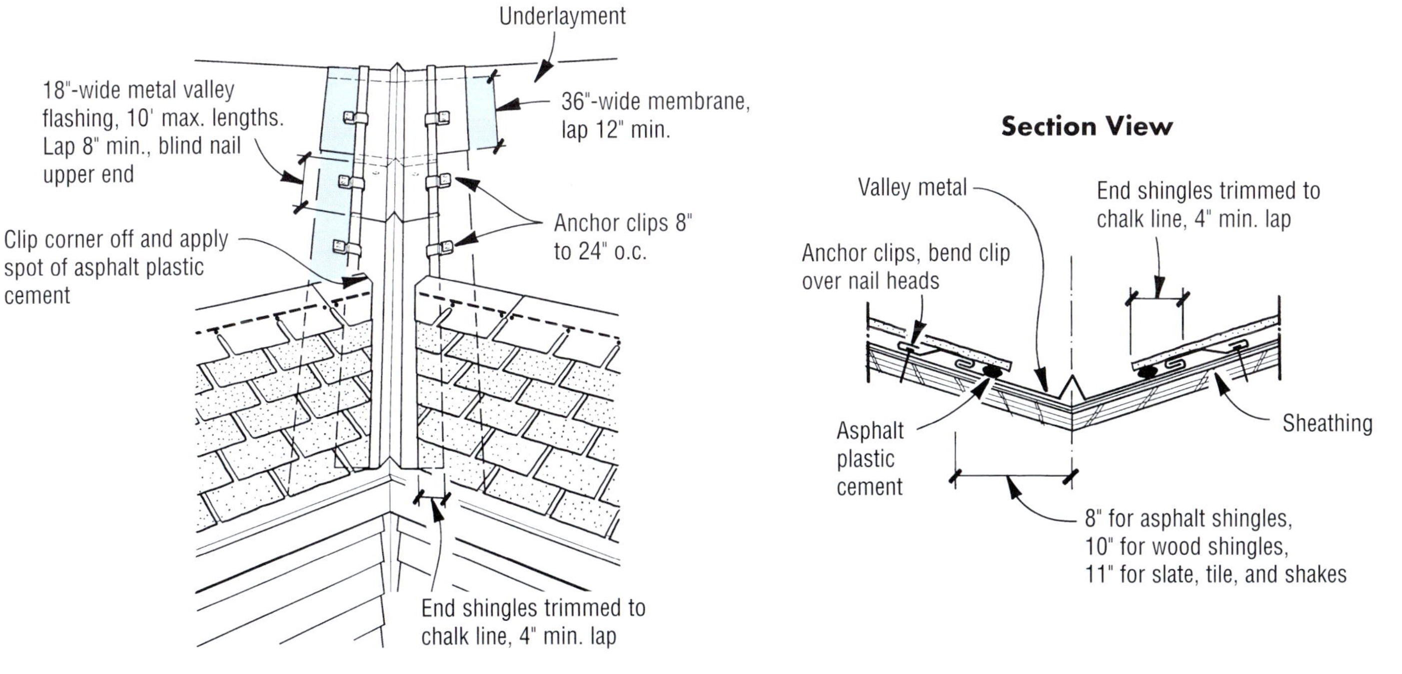

For asphalt shingles, a double layer of roll roofing is adequate for flashing. With other roofing materials, metal flashing is recommended. The metal should be formed with a "W"-shaped splash diverter, especially in valleys where the adjoining roof areas are of unequal size and slope.

Length. To avoid excessive expansion and contraction, use 8- to 10-ft. lengths of metal, maximum, for valley flashing.

- Avoid using aluminum, which expands at a greater rate than other metals.
- Lap sections 8 in. minimum.

Centerline. When a steep roof intersects a shallow one, water cascading down the steep roof can overshoot the valley and run up beneath the shingles on the shallow one. Roofs like this need a valley flashing with a crimped centerline. The steeper the roof, the more critical the crimp will be.

Edge treatments. Like all roof flashings, valley flashings should have hemmed edges. Not only do these shed water more effectively, but by holding the shingles away from the flashing, they prevent capillary suction from drawing water up under the shingles.

Figure 3-75. Valley Cleats

To allow the aluminum flashing to expand and contract, use a site-bent hem-and-clip system to secure valley flashing. Place 2 ft. o.c. at the edge of the flashing.

For more information: "Flashing a Leak-Proof Valley," JLC 3/98

Ridge treatment. Valleys that meet at a ridge should be soldered to one another. However, this connection may break when the valley metal expands and contracts. Best practice calls for covering the connection at the ridge with a moldable piece of lead flashing.

Fasteners. Metal valleys can be fastened with nails or cleats at 2 ft. o.c. (**Figure 3-75**).

- Cleats are better; they won't become fatigued and shear when the metal expands and contracts.
- When using nails, place them at least 5 in. away from the valley's center to prevent leaks.

The Code on Valleys

CABO requires that metal valleys be fabricated from no less than No. 28-gauge corrosion-resistant sheet metal.

- For asphalt shingles, valley flashing must extend at least 8 in. on either side of the centerline.
- For wood shingles, the wings of a valley must extend 10 in. each way.
- For slate, tile, and wood shakes, a valley must extend 11 in. each way.
- Sections of flashing must have an end lap of at least 4 in. for wood shingles and shakes, and at least 6 in. for slate and tile roofs.

ROOF-TO-WALL FLASHING

Front Wall Flashing Checklist

- Install shingles up the roof until you reach the base of the vertical wall. Adjust the exposure of the last few courses, if necessary, so the last course is at least 8 in. wide (**Figure 3-76**).

- Cut and bend a continuous metal flashing strip that extends at least 5 in. up the vertical wall and at least 4 in. over the last shingle course.
- When installing the metal flashing strip over the last course of shingles, embed it in roofing cement and nail it to the roof only.
- Do not nail the flashing strip to the wall.
- Apply an additional row of shingles — trimmed to the width of the flashing — over the flashing and embedded in asphalt plastic roofing cement.

Front-wall Flashing

Sidewall Flashing Checklist

Shingle roofs (asphalt, slate, or wood) that butt against vertical walls are best protected by metal step flashing placed over the end of each shingle course (**Figure 3-76**).

- Cut step flashing pieces to be 10 in. long and 2 in. wider than the shingle exposure (e.g., 7 in. for roof shingles with a 5-in. exposure).
- Bend the 10-in. length to extend 5 in. over the roof deck and 5 in. up the wall surface.

Sidewall Flashing

Figure 3–76. Flashing Against Walls

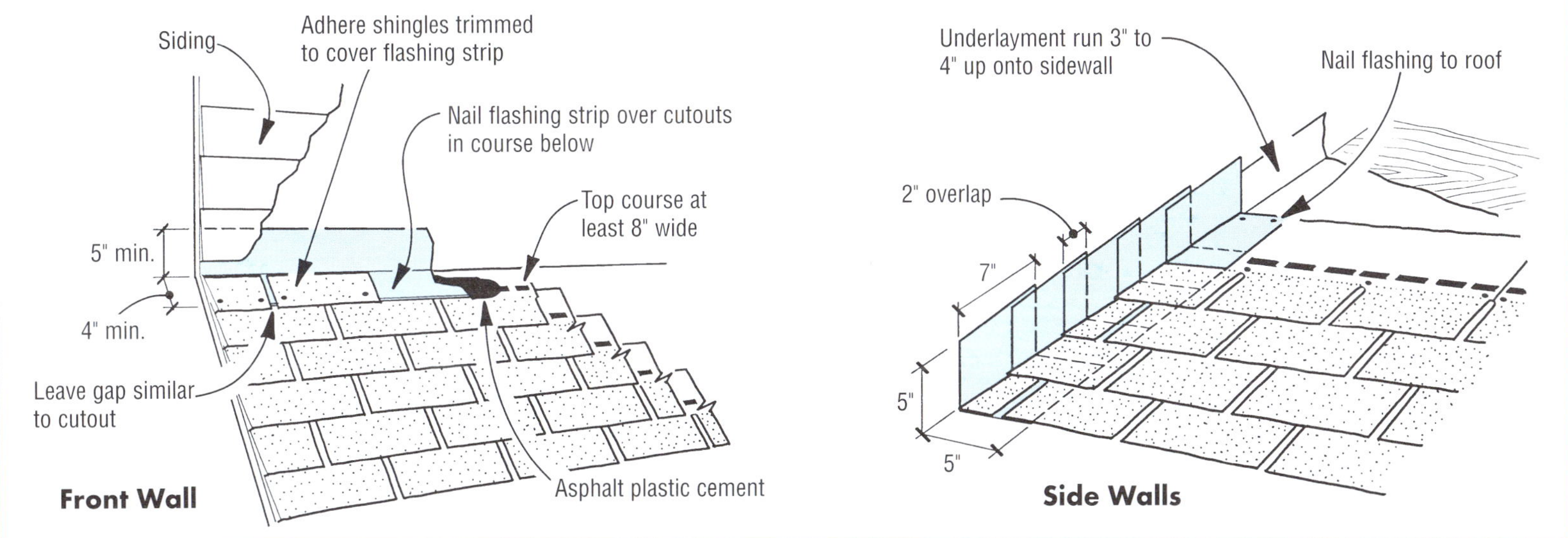

Front wall: Plan shingle layout so that the last course is at least 8 in. wide. Install continuous metal flashing over the last course and set in asphalt plastic cement. Do not nail the strip to the wall. Apply an additional row of trimmed shingles set in asphalt plastic cement. Side walls: Step flashing should be 2 in. longer than the exposure of the roof shingles in order for the step flashings to overlap approximately 2 in. In regions with severe weather, headlaps of 3 in. are common. The headlap should be made from compatible material that has a design life equal to or longer than that of the roofing material. Use 26-gauge (min.) metal for step flashing.

Figure 3-77. Counterflashing

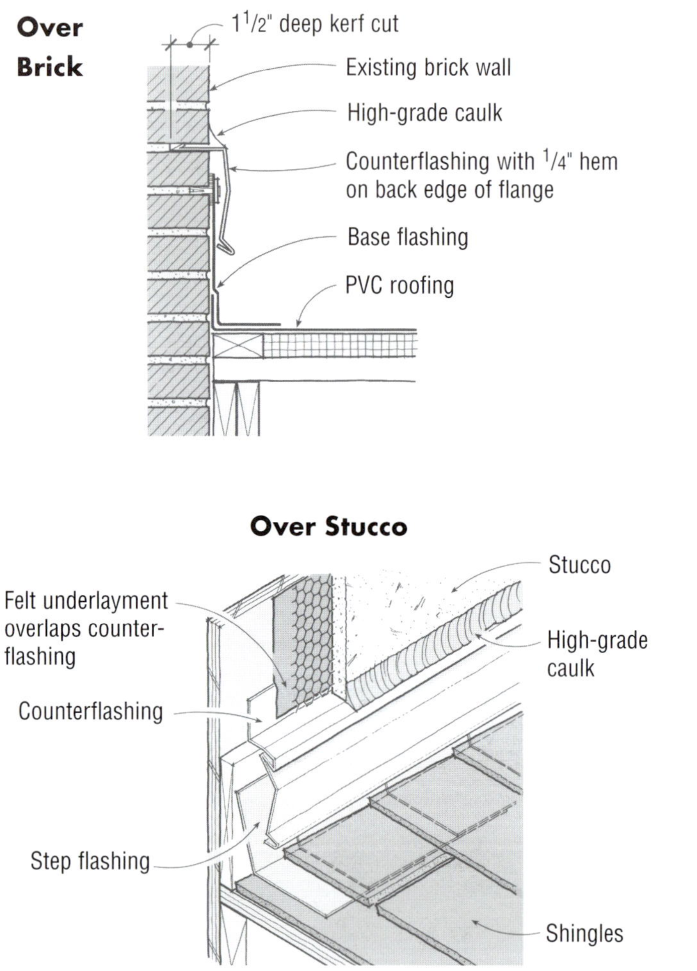

Over brick: Counterflashing should be set at a minimum of 1 1/2 inches into the mortar joint. ***Over stucco:*** *Caulk is never enough to seal between stucco and metal sidewall flashing. Use a drip screed as a counterflashing instead.*

For more information: "Building Dog House Dormers," JLC 10/94

- Place each step flashing piece about 1/4 in. above the bottom edge of the shingle that will overlap it — just enough so that the flashing piece is not visible when the overlapping shingle is in place.
- Because the metal strips are 2 in. wider than the exposure of the shingles (e.g., 7 in. for roof shingles laid with a 5-in. exposure), each step flashing piece will overlap the one on the course below by 2 in.
- Nail flashing to the roof deck only. Do not nail flashing to the wall.
- Bring wood siding down over the vertical sections of the step flashing to serve as counterflashing. Do not nail siding into step flashing.

COUNTERFLASHING

Stucco and brick siding must have a counterflashing over a step flashing or continuous wall flashing strip (**Figure 3-77**).

Counterflashing Brick

The most common failure of flashing over brick is from not setting the flange deep enough. Bring the flashing all the way through the brick in new construction. Over existing brick, set the counterflashing a minimum 1 1/2 inches into the mortar joint, as shown in **Figure 3-77**.

Counterflashing Stucco

In the case of stucco siding, the drip screed functions as the counterflashing.

FLASHING GABLE DORMERS

A gable, or doghouse, dormer includes base, step, counter, and valley flashings, as shown in **Figure 3-78**.

Flashing Gable Dormers

Figure 3–78. Gable Dormer Flashing

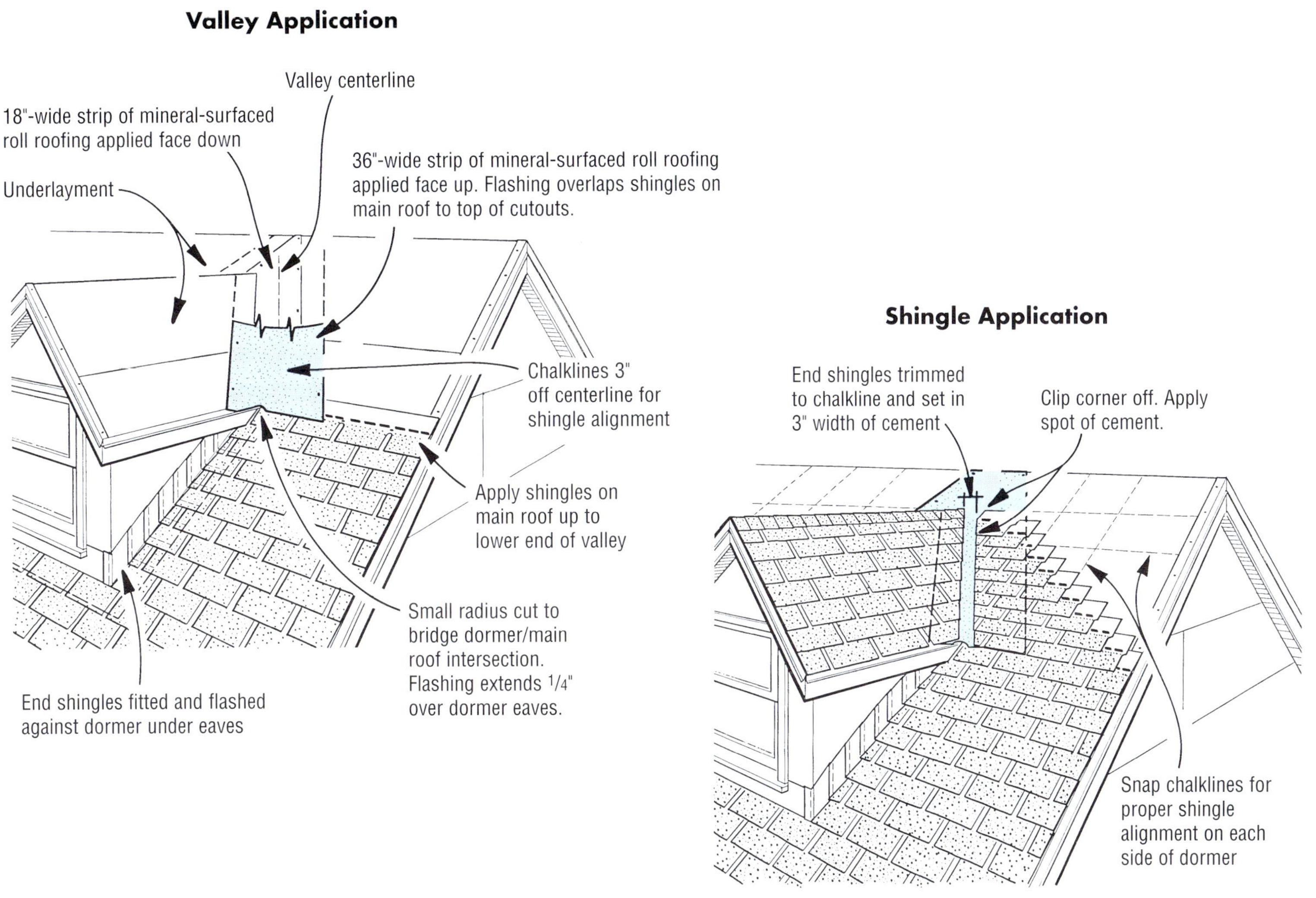

To flash a gable dormer, follow the rules for step, base, and valley flashing. At the peak, complete the last shingle course on the main roof before applying the dormer ridge shingles.

- When using metal open valleys on a gable dormer, make sure they're soldered together at the peak, then cover with a lead patch.
- Shingle the main roof before shingling over the dormer ridge.

FLASHING SHED DORMERS

Shed dormers require a continuous strip of flashing along the top of the lower roof plane where it meets the steeper roof plane. Also, they require sidewall flashings. A critical juncture that often fails is where these two flashings meet. Use taller pieces of step flashing towards the top, so the pieces can be bent over the dormer roof to help keep out snow and wind-blown rain (**Figure 3-79**).

FLASHING CHIMNEYS

Chimneys should be surrounded by base, step, and counter-flashings, as shown in **Figure 3-81**.

Chimney Cricket

Where a chimney breaks through the field of a roof, best practice calls for using a cricket to divert water away from the upper edge on the chimney. Crickets can be covered with the same material as the rest of the roof, using metal valley flashing where it meets the main roof. Or better still, fashion a metal cricket cover with 20-gauge metal as shown in **Figure 3-80**.

Figure 3–79. Shed Dormer Flashing

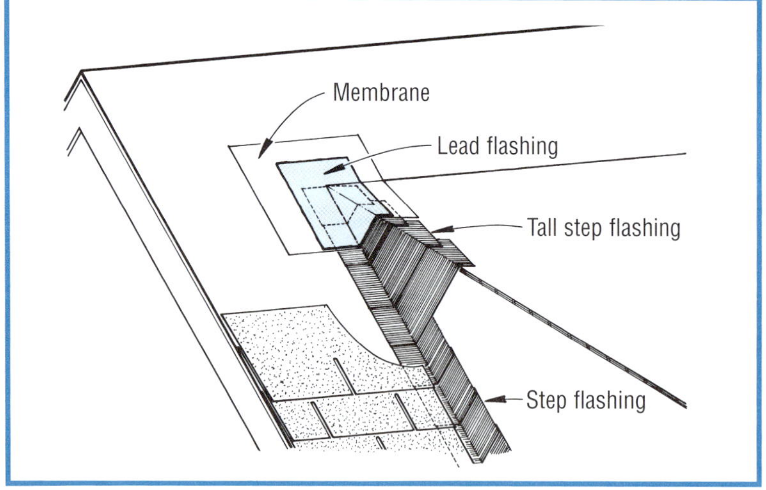

The top corners of a shed dormer are particularly susceptible to leaks. Use tall pieces of sidewall flashing, combined with a lead flashing patch.

Figure 3–80. Cricket Construction

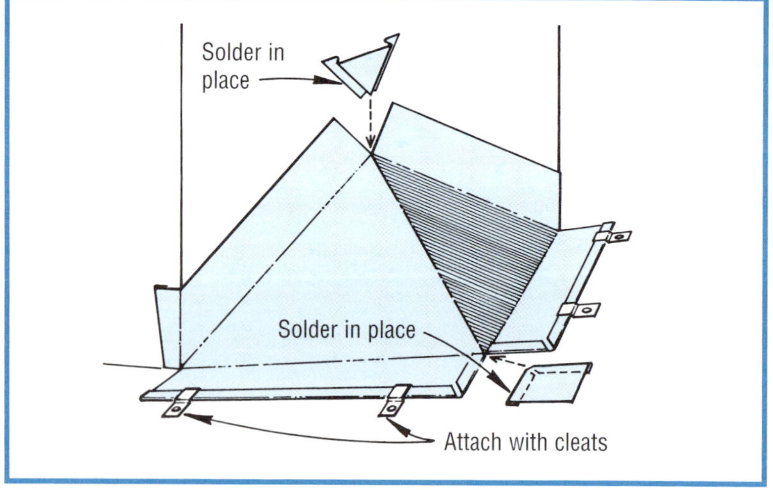

A cricket should be built of plywood, then covered by a single piece of metal. To permit expansion and contraction of the large piece of metal, attach it to the roof deck with cleats.

For more information: "Flashing the Tough Spots," JLC 5/95

"Flashing a Chimney the Right Way," JLC 11/92

Figure 3-81. Chimney Flashing

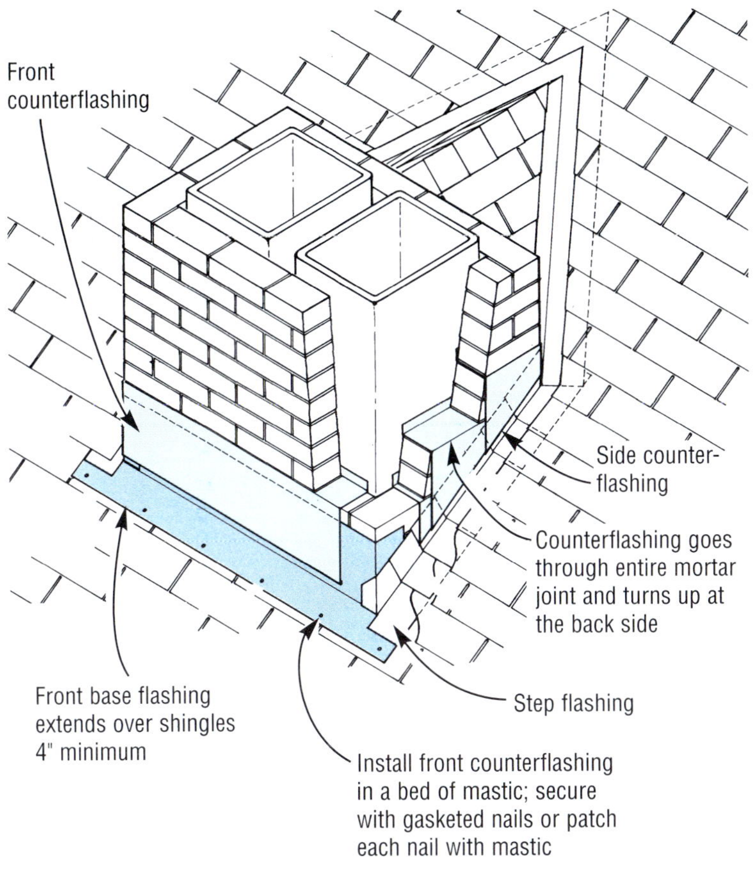

Never attach a flashing to both the chimney and the roof. The base flashing is attached to the roof only, the counterflashing to the chimney only. This allows the house to move independently of the chimney.

Figure 3-82. Skylight Opening

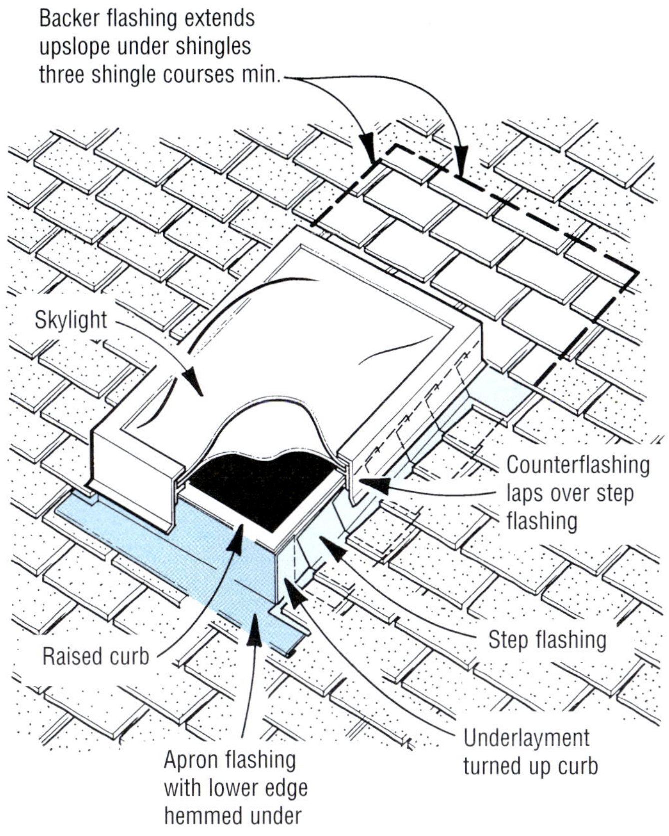

Unless it comes with a premanufactured flashing assembly, a skylight should rest on a curb or box that protrudes at least 6 in. above the roof surface.

Flashing Shed Dormers

Flashing Chimneys

Flashing Skylights

Figure 3–83. Vent Pipe Flashing

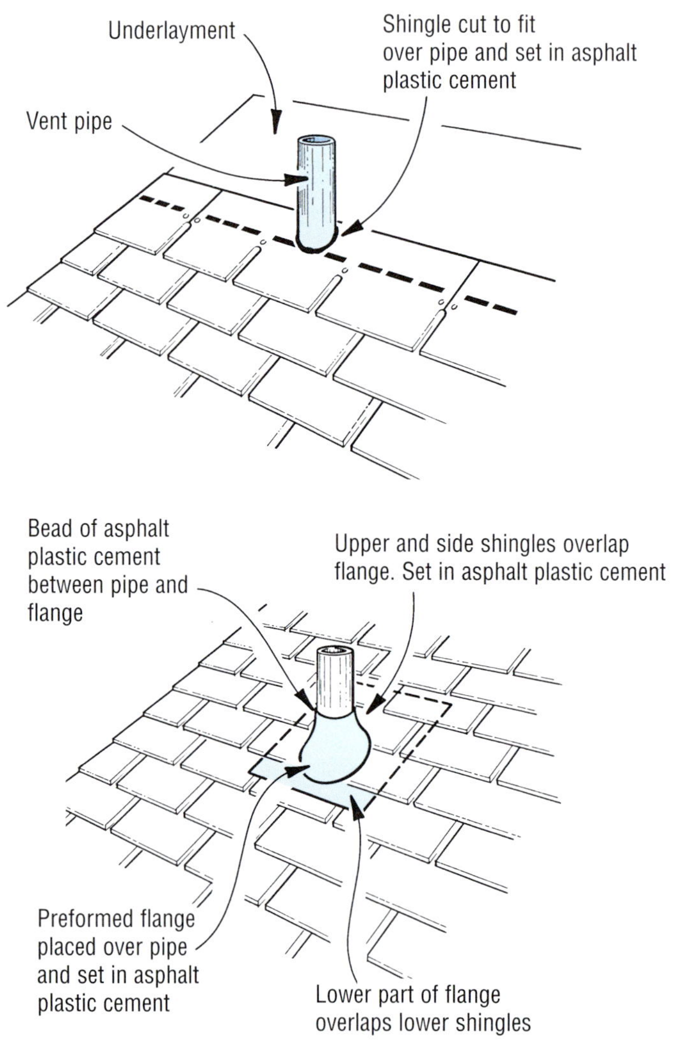

Vent pipes are sealed with a neoprene boot and a metal base plate that's woven into the roof shingles.

For more information: "Trouble-shooting Roof Leaks," JLC 10/99

FLASHING SKYLIGHTS

Most skylight flashings come as a kit from the manufacturer. Skylights without flashing kits should be mounted on a curb and flashed as shown in **Figure** 3-82.

FLASHING VENT PIPE PENETRATIONS

Vent pipes on wood and asphalt roofs are usually flashed using a preassembled metal pan with a rubber boot. Order the unit according to the outside diameter of the vent pipe and install it as shown in **Figure** 3-83. For slate and tile roofs, a malleable lead jacket will last longer.

Some cold-climate builders avoid gutters because of their reputation for getting caked with ice. The ice bends them, pops their seams, and sometimes pulls them off the house. Yet without gutters, any water not handled at the roof line must be managed on the ground.

Where to Use Gutters

- Gutters do the most good in heavy rainfall areas and on buildings with little or no overhang.
- Gutters should always be used where houses are built on clay and other frost-susceptible soils (some expansive clays can crack a foundation without freezing; they need only be wet).
- Use gutters wherever roof run-off creates a splashback problem that might accelerate the deterioration of entry steps, decks, and siding.
- Gutters go a long way toward solving basement water problems. For this reason they are required by code in some places.

Without Gutters

If you don't use gutters, it's best to cap the backfill with an impermeable soil layer around the house and slope the grade away from the house at least one inch per foot. Rectangular flagstones laid below the roof's drip line are another option: they'll prevent erosion and cause most of the water to splash away from the foundation.

GUTTER INSTALLATION

Old-style wood and yankee gutters caused a lot of damage because when they leaked, the water backed up into the eaves. This promotes decay in the rafter ends and the soffit. Modern metal and vinyl gutters do a much better job, but there is still the potential for problems if the gutters are not installed with the proper clearances (**Figure 3-84**).

Flashing Vent Pipes

In cold climates, the outside edges of all gutters should fall below the roof's slope line so snow and ice can slide clear. The shallower the roof, the more clearance is needed.

Hanger Types

There are three common hanger types (**Figure 3-85**). Of these, strap-type hangers that hold the gutter away from the fascia provide the most protection against decay.

Gutter Installation

Figure 3–84. Gutter Location

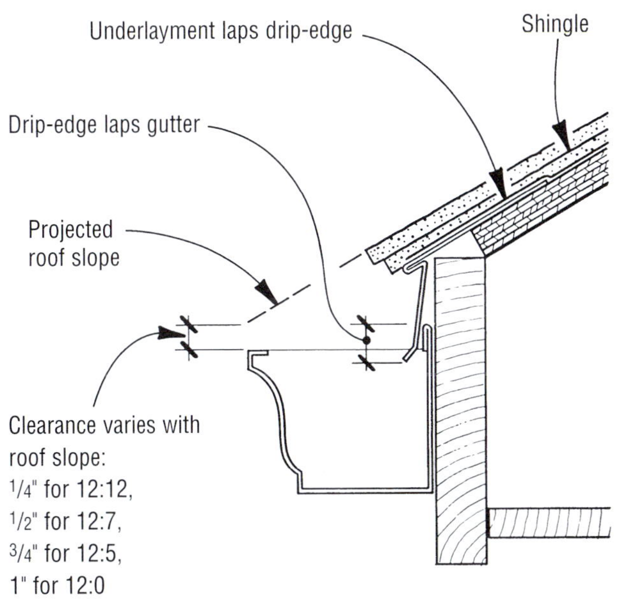

Place gutters below the roof's slope-line so that ice and snow can slide by.

Installation Details

- Aluminum gutter runs should have an expansion joint every 40 ft. of straight run since the gutter will expand and contract 1/8 in. or more over that span.
- Gutters should slope at least 1/16 in. per foot of run. It is important to move the water quickly away from the foundation with either a subsurface pipe or splash blocks and by sloping the grade at the surface.
- The number of downspouts a roof needs will depend on the size of the conductor pipe. Allow 1 sq. in. of downspout cross-section for every 100 sq. ft. of roof area. Place the downspouts at least 20 ft. apart but no more than 50 ft. apart.

Figure 3-85. Hanger Types

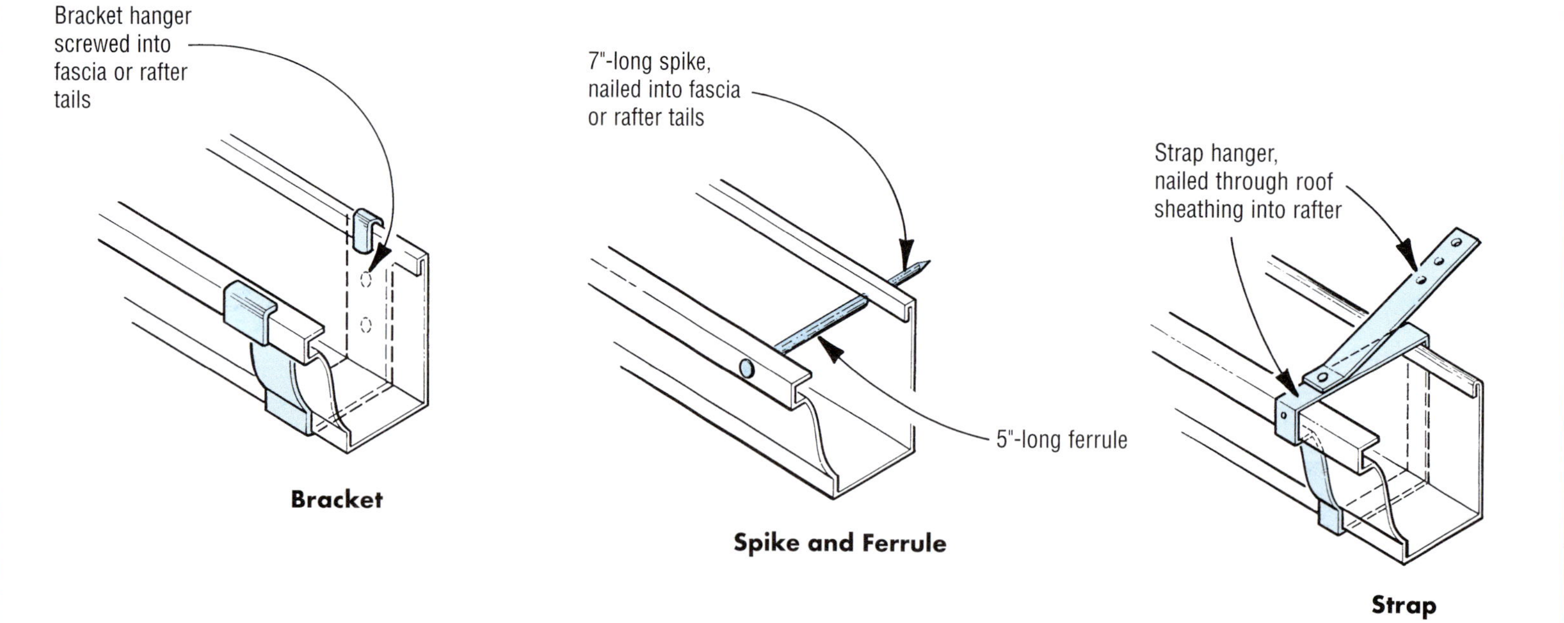

The "ogee" gutter profile is the most commonly used in residential construction. The hangers shown here are readily available for the "ogee" gutter. Expansion joints are needed for every 40 ft. (max.) of run.

For more information: "Durable Exterior Details," JLC 6/99

Roof ventilation is required on sloped roofs by all building codes and some roofing material warranties.

Why Ventilate?

Ventilation does several things:

- Ventilation reduces attic temperatures in the summer by as much as 30°F. This extends shingle life and cuts cooling bills.
- Ventilation helps rid the attic of any excess moisture.
- By keeping the roof uniformly cold in winter, ventilation helps prevent ice damming.

Attic ventilation works best where there's an airtight ceiling plane — that is, one that prevents warm, moist interior air from rising up from the house below.

SIZING VENTS

Sizing Roof Vents

Most codes specify a net-free vent area (NFVA) of 1/150 of the attic area if the ceiling has no vapor retarder, or 1/300 with a vapor retarder. The ventilation area should be split 50/50 between low and high vents, preferably soffit and ridge vents.

NFVA

NFVA is the combined area of all unobstructed ventilation openings. To calculate the actual vent area required, start with the required NFVA, then enlarge it to compensate for screens and louvers. For example, vents with 1/8-in. mesh insect screen should be enlarged by 25% (**Figure 3-86**).

Figure 3-86. Typical Net-Free Vent Areas (NFVA)

Vent Covering	Vent Area Multiplier
1/4-inch mesh hardware cloth	1.00
1/8-inch mesh screen	1.25
#16 mesh screen, with or without plain metal louvers	2.00
Wood louvers and 1/4-inch mesh hardware cloth	2.00
Wood louvers and 1/8-inch mesh screen	2.25
Wood louvers and #16 screen	3.00

This multiplier tells how many square feet of vent area is needed to give 1 sq. ft. of net-free vent area.

Figure 3-87. Roof Vent Location

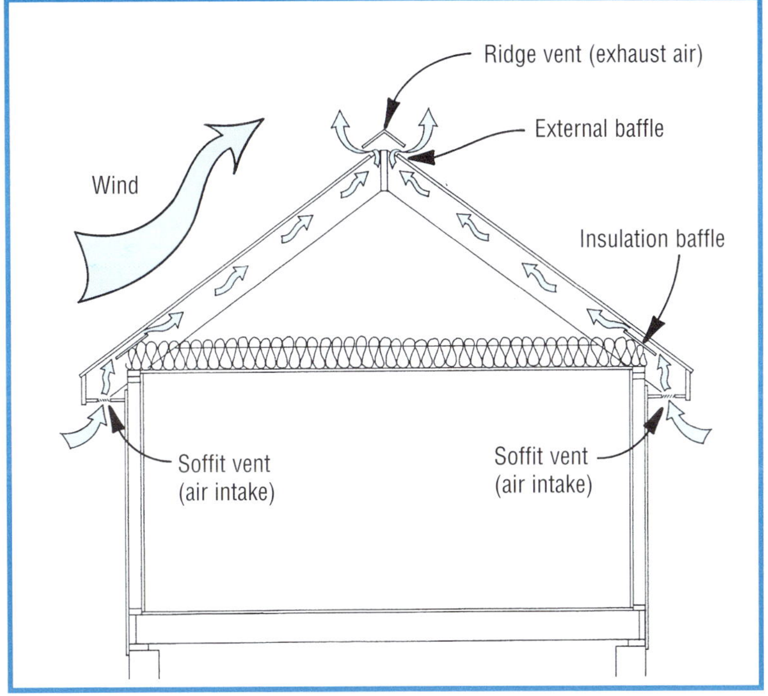

Roof ventilation should be divided about 50/50 between low and high locations. For best performance, use a soffit vent and baffled ridge vent to allow for both intake and exhaust.

Calculate vent area:

NFVA = Area of Vent Opening X Vent area Multiplier

Compare to attic area:

$$\text{Actual Vent Area} = \frac{\text{Sq. Ft. of Attic Floor}}{\text{NFVA}}$$

Actual vent area must equal .0066 or more (without ceiling vapor barrier) or .0033 (with ceiling vapor barrier).

VENT TYPES AND LOCATION

For good attic ventilation, the vents should be evenly divided between intake and exhaust (**Figure 3-87**).

Soffit and Ridge Vents

For best performance, locate vents at the soffit (intake) and ridge (exhaust).

Do not use a ridge vent without soffit vents. A ridge vent without soffit vents could make matters worse in windy

Figure 3–88. Venting Cathedral Ceilings

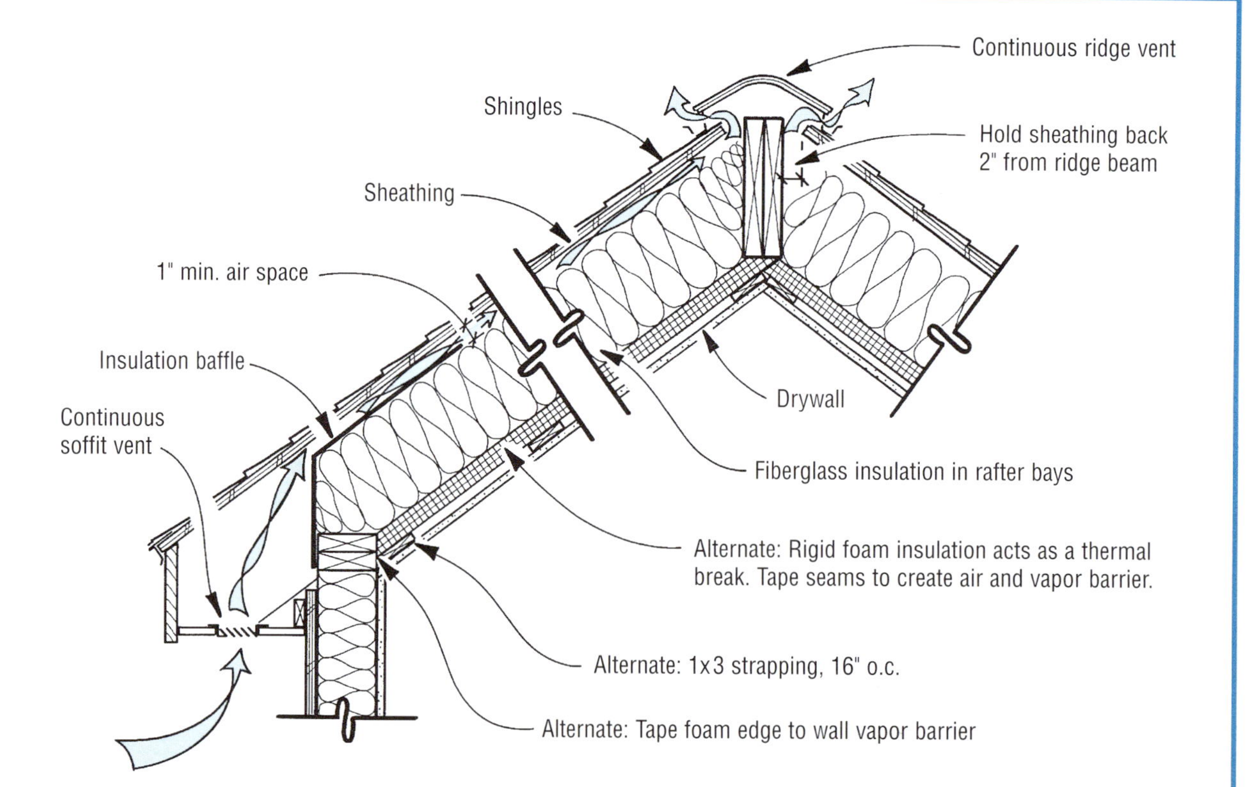

For more information: "Low-Profile Ridge Vents," JLC 5/92

"Venting Details for Cathedral Ceilings," JLC 12/96

In cathedral ceilings, leave a minimum 2-in. space for airflow between the insulation and the underside of the sheathing.

conditions by pulling moist air from inside the house, adding moisture to the attic. Research suggests that somewhat oversized soffit vents are a good idea because they provide insurance against this type of leakage.

Gable-End Vents

Gable-end vents do not pull air evenly from all parts of the attic. Use gable-end vents only when soffit vents cannot be installed.

Use a Baffle

Choose ridge vents with some type of a baffle to deflect windblown rain and snow from infiltrating the vent. Some vents use other mechanisms to restrict the entry of snow and water, although some of these restrict airflow as well.

Cathedral Ceiling Venting

Use the same combination of ridge and soffit vents in a cathedral-type ceiling, leaving a minimum ventilation channel of 2 in. between the top of the insulation and the underside of the sheathing, as in **Figure 3-88**.

Roof Vent Interruptions

Where roof penetrations such as skylights or chimneys interrupt the airflow from soffit to ridge, cut notches or drill holes in the rafters to let air flow around the obstruction (**Figure 3-89**).

ICE DAMS

When warm air escapes from a structure's conditioned space and finds its way to the back of the roof sheathing, it can melt snow that has collected on the roof. When the melted snow reaches the eaves or other unheated area, it can refreeze and cause an ice dam (**Figure 3-90**). Melted snow from farther up the roof will then pond behind this dam, back up beneath the shingles, and find its way into the structure.

Vent Types and Location

Ice Dams

Ice Dam Prevention

The way to prevent ice dams is to keep the roof sheathing uniformly cold and close to the outside temperature. The best way to do this is to:

- Install good attic insulation and ventilation, along with a tight ceiling plane.
- Detail the eaves so they have a full thickness of insulation with ventilation space above. Do not allow insulation to plug the ventilation channel near the eaves.

Figure 3–89. Venting Skylight Rafter Bays

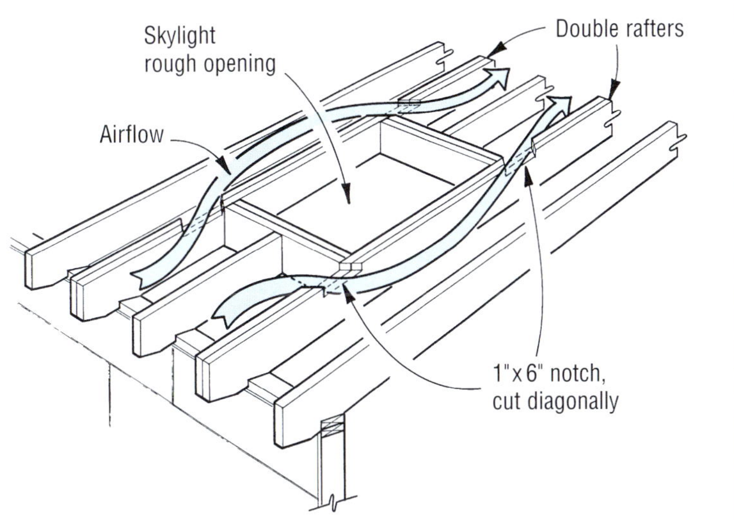

Cut notches or holes in the rafters to let air flow around skylights or other roof penetrations.

- Keep heating ducts out of the attic space, since most heating ducts leak at the joints, and will increase attic temperatures.

THE COLD ROOF

For cathedral ceilings in very cold and snowy regions, some builders have found it necessary to take roof ventilation a step further, building what's called a cold roof. It consists of two distinct roof layers separated by a 2- to 3-in. air space. A cold roof is capped by a "Boston Ridge" — a large, site-built raised ridge vent with an overhang large enough to keep out snow (**Figure 3-91**).

Figure 3-90. Ice Dams

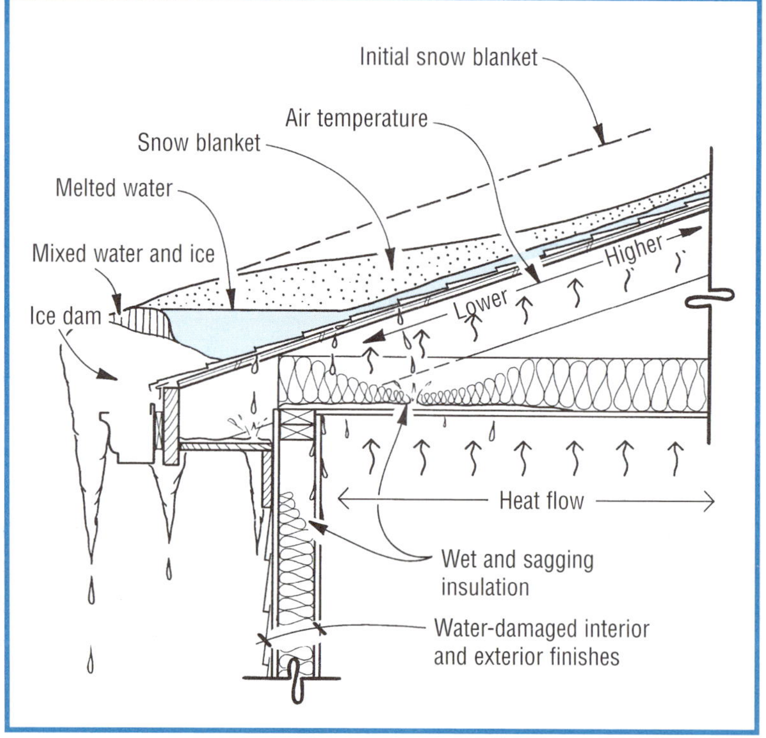

Heat leaks into the attic are the main cause of ice damming. Good insulation and ventilation offer the best protection.

Figure 3-91. Cold Roof

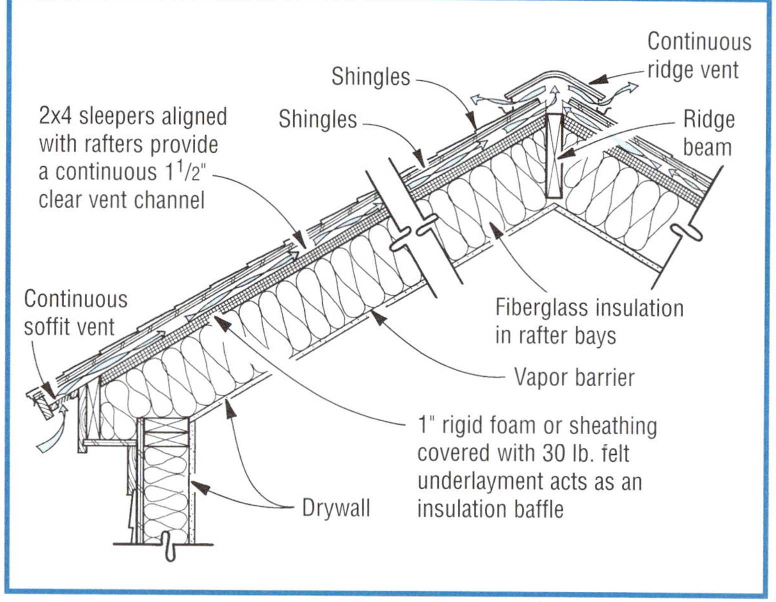

Cathedral or vaulted ceilings in severely cold, snowy climates may require a double-layer "cold" roof to prevent ice dams. The upper vented layer removes warm air that escapes through the insulated lower layer. The outside roof surface remains at the same temperature as the outside ambient air, preventing the freeze-thaw cycle that normally occurs on standard roofs.

For more information: "Preventing Ice Dam Leaks," JLC 3/95

CALCULATING WALL AREA

All siding quantities are based on the square footage of the house walls to be covered. Formulas to calculate the square footage of different shaped wall sections are shown in **Figure 4-1**.

To provide allowances for waste, apply the following:

- Do not deduct area of standard-size windows and doors;
- For openings over 50 sq. ft. (such as garage doors or large window walls), deduct only 75% of the total opening area;
- For triangular wall areas, add 1 foot in height to the original measurements.

ESTIMATING SHEATHING WRAP

- Asphalt felt typically comes in rolls 36 in. wide by 144 ft. long (432 sq. ft.), which will cover about 4 squares with overlap (see "Installing Sheathing Wrap," **page 298**).
- Plastic housewrap typically comes in rolls 9 ft. wide by 111.1 ft. long (999.9 sq. ft. or 9.2 squares with overlap) or 195 ft. long (1,755 sq. ft. or 16 squares with overlap).
- Smaller rolls of plastic housewrap are available at 4.5 ft. wide x 222 ft. long (999.9 sq. ft. or 9.2 squares with overlap) and 3 ft. wide x 111.1 ft. long (333.3 sq. ft. or 3 squares with overlap).

Calculating Wall Area

Estimating Sheathing Wrap

Figure 4–1. Wall Area Formulas

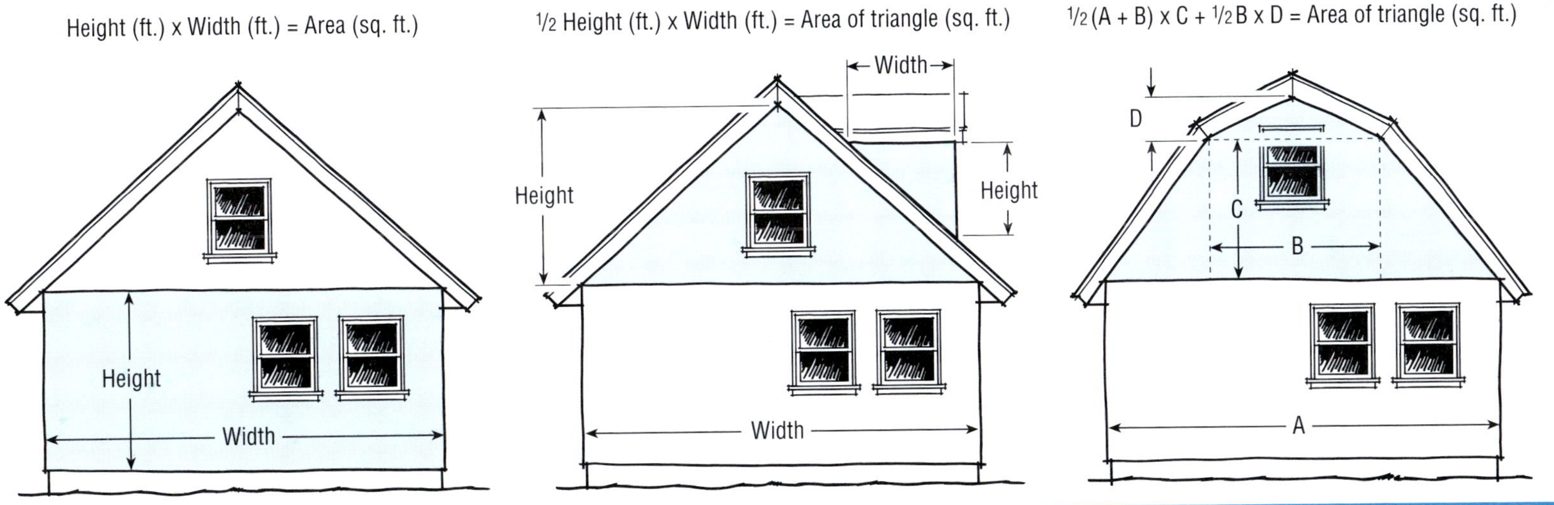

To calculate siding coverage area, add the areas of all wall sections. Include dormer and bay walls, porch areas, and entry alcoves.

ESTIMATING BOARD SIDING COVERAGE

Standard board (wood) siding usually is sold either by the linear (running) foot or by the board foot. Depending on the pattern, more or less material will overlap when installed. To estimate the amount of material to buy:

- Calculate coverage area by adding all wall sections, as described in "Calculating Wall Area" (previous page), then multiply by the appropriate factor from the table in **Figure 4-2.**
- Add 5% to 10% to linear or board-feet totals for waste. Premium-grade material will have less waste from fewer culls, but the more inside and outside corners there are, the more trimmed waste there will be.

Multipliers (factors) do not include any allowance for trim or waste and do not apply to diagonal installations. When multiplying factors by the square footage to be covered, add 10% allowance for trimmings and waste. See ***Figure 4-36*** *for examples of siding patterns.*

Figure 4-2. Board Siding Coverage

Pattern	Nominal Width	Dressed Width	Exposed Face Width	Multiplier for Linear Feet(1)	Multiplier for Board Feet(2)
Bevel &	4 in.	3 1/2 in.	2 1/2 in.	4.8	1.60
Bungalow	6	5 1/2	4 1/2	2.67	1.33
	8	7 1/4	6 1/4	1.92	1.28
	10	9 1/4	8 1/4	1.45	1.21
Dolly	4 in.	3 1/2 in.	3 in.	4.0	1.33
Varden	6	5 1/2	5	2.4	1.2
	8	7 1/4	6 3/4	1.78	1.19
	10	9 1/4	8 3/4	1.37	1.14
	12	11 1/4	10 3/4	1.12	1.12
Drop, T&G	4 in.	3 3/8 in.	3 1/8 in.	3.84	1.28
and Channel	6	5 3/8	5 1/8	2.34	1.17
Rustic(3)	8	7 1/8	6 7/8	1.75	1.16
	10	9 1/8	8 7/8	1.35	1.13

(1) To calculate the total linear feet of a particular board siding, multiply the factor in this column by the square feet of wall to be covered. This linear-foot factor is derived by dividing 12 in. by the exposed face width.
(2) To calculate the total board feet of a particular board siding, multiply this factor by the square feet of wall to be covered. This board-feet factor is derived by dividing the nominal width by the exposed face width, and assumes normal 1x stock.
(3) The exposed face width on board-on-batten and board-on-board applications will vary depending on material sizes selected and the overlap applied. Determine exposed face, then use appropriate factor. (Minimum overlap is 1/2 in.)

Starter Strip for Lap Siding

- For clapboards (beveled wood siding), as well as plywood, hardboard, and fiber-cement lap siding, include at least 100 lin. ft. of starter strip for every 1,000 sq. ft. of siding material (see **Figure 4-60, page 341**).

ESTIMATING SHINGLE AND SHAKE COVERAGE

The same shingles and shakes used to cover roofs may be used as siding; however, different maximum exposures apply (see **Figure 4-43, page 328**). Also, manufacturers package their shingles for different applications — either in four-bundle "roof" packs or two- bundle "wall" packs. Both packs contain a square of shingles, and both may be used for siding applications.

Estimating Shakes and Shingles

As with roofing shingles and shakes, actual coverage on sidewalls depends on the exposure at which the shingles and shakes are applied. Refer to **Figures 4-3** and **4-4** to calculate actual coverage.

In addition to field coverage found in tables:

- Add one square for every 100 vertical ft. of corners. (For smaller jobs, figure one bundle for every 25 ft. of corners if material comes in a four-bundle square, or

Figure 4–3. Single-Course Shingle Coverage

Approximate Coverage (sq. ft.) of One Square (4-bundle pack) of Shingles at Indicated Weather Exposures

Length and Thickness (in.)	Weather Exposure (in.) 3½	4	4½	5	5½	6	6½	7	7½	8	8½	9	9½
16x5/2	70	80	90	100	110	120	130	140	150	160	170	180	190
18x5/2¼	–	72½	81½	90½	100	109	118	127	136	145½	154½	163½	172½
24x4/2	–	–	–	–	73½	80	86½	93	100	106½	113	120	126½

	10	10½	11	11½	12	12½	13	13½	14	14½	15	15½	16
16x5/2	200	210	220	230	240	–	–	–	–	–	–	–	–
18x7/2¼	181½	191	2002	209	218	227	236	245½	254½	–	–	–	–
24x4/2	133	140	146½	153¾	160	166½	173	180	186½	193	200	206½	213

Adapted from Cedar Shake and Shingle Bureau

*For a description of shake grades, see **Figure 3-15, page 222**. To calculate quantity of shingles needed: 1) calculate total area to be sided (see **Figure 4-1, page 289**); 2) subtract area of openings greater than 50 sq. ft.; 3) add 5% of adjusted total area for waste; 4) select nearest approximate coverage for the shingle size to be used; 5) divide by 100 to find number of squares of material needed for the job. For double-course applications, divide by 50 to calculate actual squares needed.*

one bundle for every 20 ft. of corners if material comes in a five-bundle square.)

- Do not subtract area of openings unless their area is greater than 50 sq. ft.
- Add an additional 5% for waste.
- For starter courses, add one square for every 120 lin. ft. of wall skirt for single-course applications, or one square for every 60 lin. ft. of wall skirt for double-course applications.

Starter Course

The underlying course can be either the No. 1 grade product used on the outer course or a lower-grade product.

Figure 4–4. Single-Course Shake Coverage

Adapted from Cedar Shake and Shingle Bureau

Weather Exposures (in.):	**7**	**8 1/2**	**10**	**11 1/2**	**14**	**16**	**18**	**20**
Shake Type, Length and Thickness (in.)								
18 x 1/2 Handsplit and Resawn Mediums(1)	70	85	100	115	140(3)	–	–	–
18 x 3/4 Handsplit and Resawn Heavies(1)	70	85(2)	100	115	140(3)	–	–	–
No. 1 18 Handsplit and Resawn	70	85(2)	100	115(2)	140	160	180	200(3)
24 x 1/2 Handsplit and Resawn Mediums	70	85	100	115(2)	140	160	180	200(3)
24 x 3/4 Handsplit and Resawn Heavies	70	85	100	115(2)	140	16	180	200(3)
No. 1 24 Taper-Sawn	70	85	100	115(2)	140	160	180	200(3)
24 x 1/2 Tapersplit	70	85	100	115(2)	140	160	180	200(3)
18 x 3/8 True-Edge Straight-Split(5)	50(2)	60	71	82	100(3)	–	–	–
18 x 3/8 Straight-Split(6)	82	100(2)	118	135	165(3)	–	–	–
24 x 3/8 Straight-Split	70	85	100	115(2)	140	160	180	200(3)

(1) 6 bundles will cover 100 sq. ft. wall area at 8 1/2-in. exposure. 7 bundles will cover 100 sq. ft. at 7 1/2-in. weather exposure.
(2) Maximum recommended weather exposure for single coursed wall construction.
(3) Maximum recommended weather exposure for double coursed wall construction.
(4) All coverage based on 1/2-in. spacing between shakes.
(5) Four-bundle square.
(6) Five-bundle square.

For a description of shake grades, see ***Figure 3-15, page 222****. To calculate quantity of shakes needed: 1) calculate total area on building to be sided (see* ***Figure 4-1, page 289****); 2) subtract openings greater than 50 sq. ft. from total area; 3) add 5% of adjusted total area for waste; 4) select nearest approximate coverage for the shingle size to be used; and 5) divide by 100 to find number of squares of material needed for the job. For double-course applications, divide by 50 to calculate actual squares needed.*

Starter Course for Shingles and Shakes

Starter strip is not recommended for shingle and shake siding. Use a starter course of shingles instead (see "Starter Course," **page 328**). A lower-quality shingle grade, such as No. 3 Black Label or No. 4 undercoursing, is a suitable starter course for sidewall shingles and shakes. Note: This means that double-course shingle and shake applications start with three courses.

ESTIMATING HARDBOARD

Coverage for hardboard and plywood lap siding varies with board width and exposure. Follow the manufacturer's guideline. An approximate guide to the amount of lap material needed for a given area of wall is shown in **Figure 4-5**.

Figure 4-5. Hardboard Siding Coverage

Siding Width (in.)	Exposure (in.)	Siding Qty. per 1,000 sq. ft. of Wall
12	11	1,150 sq. ft.
12	10 1/2	1,200 sq. ft.
9	8	1,185 sq. ft.
9	7 1/2	1,260 sq. ft.
6	5	1,320 sq. ft.
6	4 1/2	1,460 sq. ft.

Hardboard and plywood lap siding varies by width and overlap. For various board sizes and exposures, order the square feet of siding shown for every 1,000 sq. ft. of wall area.

ESTIMATING NAIL QUANTITIES

Estimating Hardboard

Estimating Nails

Horizontal Siding

Order 1 lb. for each nail penny (d) per 1,000 sq. ft. of siding required:

- For 6d nails, order 6 lb. of nails per square of material
- For 8d nails, order 8 lb. of nails per square
- For 10d nails, order 10 lb. of nails per square

Vertical Board-and-Batten

Order 2 lb. for each nail penny (d) per 1,000 sq. ft. of siding required:

- For 8d nails, order 16 lbs. nails per square of material
- For 10d nails, order 20 lbs. nails per square

Figure 4–6. Estimating Siding Accessories

	Amt. Required per 1,000 sq. ft. of Siding
Starter strip	100 lin. ft.
J-Trim(1)	200 lin. ft.
Utility trim(1)	30 lin. ft.
H-Trim(2), 12 in.	60 pieces
H-Trim(2), 9 3/8 in.	80 pieces

(1) Typically available in 12-ft. lengths
(2) Available to cover butt joints between hardboard panels (**see Figure 4-64, page 344**)

Vinyl, metal, and hardboard siding require trim accessories, such as those listed in this table. Check with the siding manufacturer for its suggested siding trim package.

Shingles and Shakes
Order 2 lb. for each nail penny (d) per 1,000 sq. ft. of siding required:
- For 6d nails, order 12 lbs. nails per square of material
- For 8d nails, order 16 lbs. nails per square

ESTIMATING VINYL SIDING COVERAGE

When estimating vinyl siding quantities, add up square footage (see "Calculating Wall Area," **page 289**), and add 10% for waste. As with wood siding estimates, do not deduct the area of windows and doors. This will provide an allowance for waste. If the window and door areas are over 50 sq. ft. (such as garage doors or sliding glass doors), deduct 75% of the total openings. For dormers and gable-end walls, add 1 ft. in height to the original measurements to allow for waste.

Estimating Siding Accessories
Add up lin. feet of outside and inside corners needed, and add 10% for waste. Include other trim accessories, as shown in **Figure 4-6.**

ESTIMATING STUCCO

Three-coat stucco requires lath as well as cement-mortar stucco. **Figure 4-7** provides some rough measures of the

Figure 4–7. Estimating Stucco Materials

Materials for 100 sq. yd. of Wall		
Type of Construction	**Lath (sq. yd.)**	**Other Materials**
Metal lath on wood studs	105	8 lbs. self-furring nails or 15 lbs. staples
Metal lath on steel studs	105	10 lbs. tie wire, 1,000 lin. ft. of 3/4-in. channel

Stucco for 100 sq. ft. of Wall			
		1:3 Mortar Materials	
Stucco Thickness (in.)	**Stucco Amt. (cu. ft.)**	**Cement (cu. ft.)**	**Sand (cu. ft.)**
1/4	2.08	0.68	2.06
3/8	3.13	1.03	3.10
1/2	4.17	1.37	4.12
5/8	5.21	1.71	5.15
3/4	6.25	2.06	6.18
1	8.33	2.74	8.24

materials required for different construction elements and stucco thicknesses.

ESTIMATING BRICK VENEER

Estimating Brick Veneer

Brick sizes vary widely, and so do the costs. When estimating and pricing brick, keep in mind that larger brick may cost more per unit, but it takes far fewer units to complete a job, and the job will require less sand and mortar.

To estimate the number of brick per 100 sq. ft. of wall, refer to **Figure 4-8**. Or use the following formula (for standard 8x2 1/4x3 3/4 brick with a standard 3/8-in. mortar joint):

1 SQUARE FOOT OF WALL = 6.8 BRICK

WALL HEIGHT (FT.) × WALL LENGTH (FT.) × 6.8 =

TOTAL BRICK NEEDED

Figure 4–8. Brick Veneer Coverage (per 100 sq. ft.)

Mortar Joint (in.)	Brick	Wall Ties	Mortar Required (cu. ft.)
1/4	698	23	4.48
3/8	680	22	6.56
1/2	635	21	8.34
5/8	590	19	10.52
3/4	549	18	12.60

"Mortar Required" assumes 20% waste for all head and bed joints. The brick size assumed is 8x2 1/4x3 3/4 in. No waste is included for brick.

Figure 4–9. Drainage Plane

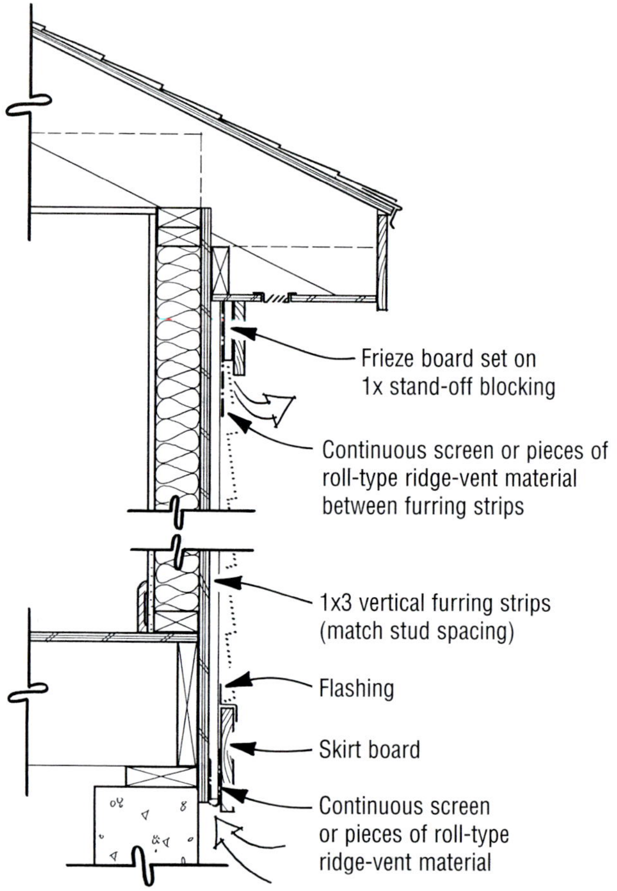

Siding that is pressed up tight against sheathing (even with felt or plastic housewrap) will retain moisture, causing the siding and finish to fail prematurely and water to move into walls. All siding should be installed over battens that create an air space behind the siding so water can drain down and out.

For more information: "Water-Managed Wall Systems," JLC 3/03

"Making Walls Watertight," JLC 12/95

All sidings will leak. Most water that leaks behind the siding is driven by rain or wicks into small cracks. Caulks and sealants are not an answer to stopping water penetration. Instead, all siding materials should be applied over a drainage plane (also called a rain screen) to allow water to fall down and out. This drainage plane is created by first applying sheathing wrap and flashings — lapped to shed water — and then by creating an air space and weep holes so water can drain away from the building (**Figure 4-9**).

SHEATHING WRAP MATERIALS

Sheathing wraps that work well include asphalt-saturated felt and plastic housewraps.

Type D kraft paper (a lighter black paper than asphalt felt), sisal kraft paper (asphalt sandwiched between two layers of kraft paper), and cross-laminated plastic housewraps all perform poorly in rainy and snowy climates. However, two layers of Type D paper work well beneath stucco in most climates (see "Stucco Drainage Plane," **page 346**).

All sheathing wrap must be integrated with flashing materials. It's not enough to wrap a house, and then to simply X-cut the openings and fold the flaps inside. Follow the guidelines in **Figure 4-10** when installing any sheathing wrap.

Asphalt-Saturated Felt

Commonly available in two weights, asphalt felt, or "black paper," is made from recycled corrugated papers mixed with sawdust and an asphalt resin. When it actually was made from cotton felt, 15-lb. felt weighed 15 pounds per square, and 30-lb. felt weighed 30 pounds per square. These days, asphalt felt weighs considerably less (between 6 and 12 pounds per square) and is now simply designated No. 15 and No. 30.

When choosing No. 15 felt, look for material rated ASTM D4869 or ASTM D229. The better material is rated by the more stringent ASTM D229 standard, which ensures a weight of at least 11.5 lbs. per square.

The biggest drawback to using asphalt felt as a sheathing wrap is that it's difficult to apply well on walls. Generally, wrinkles are not a problem, but tears — particularly at inside corners — should be avoided. Also, try to prevent sags between the relatively narrow (36-in.) courses.

Plastic Housewrap

In cold temperatures, plastic housewraps stay more flexible than paper and felt wraps. Also, plastic housewraps resist tearing, so they are easier to apply.

Drainage Planes

Sheathing Wrap

Plastic housewraps typically are made of polyolefin fabric — either polyethylene or polypropylene. The primary difference in housewraps is whether or not they are perforated:

- **Perforated** products are made from a vapor-tight plastic, and then needle-punched to allow the wrap to "breathe"

Figure 4–10. Sheathing Wrap Details

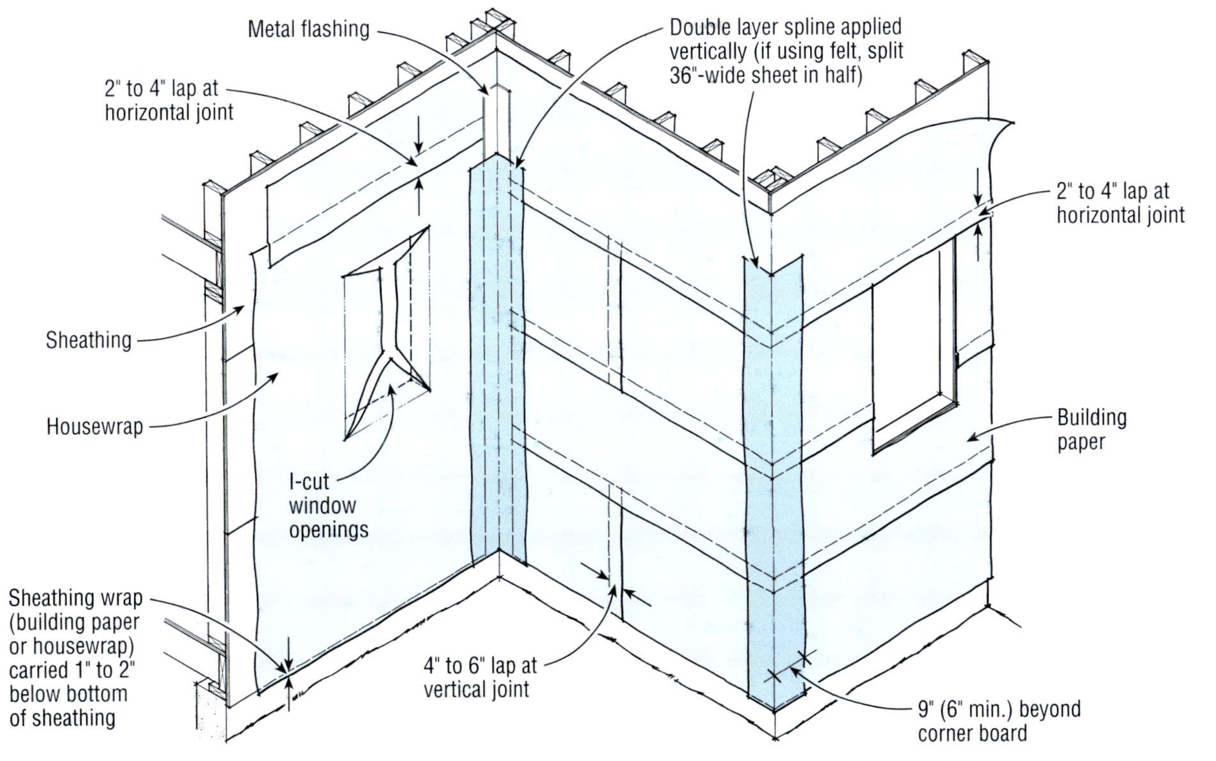

All sheathing wraps must be applied from the bottom to the top, so upper courses lap over lower courses and shed water away from the house. Always carry the sheet around corners, then reinforce the barrier at the corner by applying a second layer, or spline. This spline should extend at least 6 in. beyond the joint between the siding and corner boards.

(allow vapor to pass through). Unfortunately, these holes also allow liquid water to seep through, particularly when the siding is pressed tight against it. The housewrap will slow the seepage but will not block it entirely.

- **Non-perforated** housewraps are made from a plastic fiber that allows vapor to pass between the fibers. These generally perform much better than perforated products at resisting liquid water, which should be considered their primary task. Relatively few houses suffer from water vapor penetrating from outside, but many suffer from rain and snow infiltration.

Leaching

Housewraps are susceptible to damage from extractives that can leach from wood siding materials, particularly cedar and redwood. This can be prevented in two ways:

- Install siding on battens to create an air gap, which reduces contact between the sheathing wrap and siding (**Figure 4-9, page 296**).
- Back-prime all wood siding materials. This reduces leaching and blocks moisture from moving through the siding and blistering the finish (see "Finishes," **pages 373-380**).

For more information: "Choosing a Sheathing Wrap," JLC 8/00

"Wrapping the House," JLC 3/00

INSTALLING SHEATHING WRAP

The primary purpose of housewrap is to stop water, not air. Air-sealing and vapor control are secondary. See **Figure 4-10,** previous page, for details about how to lap sheathing wrap to shed water.

- Use sheathing wrap or felt paper on all houses, no matter what kind of siding is used.
- The wrap should be continuous; avoid a patchwork of small pieces. Review flashing details (**pages 302-306**), as some flashings (such as those around electrical fixtures, mechanical penetrations, etc.) should be installed concurrent with sheathing wrap.
- Lap upper courses of housewrap over lower courses 4 in. and lap vertical seams 12 in. Code requires a horizontal lap of only 2 in. and a vertical lap of 6 in., but water may still back up beyond these minimal overlaps.
- Do not X-cut sheathing wrap at openings. Instead, I-cut the sheathing wrap, running the knife along the rough opening at the head and sill, and then slitting vertically in the center of the opening (see **Figure 4-10,** previous page).
- If the walls are wrapped while they are lying on the deck before standing them up, be sure to add a course of sheathing wrap to cover the rim joist. Remember, this course must lap under the sheathing wrap above it.
- Always wrap gable ends. This part of a house sees a lot of exposure to wind-driven water, but often is neglected when contractors assume that the sheathing wrap is meant to stop air.
- Add a second layer of sheathing wrap along inside and outside house corners, lapping at least 9 in. (an even multiple of 36-in.-wide asphalt felt) on each wall.
- Control splashback by sloping the grade around the house and keeping siding at least 8 in. above grade. At decks and stoops where splashback may be a problem, use a wide peel-and-stick flashing membrane (see "Flexible Flashing Materials," opposite page). Make sure this membrane laps under the sheathing wrap above it.
- All flashing (for windows, doors, deck ledgers, electrical fixtures, vents, pipes, and cantilevered beams) must lap

under the sheathing wrap above it. Do not rely on tape to seal the flashing.

To Tape or Not

Manufacturers of plastic housewrap recommend taping all seams to block air leakage through walls. However, typical plywood and OSB sheathing are impermeable to air, so there is no need to tape all the seams.

Some tape is useful for stopping water leaks by repairing rips (particularly tears around staples) and for securing the wrap to flashing before the siding is in place. For best results, use a polypropylene tape intended for sheathing wrap or a flexible, peel-and-stick flashing. A high-quality polyurethane rubber caulk may also be used to seal plastic housewrap (see "Caulks and Sealants," **page 371**).

Do not patch over housewrap and rely on the stickiness of tape to seal the patch. Always lap housewrap and flashings under the piece of sheathing wrap above.

FLEXIBLE FLASHING MATERIALS

Flexible flashing — a self-adhering bituminous tape — should be used to flash all the following: exterior door sills; deck ledgers; inside and outside corners of wall sheathing; stucco shelves and parapets; below-grade concrete cracks; and wall areas subject to splashback.

Flexible flashings come in several varieties, including "peel-and-stick" products, as well as non-stick products.

RUBBERIZED-ASPHALT FLASHINGS

Self-adhesive, rubberized asphalt flashings are made from the same materials as the eaves membranes used to prevent ice dam leaks. Typically available in rolls from 4- to 12-in. wide (but sometimes available up to 36 in. wide), these flexible flashings can be applied directly to the sheathing, as long as the surface is warm and relatively clean.

Flexible Flashing

Thickness varies from 25- to 60-mils. The thinner the material, the easier it can be folded and tucked beneath siding, but also the easier it will tear and degrade in sunlight.

Climate Cautions

- **In warm temperatures,** when rubberized asphalt is the stickiest, it is nearly impossible to reposition it once the flashing has touched the surface. Make sure to dry-fit before pulling the release paper. It is also very difficult to handle alone when it is this sticky.
- **In hot temperatures,** rubberized asphalt tends to ooze, making it messy to handle. In sunbelt areas (or wherever surface temperatures might exceed 180°F), avoid using with metal flashings; the flashings will cook and the asphalt soon dries out, usually after creating a dripping mess.
- **In cold weather,** self-adhering flashings often don't stick well. Check the manufacturer's minimum application temperature. Some stick in temperatures below 25°F, but most require surface temperatures of 40°F or greater.

Compatibility

The solvents in rubberized asphalt will stain rolled vinyl flashing — the flexible coiled trim stock — but generally will not hurt the more rigid vinyls used for window fins and siding.

Do not use rubberized asphalt and butyl tapes together. On concrete, masonry, or OSB, rubberized-asphalt flashings may not stick well unless a primer is applied. On these surfaces, a butyl flashing works better.

Figure 4-11. Sill Flashing Details

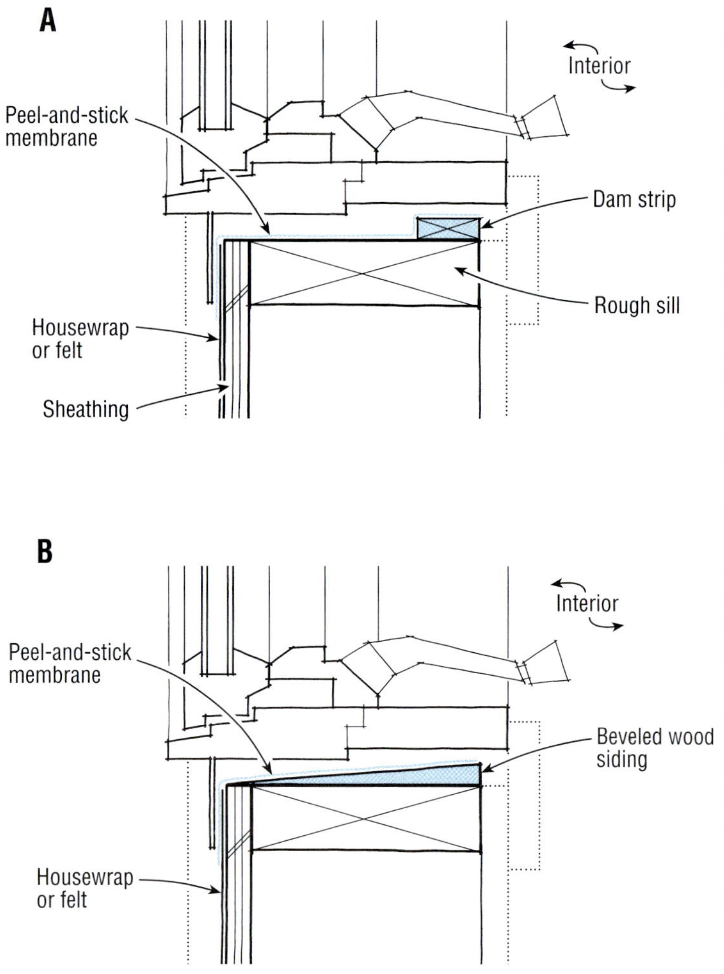

Apply a narrow dam strip or bevel cant strip to the rough sill before flashing with sheathing wrap or a flexible flashing membrane.

For more information: "Choosing Flexible Flashing," JLC 6/01

"Controlling Moisture in Mixed Climates," JLC 8/97

"Avoiding Moisture Problems in Cooling Climates," 10/97

Poly vs. foil. Rubberized-asphalt flashings typically come with a polyethylene or foil covering to protect them from sunlight. Poly-covered products need to be covered from sunlight within 30 to 60 days; foil-covered tapes have no exact exposure limitations but eventually will dry out and become brittle in full sunlight. Once the flashing has become brittle, any shift in surface movement will pull at the bond until either the bond fails or the membrane cracks.

Exposed applications. In permanently exposed applications, use an EPDM flashing tape instead of an inexpensive flexible flashing material. The drawback with EPDM tape is that it's expensive, as well as quite thick (about 70-mil), making it awkward to fold into small areas.

BUTYL RUBBER FLASHINGS

Butyl costs about twice as much as rubberized asphalt, but concedes great results: The adhesive lasts longer and sticks over a wider range of temperatures. Butyl grabs better onto rough surfaces, such as concrete and OSB, yet the bond develops slowly, allowing for easy repositioning. Also, butyl is less likely to ooze in high temperatures or to stain a finished surface.

Like rubberized asphalt products, butyl flashings are available with either a polyethylene or foil surface. Poly-covered flashings should be covered within 30 days to protect from UV light. Foil does a good job at reflecting the UV light, but it still heats up in exposed sunlight. It will hold out longer, but when left exposed, foil-faced flashing eventually will degrade.

Compatibility. Do not use butyl flashing near any

asphalt-based products. The butyl and the asphalt products will both dry out faster and loose their adhesion.

NONSTICK FLASHINGS

Not all flexible flashings are self-adhering. Some must be secured with staples or roofing nails. While generally thinner and the quickest to deteriorate in sunlight, these are often the best choice for applying beneath siding and trim in consistently cold temperatures, or when the surface is rough or dirty — factors that can diminish the bond of

Figure 4-12. Window Splines

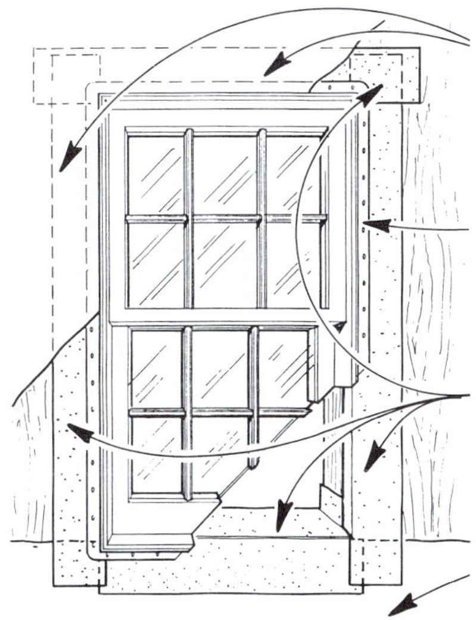

Flash windows and doors with window splines cut from felt paper or flexible flashing. Make sure the sheathing wrap laps over the splines at the top and sides. At the sill, however, the bottom end of the spline and the sill flashing should overlap the sheathing wrap.

peel-and-stick flashings. Cover nonstick flashing as soon as possible to protect it from sunlight.

DRAINAGE PLANE

Water will always get behind siding. Wind blows in through gaps, and when the sun beats down on wet siding, it drives

Figure 4-13. Window Head Flashing

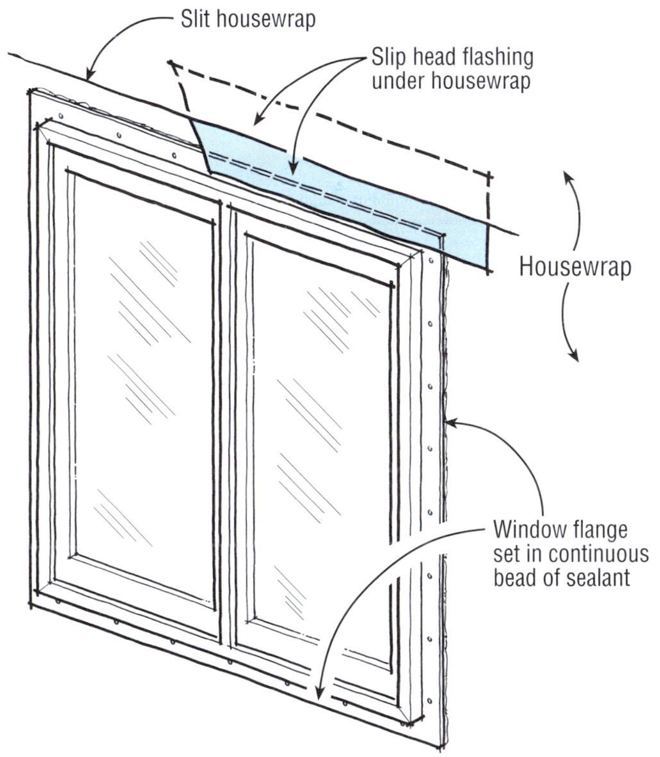

If the window flange is not tucked under the sheathing when the window is installed, slit the housewrap and then slip in a piece of flexible flashing and drape it over the flange.

Figure 4-14. Flanged Window Head

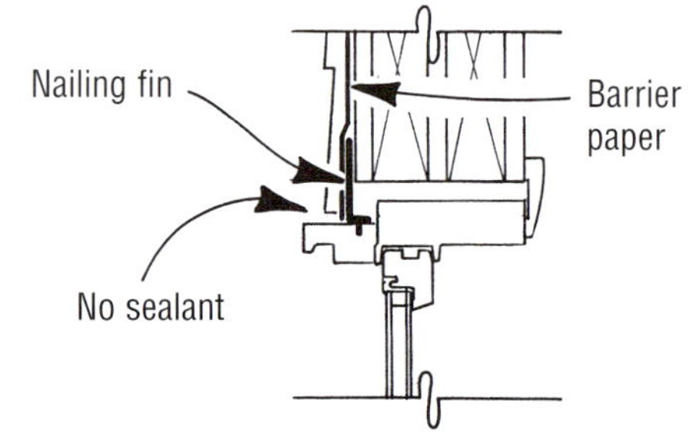

With flanged windows, make sure the sheathing wrap laps over the top fin. Leave a 1/4-in. gap between the siding and the flange, but don't caulk this gap. Water draining from above must have a place to exit.

Figure 4-15. Unflanged Window Head

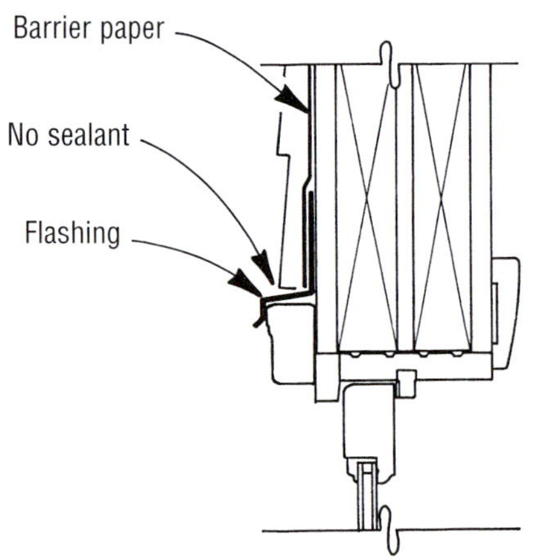

With unflanged windows, install metal Z-flashing under the sheathing wrap and over the window head trim.

For more information: "Water Managed Wall Systems," JLC 3/03

"Making Walls Watertight," JLC 12/95

moisture inward. No siding material is impervious to water, and if the siding is pressed tight against plastic housewrap or applied over foam sheathing, the front of the siding tends to dry while the back stays wet. This stresses all types of siding: It cups boards, curls shingles, accelerates stucco cracks, and weakens mortar bonds in brick veneer. With all siding, an air space behind the siding (1/4- to 3/8-in. is enough) will allow water to drain and the backside of the siding to dry.

Well-Drained Wall Systems

Different sidings call for different drainage details. Typical details for well-draining walls are shown for different types of siding:

Board siding, **Figure 4-37, page 322**;

Shake and shingle siding, **Figure 4-41, page 326**;

Vinyl siding, **Figure 4-51, page 333**;

Stucco, **Figure 4-66, page 346**.

In each case, an air space must be created behind the siding and an exit provided for water to drain out.

FLASHING DETAILS

- Never rely on the stickiness of peel-and-stick flashings for sealing; the bond is unlikely to last over the life-span of the house.
- Always lap flashings under the sheathing wrap above.

FLASHING WINDOWS

Do not X-cut sheathing wrap at openings. Instead, I-cut the sheathing wrap, running the knife horizontally along the rough opening at the head and sill, then slitting vertically in the center of the opening (**Figure 4-10, page 297**). This way, the window flange can be slipped under the sheathing wrap when installing the window.

Sill details: Install a beveled cant strip or narrow dam strip to the rough opening sill, then wrap the sill with sheathing wrap or a flexible flashing membrane (**Figure 4-11, page 300**). Make sure this membrane drapes over the sheathing wrap below.

All sheathing wrap must be integrated with flashing materials. It's not enough to wrap a house, and then to simply X-cut the openings and fold the flaps inside. Follow the guidelines in **Figure 4-10 (page 297)** when installing any sheathing wrap.

Flashing Windows and Doors

Window and Door Splines

Water will leak through the gap where the siding butts the casing. Even if this joint is caulked, the sealant won't last as long as the paint job, and water will eventually work itself inside. Windows and doors should be flashed with felt paper or flexible flashing splines (**Figure 4-12, page 301**).

Figure 4-16. Flashing Door Thresholds

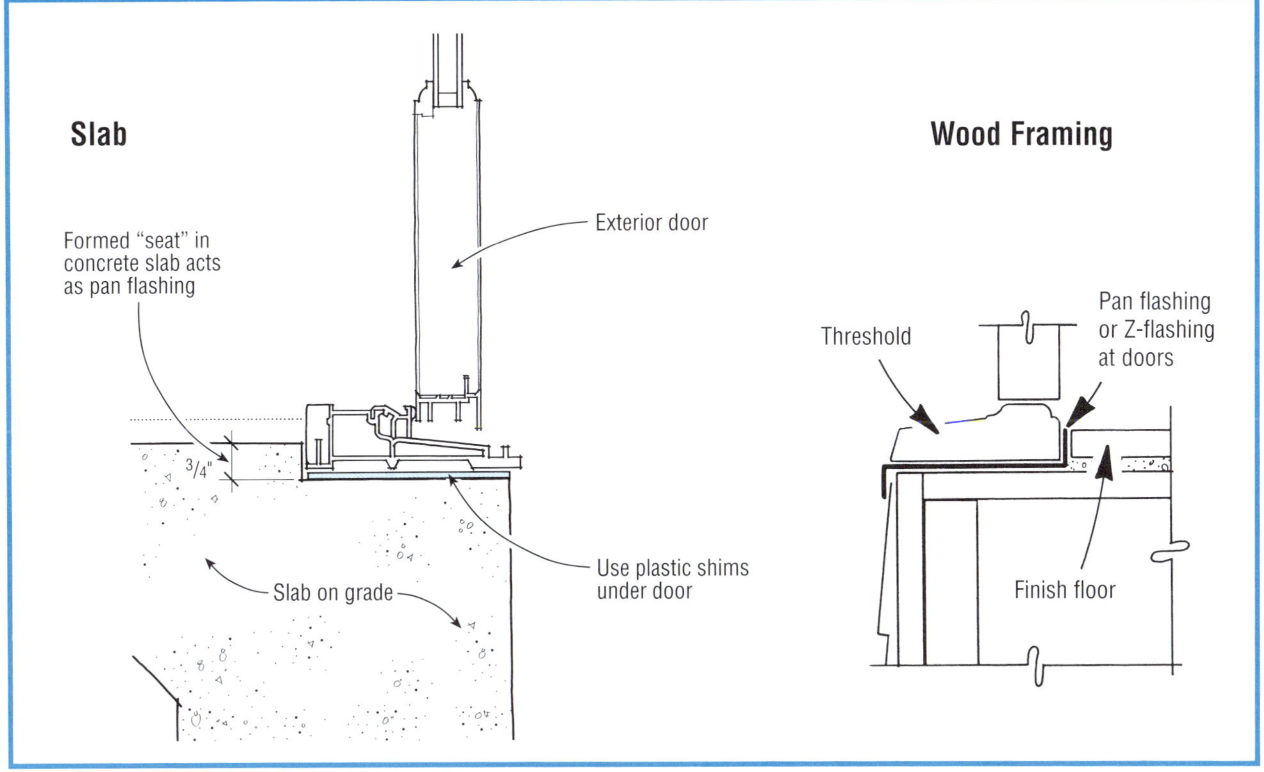

Mulled entry doors and sliders are most prone to leaks at the threshold. Provide a formed seat in the slab (far left) or a metal pan flashing to keep water off the floor.

The splines should be installed after the sill flashing, and the sheathing wrap should go on afterwards or be slit so it laps over the splines.

Flanged window heads do not need additional flashing. The nailing fin will do a good job sealing the head if it can be tucked beneath the sheathing wrap when the window is installed (**Figure 4-14, page 302**). Leave a 1/4-in. gap between the siding and the window along the head, but do not caulk this gap: Otherwise, water that runs behind the siding from above won't have a place to drain.

If the flange cannot be tucked under the sheathing wrap when the window is installed, slit the wrap and add a head flashing, as shown in **Figure 4-13, page 301.**

Unflanged windows will need an additional metal Z-flashing above the head casing or brick mold, as shown in **Figure 4-15, page 302.** Make sure this Z-flashing laps underneath the sheathing wrap.

Mulled windows and window walls must have a continuous head flashing; the manufacturer's mullion cover will not prevent water from running between the window units from above.

Figure 4-17. Flashing Deck Ledgers

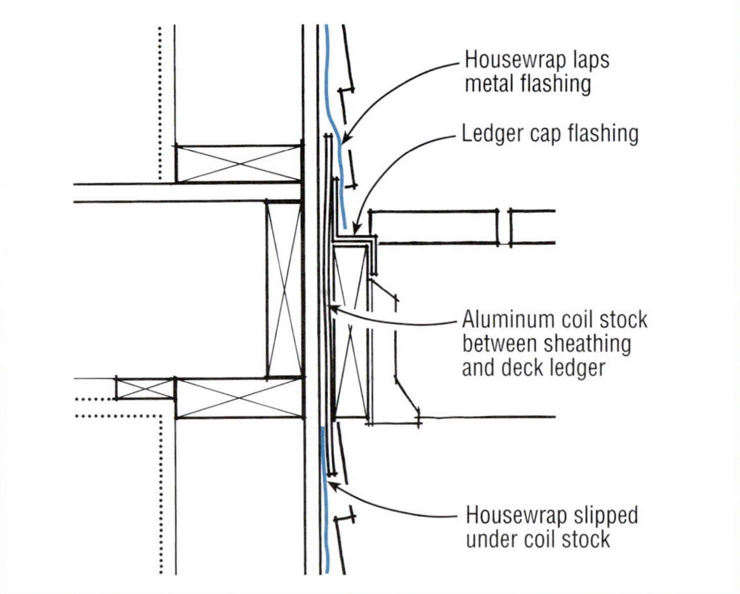

To properly flash a ledger, run the flashing behind the ledger to a drip-edge.

Figure 4-18. Flashing Cantilever Beams

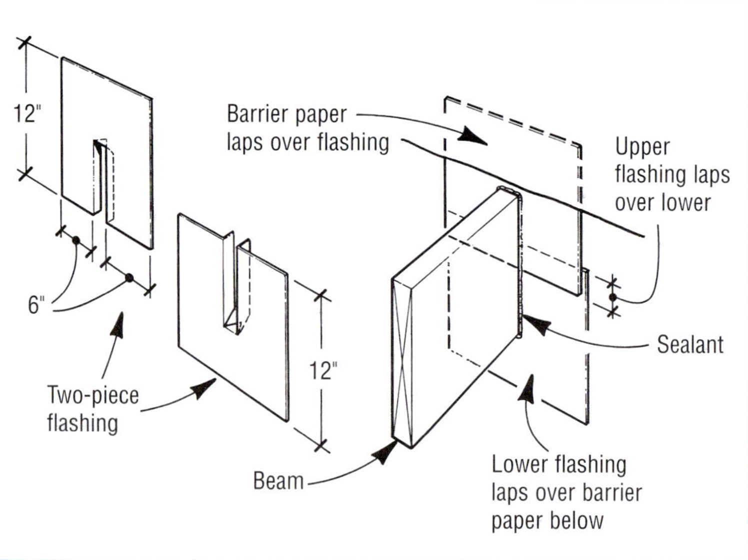

Protect cantilevered balcony and deck beams with a two-part flashing sealed with urethane. The top flashing piece must lap over the bottom one.

Round-top windows should be flashed with a peel-and-stick flexible flashing. Slit the sheathing wrap and slip it underneath, letting it drape over the window top. Then cut the flashing to conform to the nailing fin. Rely on proper lapping of the flashing, not on caulk.

FLASHING DOORS

Water that gets past siding and trim around doors can run along the floor, ruining finish flooring. This is a particular problem where splashback from dripping eaves and overhangs soaks the door. Over wood floors, door thresholds should be flashed with a metal pan or Z-flashing. Slabs should be formed with a "seat" to hold the threshold (**Figure 4-16, page 303**).

Figure 4-19. Flashing Electrical Fixtures

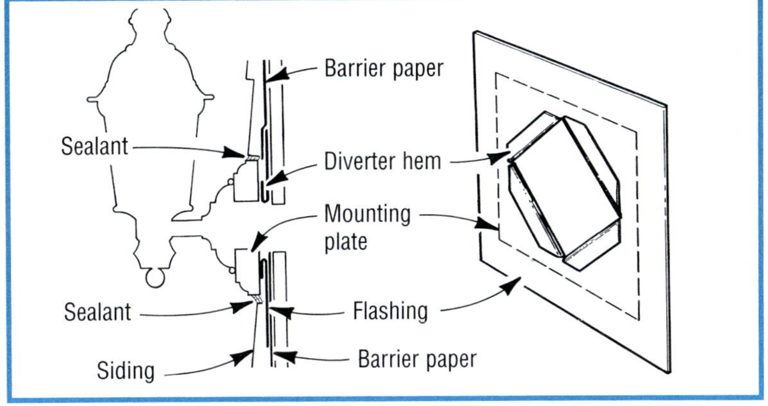

Exterior electrical boxes and fixtures can be sealed with a hemmed metal plate. Slit the sheathing wrap above the fixture and slip the plate beneath.

Figure 4-20. Flashing Penetrations

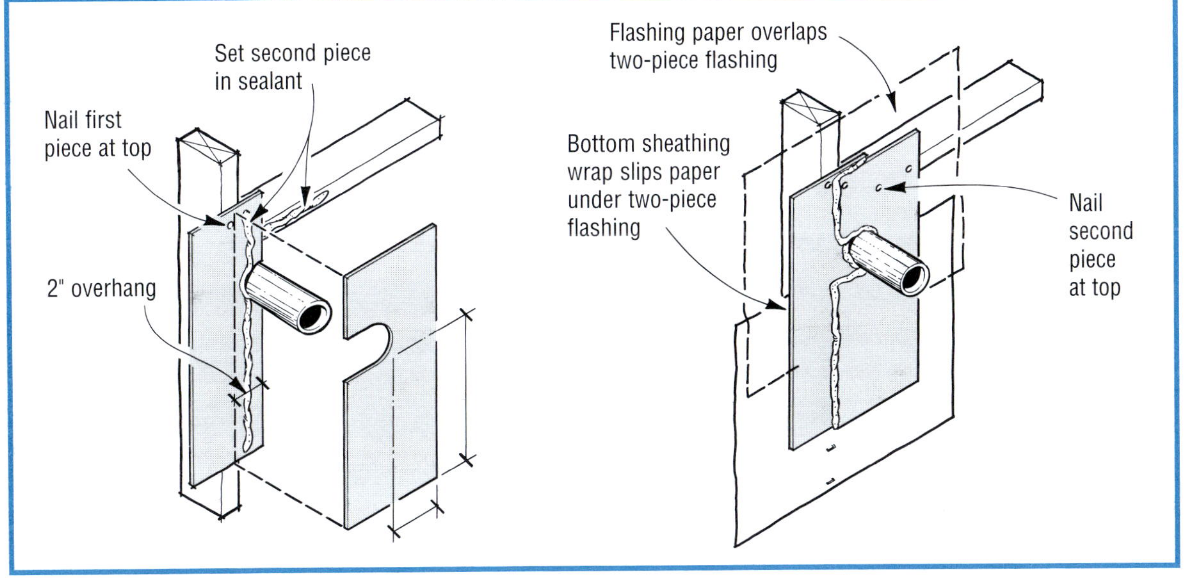

To seal around wall penetrations, use a two-piece metal plate bedded in urethane sealant. Sheathing wrap above the metal plate should lap over the plate while sheathing wrap below it should be slipped beneath.

FLASHING LEDGERS AND BEAMS

Ledger flashing needs to be more than a Z-flashing running along the top edge of the ledger board. Use metal coil stock behind the ledger, as shown in **Figure 4-17**. Make sure the coil stock slips under the sheathing wrap above the ledger, and drapes over the sheathing wrap below the ledger.

Figure 4-21. Kickout Flashing

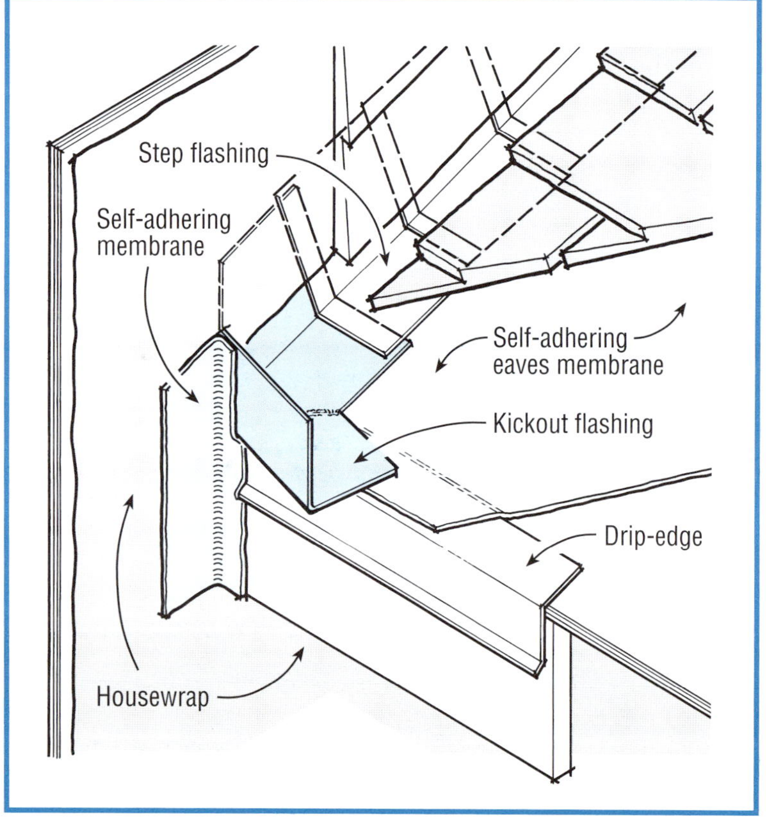

To prevent water from dumping behind the siding at the end of a roof intersection, bend a small "kickout" from metal flashing to divert water running down the roof away from the siding.

Alternatively, build a deck completely free-standing from the wall, or insert spacers between the ledger board and the wall, so it stands out and allows water to drain freely.

Cantilevered deck or balcony joists that project through the wall framing are trouble spots. Flash with a two-piece flashing membrane, as shown in **Figure 4-18**.

Flashing Wall Penetrations

Electrical fixtures. Install electrical boxes and lighting fixtures on a hemmed metal flashing plate, as shown in **Figure 4-19**. Do not tape the plate over the sheathing wrap; slit the sheathing wrap and tuck it beneath.

Mechanical penetrations. Gas line, conduit, water spigots, and other small penetrations should be flashed with a two-piece metal flashing plate (**Figure 4-20**). Seal the two plates with a "low modulus" urethane caulk (see "Caulks and Sealants," **page 371**).

Sidewall Flashing

Because of the volume of water that can pour down a sloped roof, one of the most critical flashing details occurs where a roof intersects a sidewall. The roof should be flashed with a step flashing (see **Figure 3-76, page 277**). Where the roof terminates, install a kick-out to deflect water away from the siding (**Figure 4-21**).

Though not always possible in the schedule, it's best to install a full 36-in. piece of rubberized asphalt flashing on the wall before the subfascia and trim boards are nailed in place, and then come back to install the kickout.

Use only galvanized or stainless-steel nails for exterior siding and trim. Stainless-steel nails will resist corrosion the longest, but they are very expensive. Double hot-dipped galvanized nails are the next best choice. High-tensile aluminum nails can be used for a low-budget job.

GALVANIZED NAIL COATINGS

Galvanized coatings vary widely. The following coating types are available (in order from best to worst):

Hot-dipped zinc-coated nails offer the best protection among the galvanized nails. Because nails are submerged in molten zinc, they are coated uniformly and the high heat causes the zinc and steel to fuse. Nails can be double-dipped for heavier plating. True hot-dipped nails are hard to distinguish from hot-galvanized nails; for verification, look for ASTM-153, the standard for hot-dipped galvanized hardware.

Galvanized Nails

Electroplated nails work best in nail guns. The nails are immersed in an electrolytic solution that deposits a thin film of zinc on them when an electric current is run through the solution. Because the coating is thin, it oxidizes in harsh exposures to salt air and pollution.

Figure 4-22. Exterior Nail Types

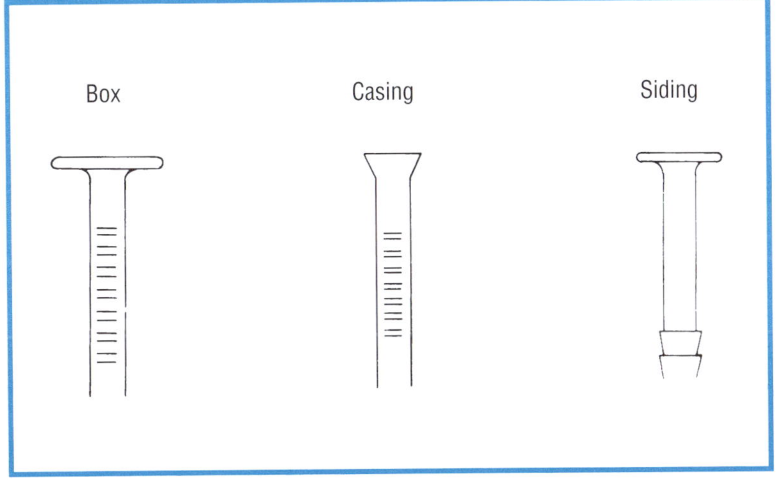

Figure 4-23. Extractive Bleeding

Bleeding is caused by excessive moisture in siding, which reacts to the iron in nails — even when the nail is intact. To prevent bleeding, keep siding dry before installation and seal it soon afterwards.

Figure 4-24. Selecting Siding Nails

For Cedar, Redwood, and Other Premium Board Sidings

Features	Advantages
High-Grade Steel	Under the galvanized coating, high-carbon steel minimizes bending, even with thin-shank nails.
Shank	A ring-shank nail provides 50% to 100% greater holding power than a smooth-shank nail. Recommended penetration is $1\frac{1}{2}$ in. into the solid wood stud. A thin-shank nail will reduce wood splits that occur with standard-thickness siding nails.
Point	A blunt point punches its way though siding material rather than wedging though wood fibers, reducing the chance of splits.
Head Size	A small head helps to make the nail unobtrusive on fine wood siding. For painted sidings, choose a checker-head nail, which greatly increases paint adhesion.

For Hardboard, Plywood, and Fiber-Cement Siding

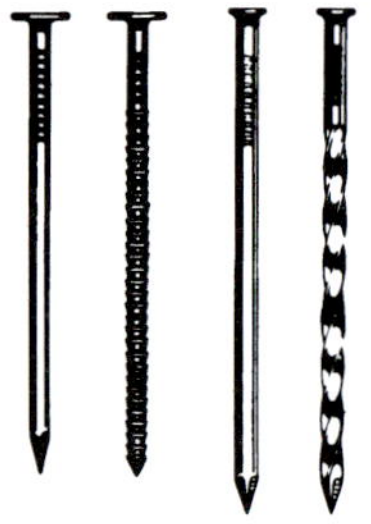

High-Grade Steel	Under the galvanized coating, high-carbon steel minimizes bending when the nail is driven through two laps of siding.
Threaded Shank	A ring or a spiral shank provides much better holding power and reducing nail "pop-outs."
Head Size	Use a minimum $\frac{1}{4}$-in.-diameter head for blind nails; use a larger roofing nail head for face nailing. A larger, flat head will help prevent the head from breaking through the surface.

For Vinyl Siding

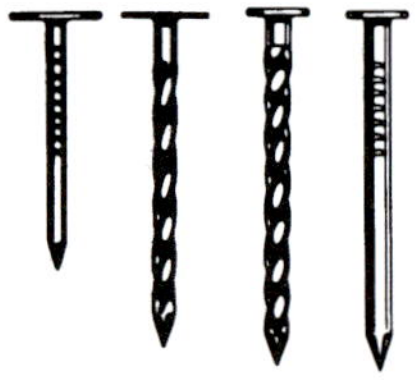

High-Grade Steel	Drives much better than aluminum nails.
Threaded Shank	Spiral shanks hold much better than smooth shanks.
Head Size	The large head allows the siding to hang properly without slipping though the expansion slot.

For long-lasting results, all exterior siding and trim nails should be double hot-dipped galvanized, or Type 304 or 316 stainless steel.

For more information: "Guide to Galvanized Nails," JLC 11/94

Mechanically plated nails are rotated when cold in a barrel of zinc dust. Glass beads hammer the zinc onto the nail, and then the nails are immersed in a chromate solution, which gives them a gold or greenish color. This process leaves threaded nails clean, but the coating is thin.

Hot-galvanized nails are the least effective. The nails are rotated in a barrel in a furnace to melt the zinc. They may not be evenly coated and threads may fill up.

SELECTING GALVANIZED NAILS

For best results, choose nails with an appropriate head, shank, and point, as shown in **Figure 4-24**.

EXTRACTIVE BLEEDING

Natural-finished siding may stain around nails, as shown in **Figure 4-23**. Even painted and stained siding may become discolored. This staining can occur even when the galvanized coating of the nail is intact, and with stainless-steel and aluminum nails. The staining is caused when a high moisture content in wood dissolves the wood's naturally occurring extractives — pigments, tannins, oils, and resins. These extractives then react with the iron in the nail to form a blue-black stain around the nail head. A poor quality nail makes the problem worse because gaps in the galvanizing expose the extractives to unprotected iron.

Selecting Siding Nails

To prevent extractive bleeding, condition the siding and protect it from water before installation (see "Conditioning Siding," **page** 318). Back-prime the siding to help seal extractives in the wood, and install the siding on a rain screen to promote drying. Seal or paint the wood when it is dry and maintain the finish to keep wood dry over time (see "Exterior Finishes," **page** 373).

EXTERIOR TRIM SERVES BOTH A FUNCTIONAL AND AN AESTHETIC purpose. Its primary function is to divert water away from vulnerable areas of walls, such as roof intersections, window and door openings, and wall corners — all places where wind-driven rain can seep into the wall cavity (**Figure 4-25**). Aesthetically, exterior trim provides a visual transition between wall surfaces and openings, edges, and corners.

WOOD SELECTION

Rot-resistant woods include redwood, red cedar, Alaskan yellow cedar, and Port Orford cedar, all of which may last a century or more without protection. These woods are easy to work with and they take paint well, but they are soft and dent easily.

Depending on the region of the country, other durable woods may be available. These include high-quality cypress, white oak, and locust. These species offer nearly the same rot resistance as redwood but are much stronger.

Moderately rot-resistant woods include eastern white pine, southern longleaf pine, larch, and swamp oak. These woods may last for many years if they are installed properly and are well-protected with exterior finish.

Slightly resistant or nonresistant woods include alder, poplar, cottonwood, hemlock, the spruces, the maples, red oak, and all other pines and true firs. Do not use these species for exterior trim unless it can be guaranteed that no water ever reaches the raw wood.

For more information: "Exterior Trim that Lasts," JLC 4/94

Cull Out Defects

Regardless of species, avoid any board containing a large percentage of sapwood, as sapwood is less durable than heartwood. Also avoid boards with splits, surface cracks, or loose knots. These defects will always get worse with exposure to the elements, and provide entry for bugs and fungi.

DESIGN PRINCIPLES OF EXTERIOR WOODWORK

Exterior joints should be designed to shed water (**Figure 4-25**). Trapped water will cause excessive wood movement and splitting, and eventually will invite rot, bugs, and fungal infestations.

Avoid Miter Joints

Only profiled moldings should be mitered; all other trim boards should have lap joints. Miter joints inevitably open as the wood shrinks, exposing the end grain of both boards to the weather.

While all trim stock should be dry (less than 12% MC at installation), this is especially important for profiled moldings to keep miters tight. Pre-prime all sides to help keep the wood dry.

Seal and Flash Edges

To keep joints dry, first make them as tight as possible:

- Make sure that adjoining surfaces meet uniformly at all points; no amount of glue or fasteners can compensate for a sloppy joint.
- Seal all joints that are likely to absorb water. Horizontal joints are the most vulnerable. Apply a coat of latex paint or a wax-based sealant such as Thompson's Water Seal® before the pieces are assembled.

Second, use metal flashing to help water find a path away from horizontal trim pieces.

Figure 4-25. Durable Exterior Trim

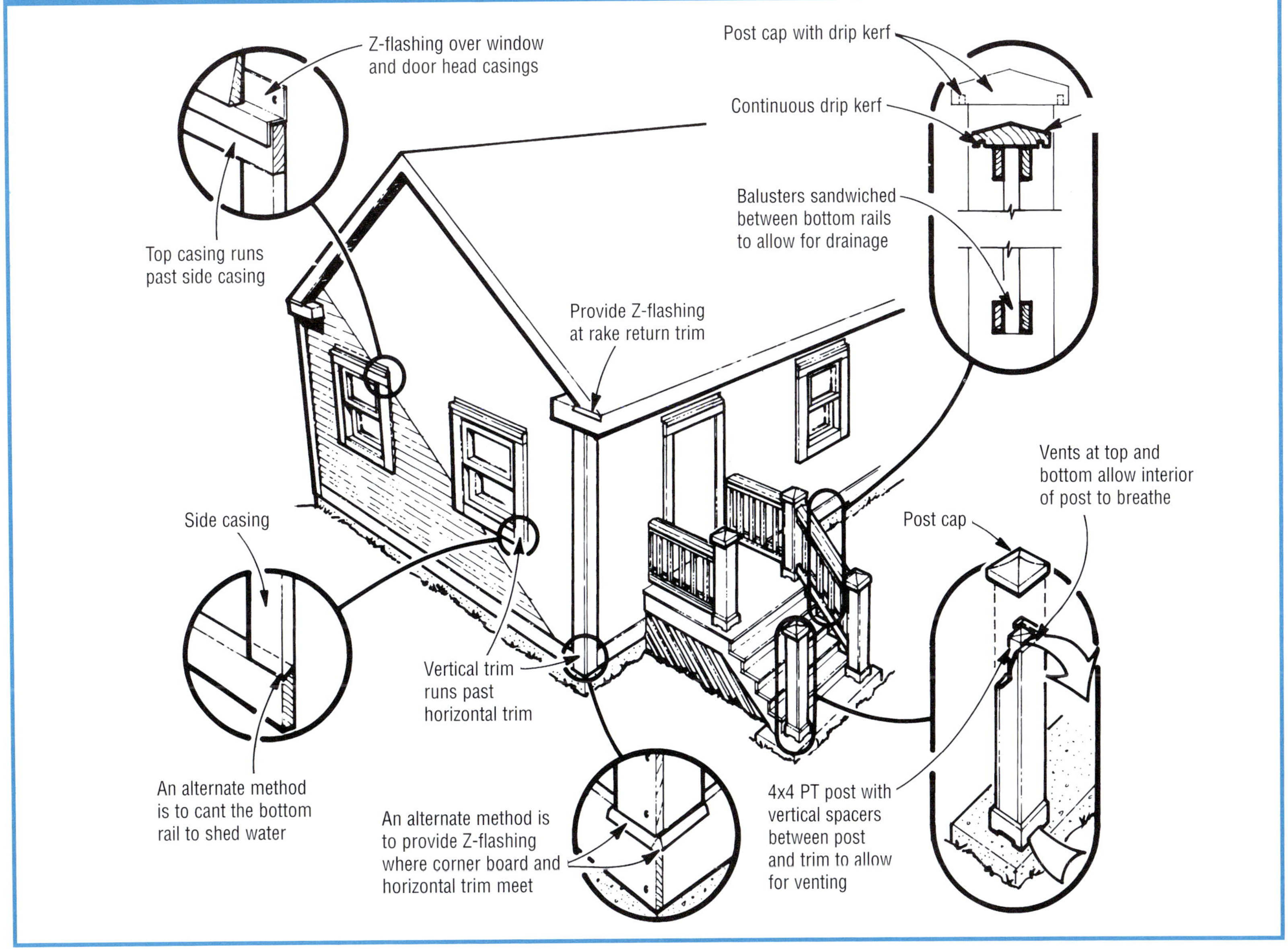

Exterior Trim Design Principles

Figure 4-26. Exterior Casings over Flanged Windows

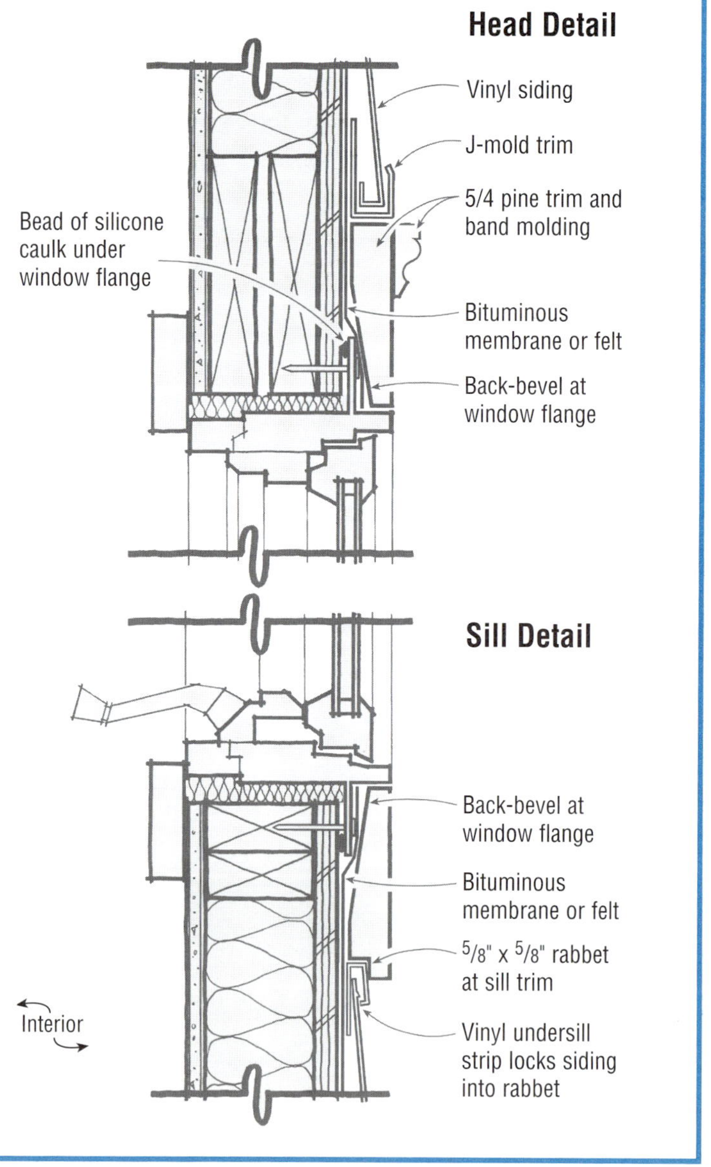

For more information "Exterior Trim Design Basics," JLC 7/95

Canted Surfaces and Drip-Edges

Railings, thresholds, and other flat surfaces can be made to shed water by bowing or sloping top surfaces, and by sawing or routing a drip groove along their undersides. Around posts and columns, build ventilation channels that will allow built-up pieces to dry.

EXTERIOR CASING

Always apply Z-flashing over the head casing, and lap the head casing over the side casings. Either extend side casings past the apron or cant at the bottom on a sloped sill (**Figure 4-25**, previous page). Keep sills as narrow as possible to limit exposure.

To keep casings flat and joints aligned, back-bevel casing stock to fit over window flanges (**Figure 4-26**). Rabbet the bottom edge of window aprons to divert water from the joint between the window trim and siding.

Figure 4-27. Corner Reveal

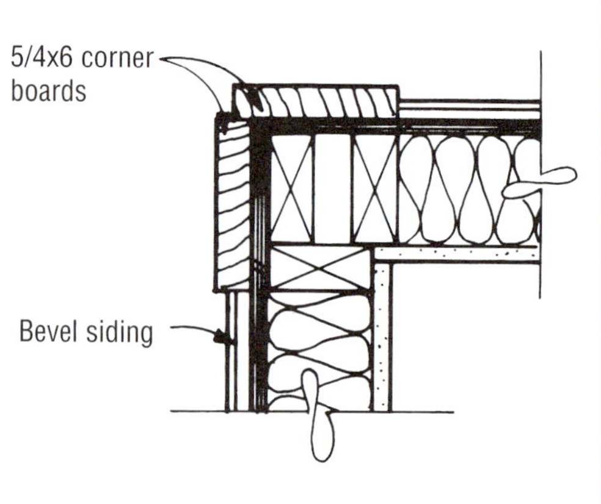

Keep corner boards plumb so siding cuts can be cut square. A reveal at the outside corner will help compensate for out-of-plumb walls.

Exterior Casing

Corner Boards

CORNER BOARDS

All corners should be double-flashed with sheathing wrap or flexible flashing (see **Figure 4-10, page 297**).

Outside Corners

Mitered outside corners should be avoided unless metal corner flashing is used (see **Figure 4-65, page 345**). The best way to handle outside corners with most sidings is with corner boards (**Figure 4-28**). Rabbeted edges will work the best to protect the joint between the corner board and the siding. Do not miter or caulk either side of the corner boards. The caulk will fail faster than the paint job and could trap water. Rely on careful corner flashing instead.

Tall walls may not always be framed perfectly plumb. To compensate (so siding cut can be level and square), build-in a reveal, which will help to hide slight discrepancies (**Figure 4-27**).

Inside Corners

Inside corners are best handled with a square corner post (**Figure 4-28**).

Figure 4-28. Corner Treatments

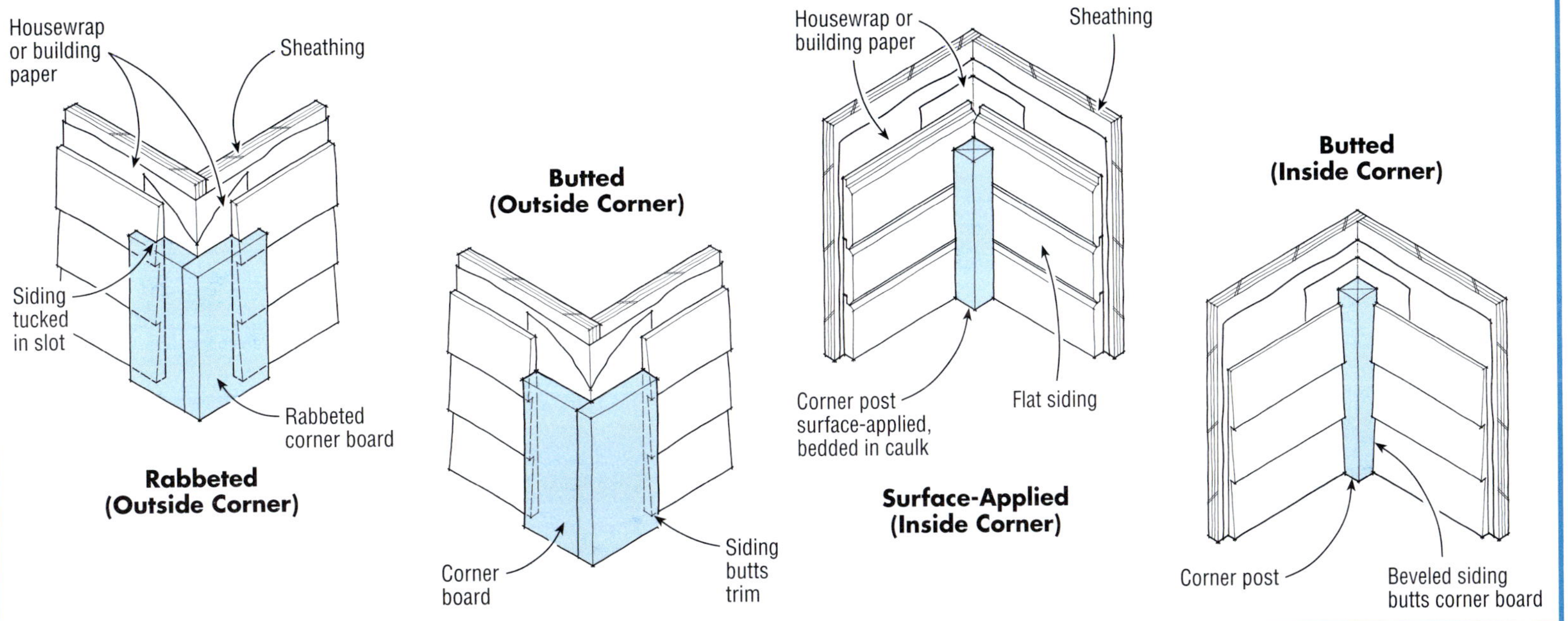

A rabbeted corner board will work best to protect siding at outside corners. This detail is critical for vulnerable siding such as hard-board. On flat siding profiles, an inside corner post bedded in caulk can overlay the siding, but beveled siding should butt.

WOOD CORNICES

A wide cornice is one of the best ways to protect walls from water. Conversely, walls without overhangs will always leak. Rake overhangs are just as important as eaves overhangs. Remember, water doesn't flow just downhill; it is driven in all directions by the wind, and a rake overhang will protect the vulnerable roof-to-gable intersection.

A cornice includes frieze board, soffit, and fascia. Several different width overhangs are shown in **Figure 4-29**. With an elaborate cornice over brick veneer, frame-in short pony walls to support the soffit and frieze (**Figure 4-30**).

Figure 4-29. Simple Wood Cornices

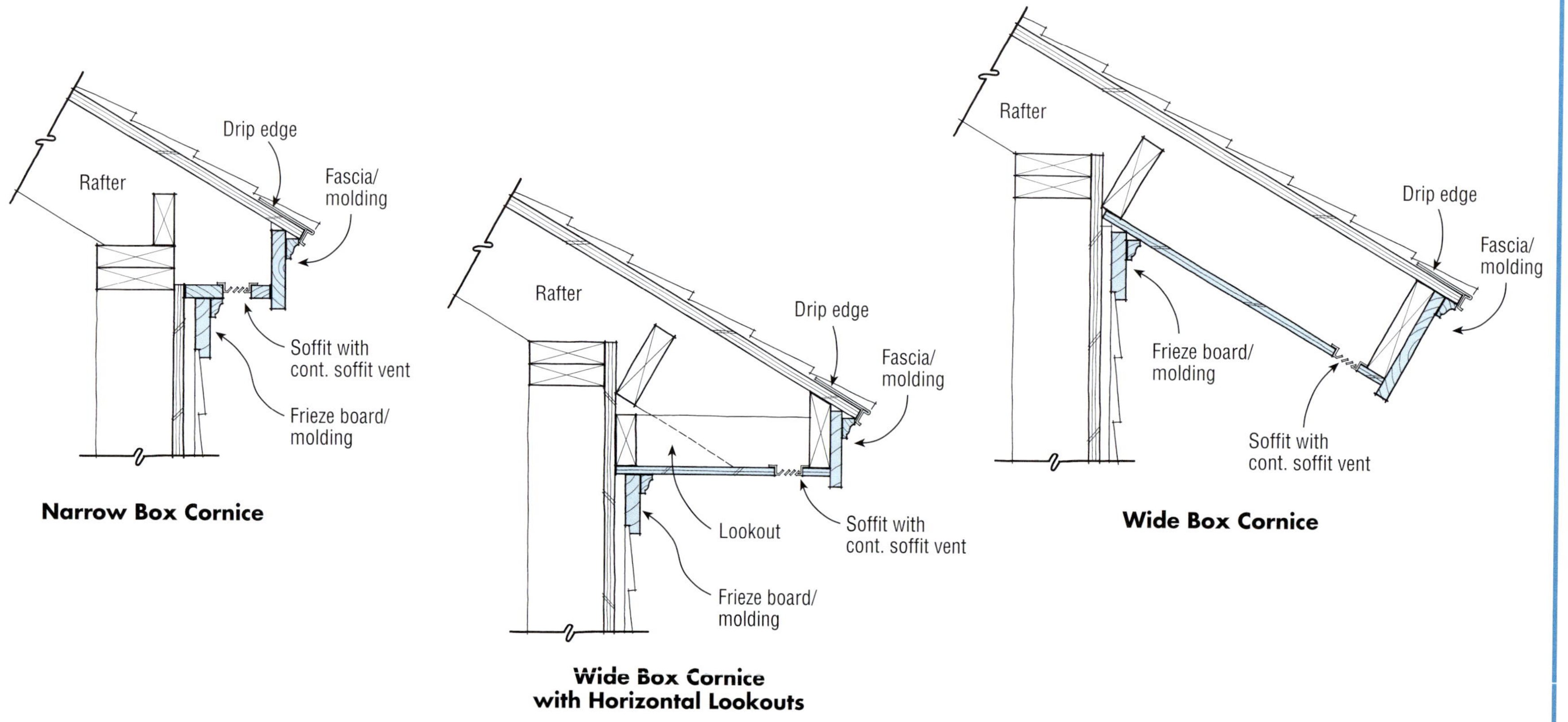

Eaves Crown

Crown molding is a common detail on traditional Colonial, Georgian, and Federal period houses. The most difficult detail is at the eaves-rake transition. There are only two ways to turn the corner with the crown and match the profile: Either cut one piece of trim for the eaves and one piece for the rake (called matched profiles) or return the crown horizontally and terminate the rake crown on top of the cornice return (**Figure 4-31**, next page).

GABLE-END TRANSITIONS

Just as a fascia is protected by the roof's drip-edge, other horizontal runs, such as head casing and skirt boards, should be protected by a drip-cap and Z-flashing. This is especially important on transition boards, such as the joint between gable siding and the wall below (**Figure 4-32, page 317**).

TUDOR TRIM DETAILS

The key to leak-free details in a traditional Tudor exterior is to install "build-out boards" behind the 1x6 cedar trim (**Figure 4-33, page 317**).

Vertical and Angled Trim

Install vertical and angled trim over 4-in.-wide plywood "build-out" boards. The narrower build-out board allows the scratch and brown coats of stucco to be troweled in behind the 1x6 (see "Three-Coat Stucco," page 00). Then, run the finish coat tight to the edge of the overlapping 1x6 trim board. Angled board details are problematic since they channel water towards the interior of the wall. Always flash behind diagonal boards with flexible flashing membranes.

Horizontal Trim

Unlike vertical trim, horizontal trim is held flush with the upper edges of the build-out board. Run Z-flashing continuously across the tops of horizontal boards, making sure to lap sheathing wrap over the Z-flashing so water drains down and out.

Wood Cornices

Seal Trim Before Installation

Tudor trim boards should be sealed with paint or stain on all sides (including field cuts) before installation.

Figure 4-30. Traditional Cornice with Brick Veneer

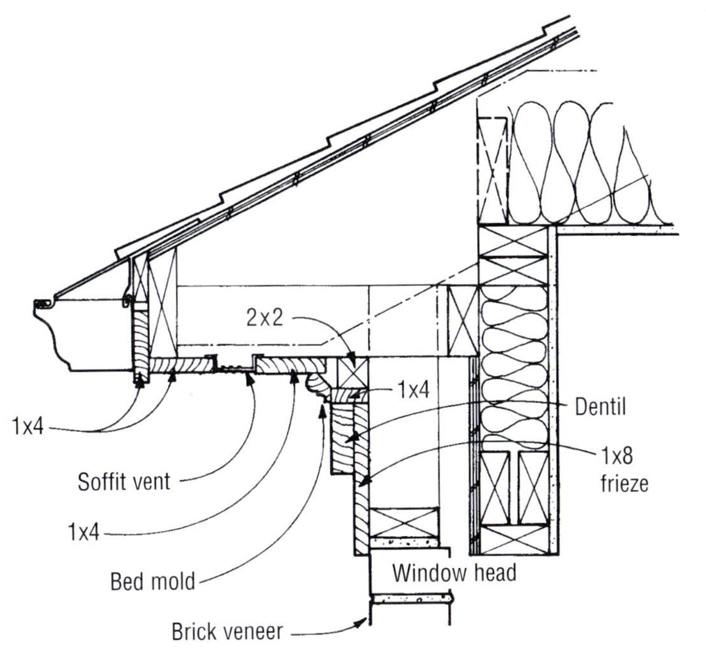

Figure 4-31. Crown Return

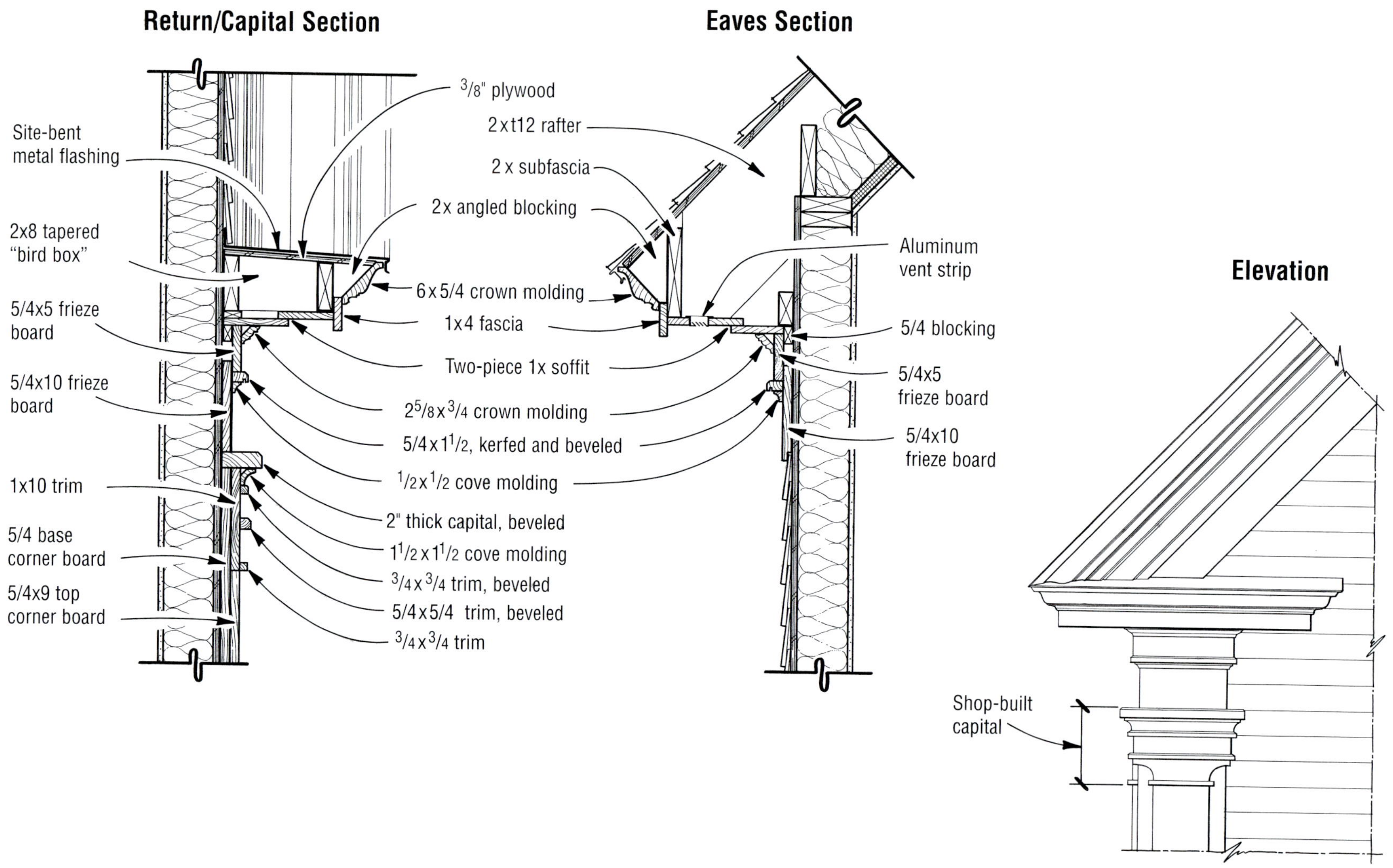

The rake of a traditional Greek Revival cornice should die into a horizontal return; otherwise, the crown molding profiles on the eaves and rake will not align.

Figure 4-32. Siding Transition

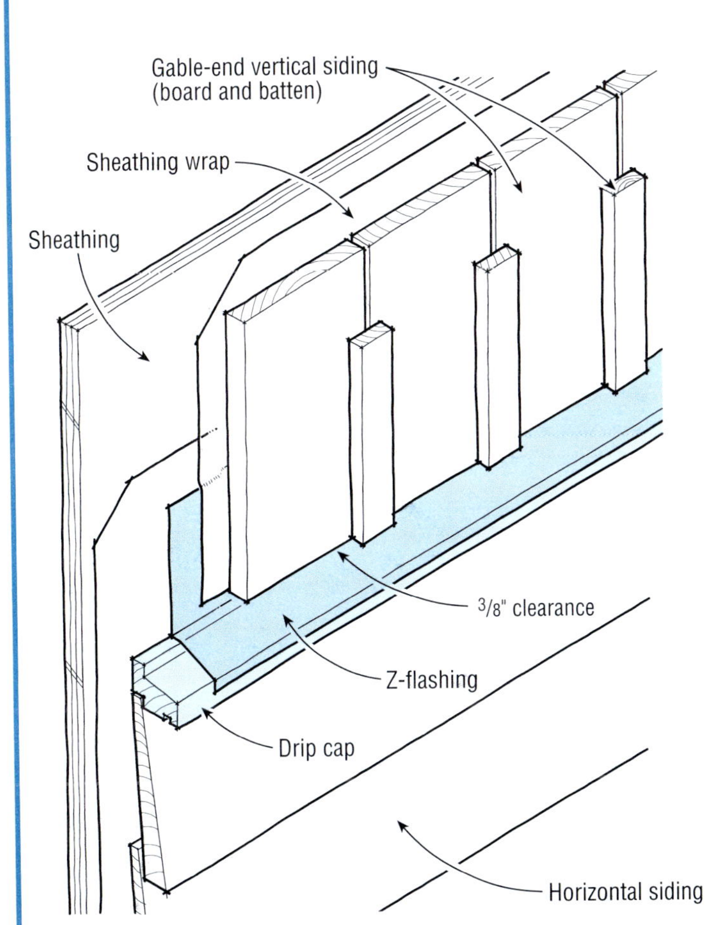

To divert water away from horizontal siding when vertical siding is used on gable ends, install a transition board with a drip-cap and Z-flashing.

Figure 4-33. Tudor Trim Details

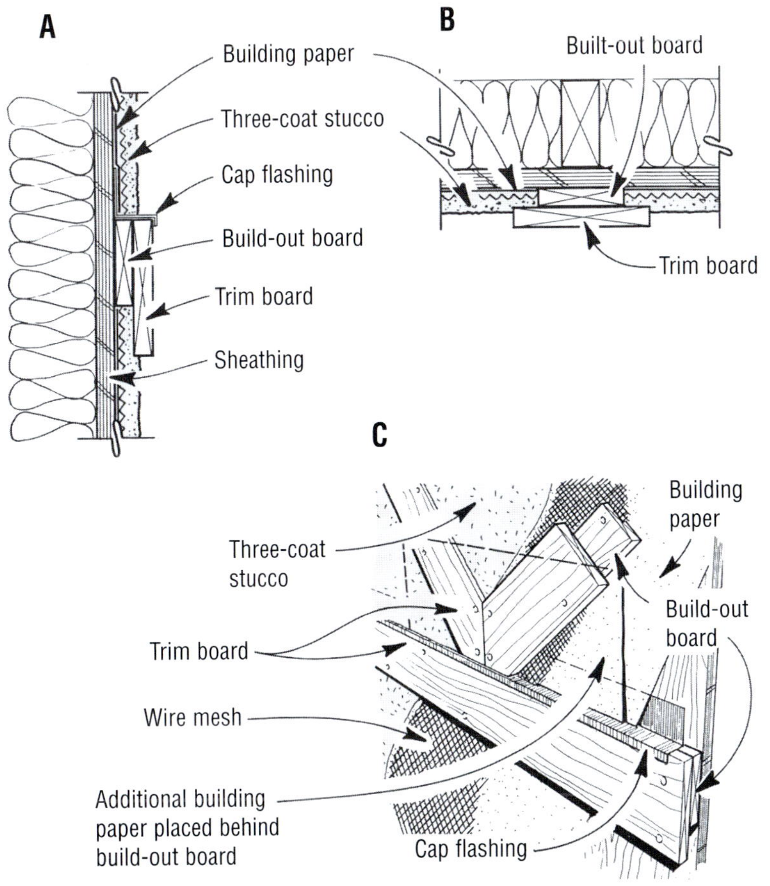

Tudor trim set in stucco must be carefully flashed. Avoid diagonal trim if possible; the diagonals channel water down. If diagonals can't be avoided, flash the wall with peel-and-stick flashing membrane where the boards end, and provide a metal Z-flashing on horizontal trim pieces.

Eaves Crown

Siding Transition

Tudor Trim

SIDING GRADES

Problems occur with siding grades that are unsuitable for the job. To avoid problems, use select grades. Lumberyards may use other terms for their siding grades, but there's little chance of knowing what will be delivered or how it will perform. Ask for equivalents to the standard grades established by industry trade associations shown in **Figure 4-34**. Cheap or ungraded siding usually warps, shrinks, splits, and otherwise deteriorates far quicker than premium-grade material.

Figure 4-35. Storing Siding On Site

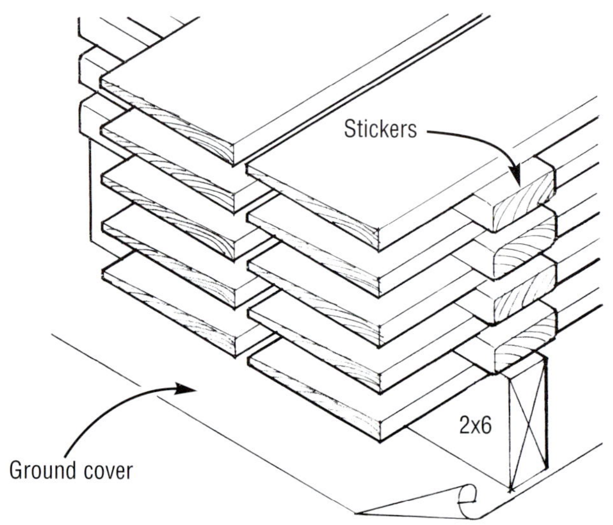

It's best to get any wood siding at least a week early and store it at the job site to let it adjust to the site's humidity. Keep siding out of weather in the driest space available. Break apart bundles and restack the material on evenly spaced, vertically aligned stickers to allow air to circulate freely.

For more information: "Economical Wood Siding," JLC 6/91

MOISTURE CONTENT

Nearly all wood siding shrinks somewhat after installation. If the wood is too wet when installed, excessive shrinkage can lead to splitting, warping, cupping, or paint checking. Problems can be minimized by specifying "S-Dry" material, which contains no more than 19% moisture, or by using premium-grade materials.

MC on Delivery

Unseasoned or green wood has a moisture content (MC) of more than 19%. Dry wood has a maximum moisture content of 15% or 19% depending upon grade. Premium grades are dried to MC 15, meaning that the wood has 15% moisture content or less (and that 85% of the pieces are dried to 12% or less). Dry siding will acclimate more quickly to its final surroundings and will have less dimensional change, before or after installation, than unseasoned or green siding.

MC at Installation

Condition all board siding. Moisture content will vary from piece to piece, but the material should average about 12% MC at the time of installation (individual pieces ranging from 9% to 14%). In dry southwestern states, siding should average 9% MC.

Conditioning Siding

Wood siding picks up and loses moisture in transit and storage, so it is important to allow it to acclimate to the site before installation (**Figure 4-35**):

- "Splinter" bundles by breaking them up and stacking them with stickers (spacers between every layer).

Figure 4-34. Wood Siding Grades

Product	Grade	Description	Moisture Content
Standard Clear Grades — Western Red Cedar			
Bevel Siding	Clear VG (vertical grain)	Free of knots and imperfections; for use where the highest quality appearance is desired.	MC-15 (15% or less – most pieces 12% or less)
	A Grade	Includes some mixed grain and minor growth characteristics.	
	B Grade	Includes mixed grain, limited characteristics and occasional cutouts in longer pieces.	
	Rustic	Similar to A grade, but graded from sawn face.	
	C Grade	Admits larger and more numerous characteristics than A or B grades.	
Boards (Finish, Trim)	Clear	Finest appearance with clear face, few minor characteristics.	MC-15 (15% or less – most pieces 12% or less)
	A Grade	Recommended for fine appearance. May include minor imperfections or growth characteristics.	
	B Grade	Permits larger and more characteristics, but may have short lengths of fine appearance.	
Standard Knotty Grades — Western Red Cedar			
Bevel Siding, Boards, Channel, T&G, etc.	Select Knotty	For fine knotty appearance.	19% or less
	Quality Knotty	Permits more pronounced characteristics and has occasional cutouts in longer pieces.	
	Select Merchantable	Has fine appearance and includes knots and minor markings.	Unseasoned
	Construction	Limited characteristics allowed to assure high degree of serviceability.	
	Standard	Allows more characteristics than construction.	
Standard Softwood Grades (all species except redwood)			
All board patterns	C select	Mixed grain, a few small knots allowed. For uses where a fine finished appearance is desired.	MC-15 (15% or less – most pieces 12% or less)
	D select	Mixed grain, slightly larger knots than allowed in C Select.	
All board patterns	#2 common	Has fine appearance and includes knots and minor markings.	19% or less
	#3 common	Limited characteristics allowed to assure high degree of serviceability.	
	#4 common	Allows more characteristics than #3. Used chiefly for serviceability rather than appearance.	

Note: These grades apply to all lumber graded under the rules of Western Wood Products Assoc. (WWPA), West Coast Lumber Inspection Bureau (WCLIB), or National Lumber Grades Authority (NLGA) of Canada. The term "characteristics" refers to knots, wane, pitch pockets, irregular grain, etc.

- If stacked over the ground or concrete, lay poly down first, and then use 2x6s to elevate the first layer.
- Stack siding in a dry area where plenty of air can flow through the stack, and protect it from weather.
- Store on site 7 to 10 days before installation. Extend drying period to 30 days for unseasoned siding or during very humid seasons.

BOARD SIDING PATTERNS

Pattern is more than just an aesthetic choice. Certain patterns respond better than others to climate changes (**Figure 4-36**).

Pattern Width

Narrow patterns perform best because there is less movement from wet to dry periods and with seasonal climate changes.

Figure 4-36. Board Siding Patterns

Siding Patterns	Nominal Sizes (thickness and width)	Nailing (Do not nail where siding pieces overlay): 6 in. and narrower	Nailing (Do not nail where siding pieces overlay): 8 in. and wider
Bevel or Bungalow Bungalow (Colonial) is slightly thicker than Bevel. Either can be used with the smooth or rough-faced surface exposed. Patterns provide a traditional-style appearance. Horizontal applications only. 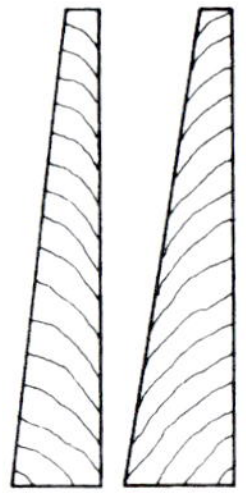	1/2 x 4 1/2 x 5 1/2 x 6 5/8 x 8 5/8 x 10 3/4 x 6 3/4 x 8 3/4 x 10	Recommend 1" overlap. One siding nail or box nail per bearing, just above the 1" overlap. 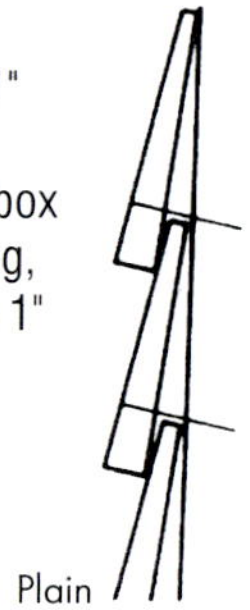	Recommend 1"overl: One siding nail or bc nail per bearing, just above the 1" overlap 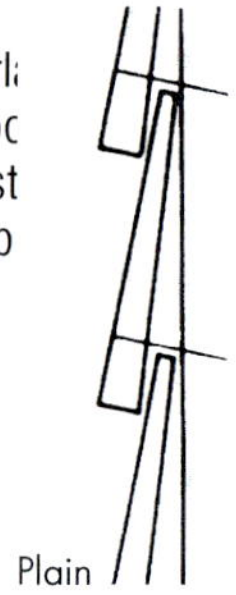
Tongue & Groove T&G siding is available in a variety of patterns. Vertical or horizontal applications. Install horizontal T&G with the tongues up so the joint will drain. 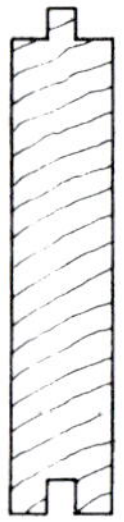	1 x 4 1 x 6 1 x 8 1 x 10 Available with 1/4", 3/8", or 7/16" tongues. For wider widths, specify the longer tonguc.	Use one casing nail per bearing to blind nail. 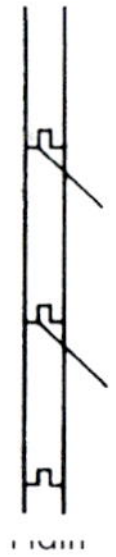	Use two siding nails or box nails 3" to 4" apart to face nail. 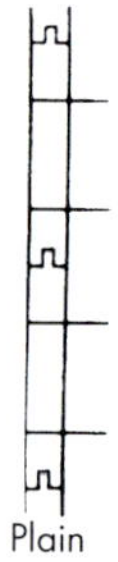

Wide patterns will cover an area faster, thus reducing labor costs, but any savings may be offset by callbacks due to warping, cupping, or checking. Avoid patterns over 8 in. in width.

DRAINAGE PLANE

The easiest way to create a drainage plane is to use a self-draining rainscreen mat, such as Home Slicker® (Benjamin Obdyke; 800/346-7655; www.obdyke.com). This material is thin enough that windows, doors, and corners don't need to be packed out when used with 5/4 trim stock.

Also, siding may be installed over vertical battens spaced

Selecting Board Siding

Drainage Plane for Board Siding

Figure 4-36. Board Siding Patterns

Siding Patterns	Nominal sizes (thickness and width)	Nailing (Do not nail where siding pieces overlay): 6 in. and narrower	Nailing (Do not nail where siding pieces overlay): 8 in. and wider
Channel Rustic Channel Rustic has 1/2" overlap and a 1" to 1 1/4" channel when installed. The profile allows for maximum dimensional change without harming appearance. Available smooth, rough or saw-textured. Horizontal or vertical applications. 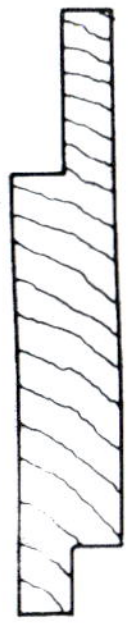	3/4 x 6 3/4 x 8 3/4 x 10	Use one siding nail or box nail to face nail once per bearing, 1" up from bottom edge. 	Use two siding nails or box nails 3" to 4" apart per bearing. 
Board-and-Batten Boards are surfaced smooth, rough or saw-textured. Rustic ranch-style appearance. Requires horizontal nailers. Vertical applications only. 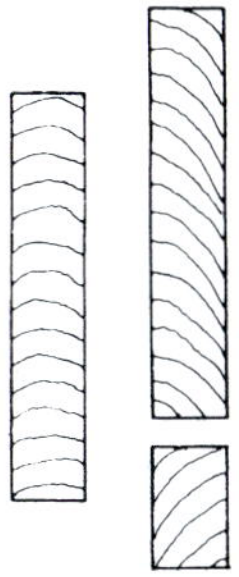	(4/4) 1 x 2 1 x 4 1 x 6 1 x 8 1 x 10 1 x 12 (5/4) 1 1/4 x 6 1 1/4 x 8 1 1/4 x 10 1 1/4 x 12	Recommend 1/2" overlap. One siding or box nail per bearing. 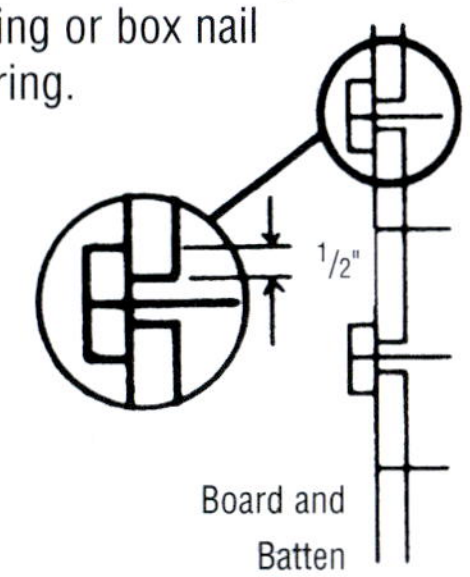	Increase overlap proportionally. Use two siding nails or box nails, 3" to 4" apart. 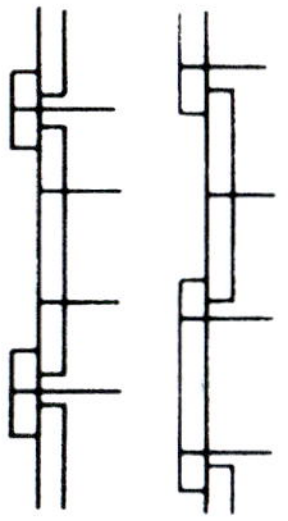

16- to 24-in. apart to create the air space (**Figure 4-37**). In this case, door and window jambs must be extended to accommodate the added thickness, and a vent strip or screen should also be included over weep channels to keep out insects (**Figure 4-38**).

TRIM FOR BOARD SIDING

For information about detailing exterior wood trim for board siding, see "Wood Trim," **pages 310-317**.

Figure 4-37. Drainage Plane for Board Siding

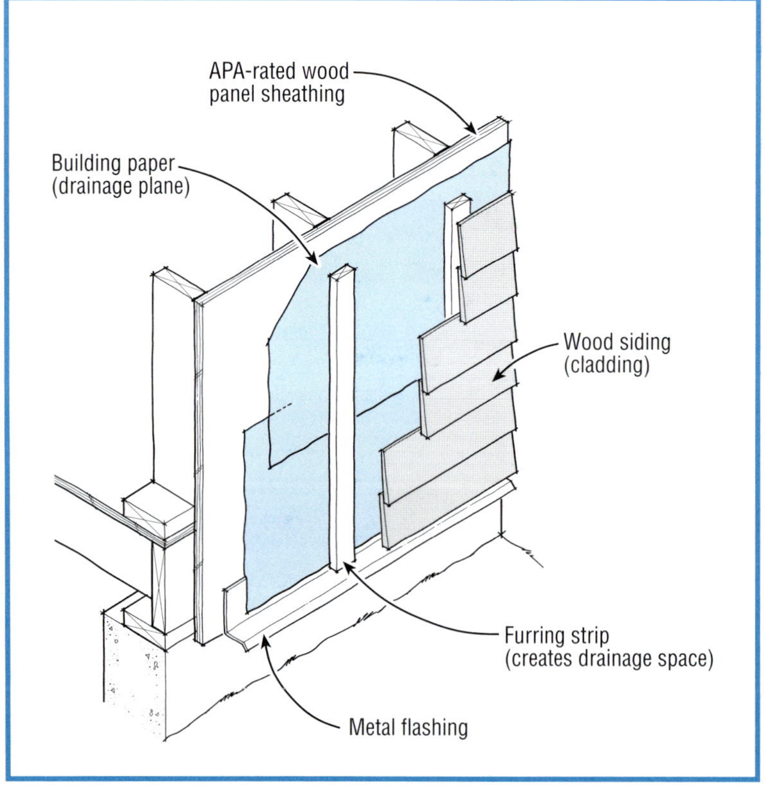

For more information: "Water-Managed Wall System," JLC 3/03

"Rain Screen Siding Retrofit," JLC 4/98

INSTALLING HORIZONTAL SIDING

Bevel siding is the most forgiving pattern because it's designed to shed water, and cross-grain shrinkage will not expose the wall beneath. See **Figure 4-39** for horizontal installation details.

Problems can develop from too much overlap. Buy bevel siding only 3/4- to 1-in. wider than the desired reveal. A large overlap may require double-nailing (which can cause splitting) or nailing through the thinnest part of the board (which can lead to excessive cupping).

INSTALLING VERTICAL SIDING

Installing vertical siding over battens to create a rain-screen can be problematic. Battens must be run horizontally to hold the siding boards, but then the channels won't drain. Install vertical siding over a self-draining rainscreen backing material. See **Figure 4-40, page 325**, for vertical installation details.

Vertical T&G siding is more weatherproof than board-and-batten or edged-butted boards. However, all vertical siding is more prone to leaks than most horizontal board siding. Take extra care with sheathing wrap and flashing behind all vertical siding.

Extend the lower ends of vertical siding below the sheathing to allow water to drain. To prevent wicking and water staining, seal the bottom end-grain of vertical siding with water repellent.

Diagonal Board Siding

Diagonal siding is not recommended. The joints in diagonal siding act like gutters, collecting and channeling water downward at an angle against the sides of windows, doorways, corner boards, and any horizontal or vertical piece that stands in its path. Without very careful flashing and drainage plane preparation, the runoff will gush into the building at these points.

Installing Board Siding

NAILING CHECKLIST

Proper nailing is essential to the performance of any wood siding.

- Always use non-corrosive fasteners (see "Siding Nails," page 307).

Figure 4-38. Drainage Plane Details

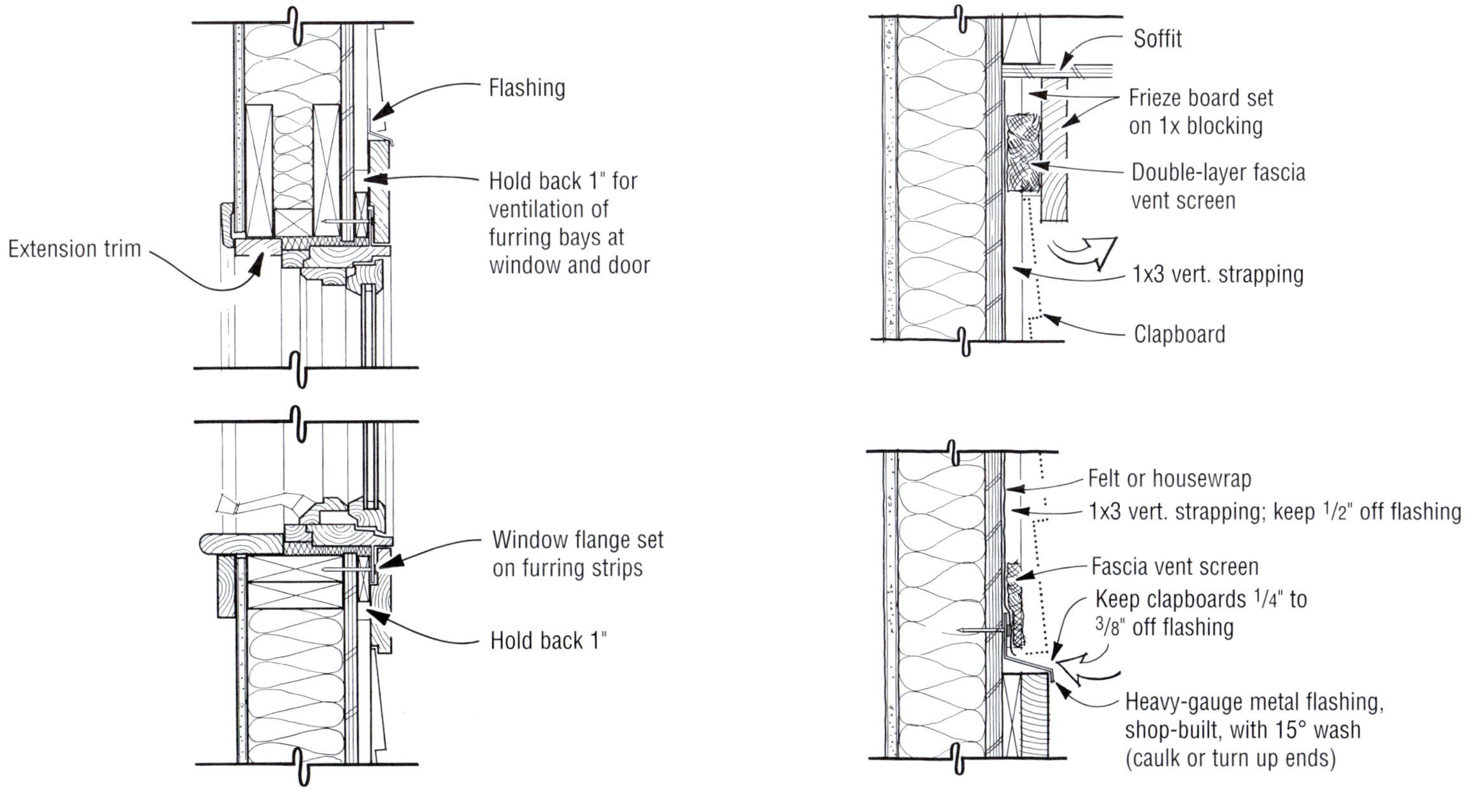

If siding is installed on battens, window and door jambs must be extended (left), and the gap created by the strapping should be screened at the top and the bottom (right) to keep out insects.

- Drive siding nails flush with the surface of the siding. Nails that are set and filled are prone to popping with any movement or shrinkage in the wall framing, and may promote cupping and splitting. However, casing nails used to fasten exterior trim should be set and filled (see "Filling Nail Holes," **page 377**).
- Nails should penetrate at least 1- to 1 1/2-in. into studs or blocking.
- Ring-shank or spiral-shank siding nails are recommended for increased holding power (**Figure 4-24, page 308**).
- Never double-nail solid wood siding materials. If the siding is pinned along both edges, it is likely to split. See **Figure 4-36, pages 320-321** for proper nailing configurations for different siding patterns.

BOARD SIDING OVER FOAM

Siding applied directly over rigid insulated sheathing has a history of failures. Since the foil behind the siding is impermeable, moisture collects on the back of the siding. The back of the siding absorbs moisture at a faster rate than

Figure 4-39. Horizontal Siding Details

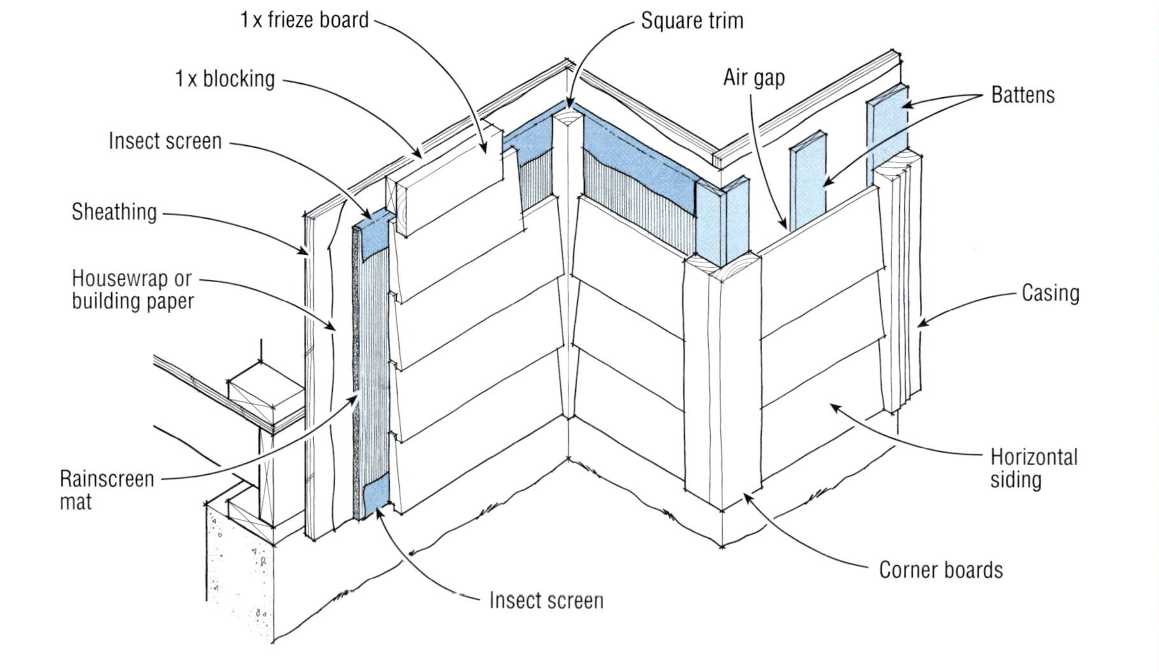

For more information: "Trouble-shooting Wood Siding," JLC 6/93

"Working with Preprimed Wood Siding," JLC 2/00

To avoid siding failures, install all wood siding over an air gap created by battens or another spacing material.

the finished front side, causing boards to cup inwards. That same moisture will be drawn through the siding when the sun comes out, causing the finish to blister. This cycle of wetting and drying is accelerated because the insulation prevents the wall framing from absorbing any of the heat generated by exposure to direct sun.

To prevent siding problems over foam:

- **Rainscreen.** Install siding over furring strips or rainscreen backer, creating a drainage plain or "rain screen" (see "Drainage Plane," next page). The resulting air space allows both heat and moisture to dissipate.
- **Back-prime.** All wood siding should always be back-primed (on natural finish material, use a clear water-repellent preservative; see "Finishes," **page** 373). This will slow the cycle of wetting and drying. Field-cut ends should also be coated with stain or preservative.
- **Foil caution.** Foil-faced foam is a moisture barrier, so it creates a wrong-side vapor barrier. Foam should be used only with a continuous-film vapor retarder on the inside wall to keep condensation out of the wall cavity.

Siding Over Foam

Vertical and Horizontal Siding Details

Figure 4-40. Vertical Siding Details

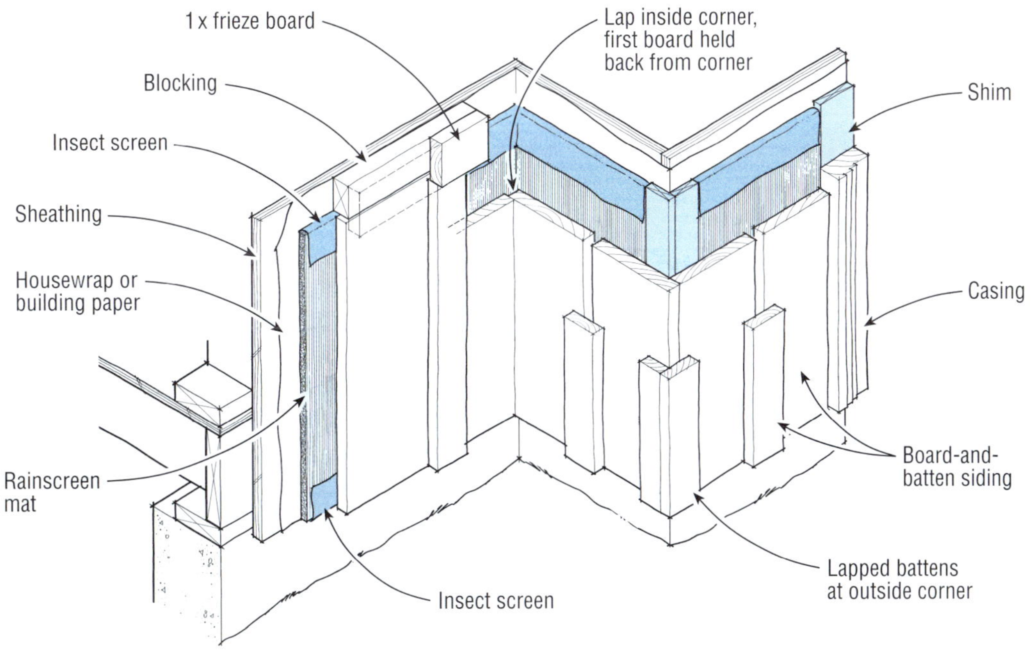

Do not nail through board when attaching battens or the boards may split.

SHINGLE AND SHAKE GRADES

Wood shingles and shakes are made from redwood, western red cedar, northern white cedar, and cypress. Most shingles are sawn smooth on both faces. Shakes are thicker and have at least one split surface, which leaves a rough texture.

Figure 4-41. Drainage Plane for Shingle Siding

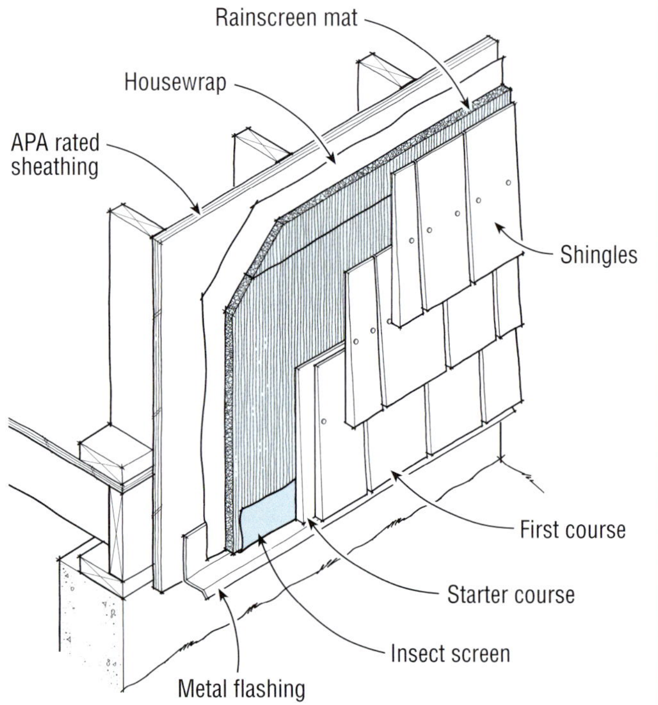

To create a drainage plane for wood shakes and shingles, install the siding over a drainage mat that will provide a channel to allow water down and out.

Shingle and shake grades for sidewalls are largely the same as those used for roofs (**Figure 3-15, page 222**). The exception is machine-grooved shakes (also called "rebutted/regrooved shingles"). These are sawn into 16-, 18-, and 24-in. lengths (butts measuring .4 in., .45 in., and .5 in., respectively) with parallel edges, and the face is grooved with striations. Machine-grooved shakes should be double-coursed (see "Double Coursing" **page 328**).

DRAINAGE PLANE FOR SHAKES AND SHINGLES

The most practical way to create a drainage plane for shingle and shake siding is with a self-draining rainscreen mat (**Figure 4-41**).

If the wall is furred out, horizontal battens will block drainage unless they are installed over opposing vertical battens (spaced 16- to 24-in. o.c.), creating a grid of strapping. Space horizontal battens the length of the shingle, and space vertical battens 16- to 24-in. o.c. With double thickness 1x battens, pack out windows, doors, and corners with 2x material.

Prior to installing strapping or a rainscreen mat, wrap and flash wall sheathing (see "Sheathing Wrap and Flashing," **pages 296-306**).

TRIM FOR SHINGLES AND SHAKES

Like other wood and wood-based sidings, most exterior trim for shingles and shakes can be handled with wood. For information about detailing exterior wood trim, see "Wood Trim," **pages 310-317**.

SHINGLE AND SHAKE CORNER TREATMENTS

Corner boards are commonly used with white cedar shingles (see "Corner Boards," **page 313**), but in traditional style, shingles or shakes are joined to each other at the corners.

Outside corners. With shakes and red-cedar shingles, corners are often woven with two courses on one side alternating with two courses on the other, each hand-planed to the corner profile. This is labor-intensive and requires small, galvanized finish nails to hold the butt joints closed (**Figure 4-42**).

Mitering shakes and shingles is not recommended. These corners are likely to open when the siding shrinks, even if nailed.

Inside corners. Flash inside corners with a right-angle flashing extended at least 3 in. under the shingle courses. At a minimum, they should receive an extra layer of building paper.

The simplest inside corner trim is a 1x1 or 2x2 wood strip installed in the corner with the shingles butted against it (**Figure 4-42**). Inside corners also may be woven in a manner similar to outside corners — two courses on one side, followed by two courses on the other.

WEATHER EXPOSURE

Shingles and shakes may be applied in single or double courses. Double-coursing allows for a greater weather exposure (**Figure 4-43**).

Figure 4-42. Inside and Outside Corners

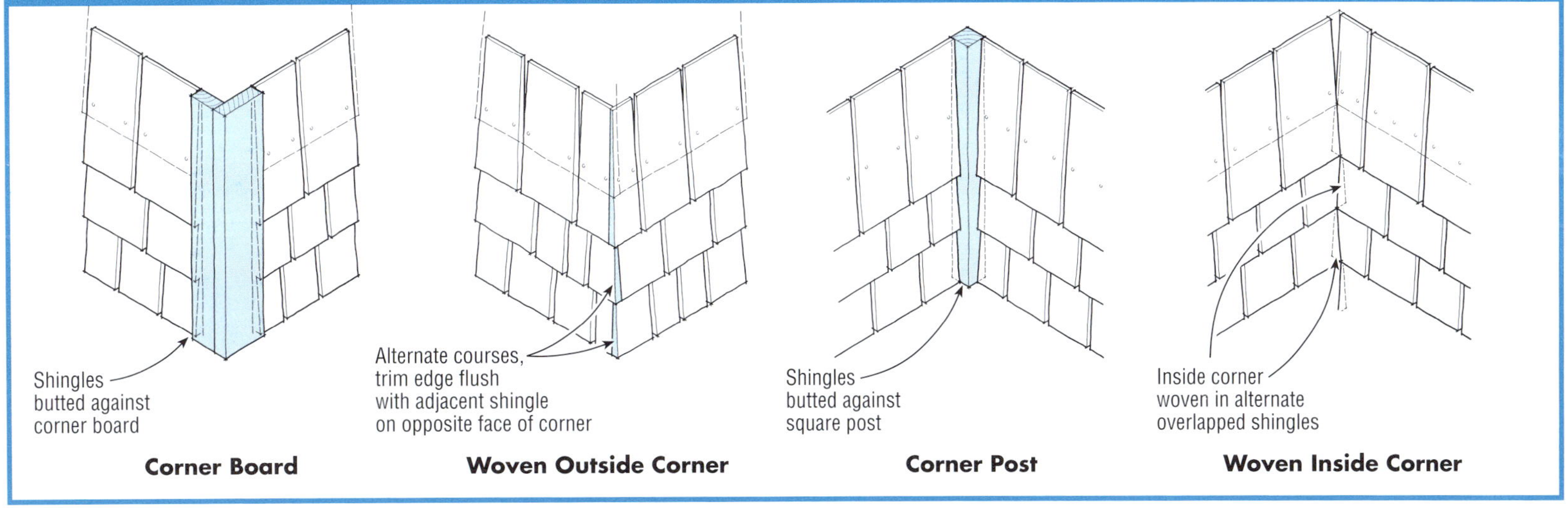

A woven corner, typically accomplished with a block plane, is made by alternately lapping a shingle piece on each course. Inside corners may also be woven, but the easiest and cleanest treatment is to butt shingles against a 1- to 2-in.-square post.

DOUBLE-COURSING

Double-coursing uses two layers of shingles (or one layer of shakes over a layer of No. 3 or No. 4 shingle). This application allows for wider weather exposures, and creates deep shadow lines at the butt ends (**Figure 4-44**).

Figure 4-43. Shingle and Shake Maximum Exposures

Shingle Length	Single Course	Double Course (No. 1)	Double Course (No. 2)
16″	7 1/2″	12″	10″
18″	8 1/2″	14″	11″
24″	11 1/2″	16″	not recommended

Shake Length & Type	Single Course	Double Course (No. 1)
16″ machine-grooved	7 1/2″	12″
18″ machine-grooved	8 1/2″	14″
18″ resawn	8 1/2″	14″
24″ resawn	11 1/2″	18″
18″ taper-sawn	8 1/2″	14″
24″ taper-sawn	11 1/2″	18″
24″ tapersplit	11 1/2″	18″
18″ straight-split	8 1/2″	16″

The maximum weather exposures for shingles and shakes varies, depending on the length and type of material, as well as on the application (single- or double-course).

Starter Course

Begin with a double underlay course or a single underlay course shimmed out the thickness of a shingle butt with wood lath. Nail undercourse shingles with a single nail in the center.

Figure 4-44. Double-Coursing

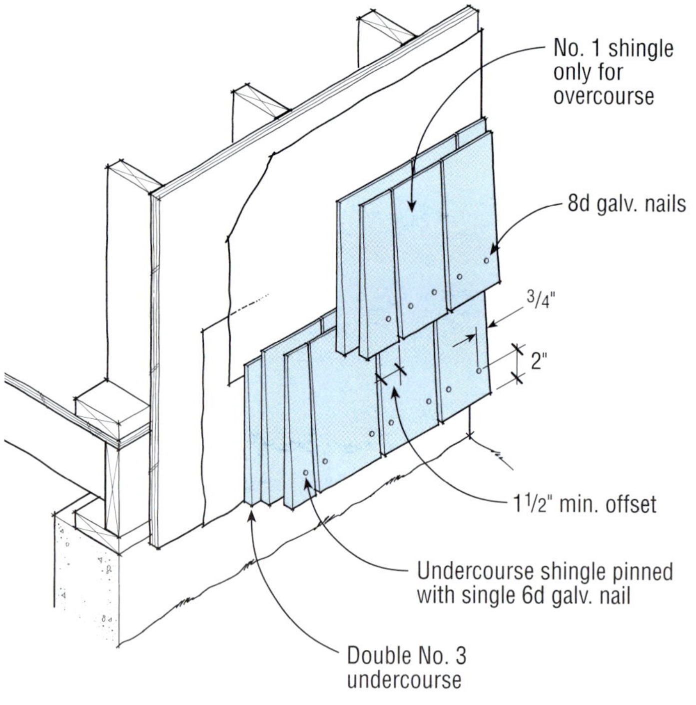

This approach to laying shakes and shingles features wide weather exposures and deep shadow lines. The outer course drops about 1/2 in. below the undercourse, which typically consists of low-grade shingles.

Face Nail

Double-course applications require face-nailing. Face-nail outer shingles with two nails placed about 2 in. above the butt line and 3/4 inch in from each end. Shingles wider than 8 in. get two more nails driven near the center, about 1 in. apart. The outer course drops about 1/2 in. below the undercourse.

Double-Course Variations

Double-course shingles or shakes may also be used over non-wood sheathing (fiberboard or gypsum). In this case, nail strapping below the butt end of the undercourse with 8d box nails, spaced on-center the distance of the weather exposure. For visual effect, raising the overcourse about 1 in. provides a double shadow line (**Figure 4-45**). In this case, the undercourse should be a No. 1 grade shingle.

Double-Coursing vs. Single-Coursing

SINGLE-COURSING

Single-coursing is quicker and less expensive because less material is used for a given wall area. However, it requires a narrow weather exposure (**Figure 4-46**).

Starter Course

Double the first course using inexpensive, lower-grade shingles (No. 3 Black Label or No. 4 Undercoursing; see **Figure 3-15, page 222**).

Figure 4-45. Decorative Coursing Variations

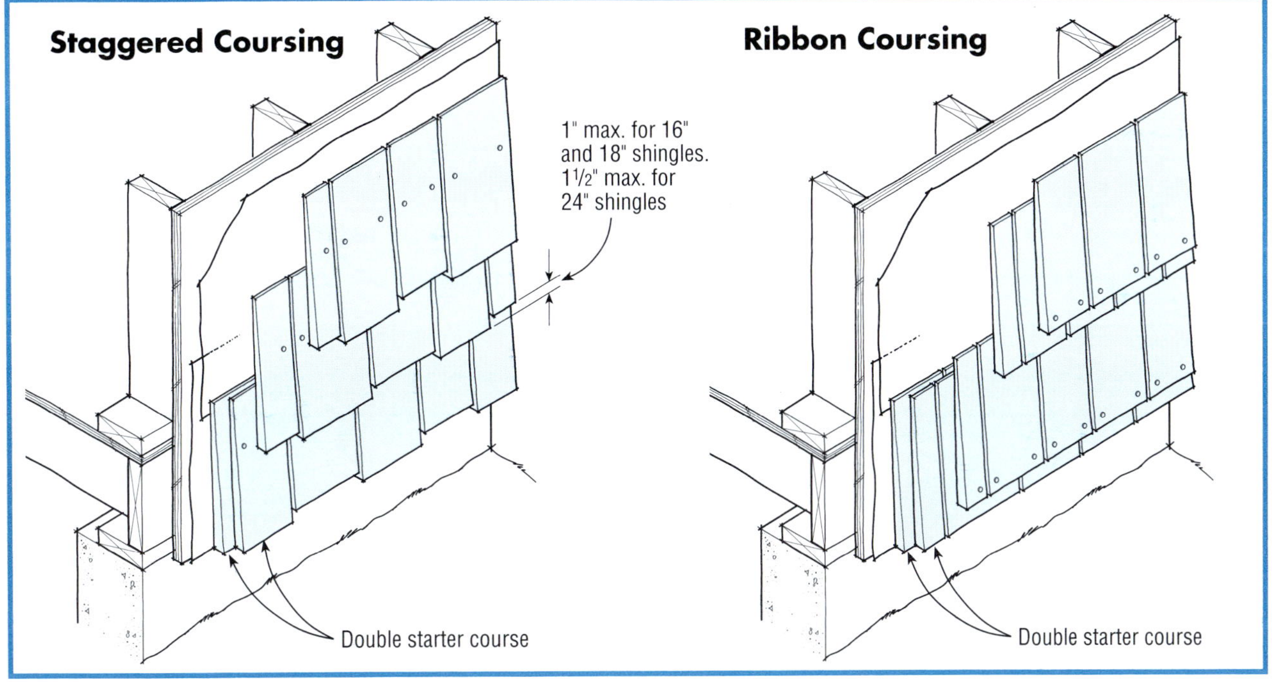

Over solid sheathing, a double shadow line can be achieved by raising the outer course about 1 in. (near left). Single-course applications can be staggered to create an irregular appearance (far left).

Blind Nail
Drive nails about 1 in. above the butt line of the course above. For shingles up to 8 in. wide, use two nails per shingle about 3/4 in. from each edge. For wider shingles, drive two more nails about 1 in. apart near the center (**Figure 4-47**).

Staggered Coursing
For visual effect, butts can be staggered below (not above) the horizontal line (**Figure 4-45**). The maximum distance to stagger the butts varies with shingle or shake length: Stagger 1 in. for 16's and 18's or 1 1/2 in. for 24's.

NAILING CHECKLIST

- Install sidewall shingles or shakes with rust-resistant box nails or special shingle nails long enough to penetrate the sheathing (or nailing strips) at least 1/2- to 3/4-in. Nails should be hot-dipped galvanized, Type 304 or 316 stainless steel, or high-tensile aluminum.

Figure 4-46. Single-Coursing

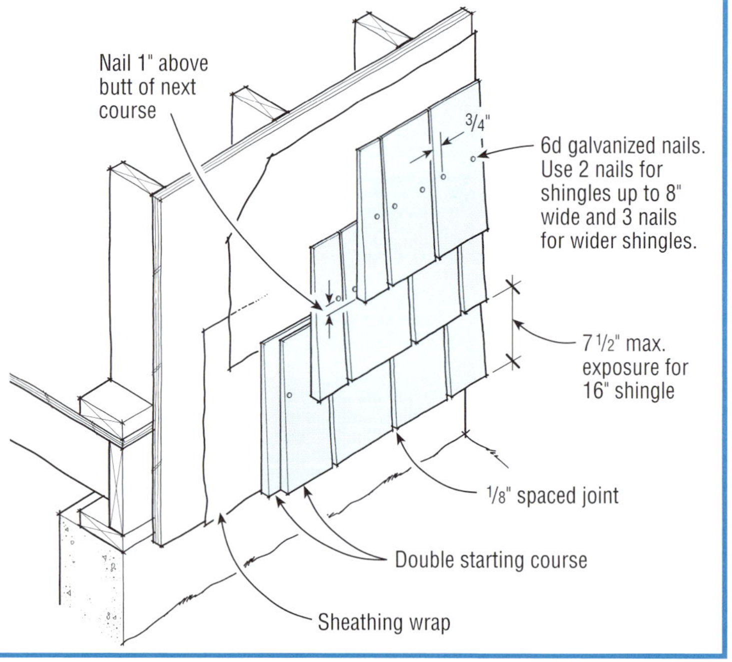

Each course is blind-nailed about 1 in. above the butt line, using two nails for shingles up to 8 in. wide and four nails for larger ones. Offset joints in successive courses by at least 1 1/2 in.

Figure 4-47. Preventing Splits in Wide Shingles

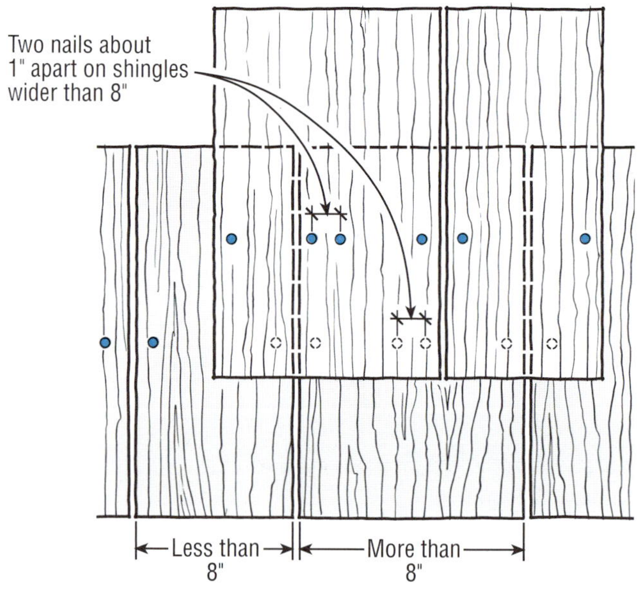

Wide shingles (over 8 in.) are prone to splitting. To avoid problems, blind-nail as shown above.

- Drive nails flush to the shingle or shake surface, but not hard enough to crush the wood fibers, which could split the wood.
- For single-coursing, 3d or 4d shingle nails are commonly used. For double-coursing, 3d or 4d nails are typically used on the undercourse, and 5d box or shingle nails are used on the outer exposed course.

Nailing Shakes and Shingles

SHINGLE AND SHAKE LAYOUT

Shingle Gap

Space all exposed shakes or shingles about 1/8 in. from each other. This allows room for expansion and gives pieces an individual appearance. Tightly packed shingles that get wet before finishing can buckle.

Offset joints in successive courses by at least 1 1/2 in. to avoid leakage between shingles.

Aligning Courses

To determine the number of courses, divide the height of the wall into equal parts whose length is not greater than the weather exposure for the shake or shingle. Wherever possible, align shingle butt lines with the tops or bottoms of windows or other openings. This may require adjusting a few courses by a fraction of an inch (**Figure 4-48**).

CLEANING SHINGLES AND SHAKES

Cleaning Shakes and Shingles

Mildew should be removed from shingles before refinishing since it will grow quickly through the new finish.

To remove mildew, scrub with a solution of 1 cup trisodium phosphate (TSP) dissolved in 1 quart of household bleach mixed with 3 quarts of warm water. Rinse the area thoroughly with clear water, and immediately apply a 50/50 solution of household bleach and water. Allow the last treatment to dry completely before applying any finish. Because bleach can harm skin and eyes, wear protective gloves and goggles during applications and carefully follow manufacturers' instructions.

Figure 4-48. Even Shingle Layout

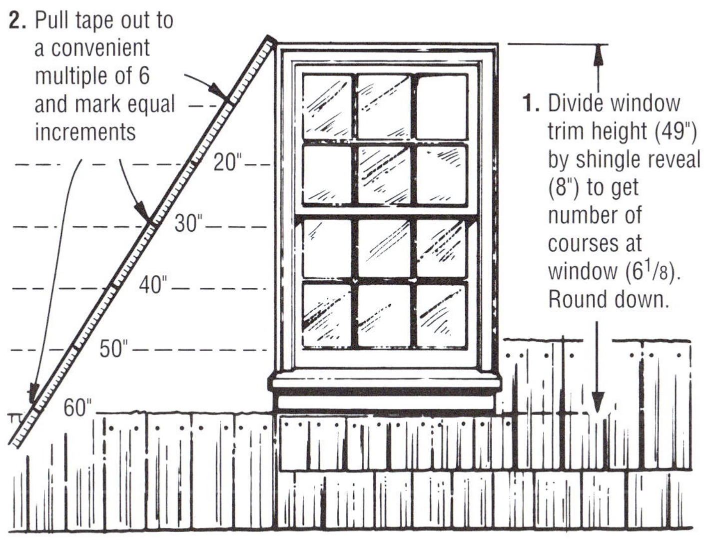

To find the number of shingle courses between the top and the bottom of a window: 1) Measure the window height plus the trim. Divide this number by the average shingle exposure. If the resulting number is a fraction, choose the nearest whole number. 2) Hold a tape measure at the corner where the head trim will be, and pull it out to an even multiple. To find the course lines, mark off the even multiples.

Vinyl siding is made from extruded PVC (polyvinyl chloride), a durable lightweight plastic. Vinyl siding doesn't dent, rot, or corrode, and it is manufactured in a wide variety of light colors and textures. Older formulations may be quite brittle from extended exposure to ultraviolet light, however most formulations nowadays are a co-extrusion of a UV-resistant capstock and a resilient substrate.

Figure 4-49. Locking Edge

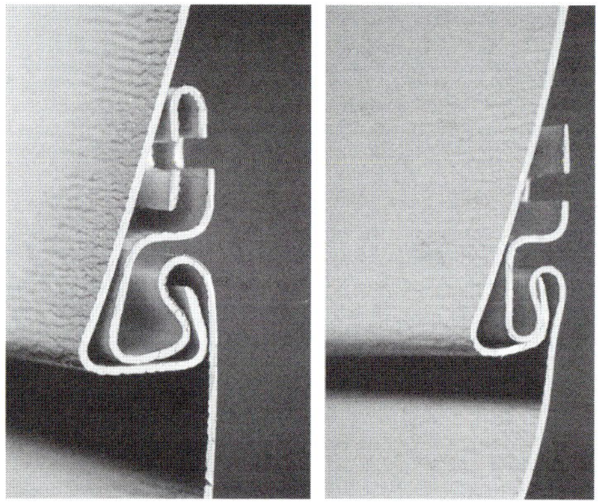

The stiffness of a vinyl panel is more affected by the butt dimension and the lock profile than by the gauge of the vinyl. A square butt (near right) will stay straighter and droop less than a post-formed edge (far right).

Figure 4-50. Nail Hem

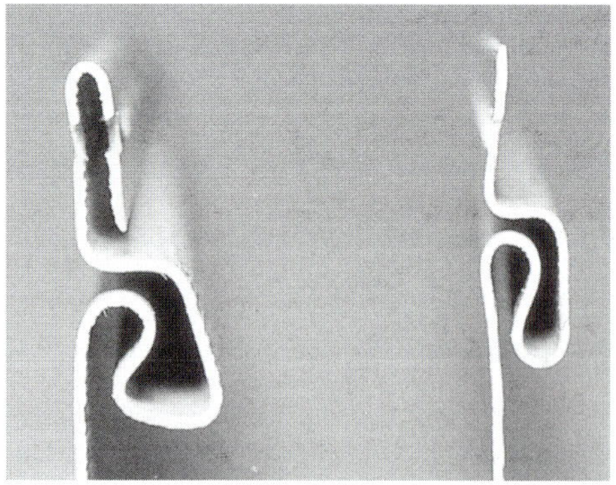

A double-thickness nail hem like the one on the left is less likely to rip in a windstorm.

For more information: "Builder's Guide to Vinyl Siding," JLC 6/00

VINYL SIDING TYPES

Typically, vinyl siding is available in horizontally applied panels that simulate lap siding — with one, two, or three siding courses per panel.

V-groove and board-and-batten panels are the most popular forms of vertical vinyl siding. Usually, they are used in combination with horizontal siding on gable ends.

Panel Stiffness

Vinyl thickness varies from about .035 in. (the minimum thickness allowed to comply with the ASTM D3679 standard) to .055 in. While thicker panels are less likely to droop and wave in hot weather, they are not necessarily the best choice, as overlaps between panel lengths are more noticeable with thicker panels. Most installers prefer a "midrange" panel thickness of .042 to .044 in.

The stiffness of panels depends much more on the shape of the panel than on the thickness of the vinyl. The thicker the butt at the locking edge, the stiffer the panel will be and the less likely it will be to droop or wave (**Figure 4-49**).

Color Choices

While advances in materials science have improved the performance and durability of darker shades of vinyl siding, vinyl is typically available in light colors to reduce surface temperatures. If vinyl gets too hot, it can droop. Also, vinyl chalks as it oxidizes, particularly in humid climates. The white film is more noticeable on darker panels. White siding will be the least likely to show any staining from oxidation.

Wind Resistance

Because vinyl siding must be applied loosely, it may be susceptible to blow-offs. Panels with reinforced nail hems are less likely to be torn off in a storm (**Figure 4-50**).

DRAINAGE PLANE

Of all the siding choices available, vinyl is the only one that does not need extra work to create a drainage gap. Properly nailed, it should hang loose over the wall, allowing water that gets through to drain down and out (**Figure 4-51**). However, because vinyl must be installed loosely to accommodate expansion, it's critical to flash and wrap walls carefully (see "Sheathing Wrap and Flashing," **pages 296-306**).

TRIM FOR VINYL SIDING

Aluminum covered trim has become the norm in the vinyl siding business. It offers a virtually maintenance-free finish, but how good it looks depends entirely on the skill of the installer.

Trim Coil

Trim is typically bent from two types of aluminum — smooth and striated. Smooth trim coil is a thin aluminum flashing finished with polyester or acrylic paint. Striated coil has a textured finish made with a vinyl paint that forms striated patterns as it dries. Vinyl-coated coil marks up a little easier, requiring a bit more care, but in general the difference is largely a matter of personal preference.

The gauge of almost all aluminum trim coil, whether smooth or striated, is a "nominal" 0.019 in., which can be as thin as .0175 in. and still meet spec. Less common, and about 20% more expensive, is a heavier gauge coil with a nominal measurement of 0.022- to 0.024-in.

Figure 4-51. Drainage Plane for Vinyl Siding

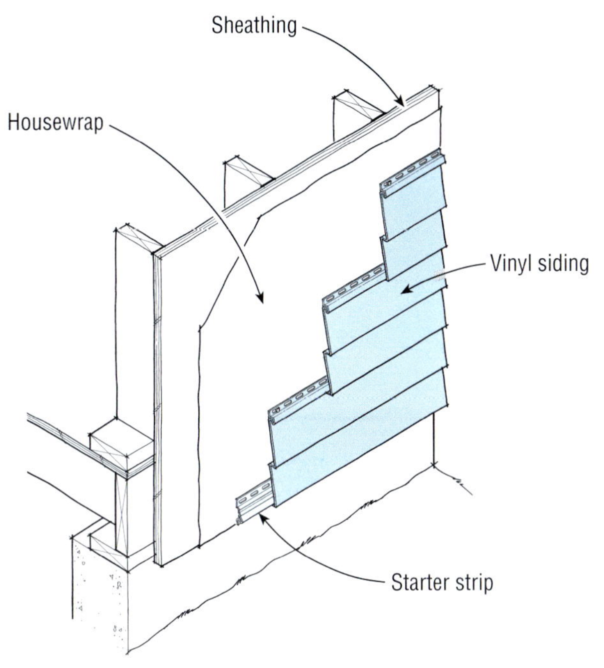

If vinyl siding is nailed loosely (as it should be to accommodate expansion), water that gets behind the siding will be able to drain. A well-detailed sheathing wrap and flashings behind the siding protect the wall from water.

Selecting Vinyl Siding

Trim coil can ripple as it expands in warm weather. This is a problem only if it is face-nailed in cold weather. In addition to limiting face-nailing, try the following to avoid ripples:

- Don't nail too snugly.
- Hem or crease the material to stiffen it.
- Drill or punch an oversized hole in the aluminum before nailing.
- Avoid nailing long fascia and rake pieces. Instead, hold them in place with utility trim and F-channel (**Figure 4-52**).
- On rakes and fascias, install a 1x2 at the top of the trim

Figure 4-52. Aluminum Soffit and Fascia Details

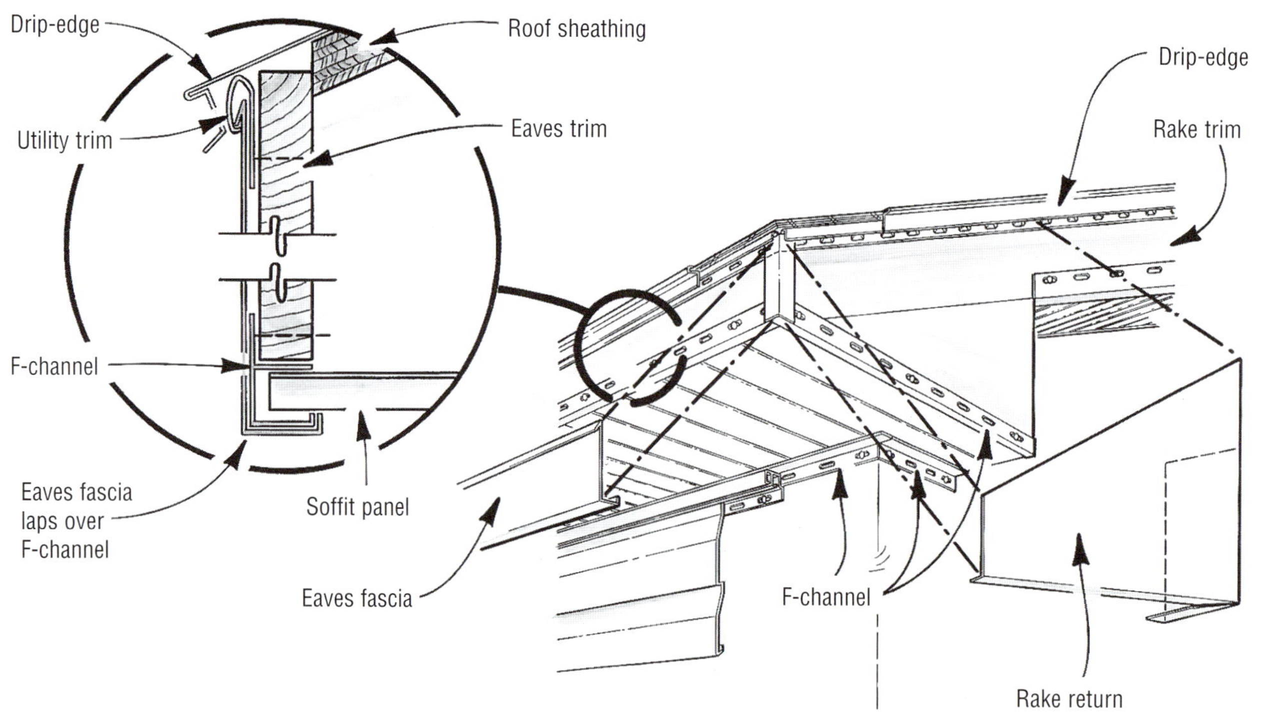

Aluminum trim coil will ripple if nailed. Although small pieces of trim coil may be nailed tight, long lengths of fascia should be secured with utility trim or nails in slotted holes.

to provide a stepped profile. This stiffens the trim and limits rippling (**Figure 4-53**).

- To avoid face-nailing a fascia, insert the top of the trim coil under the drip-edge or into a piece of undersill trim and install a 1¼-in. bend at the bottom toward the soffit. Nail up through the bottom leg every 18- to 24-in.

Vinyl Window Trim

Trim for Vinyl Siding

Window trim is typically framed with J-channel to receive the siding panel (**Figure 4-54**). The side pieces should be run long enough to receive the mitered head trim (top detail, **Figure 4-54**), and the sill corners should be flashed with trim coil to divert water running down the J-channel.

Figure 4-52. Aluminum Soffit and Fascia Details, continued

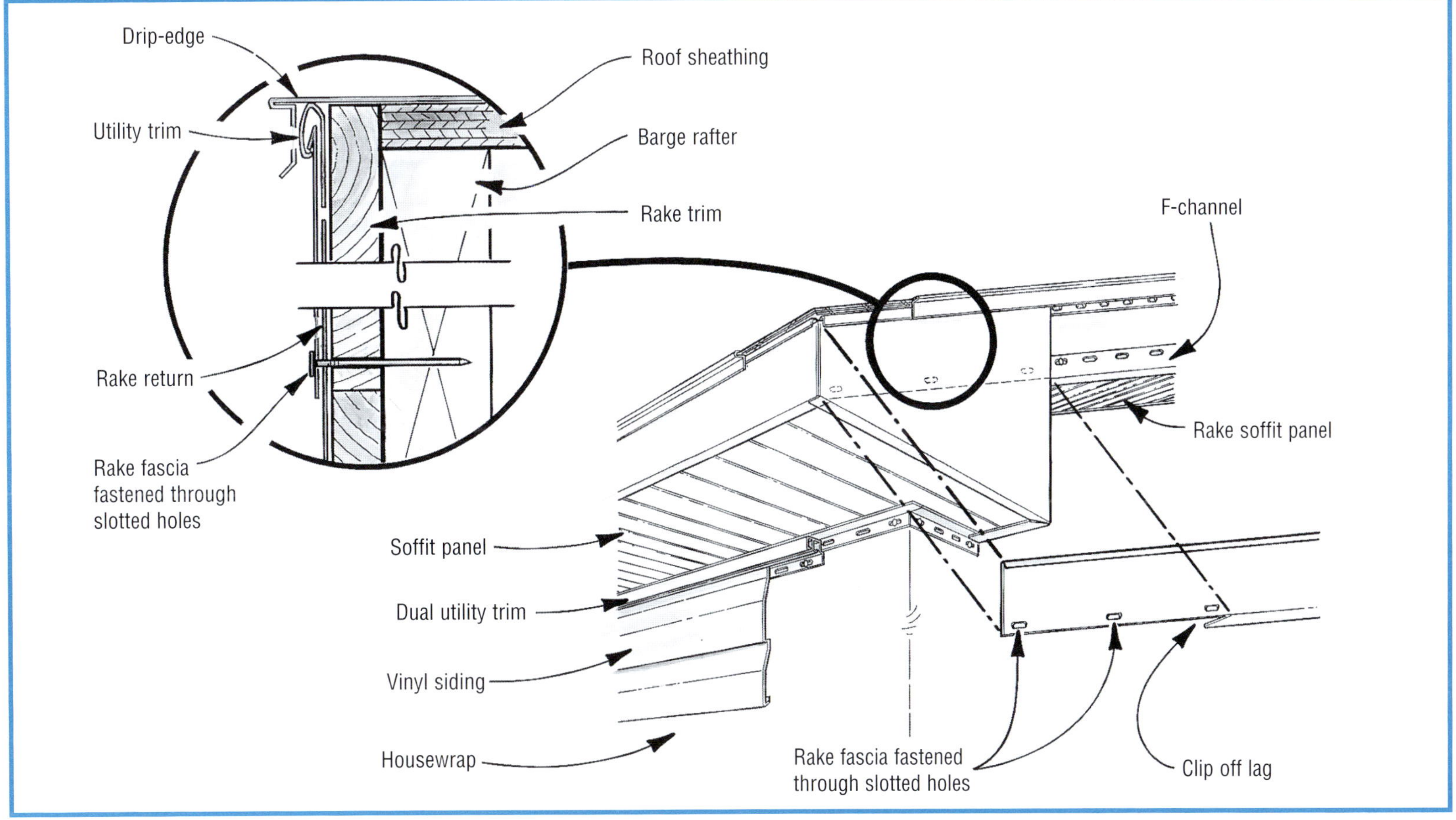

Notched wall panels, which have the nail hem ripped off, can be locked in place with dual utility trim (**Figure 4-55**).

VINYL SIDING LAYOUT

Vinyl siding has a high coefficient of thermal expansion. A 12-ft. length can expand 5/8 in. from winter to summer. To allow for expansion and contraction of the material, layout is critical.

- Leave a 1/4 in. gap between corner posts and the soffit.
- Leave a 1/4-in. gap between the siding and all corner posts and channels (**Figure 4-56**). Increase this gap to 3/8 in. when installing in temperatures below 40°F.
- Start the layout with a level starter strip. If the foundation is out of level, extend the nailing base as shown in **Figure 4-57, page 339**.
- Never overlap starter strip, channels, or utility trim. Leave at least 1/4 in. between pieces to allow for expansion.
- When panels overlap, make sure they overlap by one-half the length of the notch at the end of the panel, or approximately 1 in.
- Stagger the siding end laps so that no two courses (rows of panels) are aligned vertically, unless the laps are separated by at least three courses.
- Plan the direction of siding laps so they face away from high traffic areas. Someone looking into a lap will see a prominent shadow.

For more information: "Aluminum Trim for Vinyl Siding," JLC 4/99

"Durable Details for Vinyl Siding," JLC 6/97

To reduce ripples in aluminum fascia, install a 1x2 to create a two-piece fascia. To keep coil stock stiff, hem the top edge and slip it under the drip-edge.

Figure 4-53. Aluminum Fascia Alternatives

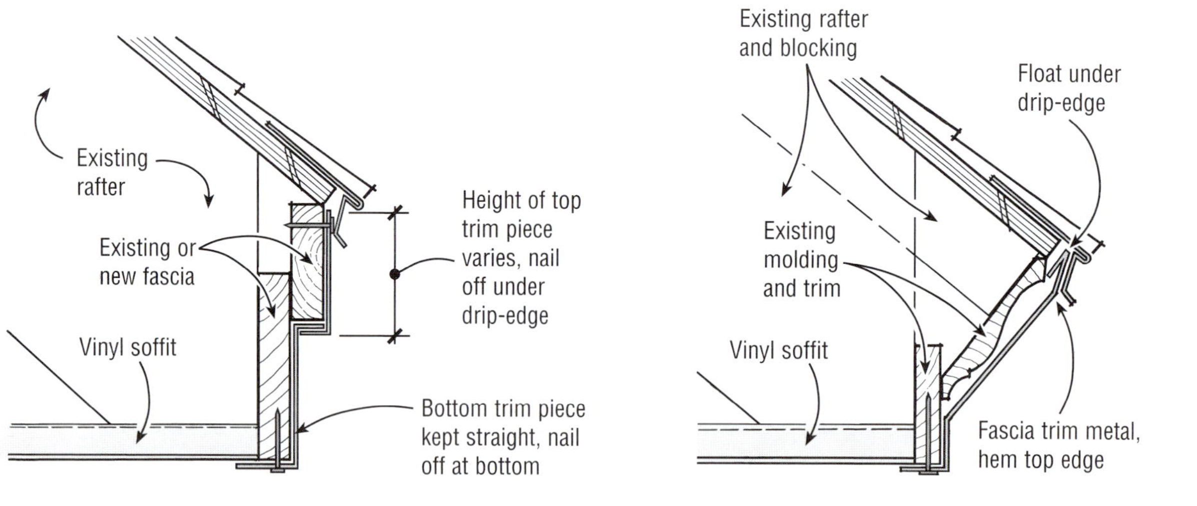

NAILING CHECKLIST

- Nail heads should be a minimum 5/16-in. in diameter, and the shank should be 1/8 in. in diameter. All fasteners must be able to penetrate at least 3/4 in. into framing or furring.
- Space nails to hold accessories such as starter strip, J-channel, and utility trim every 8- to 10-in.
- Space siding panel fasteners a maximum of 16 in. apart for the horizontal siding panels, and every 12 in. for vertical panels.
- Lock panels along the bottom edge before nailing top edge. Do not force the panels up or down when fastening.
- While the panel locks should be fully engaged, the panels should not be under vertical tension or compression

Installing Vinyl Siding

Figure 4-54. Trimming a Window

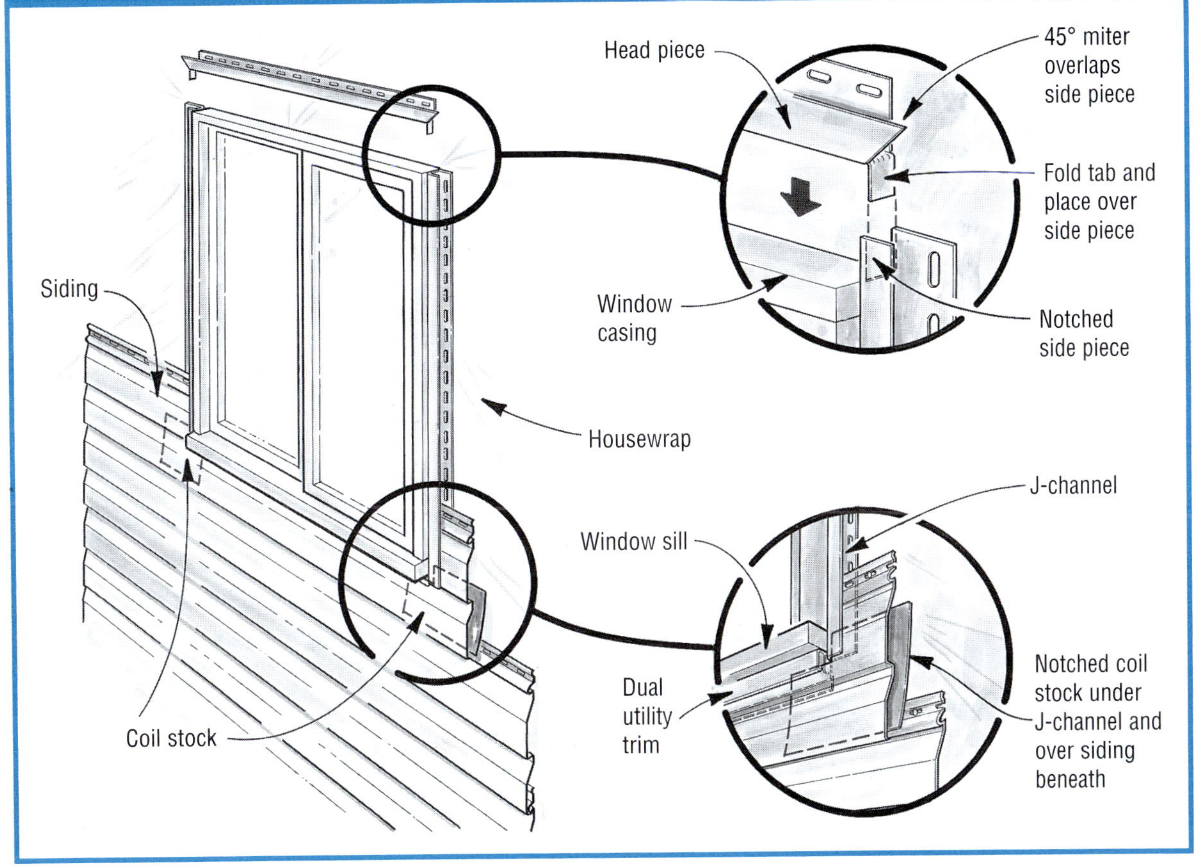

When trimming a window, first run J-channel around all four sides. Slip a notched piece of trim coil at each corner of the sill to direct water running down the J-channel over the top of the siding beneath.

when they are fastened. An exception can be made at the rim joist between first and second floors. When joists shrink, the panel may be forced down and may unlock, allowing the panel to flap in the wind. Keep some upward tension on the lock at this critical area to account for the shrinkage.

- Start nailing in the center of the panel and work towards the ends.
- Center fasteners in the slots to permit expansion and contraction of the siding. An exception to this rule can be made when fastening vertical siding and corner posts:

Figure 4-56. Corner Treatments for Vinyl Siding

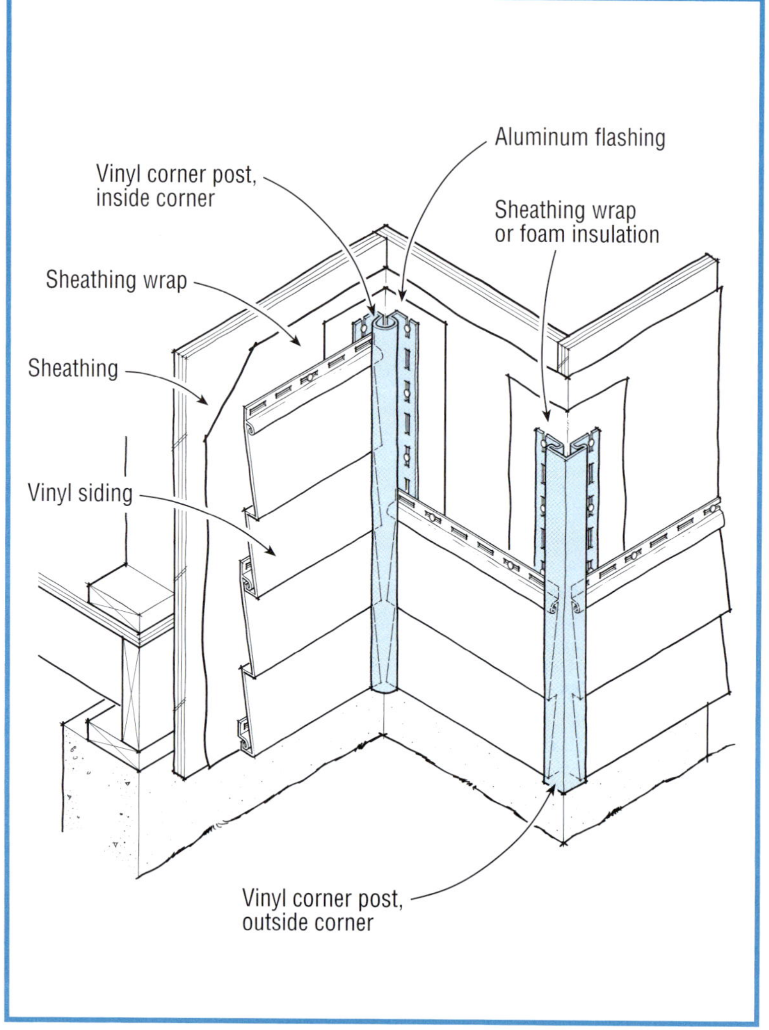

Outside corners (at right) are handled with vinyl corner posts. The corner should be wrapped with foam insulation or sheathing warp first. Inside corners (center) should be flashed with aluminum before installing the siding.

Figure 4-55. Dual Utility Trim

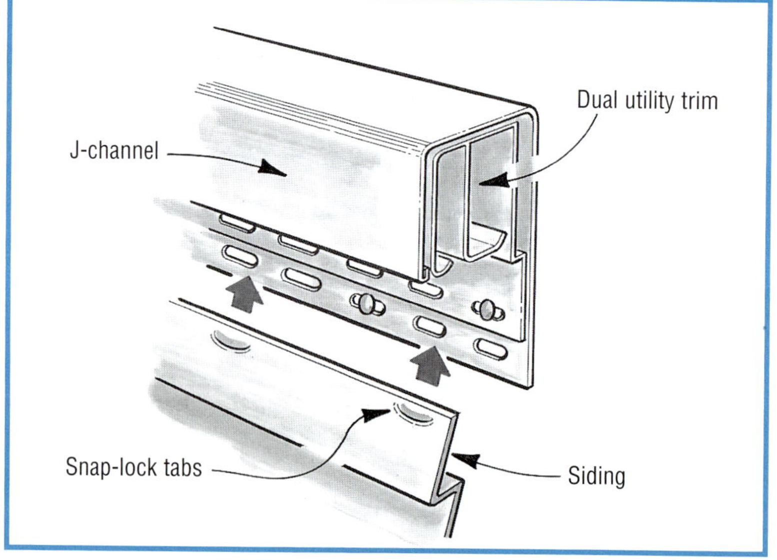

Use dual utility trim to secure notched panels without a nail hem. Use the outer channel for siding panels ripped near to the butt edge; use the inner channel when the rip is made closer to the nail hem.

In this case, start the top nail at the top of the uppermost slot to hold the piece at the proper elevation. All other nails should be centered in the slot.

- Do not drive the head of the fastener tightly against the siding nail hem. Leave a minimum of 1/32 in. (the thickness of a dime) between the fastener head and the vinyl (**Figure** 4-58).

Cutting Vinyl Siding

CUTTING VINYL SIDING

Cut vinyl panels with a fine-toothed (plywood) blade mounted backwards on the saw for a smoother, cleaner cut. This works especially well in cold weather when cutting with snips may crack the panel.

When cutting with tin snips, avoid closing the blades completely at the end of a stroke for a neater, cleaner cut.

Cleaning Vinyl Siding

CLEANING VINYL SIDING

To clean vinyl siding, use a long-handled brush with soft bristles (such as a car washing brush), and a handle that fastens onto the end of the hose. Avoid using stiff bristle brushes or abrasive cleaners, which may change the vinyl's gloss and cause the siding to look splotchy.

Figure 4-57. Starting Out Level

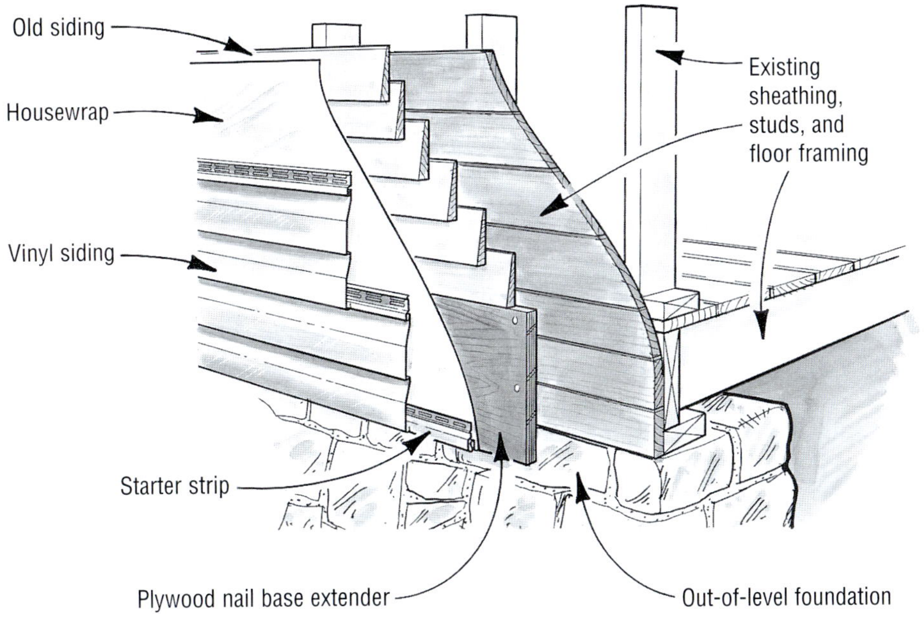

Siding should be installed level. When foundations are out of level, extend the nail base with a piece of plywood that matches the thickness of existing siding.

When cleaning soot and dust from the house, use the following cleaning solution, starting at the bottom and working up to the top in order to prevent streaking:

- 1/3 cup powdered laundry detergent, such as Fab®, Tide®, or equivalent*
- 2/3 cup powdered household cleaner, such as Soilax®, Spic &Span®, or equivalent*
- 1 gallon water

To remove mildew, add 1 qt. liquid laundry bleach to the solution above. Clean stubborn stains using the recommended cleaners shown in **Figure 4-59**.

Figure 4-58. Nailing Vinyl Siding

Do not drive nails home when installing vinyl siding. Vinyl siding needs to be able to slide back and forth as it expands and contracts with changes in temperature.

For more information: "A Builder's Guide to Vinyl Sidings," JLC 6/00

"Vinyl Siding Installation," VSI (www.vinylsiding.org/install/index.htm)

Figure 4-59. Stain Cleaners for Vinyl Siding

Stain	Cleaner*
Bubble Gum	Fantastik®, Murphy's Oil Soap®, or solution of vinegar (30%) + water (70%)
Crayon	Lestoil®
Oil-based caulk	Fantastik®
Felt-tip pen	Fantastik® or water-based cleaners
Grass	Fantastik®, Lysol®, Murphy's Oil Soap®, or Windex®
Lipstick	Fantastik® or Murphy's Oil Soap®
Lithium grease	Fantastik®, Lestoil®, Murphy's Oil Soap®, or Windex®
Mold and mildew	Fantastik® or solution of vinegar (30%) + water (70%)
Motor oil	Fantastik®, Lysol®, Murphy's Oil Soap®, or Windex®
Oil	Soft Scrub®
Paint	Brillo® Pad or Soft Scrub®
Pencil	Soft Scrub®
Rust	Fantastik®, Murphy's Oil Soap®, or Windex®
Tar	Soft Scrub®
Soil	Fantastik®, Lestoil®, or Murphy's Oil Soap®

*Cleaning materials are listed in alphabetical order.

Courtesy of Vinyl Siding Institute

Selecting and Handling Fiber-Cement

THE MAIN INGREDIENTS IN FIBER-CEMENT ARE PORTLAND cement, sand, and wood fiber; essentially, it's a masonry product, and nearly as durable. It won't burn, rot, or be eaten by insects, and it is dimensionally stable. Unlike wood and wood-composites, such as hardboard, it won't cup, shrink, or swell, and it holds paint exceptionally well. Fiber-cement siding typically carries a 50-year warranty.

FIBER-CEMENT TYPES

Fiber-cement siding is available as lap siding (12-ft. lengths from 5 1/4- to 12-in. wide), shingle panels (16x48-in.), and vertical or stucco-like panels (4x8, 4x9, and 4x10).

HANDLING FIBER-CEMENT

Fiber-cement siding is a lot heavier than wood of comparable thickness. The 5/16-in.-thick material weighs about 2 1/4 pounds per sq. ft., making it difficult to handle when working alone. It's also quite flexible, and full-length pieces can snap under their own weight. Always carry boards on edge rather than on the flat.

Aluminum/Cement Conflict

Any cementitious material, including fiber-cement siding, can attack aluminum if the aluminum is not protected. Aluminum trim should be anodized. Most window manufacturers anodize their aluminum trim. To be safe, prime the ends of the siding at field cuts near aluminum windows or flashing.

TRIM FOR FIBER-CEMENT

Fiber-cement trim and soffit panels are available. Because the material is so heavy, trim stock is typically only 7/16-in.-thick. It does not provide a good nail base for built-up trim and cannot provide deep shadow lines, so it requires the use of wood trim. For information about detailing exterior wood trim, see "Wood Trim," **pages 310-317**.

LAYOUT FOR FIBER-CEMENT SIDING

- Apply a level starter strip under the first course of lap siding (**Figure 4-60**).
- Overlap courses at least 1 1/4 in.

Figure 4-60. Starter Strip for Lap Siding

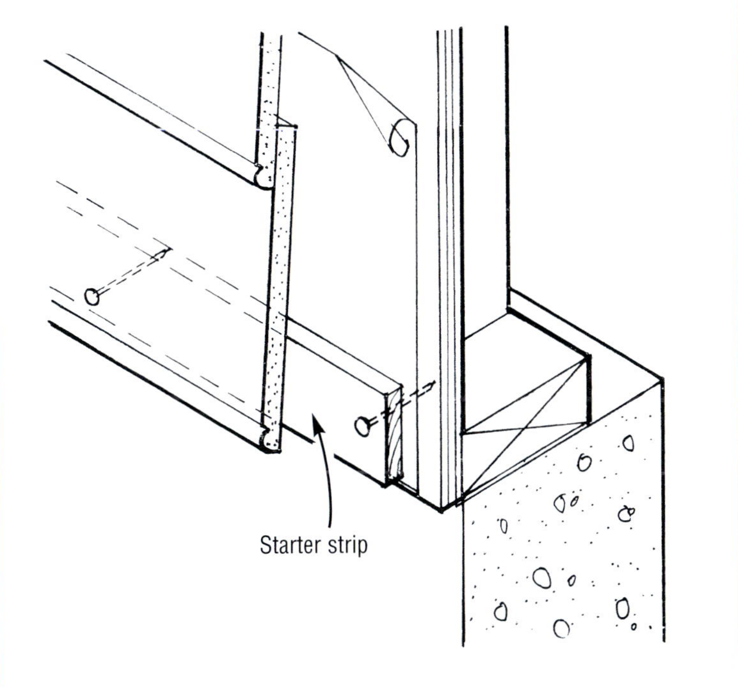

The first course of fiber-cement lap siding should be applied over a 1/4-in.-thick starter strip. Extend the siding 1/2 in. below the starter strip.

- Butt lap siding against vertical trim pieces at inside corners. Wood 2x2s can be used, but vinyl corner trim, available from Tamlyn and Sons (800/334-1676, www.tamlyn.com), is preferred (**Figure 4-61**).
- Join lap siding over a stud. If the butt joint falls between studs, use an off-stud joiner plate (**Figure 4-62**).
- Leave a 1/8-in. gap between the siding and the edge of wood casings and corner boards. This gap is necessary to allow for settlement and dimensional changes in wood framing. The joint should be sealed with caulk.

NAILING CHECKLIST

- The most important caution when nailing fiber-cement is to avoid overdriving fasteners. The siding is only 5/16 in. thick, so setting nail heads below the surface decreases holding power and voids the warranty.
- Narrower lap siding (up to 9 1/2 in. wide) can be blind-nailed with a wide-head, .121-in.-shank roofing nail. While this looks better, it is not recommended in high wind areas.
- Wider panels must be face-nailed at the bottom edge with 6d nails (minimum .093-in. shank with a .267-in.

Figure 4-61. Inside Corners for Fiber-Cement Siding

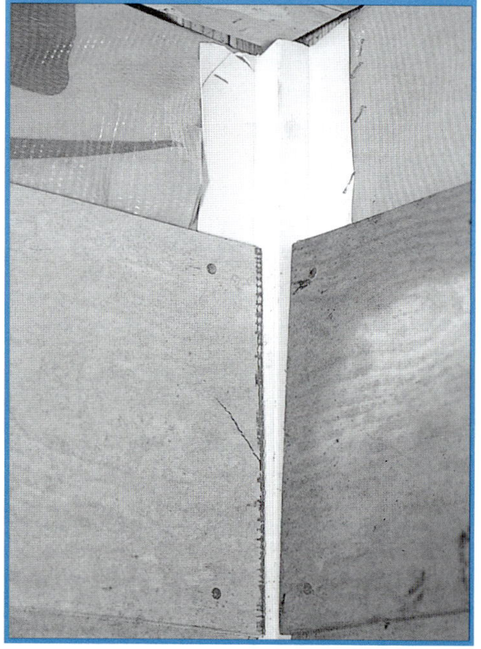

Figure 4-62. Off-Stud Joiner Plates

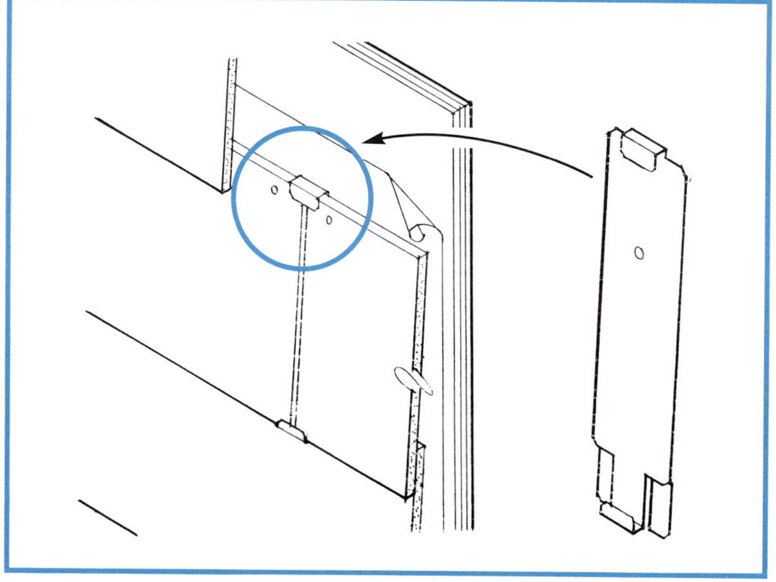

For more information: "On Site with Fiber Cement Siding," JLC 1/98

"Cutting Fiber-Cement Siding," JLC 4/03

Where joints fall between studs, apply metal joiner plates onto the butt ends of the siding. These should not be fastened to the wall.

Vinyl inside corner trims, which are similar in profile to a W valley flashing, are designed specifically for use with fiber-cement lap siding. Once painted, they look just like wood.

head diameter). This effectively pins each panel down with nails along both edges since the face nail penetrates the underlapping piece as well (**Figure 4-63**).

- Fasteners must penetrate 1- to 1 1/4-in. into framing (including sheathing).
- Use only corrosion-resistant roundhead nails (not clipped-head nails or staples).

CUTTING FIBER-CEMENT

Fiber-cement can be cut with carbide or abrasive circular saw blades. Fiber-cement is hard on saws and produces serious clouds of dust. Toothed carbide blades cut faster and produce less dust than abrasive blades, but the carbide tips dull fairly quickly.

An alternative is to use Pacific International's Snapper Lite shears (Pacific International Tool and Shear, Ltd., 800/297-7487, www.snappershear.com). These work like motorized tin snips, and while they are slower than a circular saw, they can make straight and curved cuts without producing much dust.

Figure 4-63. Nailing Fiber-Cement Siding

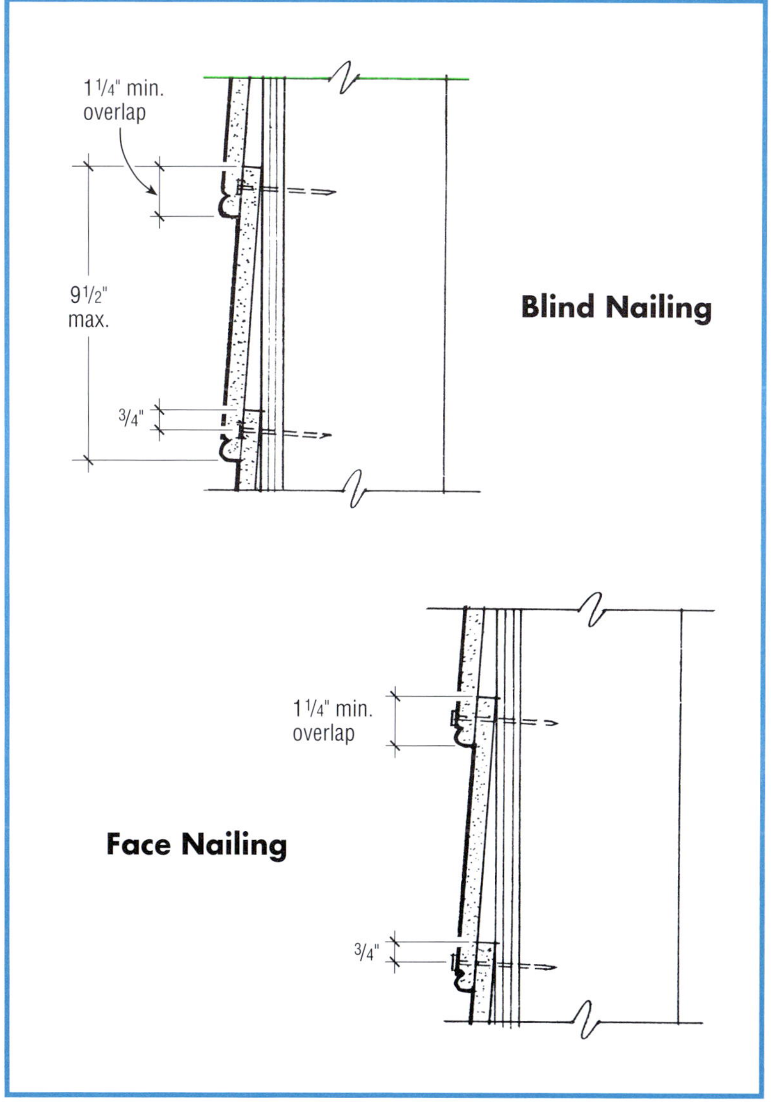

Fiber-cement lap siding that's under 9 1/2 in. wide can be blind-nailed. Wider lap siding must be face-nailed.

Cutting Fiber Cement

HARDBOARD IS A HOMOGENIZED WOOD PRODUCT MADE FROM a wood fiber pulp formed into panels or lap pieces under high heat and pressure.

HARDBOARD GRADES

Hardboard siding comes in 3/8-, 7/16-, and 1/2-in. thicknesses. In general, the 1/2-in. product performs better than the thinner versions.

At least five different grades are available. The better quality sidings are of "tempered" grade, which is impregnated with additives and/or is heat-treated to make it stiffer, harder, and more resistant to water and abrasion.

Figure 4-64. H-Trim for Hardboard Siding

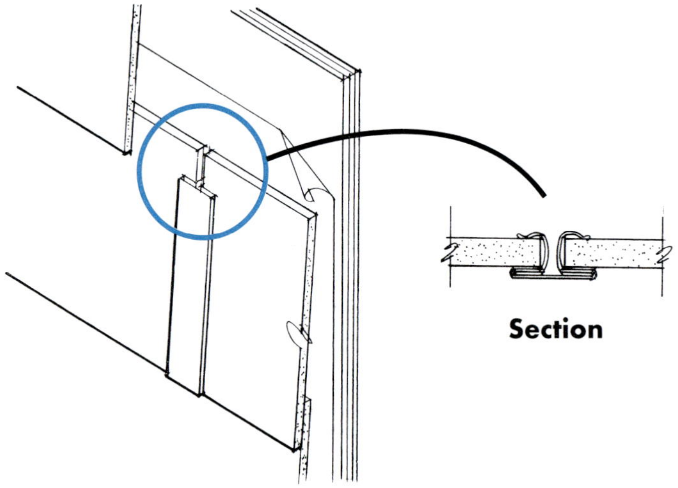

Butt joints between siding panels should be joined with color-matched, metal H-trim. This aluminum trim covers the vulnerable ends of the material without relying on caulk.

For more information: "A Hard Look at Hardboard Siding," JLC 6/90

Nearly all hardboard siding today comes primed and ready to paint or stain. About 10% of hardboard siding is sold pre-finished for quicker installation.

HARDBOARD SIDING PATTERNS

Hardboard is available to mimic nearly every siding style, but it falls into two basic categories: lap siding and panel siding.

Lap profiles, designed to look like cedar, redwood, or fir clapboards, account for 60% of hardboard sales. Boards come in standard 16-ft. lengths, ranging in width from 4- to 16-in.-wide, as well as wider "multilap" versions that will take care of several "rows" of siding at once.

Hardboard siding also comes in 2x8, 2x10, 4x8, and 4x9 panels, which install quickly. The panels imitate everything from board-and-batten construction to stone and stucco.

DRAINAGE PLANE

Because hardboard is so vulnerable to moisture, it's imperative to install it over a drainage plane — either a rain-screen mat or battens. Follow the guidelines outlined for "Board Siding, Drainage Plane," **page 321**.

INSTALLING HARDBOARD SIDING

Hardboard comes from the factory at a low moisture content (below 10%), so it must acclimate to job-site conditions. Otherwise, it will pick up moisture and expand (**Figure 4-35, page 318**). Also, because hardboard is a homogenized wood, it expands more in length than a piece of solid wood siding does, and it will buckle if it's nailed down and then expands.

Avoid installing hardboard siding when a building's concrete foundation is still green, or when the building is excessively wet from fresh plaster, wet-spray insulation, or other wet materials.

Selecting and Installing Hardboard

Hardboard Siding Checklist

Because hardboard is vulnerable to moisture damage, quality of installation is critical to its performance. To stray from the manufacturer's installation guidelines is asking for trouble and can void the warranty. Installation guidelines vary only slightly from one manufacturer to another.

To avoid problems caused by moisture:

- Leave gaps of about 1/8 in. between butt ends of boards and between boards and trim. This gap will provide maximum hold for an exterior caulk (see "Caulks and Sealants," **page 371**). Caulk should be primed and painted with two topcoats. Keep caulk joints to a minimum; they will be the first point of failure in the paint job.
- To limit caulk joints, cover butt joints between panel lengths with a special "H-trim" (see **Figure 4-64**).
- Use galvanized box nails driven flush with the board's surface. Do not break the siding, or the hole will absorb moisture and will pucker.
- Cut with a fine-tooth saw blade. Always cut into the face of the board: Place boards face up when using a hand saw, or face down when using a circular saw.
- Outside corner boards and trim should be rabbetted. If a mitered corner is desired, cover the joint with color-matched metal corners (**Figure 4-65**). At the very least, use 5/4 trim that is thick enough to completely cover the ends and edges of the siding.
- Do not install hardboard siding closer than 6- or 8-in. to the ground, or closer than 2 in. to roofs. In areas where it snows a lot, leave a 4- or 6-in. gap above roofs intersecting the walls.
- Paint all field cuts to help seal the ends. This is particularly important above intersecting roofs.
- Apply finish to the siding within 30 to 90 days after installation (see "Finishes," **page 373**).

Figure 4-65. Corner Treatments for Hardboard Siding

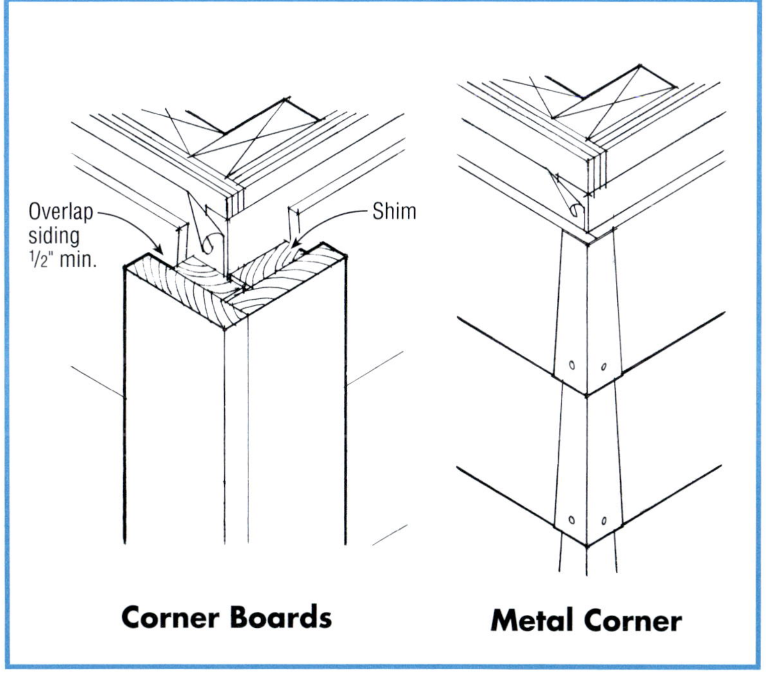

To protect the cut ends of hardboard panels, use rabbeted corner boards or color-matched metal corners. Do not leave cut ends exposed or they will absorb moisture and swell.

STUCCO IS MOSTLY SAND AND PORTLAND CEMENT. Traditionally, it is applied in three separate coats that add up, more or less, to 7/8 in. for one-hour fire-rated construction.

Figure 4-66. Drainage Plane for Three-Coat Stucco

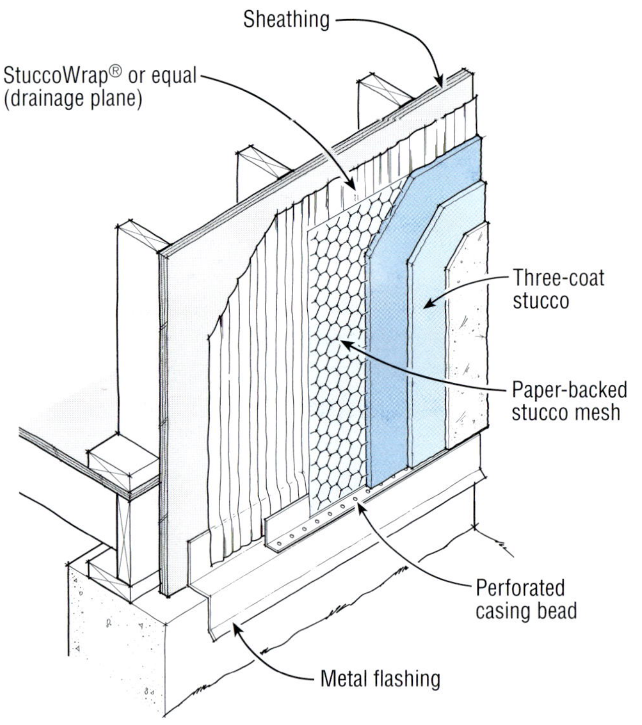

Stucco is a highly porous material and will absorb water. To protect the structure against water damage, apply a "wrinkle wrap" type drainage plane and proper flashings so water can drain down and out.

For more information: "Water Managed Wall Systems," JLC 3/03

STUCCO TYPES

Traditional Three-Coat Stucco

Three-coat stucco consists of three separate layers of cement plaster materials: a trowel-applied scratch coat, which is applied over lath and grooved to provide a key way for the next coat; a sand-finished brown coat, which keys into the scratch coat; and a cementitious finish, or color coat (**Figure 4-66**).

One-Coat Stucco

Often known as thin-coat stucco or fiberglass reinforced stucco (FRS), one-coat systems have some distinct advantages. "One-coat" is somewhat of a misnomer as it actually requires two coats: a single, thin base coat, followed by a cementitious color coat or one of the newer acrylic-based synthetic color coats.

EIFS

Exterior insulation and finish system (EIFS), is an acrylic stucco finish applied over a rigid insulation board, such as extruded polystyrene sheathing. The original intent of this material was to create a watertight skin using sealants and control joints. In practice, this system proved impractical without impeccable detailing, and many problems — leading to many lawsuits — ensued. Most successful synthetic stucco systems used today are applied over a drainage plane using cementitious backerboard, rather than foam insulation, as the substrate (**Figure 4-67**).

DRAINAGE PLANE

Stucco of any type will crack, even under the best of conditions. Even synthetic stucco, which is more elastic than other stucco types, is still intolerant to movement in the underlying surface. Movement is a bigger problem on wood-framed walls than on block, insulated block, and poured concrete. However, no structure is immune to settlement or movement caused by changing climate conditions. All stucco systems should be applied over a drainage plain (**Figure 4-66** and **Figure 4-67**).

Sheathing Wrap for Stucco

All stucco systems require sheathing wrap — either two layers of Type D paper (the minimum allowed by code) or No. 15 felt paper, or a single layer of "wrinkle wrap," such as Stuccowrap®. Stuccowrap bends more easily than black paper and is less likely to tear during installation. Paper-backed lath should not be considered a substitute for well-detailed sheathing wrap.

Lap lower courses of sheathing wrap under upper courses so water will drain down and out, and flash all openings, following the flashing guidelines found in "Flashing Details," **page 302**.

WEEP SCREEDS AND CONTROL JOINTS

A perforated weep screed at the bottom of the wall and over openings serves two important purposes — it provides a screed that helps ensure a uniform coat thickness, and it provides a place for water that leaks into the wall to escape (**Figure 4-68**).

Stucco Control Joints

Control joints allow for structural movements that would otherwise crack the stucco. On a house, control joints should be applied over the metal lath at a few critical locations:

- Over the rim joist at the intersection between floors to accommodate shrinkage in the floor joists

Figure 4-67. Drainage Plane for Synthetic Stucco

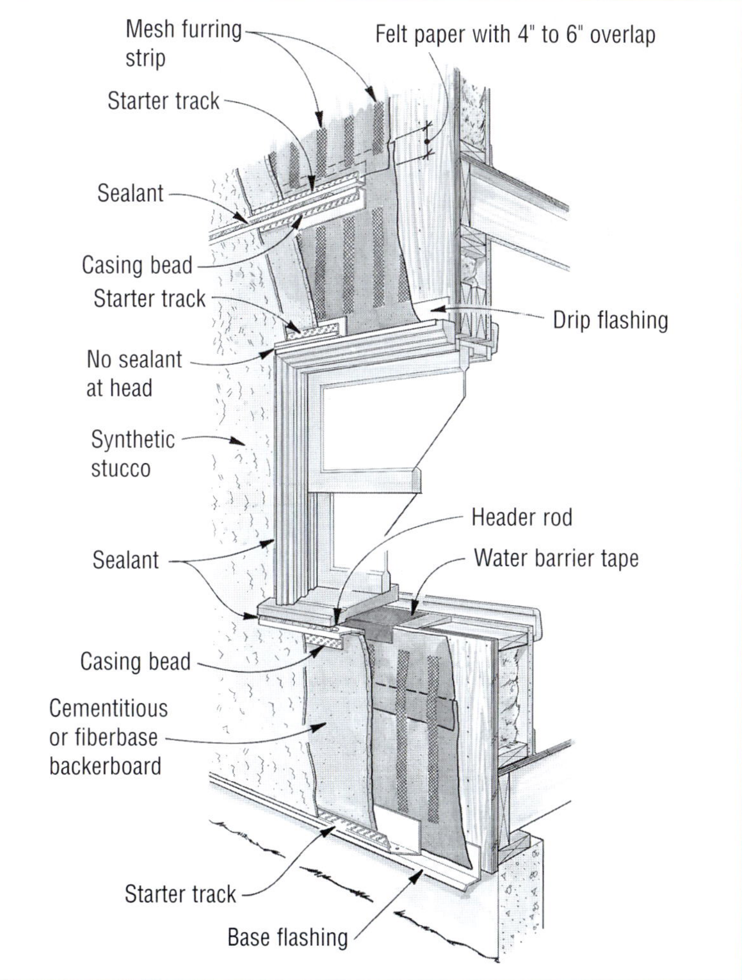

Cementitious backerboard provides a more durable substrate for synthetic stucco than foam insulation board. The backerboard must be installed over mesh furring strips to provide a drainage gap so that any water that gets through the stucco is directed to weep areas at windows and wall skirts.

Stucco Types

Drainage Plane

Weep Screeds

- Over the intersection between different types of construction, such as the joint between wood framing and a block wall (**Figure 4-69**)
- On tall walls and long runs, applied vertically or horizontally to break up wall areas into smaller sections (less than 18 ft.)

Similarly, casing bead and corner bead help to create expansion joints at critical intersections:

- Outside corners are best detailed with corner bead (**Figure 4-70, page 350**). Continuous stucco wrapped around corners is likely to crack.
- Inside corners should be flashed with two layers of sheathing wrap or flexible flashing, and then casing bead should be applied vertically to create an expansion joint (**Figure 4-70, page 350**). While the flashing serves as the most important water sealing protection, the joint between the two casing beads should be caulked with a high-quality urethane caulk, as a first-line defense.
- Apply casing bead (also called J bead) at the joint between any dissimilar materials — around window flanges (**Figure 4-67**), where stucco meets a brick chimney (**Figure 4-69**), or at the top of a wall where stucco meets a soffit or frieze board.

Figure 4-68. Weep Screed

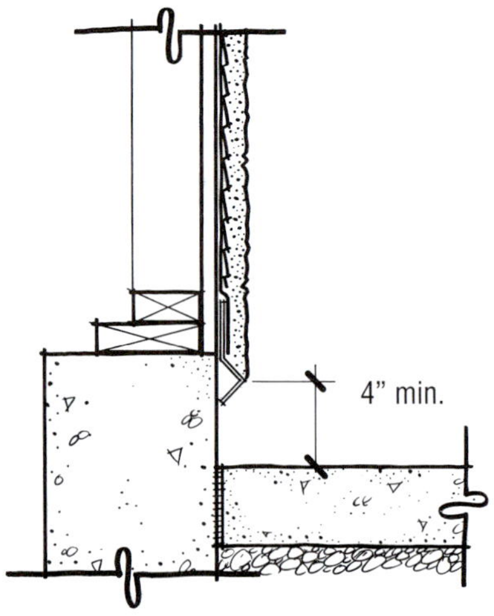

Install weep screeds to ensure a uniform stucco thickness and to provide a drainage outlet for water.

METAL LATH

Traditional three-coat systems and one-coat systems both require lath — either expanded metal lath or wire lath (**Figure 4-71, page 351**). The primary purpose of metal lath is to provide a keyway to bond the stucco to the structure, and so it must stand off the wall slightly to allow stucco to get behind it and fully encase the metal. This is done by using a self-furring material or with furring nails. Metal lath also provides some tension reinforcement for the stucco, but it will not prevent cracking. Cracking must be controlled by using the right mix, controlling the temperature during application, and allowing the stucco to cure.

Expanded metal lath comes in several varieties. Diamond lath is available in plain, self-furring (allows plaster underneath the lath), and paper-backed. Stucco mesh is similar to diamond lath but has larger openings. It is available only plain, so it must be installed with furring nails over sheathing wrap. Flat ribbed lath typically requires furring nails; it is available in two weights — 2.75 and 3.4 lbs. per yd. The lighter material is made for interior plaster and is not suitable for exterior stucco. Three-eighths-inch rib lath is self-furring, but it's difficult to encase completely in stucco and so is not recommended for wet climates.

Wire lath is not the same as chicken wire. Woven wire is available in plain, self-furring, and paper-backed versions; plain wire lath should be installed with furring nails. Use heavier 16- and 17-gauge material for three-coat stucco; lighter 20-gauge wire can be used for one-coat systems.

Furring nails. When using plain lath, use furring fasteners — which have fiber spacers — to keep the lath about 1/4 in. away from the sheathing wrap. This allows the stucco to fully encase the lath.

Installing Metal Lath

Metal Lath

- Run the long dimension of metal lath perpendicular to the framing.
- With plain (unbacked) lath, start at the top so lower courses lap over upper courses. Paper-backed sheets should be the opposite — "shingled" so upper courses lap over lower courses, like the sheathing wrap, allowing water to drain down and out.
- Expanded metal lath should be lapped 1/2 in. along the top and bottom sides and 1 in. at the ends. Wire lath should be lapped on all sides at least one full mesh (1- to 2-in.).

Figure 4-69. Stucco Control Joints

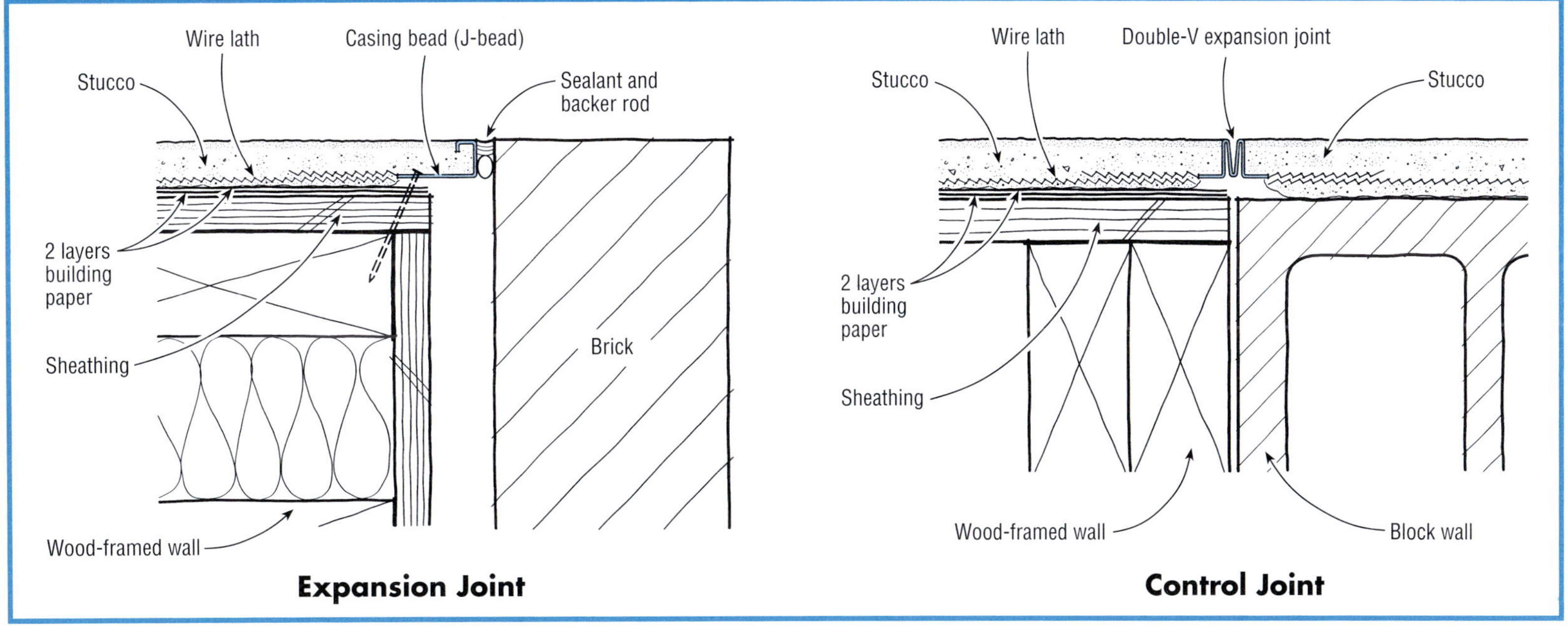

Control joints are needed at the juncture between dissimilar materials. Without an expansion joint, the stucco will crack.

- Use galvanized fasteners. Do not use aluminum fasteners; these will react galvanically with the steel lath.

THREE-COAT STUCCO

The first two coats of a three-coat system are made of a site-mixed cement plaster. Each batch contains measured amounts of Portland cement, hydrated lime, fine sand, and water. To help reduce the cracking that is common with stucco, the basic mix can be modified with liquid polymer additives, which increase strength and improve curing, and with chopped fiberglass for reinforcement.

Figure 4-70. Stucco Corner Treatments

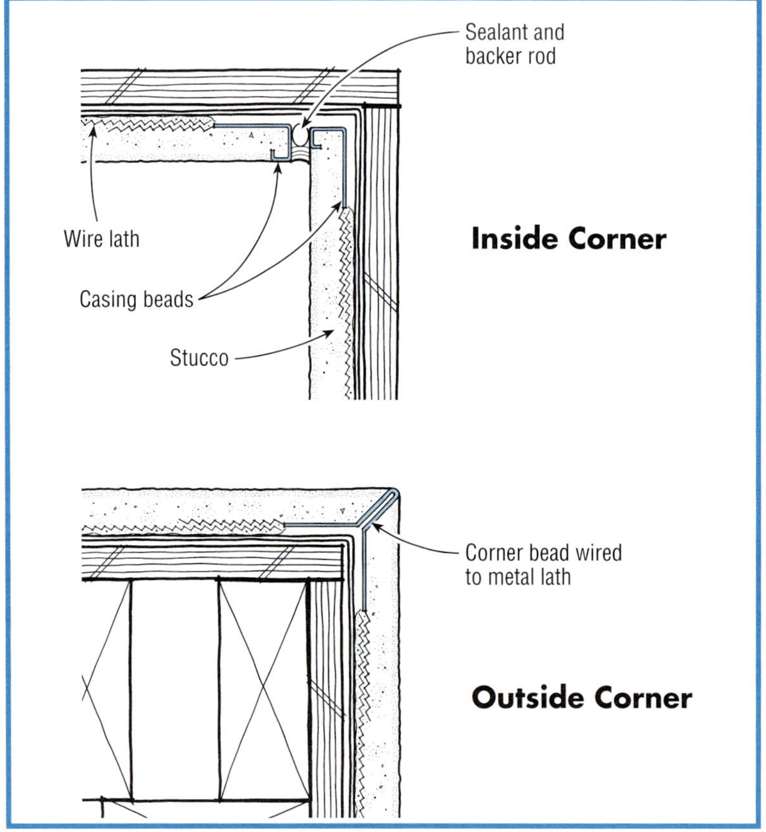

For more information: "Stucco Flashing Details," JLC 10/98

Scratch Coat

The scratch coat is applied directly to a clean block wall or, on wood structures, over "D" paper (or better) and galvanized metal lath. A metal rake is used to create "scratch" lines in the still-wet surface. The furrows create a rough surface into which the brown coat can key for a good bond.

Scratch-coat thickness. The scratch coat should be approximately 3/8- to 1/2-in. thick — thick enough to just cover the wire.

Brown Coat

The brown coat (which is actually gray, like other cement stucco) is applied the next day as a smooth coat, usually "rodded" with straightedges and wooden floats (**Figure 4-72**). In the western U.S., a steel trowel is used to produce an adobe look when the brown coat is the final coat.

Brown-coat thickness. The brown coat is composed of slightly more sand than the scratch coat mixture, and therefore it is more manageable but slightly weaker. It should be applied 1/4- to 3/8-in. thick. Apply the brown coat as evenly as possible. An uneven brown coat can lead to an inconsistent thickness in the finish coat, causing some areas of the finish coat to dry before others. The result is a splotchy or "mottled" color coat and excessive hairline cracking.

Once applied, the two coats should be allowed to cure for at least a week, preferably two weeks, before applying the finish coat (see "Curing Stucco," **page** 355). This allows the scratch and brown coats to strengthen, settle down, and do all the shrinkage cracking they intend to do.

Finish Coat
After the curing period, a finish coat can be applied (**Figure 4-73**). Color is applied to the finish mix by adding iron oxide pigments to a mix made with white Portland cement. There are many types of pigments available at different prices. The cheaper pigments tend to fade and be inconsistent from batch to batch.

Finish-coat thickness. The finish coat is only about 1/8 in. thick. It should be applied as evenly as possible to ensure uniform drying.

Three-Coat Stucco

MIXING STUCCO

Stucco (like concrete) derives its strength as a result of two factors: the relative strength of the mix and its drying time. A wet mixture will result in excessive shrinkage (often called checking), and a dry mixture — or stucco completed in very hot weather — will create a weak bond (see "Weather Precautions," **page 354**).

Figure 4-71. Metal Lath for Stucco

	Type	Weight (lb. per sq. yd.)	Opening Size (in.)	Typical Unit Dimensions
Expanded Metal	Diamond mesh*	1.75, 2.5 and 3.4	5/16 x 3/8	27 in. x 96 in.
	1/8-in. flat rib mesh*	2.75 and 3.4	5/16 x 3/8	27 in. x 96 in.
	3/8-in. rib lath (high rib)	3.4 and 4.0	5/16 x 3/8	27 in. x 96 in.
Wire Lath	Woven wire**	1.7 (18 gauge) 1.4 (17 gauge)	1 (hexagonal) 1 1/2 (hexagonal)	3 ft. x 150 ft. (rolls)** 3 ft. x 150 ft. (rolls)**
	Welded wire**	1.4 (16 gauge)	2 x 2	3 ft. x 150 ft. (rolls)**
		1.4 (18 gauge)	1 x 1	3 ft. x 150 ft. (rolls)**
		1.9 (16 gauge)	1 1/2 x 2	3 ft. x 150 ft. (rolls)**

*Available in plain, self-furring, and paper-backed
**Paper-backed and self-furring also available in sheets

Adapted from Portland Cement Association

STUCCO INGREDIENTS

Water

Use clean water that is suitable for drinking. Mineral and organic impurities in water may discolor or affect the set time, and may attack metal lath.

Sand

It is important to use bagged silica sand to ensure that there are no iron particles in the mix, which may cause staining. Do not use the less expensive "yard sand," which often has impurities.

Cement

Most stucco uses Portland cement (Types I and III, **Figure 1-34, page 34**) or masonry cement (Types M, S, and N, **Figure 1-59, page 57**). White Portland cement is used to produce a white or light colored finish coat. Plastic cement is a special cement made expressly for the plaster industry, and is commonly available in the Southwest and on the West coast. It is a blend of Portland cement and plasticizing materials, such as limestone or hydrated lime. When plastic cement is used, no lime or other plasticizer is required.

Figure 4-72. Brown Coat

A 1/4- to 3/8-in.-thick brown coat is applied over the scratch coat and "rodded" with a straightedge.

Figure 4-73. Finish Coat

A 1/8-in.-thick finish coat is applied with a trowel.

Figure 4-74. Stucco Mixes

	Mix Proportions (parts per volume)				
Group*	Portland Cement	Type II Masonry Cement	Plastic Cement	Lime	Sand
C (Cement)	1	–	–	0 to 1/4	3 to 4
C	1	1	–	–	6 to 7 1/2
C	1	–	–	1/4 to 1/2	4 to 6
L (Lime)	1	–	–	2 to 1 1/4	4 1/2 to 9
L	–	1	–	–	3 to 4
F (Finish)	1	–	–	1 1/4 to 2	5 to 10
P (Plastic)	1	–	1	–	6 to 10
P	–	–	1	–	3 to 4

	Recommended Stucco Group		
	Scratch	Brown**	Finish
Base Material			
Low Absorption (poured concrete, dense brick)	C, P	C, L, P	L, F, P
High Absorption (concrete block, clay brick, structural tile)	L, P	L, P	L, F, P
Metal Reinforcement (over all types of construction)	C, P	C, L, P	L, F, P

* Use as base coat in two-coat work

Adapted from Portland Cement Association

Mixing Stucco

Choose the stucco group for each coat, depending on the base material requirement shown in the bottom half of this table. When mixing each stucco group, keep proportions consistent between batches and follow mix proportions shown in the top half of the table.

For more information: "Extreme Stucco," JLC 2/97

"One-Coat Stucco," JLC 6/95

Plasticizer

The function of a plasticizer is to improve the workability of the brown and finish coats or the scratch coats applied over poured concrete walls. Type S hydrated lime is the most common plasticizer used in stucco.

MIX PROPORTIONS

Inconsistent mixing will result in obvious differences in color, texture, and strength between batches. Over time, mismixed batches can effloresce or fail. Make every effort to keep mix proportions consistent, following the guidelines shown in **Figure 4-74**, previous page.

MIXING PROCEDURE

Follow this procedure, using a paddle-type mortar mixer:

- Add majority of the mix water and start mixer. Keep in mind that the drier the mix, the stronger the stucco, so don't add too much water.
- Add approximately half of the sand.
- Add lime (if required), followed by cement and any admixtures required.
- Add remainder of sand.
- Add pigment (finish coats only). Allow the mixer to run until the color is dispersed throughout the entire load. Undermixing the finish will permit lumps of raw color to be left in the finish.
- Add water required to reach desired consistency (and no more). Continue mixing 3 to 5 minutes until batch is mixed uniformly.
- Keep batches to a size that can be applied within one hour after mixing.

Figure 4-75. Curing Schedule for Cement Stucco

Stucco Coat	Moist Curing	Total Setting Time*
Scratch	12 to 24 hr.	At least 48 hr. (between coats)
Brown	12 to 24 hr.	At least 7 days (between coats)
Finish**	12 to 24 hr.	At least 48 hr. (before painting)

* Air temperature min. 50°F should be maintained during this time.
** Moisten base coat immediately before applying finish coat.

Adapted from Portland Cement Association

WEATHER PRECAUTIONS

Cold Weather

Don't mix materials or apply stucco when the air temperature is below 35°F. Keep materials covered at night, and if working temperatures are low, warm the mix water and the sand with salamander heaters. Also keep freezing rain and snow away from freshly stuccoed walls. If possible, install gutters and tip up scaffold planks at the end of the day to reduce splashback.

A calcium-based accelerant can be added as an "anti-freeze." But this also increases the salinity of the mix and may cause efflorescence — the migration of salts to the surface.

Hot Weather

In hot, or warm windy weather (above 75°F), there is a danger of the stucco drying too fast, or flash curing. Keep materials out of the direct sun. Sand and water both hold heat, and if these can be kept cool, the stucco will dry more slowly. But don't moisten the sand to cool it off; this may throw off the mix ratio. If possible, start work early in the day and "chase the shadows" (working in the shade will slow drying times).

After applying brown coat, keep freshly stuccoed walls damp in hot weather, using a garden sprayer or a garden hose with a fogging head. Mist the walls every hour until the sun and wind are no longer a problem. Also, draping new work with wet burlap and keeping a soaker hose on the top of the wall to keep the burlap wet will retard evaporation as well.

Do not mist or dampen a color coat, however. This may result in splotchy walls. To be safe, the only choice is to wait for the right weather to apply the finish coat.

Cold- and Hot-Weather Stucco

CURING STUCCO

The brown and scratch coats must cure for a minimum of seven days before the color coat can be applied (**Figure 4-75**). For best results, wait four weeks to allow the brown coat to achieve full strength.

Three-coat stucco requires moist curing — fogging with fine water spray at the beginning and the end of the work day under normal weather conditions. Fogging should be delayed until the scratch and brown coats are sufficiently set to prevent erosion, and at least 12 hours after applying the finish coat. For two-coat stucco applications, follow curing recommendations for brown and finish coats.

Curing Stucco

FINISH TEXTURES

The color-coat finish is usually floated to a sand finish, but an experienced applicator can vary the texture and color with trowel techniques (**Figure 4-76**).

Finish Textures

Figure 4-76. Common Stucco Textures

Sand Float

After troweling on stucco, use a sponge float in a swirling motion. Additional water may be sprinkled lightly on the area, as needed, to roll the sand, but don't over-rub; this may create a bald spot.

Fine Lace

Use a small brush and lightly dab wet stucco onto the colored stucco base. Wait a minute for the excess water to dry out, and then flatten the ridges by lightly running a clean trowel over the dabbed spots.

Smooth Finish

Trowel on the finish stucco, continually working and smoothing the material. To reduce small shrinkage cracks, it may be necessary to retrowel a second and third time, sprinkling water on the wall with a brush, as needed, to act as a lubricant for the trowel.

Figure 4-77. Drainage EIFS Details

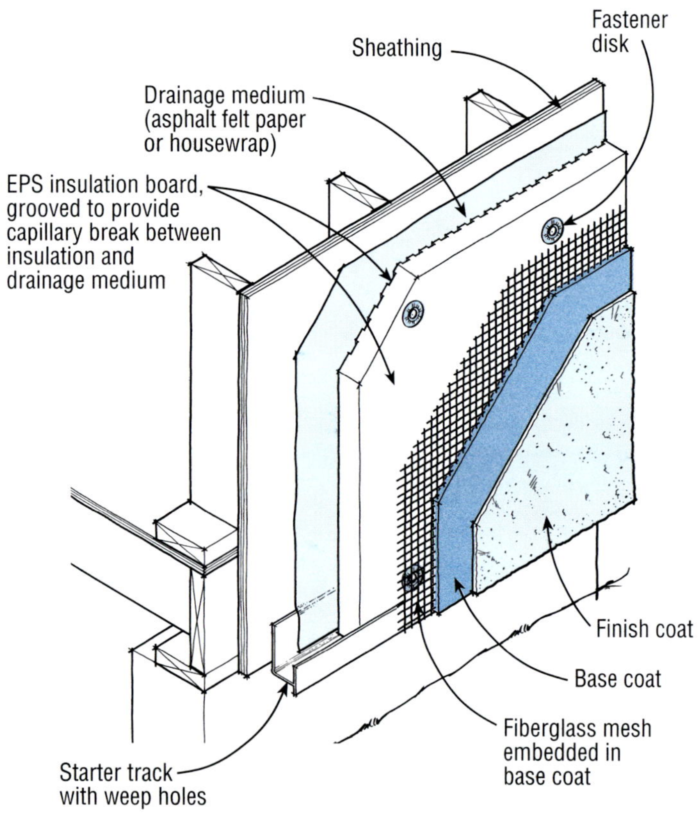

EIFS can be applied on wood-framed walls, provided the wall is designed to drain when the stucco cracks. The drainage plane relies on well-detailed sheathing wrap and flashing, and a grooved EPS insulation board.

ONE-COAT STUCCO

Base coat. One-coat stucco uses a single factory-prepared base coat, consisting of fiberglass-reinforced cement plaster with modifying polymers. Sand and water are the only additions on the job site.

Color coat. Synthetic stucco finishes come pre-mixed in five-gallon buckets. These products primarily use an acrylic polymer base (much like exterior latex paint, but more flexible and durable) mixed with fine aggregate and pigment. Synthetic finishes can be tinted to almost any color and are not subject to shade variations.

Pros

One-coat systems offer several advantages. Factory batching of the base and color coats helps eliminate much of the variation that can occur with batching three-coat mixes. It also speeds up application by eliminating the need to work to corners in warm weather because drying time does not affect the final color. In addition to eliminating one coat altogether, less curing time is required for the base coat. In a pinch, a cementitious finish coat can go on after 48 hours, or a synthetic color coat can go on after just 24 hours. However, it is usually better to wait at least seven days, if possible, to allow the base coat to reach its full strength and complete all shrinkage cracking.

Cons

On the downside, one-coat stucco is not as impact-resistant as three-coat stucco. Also, the finish appearance of a one-coat stucco job is so uniform and monochromatic that it may not perfectly capture the three-coat stucco look some clients have in mind. Nor will a one-coat finish last as long.

Most manufacturers claim a 15-year finish, whereas a traditional color coat on a three-coat stucco, properly applied, will last 20 to 30 years.

EIFS

EIFS (Exterior Insulated Finish System) was originally developed for use on concrete buildings in Europe. The original system — known as barrier EIFS — depends on an extruded polystyrene (EPS) insulation board that is glued to the wall, helping to prevent movement of the substrate. This detail was lost when it was applied to wood-framed buildings in the U.S. Stucco of any kind is bound to crack and allow water to intrude — a problem that has more serious consequences on a wood-framed structure than on a concrete one. On wood-framed structures, a drainage EIFS (also called water-managed EIFS) is required.

Drainage EIFS Design

- Provide a complete secondary weather barrier with a water-shedding drainage plane and flashings (**Figure 4-77**).
- Provide sealed movement joints at floors (**Figure 4-78**).
- All penetrations should be flashed beneath the EIFS with peel-and stick flexible flashing (see "Flexible Flashing Materials," **page 299**), as well as a surface-seal at the joint with backer rod and silicone sealant (see "Caulks and Sealants," **page 371**).
- At wall-roof junctions, provide kickout flashings at roof terminations on sidewalls (**Figure 4-79**).
- Provide Z-shaped counterflashing along sidewalls (**Figure 4-80**) and two-piece head flashings on front walls (**Figure 4-81**).

Figure 4-78. EIFS Control Joints

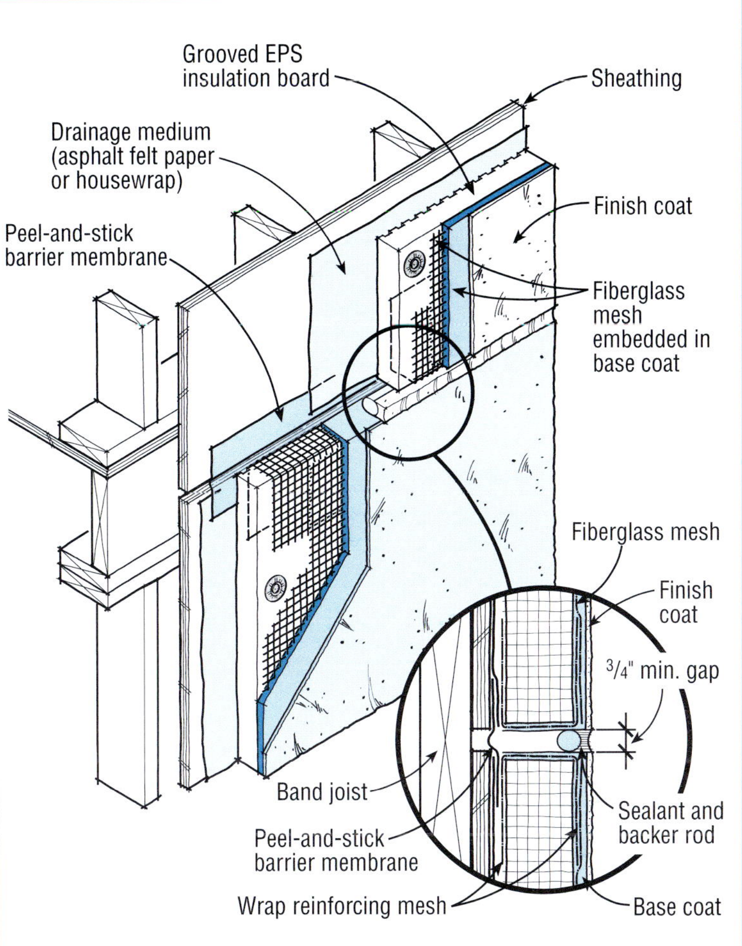

EPS board in an EIFS system can buckle when floor framing shrinks. Build a way for the board to move by creating a 3/4-in. gap between upper and lower sheathing panels and upper and lower EPS boards. Seal the gap over the sheathing with peel-and-stick flashing, and seal the gap in the insulation with backer rod and silicone sealant.

One-Coat Stucco

EIFS

- Provide sill flashings at windows (**Figure 4-11, page 300**).

Drainage EIFS Materials

- Use only high-density, low-porosity EPS board with grooves to provide drainage.
- Use wet-mix base coats, not dry mixes.
- Use 6-ounce or heavier mesh (nonwoven mesh if available).

Figure 4-79. Kickout Flashing for Synthetic Stucco

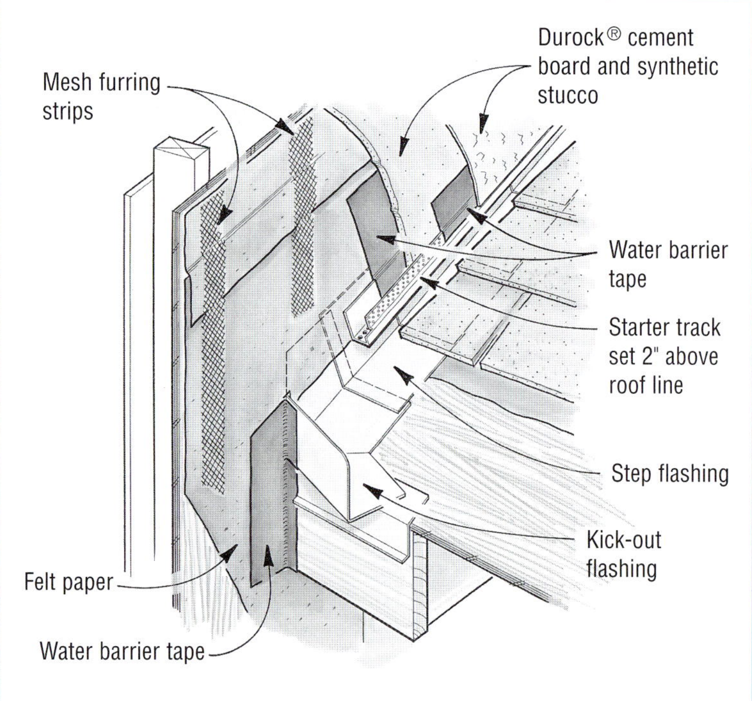

Missing kickout flashings at roof-wall junctions are a common source of EIFS failures. Use a flashing metal of the same type as the drip-edge to avoid galvanic corrosion.

For more information: "Installing Water-Managed Synthetic Stucco," JLC 9/98

"Success with EIFS," JLC 11/01

- Use high-impact mesh on the ground floor and in traffic areas.
- Use silicone sealants for joints.

Drainage EIFS Installation

- Do not break EPS board over sheathing joints. Keep board joints away from window corners.
- Apply base coat in two layers for adequate thickness; ideally, the base coat should be between 1/16- and 3/32-in. thick.
- Back-wrap all panel edges with mesh and base coat. Do not leave exposed mesh.
- Apply sealant to base coat, not to finish coat.
- Coordinate trades so that flashings and weather barriers are installed in the proper sequence, and be sure that each trade knows its responsibilities.

REPAIRING STUCCO

Hairline Cracks

It's best not to attempt patching hairline cracks with wet stucco; the patch may be more noticeable than the crack. Instead, dust the crack with stucco, as shown in **Figure 4-82**.

Small Cracks

Slightly larger cracks (up to 1/8 in.) can be filled with color-matching caulk. To fill the crack, first cover the crack with clear masking tape, and then slit the tape along the crack with a utility knife. Apply the caulk through the slit and remove the tape before the caulk sets.

For cracks 1/8- to 1/4-in. wide, use a narrow chisel or large screwdriver to scrape or chip through the color and brown coats to expose the scratch coat. Then, apply an acrylic bonding agent (masking as needed) to strengthen adhesion between the old and new materials. Finally, patch with stucco or special rapid-set patching mortar.

Larger Cracks

When patching wide cracks, wet the surrounding stucco and then apply an acrylic bonding agent. To keep the patch from sagging and shrinking, work the new stucco in layers no more than 1/2 in. thick. Screed the final layer flush with the existing stucco.

Repairing Stucco Cracks

Patching Small Holes

To repair small holes up to 3 in. across, first clean the hole of any loose material and wet the edges. Fill the hole with a rapid-setting mortar mix in one coat. Use a trowel turned on edge to rake off any excess so that the patch is

Figure 4-80. Side-Wall Flashings for EIFS

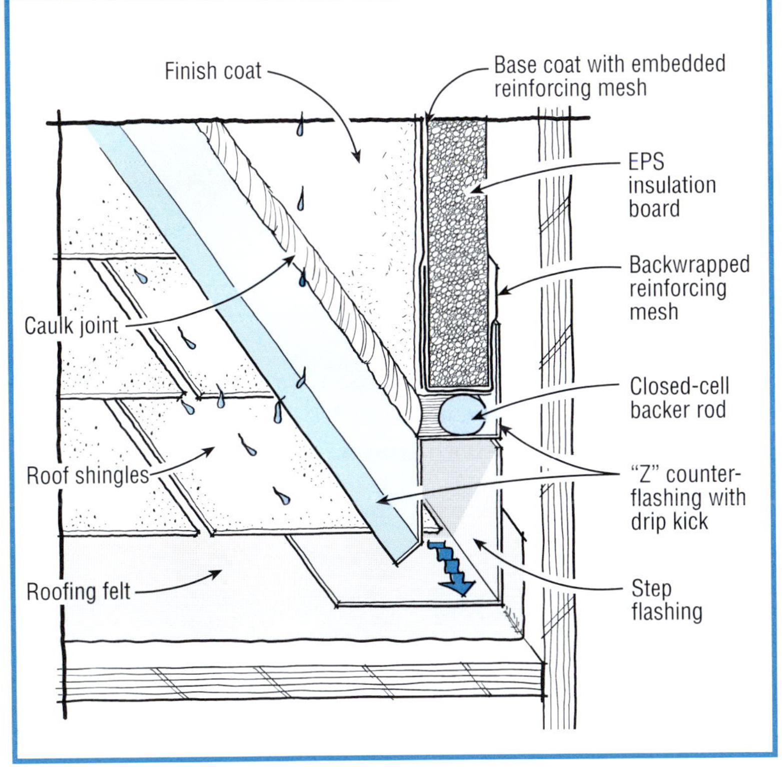

Special counterflashings at wall-roof junctions provide a critical space to caulk EIFS.

Figure 4-81. Front-Wall Flashings for EIFS

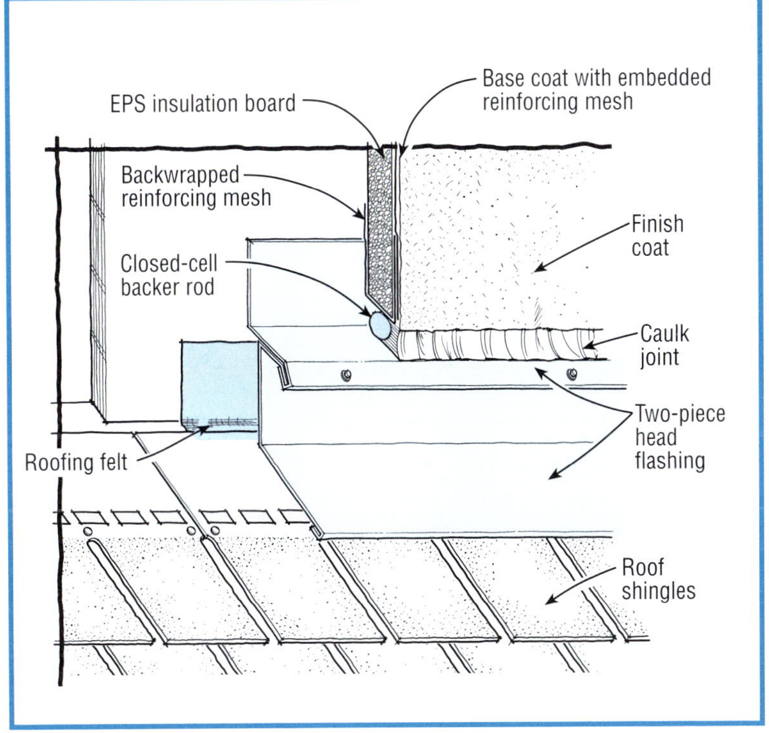

A two-piece front-wall flashing held together with sheet-metal screws allows the roof covering to be replaced without mangling the EIFS wall covering.

flush with the surrounding wall. After the base hardens, apply the matching texture.

For holes larger than 3 in., use a hammer and cold chisel to carefully chip a tapered edge around the perimeter of the hole. Tapered edges offer more surface area and produce a better bond. A back-cut taper will also lock in the patch. In addition, use wire lath to support the plaster. If possible, nail the metal lath to adjacent studs; otherwise, weave and tie it to the old lath.

Figure 4-82. Repairing Hairline Cracks

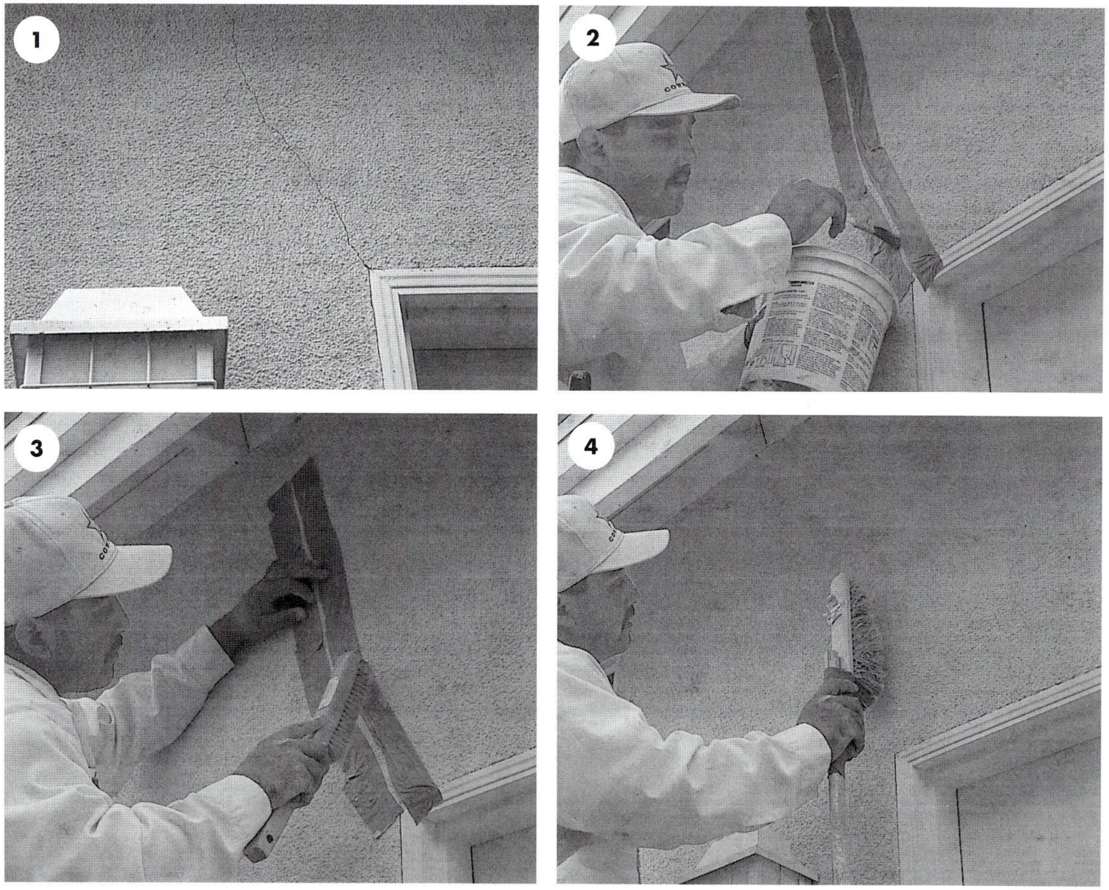

To patch hairline cracks: 1) mask both sides with tape; 2) dab the crack with dry stucco on a small brush; 3) brush off the excess to avoid ridges in the patch; and 4) remove the masking tape and brush again. Overnight condensation will provide enough moisture to cure the dry stucco.

For more information: "Patching Stucco," JLC 9/97

Patching Large Areas

To patch in a large area, the sheathing wrap and lath need to be repaired first. Use a piece of building paper slightly larger than the patch area and lap it to shed water (**Figure 4-83**).

After the scratch and brown coats have cured, repair the color coat by spraying a light color coating called a "fog coat." Fog consists of cement, color pigment, and lime, but no sand. If possible, use the original stucco manufacturer's fog coat. Since stucco darkens with age, you may have to add as much as 50% additional pigment if the stucco you're matching is older than six months.

When faced with a large patch like a patio door infill, chances are very high that both the color and texture of the new stucco will not match. Painting the stucco is the only way to assure uniformity of color.

Patching Stucco

Figure 4-83. Patching Three-Coat Stucco

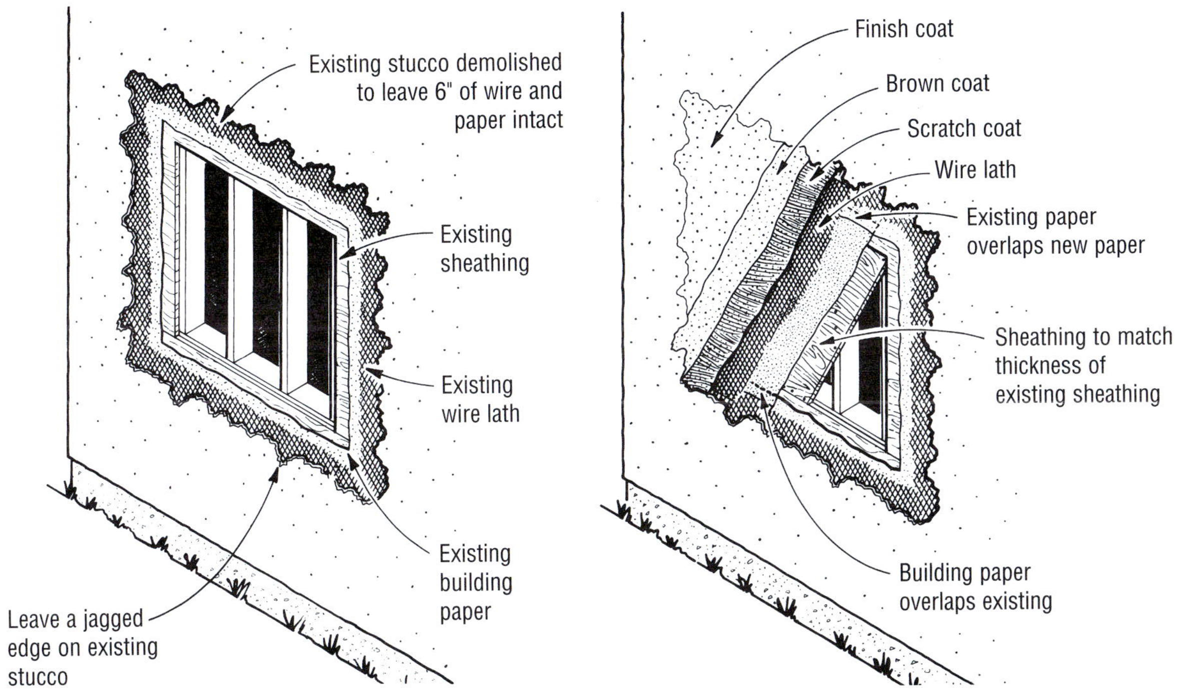

When patching stucco, chip back at least 6 in. to expose the existing lath and building paper (far left). The existing stucco should have a jagged edge to create a less noticeable transition to the new stucco. Match the existing thickness of the sheathing and make sure to lap the building paper so water can drain out (near left).

BRICK TYPES AND GRADES

Brick veneer is available in several types and grades, each of which differs in durability, appearance, and dimensional tolerances (**Figures 4-84** and **4-85**).

The quality of brick and mortar affects how watertight a wall will be. Softer porous brick absorbs water, while very hard-fired brick absorbs little water but may not bond as well with mortar. Without a good mortar bond, the wall may leak at the joints. Brick with about 6% to 9% water absorption (conforming to ASTM C216) will provide a good bond.

Avoid Used Brick

Under frost/freeze conditions, poor quality bricks will disintegrate rapidly. Old brick may also have contaminants on the surface that will prevent the mortar from bonding.

When laying brick, the mason should pull brick from several cubes at a time, not consume one cube and move onto the next. This will minimize the chance of creating blotchy walls due to normal variations in brick color.

MORTAR TYPES

Mortar bonds masonry units together and seals against air and moisture penetration. It also bonds with joint reinforcement, metal ties, and anchor bolts. For load-bearing applications, mortar performance is as critical as the brick units themselves.

No single mortar type is suited for all applications. For new brick veneer, Type N mortar provides the best combination of strength, workability, and economy (see "Mortar Types," **page** 57).

Figure 4-84. Brick Grades

	Exposure Grades	Use Conditions
SW	Severe Weathering	All exterior applications
MW	Moderate Weathering	Exterior applications in very mild climates
NW	Negligible Weathering	Interior application only
	Appearance Grades	**Tolerances**
FBX	Face Brick Extra	1/8-in. to 1/4-inch dimension variation
FBS	Face Brick Standard	3/32 to 5/32 dimension variation; minor chips, cracks possible
TBX	Thin Brick Extra	1/8-in. to 1/4-inch dimension variation
TBS	Thin Brick Standard	3/32 to 5/32 dimension variation; minor chips, cracks possible
HBX	Hollow Brick Extra	1/8-in. to 1/4-inch dimension variation
HBS	Hollow Brick Standard	3/32 to 5/32 dimension variation; minor chips, cracks possible

For more information: "Getting Started with Brick Veneer," JLC 11/97

When selecting mortar types, a good rule of thumb is to use the mortar with the lowest compressive strength that meets requirements (**Figure 4-86**).

MORTAR JOINTS

Examples of common mortar joints are shown in **Figure 4-87**. Each joint provides more than a different architectural appearance; it provides varying degrees of weather resistance, as well. Joints with ledges, such as flush, rake, and struck joints, tend to perform poorly in exterior applications

Selecting Brick

Figure 4-85. Common Brick Types

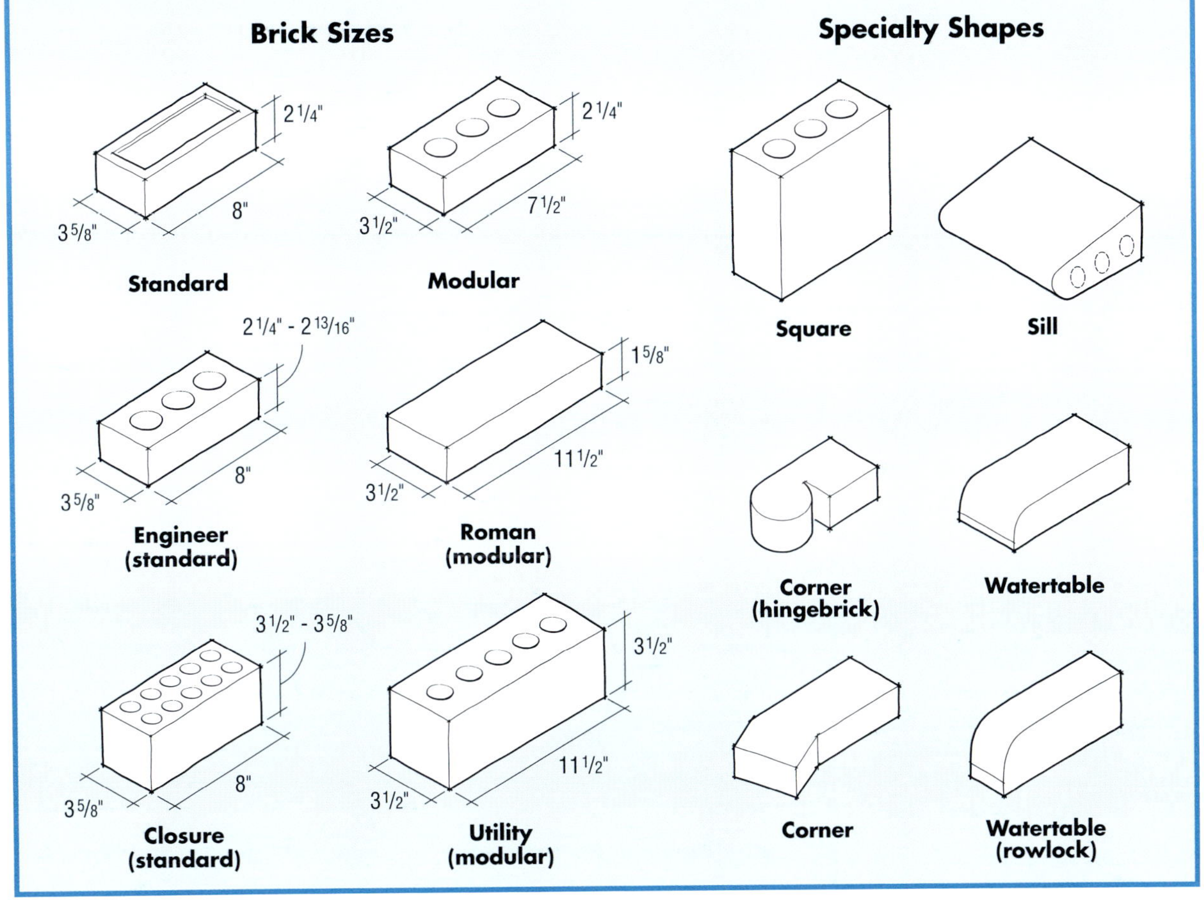

There are literally thousands of brick types to choose from in various shades and textures. Make sure that clients see and approve the brick type before ordering, and that the selected brick works well with the colors and textures of roofing, siding or stucco, window cladding, and trim.

and can allow moisture penetration. Tooled concave or vee joints are recommended for exterior application. The tooling compacts the mortar tightly, minimizing moisture penetration.

DRAINAGE PLANE

Because brick is porous, every brick-veneer wall should be designed around a drainage plane. Use well-detailed sheathing wrap over plywood, OSB, foam sheathing, or flexible flashing membrane to waterproof the framing behind it (see "Sheathing Wrap and Flashings," **page 296**).

Air Space

In commercial work, a 2-in. space is often specified. However, in residential work, a 1-in. gap is acceptable (**Figure 4-88**). With anything less, there's a good chance that mortar droppings will clog the air space. Be sure to keep the air space open at the top (**Figure 4-89**).

Weep Holes

Weep holes are an integral part of any drainage plane. In brick work, weep holes are usually formed by laying cotton rope or plastic tubing every 16 in., or by leaving an

Figure 4-86. Compressive Strengths of Mortar Types

Mortar Type	Average Compressive Strength at 28 Days (min. psi)
M	2,500
S	1,800
N	750
O	350

Figure 4-87. Mortar Joints

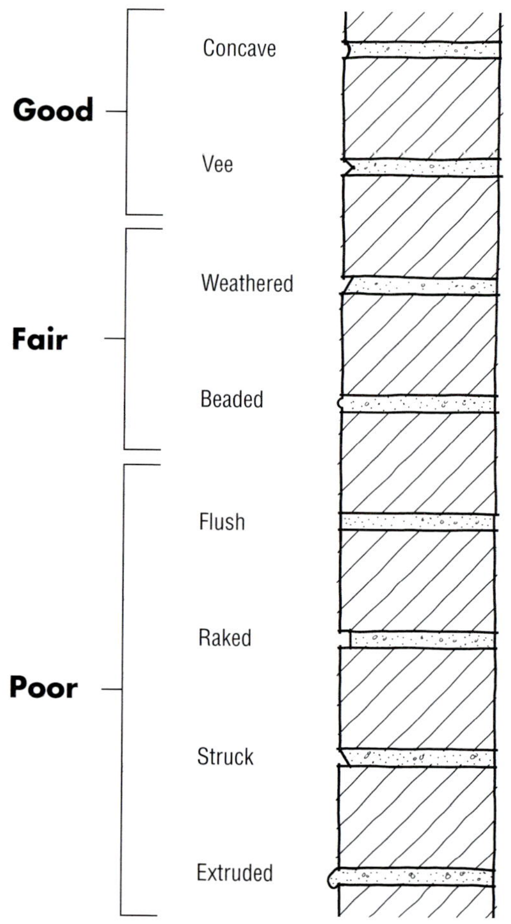

The style of the mortar joints in a brick wall influences how well the wall can repel water. Concave and vee joints provide the most weather-resistant joints in exterior walls.

empty head joint (no mortar at the end of the brick) every 24 in. (every third brick). In empty head joints, Cell Vent® (Dur-O-Wall, 630/375-6440, www.dur-o-wal.com) will prevent insects from building nests in the wall, and it looks better than an empty joint.

Figure 4-88. Drainage Plane for Brick Veneer

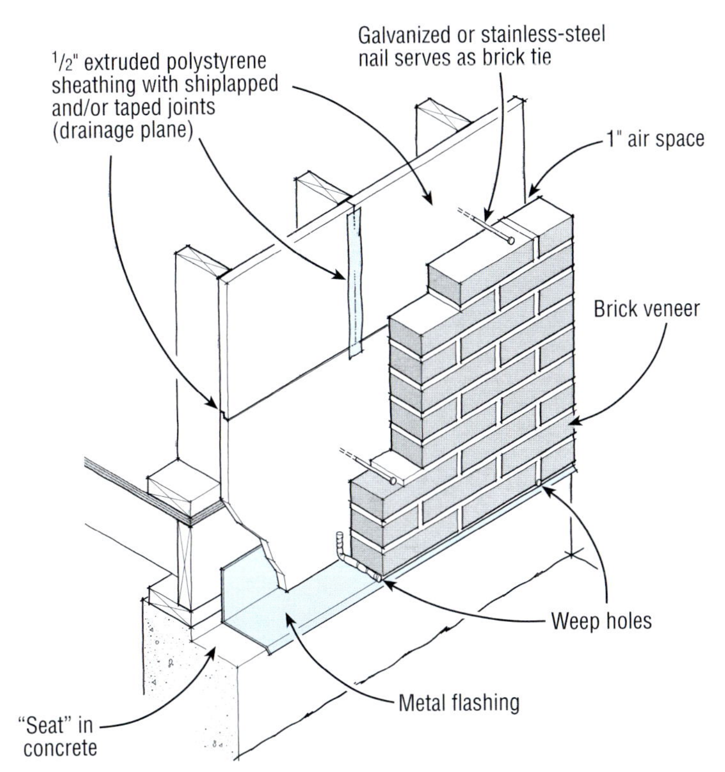

Brick is very porous and will leak. To prevent water damage to the wall, provide an air space and weep holes to drain water down and out.

STRUCTURAL SUPPORT

Brick veneer must be well supported at the foundation and lintels to prevent settlement. Problems will show up as the building settles at different rates. If differential settlement occurs, step cracks will follow the mortar joints and the

Brick Mortar

Mortar Joints

Drainage Plane for Brick Veneer

Figure 4-89. Drainage Plane Details

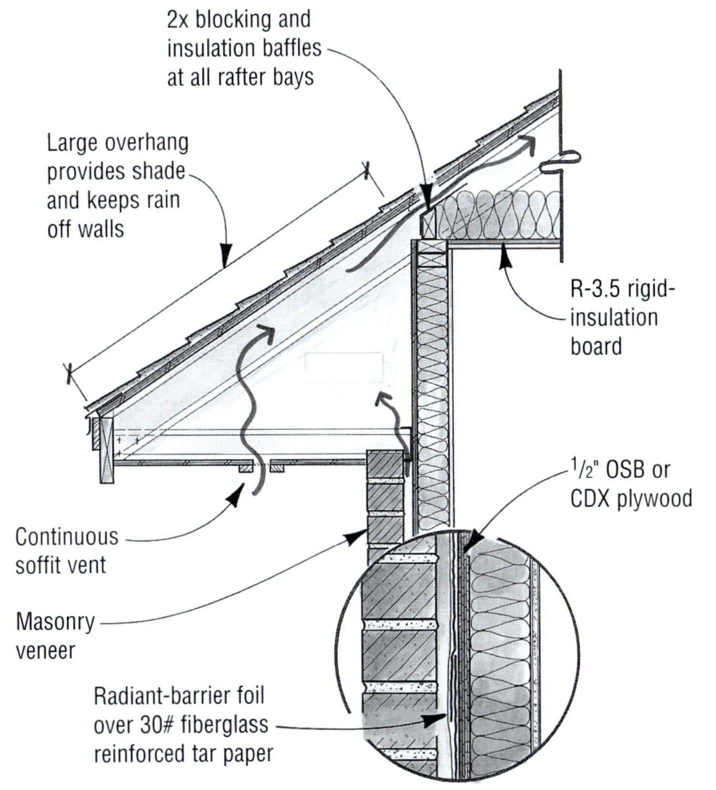

Do not block the air space behind brick veneer at the eaves. Air should be allowed to rise up and out to increase the drying potential of the drainage plane.

crack pattern will grow larger as it goes from the bottom to the top of the wall. To avoid problems:

- Support the brick on a stiff, stable ledge or angle bracket (**Figure 4-90**);
- Tie brick securely to the frame wall.

Steel Lintels

In residential work, 3x3 1/2 x1/4-in. steel angles will support brick headers for openings less than 6 ft. wide. For larger openings, such as garage doors, use a 5x3 1/2x1/4-in. angle. The angle should be laid across the opening and then flashed (**Figure 4-91**). This metal flashing should tuck under the sheathing wrap 2- to 4-in.

Generally, steel lintels are not attached to the framing. This allows the brick and wood framing, which have different rates of expansion and contraction, to move independently of one another. It is also important to leave space at the end of the lintel for expansion where it is supported on the masonry.

Brick Arches

For standard windows and doors, a segmented or a flat "jack" arch can be used without steel (**Figure 4-91**). Because the archwork transfers the weight of the wall above the opening to the walls on either side, make sure there is sufficient mass to the left and right of the opening to take the added load. A wide arch placed too near the corner of a house can blow out the corner.

Figure 4-90. Foundation Support for Brick Veneer

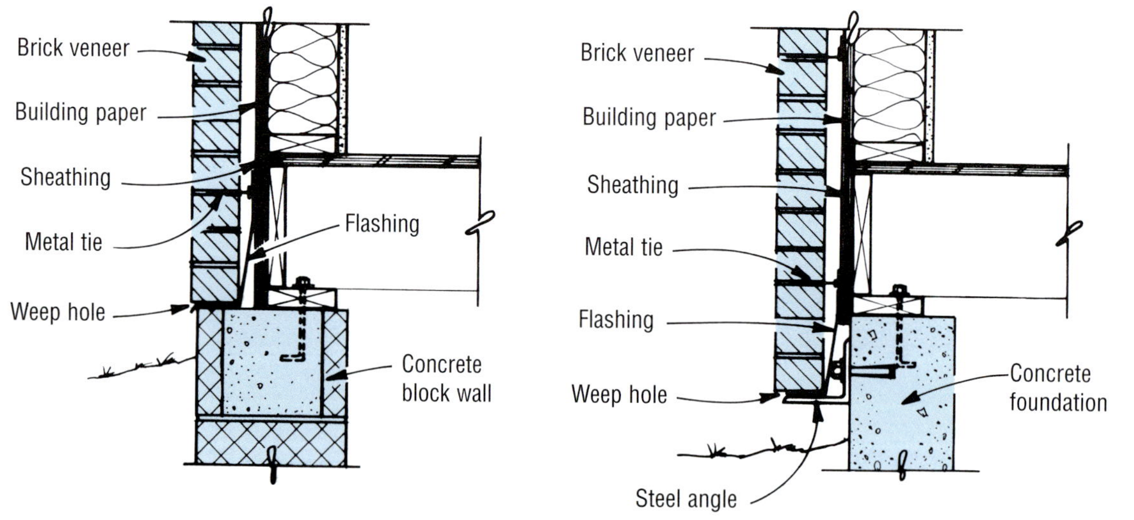

In new construction, build foundation walls wide enough to provide a 4-in. shelf for the brick (near right). A stainless-steel angle bolted onto an existing foundation (far right) can support a retrofit brick veneer wall up to 14 ft. high. The steel should be installed above-grade, and should be wide enough to support two-thirds the width of the brick.

FLASHING DETAILS

In veneer construction, flashing is required at all shelf angles, concrete foundations, and over door and window heads to interrupt the downward flow of water (Figure 4-92).

Flashing Materials

Flashings for brick were traditionally made from copper or lead sheeting. These materials perform well, but their use has steadily decreased because of the cost and the potential for staining and galvanic corrosion.

Figure 4-91. Brick Arches for Openings

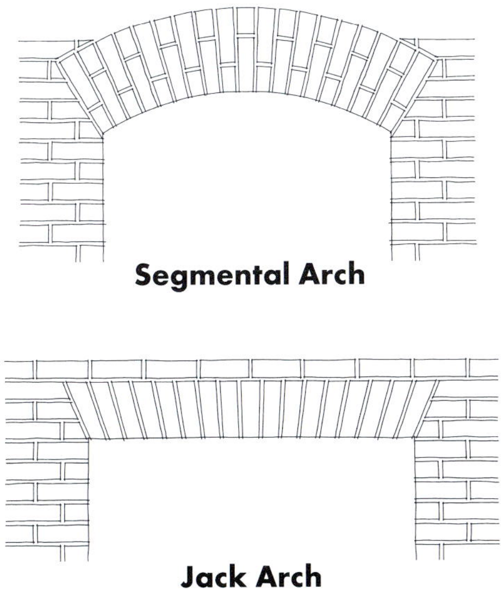

A segmented arch (top) can span an opening up to 36 in. wide. A jack arch (above) can span only about 30 in. without special engineering.

Figure 4-92. Details for Durable Brick Veneer

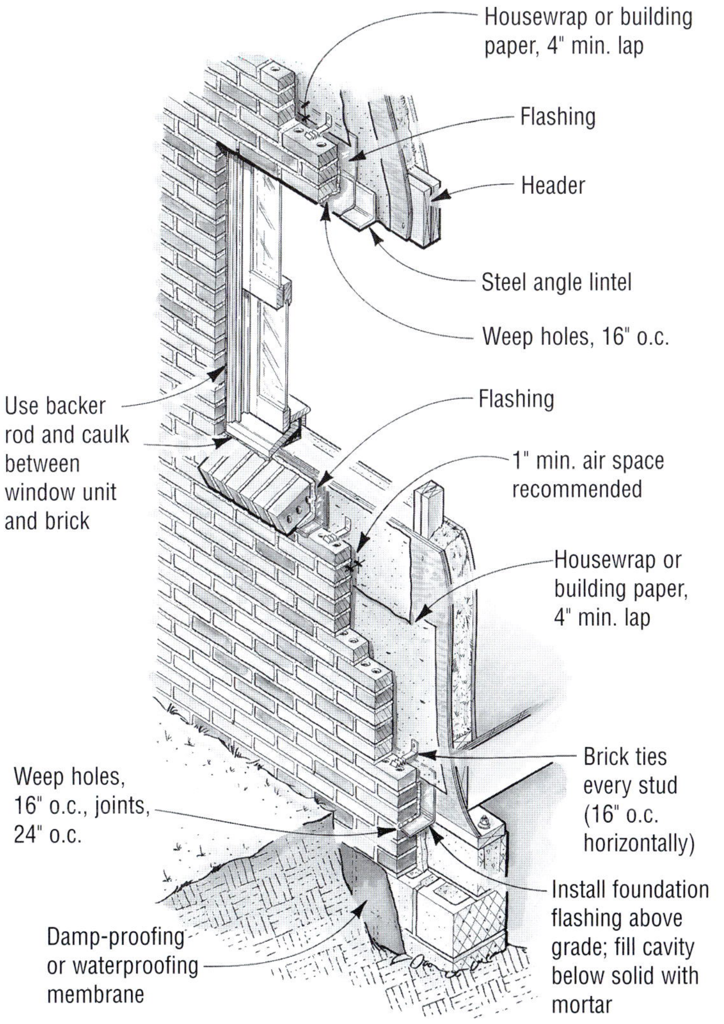

Flashing details are critical for the durability of brick veneer. For best results, use flexible flashings (see also ***"Flexible Flashing Materials," page 299****).*

Foundation Support

Window Arches

Flashing Brick Veneer

Aluminum flashings should be used only with extreme caution since the wet and alkaline environment of mortar can corrode the aluminum. Similarly, some galvanized flashings can corrode in fresh mortar, and the galvanized coating may crack during bending and handling.

Flexible flashings work the best. They are resilient, durable, and resistant to corrosion (see "Flexible Flashing Materials," **page 299**). Make sure that flexible flashing membranes are kept covered; UV light will degrade them.

Figure 4-93. Brick Ties

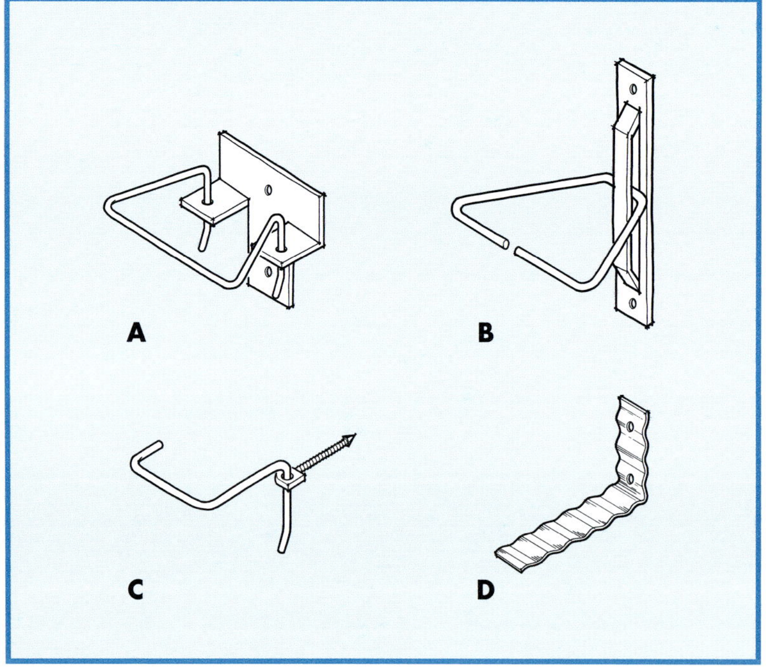

Brick ties come in a wide selection of styles. Those used most commonly in residential construction include U-tabs (A and B), Z-ties (C), and corrugated ties (D).

For more information: "Brick Veneer Basics," JLC 6/94

BRICK TIES

Brick ties transfer horizontal wind loads on the building to the structural wall framing and the foundation. The ties must be strong enough to resist tensile and compressive forces (**Figure 4-93**).

For low-rise construction, follow these rules of thumb:

Figure 4-94. Removing Mortar

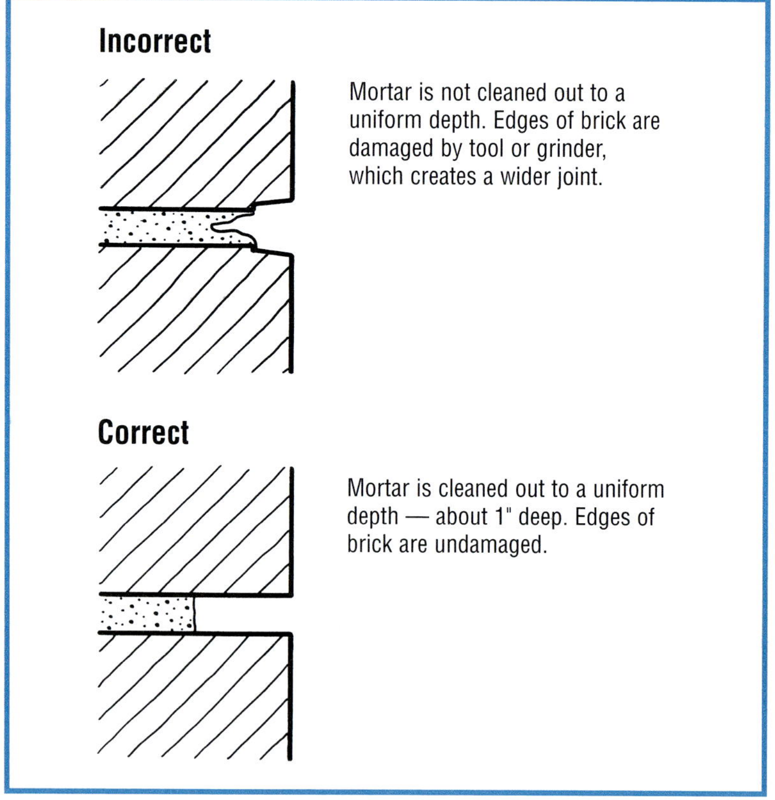

When raking joints, remove mortar to a uniform depth (above) for maximum bond of the new mortar, and take care not to damage the edges of the brick and create an uneven joint (top).

- Use one tie for every $3^1/4$ sq. ft. of wall area, with a maximum spacing of 24 in. on-center. Over a wood-framed wall, this translates to approximately one tie per stud every sixth or seventh course.
- Over a concrete masonry structure or steel-framed wall, use continuous horizontal joint-reinforcement with U-tabs or individual Z-ties.
- For brick veneer over wood framing, use corrosion-resistant corrugated metal ties, at least 22-gauge, $^7/8$-in. wide, and 6 in. long (**Figure 4-93**), or 20d galvanized spikes (**Figure 4-88, page 365**).

Brick Ties

EXPANSION JOINTS

Expansion joints in brick masonry will prevent cracking due to thermal movement, moisture-absorption, and load effects. Accumulated movements are usually not large enough in residential construction to warrant expansion joints. However, long walls (over 75 ft.) can have problems.

REPOINTING BRICK

Old brickwork should be repointed to protect it from further damage. Once mortar joints begin to fail, the deterioration can accelerate damage to the brick itself and increase the likelihood of leaks.

Repointing Brick

Raking Out Mortar

Old mortar should be removed (using a hammer, chisel, and a tuck-point grinder or a pneumatic chisel) to a minimum depth of two-and-a-half times the width of the joint. A depth of about 1 in. deep usually will ensure an adequate bond. Any loose or disintegrated mortar beyond this depth also should be removed. Clean joints carefully; damage to bricks will affect the appearance, and also can lead to accelerated weather damage (**Figure 4-94**).

Tuck-Pointing

- Dampen the brick and old mortar before filling the joint to control the rate of hardening, but avoid free water or excessive wetting. Too much water will delay the tooling

Figure 4-95. Lime vs. Cement Mortar

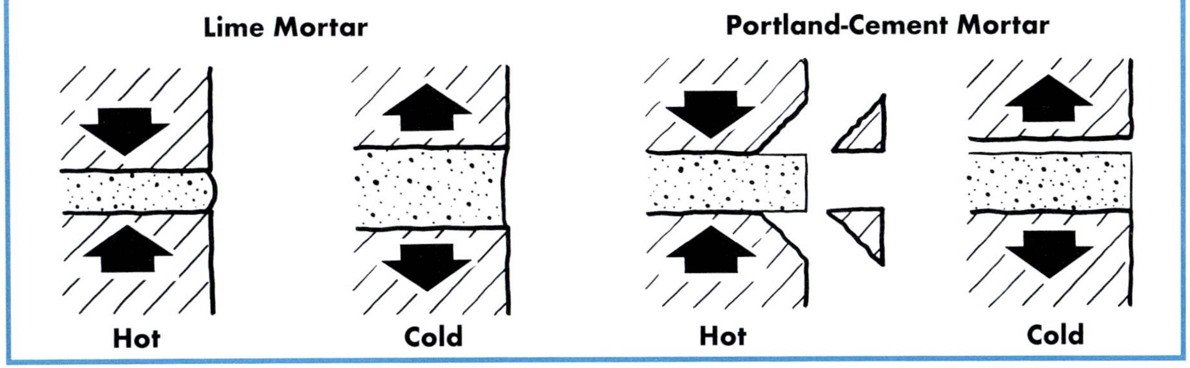

With old brick that is soft, it's important to use a soft lime mortar, which will flex with changes of temperature. If a hard, Portland-cement mortar is used, the brick will spall or joints will open up when the brick moves.

or cause excessive shrinkage; too little water will reduce bond strength.

- Use only prehydrated lime mortar for tuck-pointing. A hard, Portland cement mortar may cause the brick to spall (**Figure 4-95**). To prehydrate, thoroughly mix dry mortar ingredients. Then add just enough water to produce a very stiff, unworkable mix (it should retain its shape only when pressed into a ball). Let stand for about two hours, and then add enough water to bring it to a good, workable consistency. This will be somewhat stiffer than conventional masonry mortars and will be much easier to work into joints.
- Where existing mortar has been removed to a depth greater than 1 in., compact the new mortar in several 1/4-in.-thick layers to reduce overall shrinkage. Allow each layer time to harden before applying the next layer.
- When the mortar is thumb-print hard, the joint should be tooled to match the existing joint (**Figure 4-96**). Proper timing of the tooling is important. If tooled when the mortar is too soft, the color will be lighter than expected, and hairline cracks may occur. If tooled too hard, there may be dark streaks called "tool burning," and the mortar will not bond tightly to the brick.

Figure 4-96. Tooling Joints

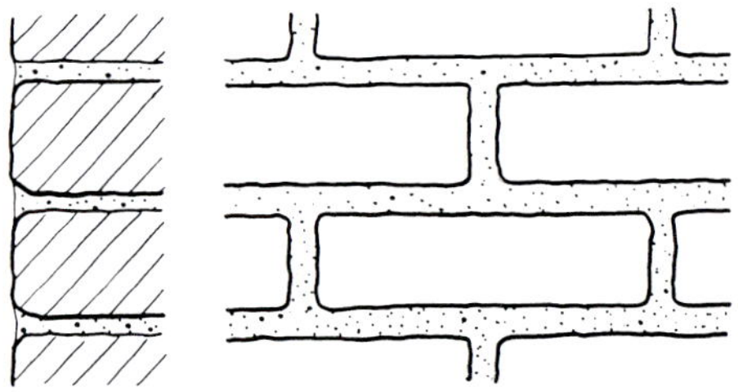

Incorrect
Joints are filled too full. Wide feather edge is susceptible to spalling.

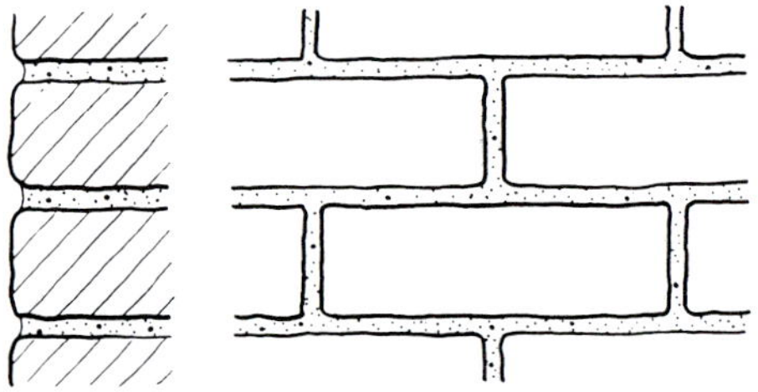

Correct
Joints are slightly recessed, which makes a durable joint.

Do not finish mortar joints flush. The thin edges of the mortar where it meets the weathered brick will be weak and may fail.

For more information "Restoration Primer: Repointing Brick Masonry," JLC 7/90

"Restoration Primer: Repointing Pointers," JLC 8/90

Cleanup

If repointing work is done carefully, the only cleaning required after tooling will be a small amount of mortar brushed from the edge of the joint with a stiff bristle brush. This is done after the mortar has dried, but before it is fully hardened (one to two hours). Mortar that has hardened can usually be removed with a wooden paddle or, if necessary, a chisel.

Efflorescence, or "bloom," may appear within the first few months after repointing, but it usually disappears through the normal weathering process. If natural processes do not remove the efflorescence, the safest way to remove it is by dry-brushing with stiff natural- or nylon-bristle brushes and water. Usually, muriatic acid is ineffective and it should be avoided. In fact, it can deposit salts, which can lead to additional efflorescence.

WATERPROOFING DETAILS FOR EXTERIOR TRIM AND SIDING should rely on sheathing wrap and flashing (see **pages 296-306**) and good design principles (see "Design Principles of Exterior Woodwork," **page 310**), not on caulk. Caulk is only a first-line defense. Caulk joints accommodate dimensional changes due to climate and settlement, but they should never be the sole means of keeping water out of the wall.

SELECTING CAULK

Selecting Caulk

There are four types of polymer used to manufacture most high-performance caulks: silicones, polyurethanes, latex acrylics, and the solvent-based "block copolymer" synthetic rubbers. The best type of caulk for the job depends largely on the materials to which the caulk must bond. For the most part, polyurethane caulks provide the best all-around performance for exterior applications, but they are more expensive than most other caulks and can be difficult to clean up (**Figure 4-97**).

JOINT DESIGN

Caulk Joint Design

If the joint sealant is too shallow or deep, it will not expand and compress with the movement of the joint. Instead, it may split or rip away from one of the joint surfaces.

Shape

Ideally, caulk joints should be configured to span a gap that is half as deep as it is wide (**Figure 4-98**). The caulk should not bond to the back (three-sided bonding), but only to two sides, so it can flex as it expands and contracts.

Corner fillets do not move freely and generally peel off the

Figure 4-97. Exterior Caulk Selection Guide

Type	Joint Movement	Life Expectancy	Comments
Silicone	50%	20 yrs.	Very low shrinkage; may be applied over the widest range of temperatures; not paintable; poor adhesion to damp surfaces and to masonry
Polyurethane	25%	20 to 30 yrs.	Best all-purpose caulk; excellent adhesion to most materials; difficult to clean up
Ethylene copolymer	25%	18 to 20 yrs.	Reasonable adhesion to most materials; can be painted
Polysulfide	12 to 25%	20 yrs.	Difficult to handle; may require primer on porous surfaces; wear gloves and maintain ventilation during application
Acrylic latex with silicon	10%	10 to 20 yrs.	Easy to work with; cleans up with water
Acrylic latex	2%	4 to10 yrs	Interior application only; easy to work with; cleans up with water
Butyl rubber	5 to 10%	4 to10 yrs	Poor adhesion on damp surfaces; attracts dirt; difficult to handle
Oil-based	1%	3 to 5 yrs.	Not recommended

two surfaces. If a joint depends on a caulk sealant, open it up slightly to allow caulk to fill the gap. If a corner is tight, don't caulk it; the caulk will fail faster than any finish applied to it and will only complicate refinishing.

BACKER ROD

Backer rod should be used to fill large gaps and control the depth (**Figure 4-98**). Most backer rod today is made out of foam and comes in diameters from 1/4 in. to 2 in. and up. Choose a backer rod that's just a little larger than the joint so it will stay in place and allow you to control the depth of the sealant.

Foam backer rod comes in open-cell and closed-cell form. Open-cell backer rod compresses more and is easier to work with, but it's like a sponge and can absorb water. In critical areas that may absorb moisture from the substrate, use a closed-cell material.

Figure 4-98. Proper Joint Design

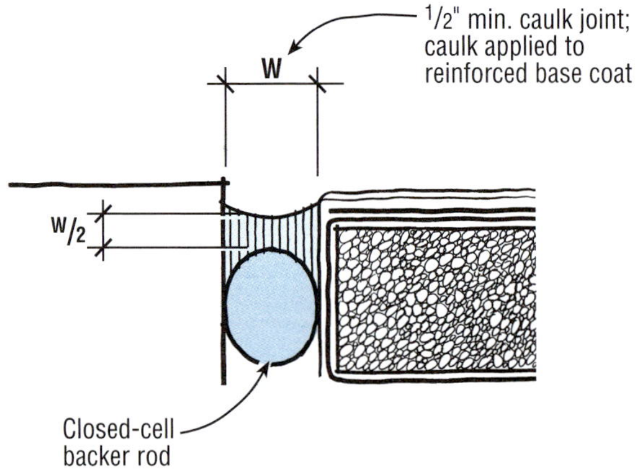

A good caulk joint should bond to only two sides, not to the back. Use backer rod to control the depth, which should be half the width of the joint.

Bond-breaker tape. The main purpose of backer rod is to keep sealant off the back of the joint, thus preventing three-sided bonding. If the joint is too shallow to fit backer rod, apply bond-breaker tape to the back of the joint.

APPLYING CAULK

The fine print on the label of most caulk tubes reads: "Apply to clean, dry surfaces. Application temperature: 40°F to 80°F." However, construction sites are never clean, seldom dry, and work doesn't stop when the temperature drops below 40°F. To promote good adhesion, prepare the surface using the following strategies:

- Scrape, chip, brush with a stiff-bristle brush, and dust the inside of the joint, or even blow it out with compressed air.
- Wipe the surface with a strong solvent, such as acetone or methyl ethyl ketone (available in hardware stores and contractor supply houses). These solvents evaporate water and dissolve greasy substances that can retard adhesion.
- Every caulk joint should be tooled. The best tool is a steel spatula (or sculptor's tool) that is slightly smaller than the width of the joint. Try to achieve a smooth, even appearance without squeeze-out on the sides.
- Cut the nozzle of the tube at an angle slightly smaller than the size of the joint.
- In cold weather, keep the caulk warm. If necessary, use an electric heat chest to store tubes. Also, it may be necessary to heat the surface with a hair dryer or a heat gun (just be careful not to strip the paint or start a fire).

Applying Caulk

Paints and Stains

SUCCESS WITH PAINTED SIDING AND TRIM STARTS WITH AN understanding of how the siding will perform under local weather conditions. Next, it's necessary to match the finish to the siding — a given finish performs differently on different types of siding.

PAINT BONDING CHARACTERISTICS OF WOOD

Avoiding paint problems begins with selecting the best wood available (**Figure 4-99**). When choosing softwood for exterior siding and trim, begin by looking at the following characteristics:

- **Wood density.** The density of wood is one of the most important factors affecting paint life for a simple reason: "Heavy" woods shrink and swell more than "light" woods (**Figure 4-99**). Excessive dimensional change constantly stresses paint or a solid-color stain, and may result in early failure. Finishes that don't form a film, such as penetrating stains, are not affected by these dimensional changes.
- **Flat-grain vs. edge-grain.** Flat-grained lumber shrinks and swells more than edge-grained lumber, so edge-grained (or vertical grain) lumber will usually hold paint better than flat-grained material. Most standard lumber grades contain a high percentage of flat grain.
- **Earlywood and latewood.** Earlywood and latewood can be seen as two distinct bands in wood. Latewood is denser, harder, smoother, and darker than earlywood. Although new paint or solid-color stain will adhere well to both earlywood and latewood, old alkyd paints and solid-color stains that have become brittle with age and weathering will peel sooner from the smooth, hard surface of the latewood.
- **Heartwood and sapwood.** Mature trees have a darker central column of wood, called heartwood, surrounded by a lighter cylinder of wood, called sapwood. Heartwood is impregnated with extractives, pitch, and oil, which give the heartwood of some species — such as redwood, cedar, and cypress — a natural resistance to decay and insects. Extractives, however, can sometimes

Figure 4-99. Paint Holding Ability of Selected Softwoods

	Weight per Cu. Ft. at 8% Moisture Content	Paint-holding Ability (I best, IV worst)
Western red cedar	22.4	I
Redwood	27.4	I
Eastern white pine	24.2	II
Ponderosa pine	27.5	III
Western hemlock	28.7	III
Spruce	26.8	III
Douglas fir	31.0	IV
Red pine	30.8	IV
Southern yellow pine	38.2	IV

Lighter-weight woods shrink and swell less than denser woods, so they tend to hold paint better. Besides being less dense, redwood and cedar have more narrow bands of latewood compared with Southern yellow pine and Douglas fir, which are higher in density and have wide bands of latewood.

cause discoloration problems as they dissolve in water and are transported to the wood surface.

- **Knots and other irregularities,** such as bark, splits, and pitch pockets also affect paint adhesion. Knots are mainly exposed end grain, which absorbs more finish than flat- and edge-grained lumber. In pine, knots often contain a high percentage of resin, which may cause the paint over the knot to discolor. Large knots usually check and crack. To reduce the brown stain over knots, apply a primer, such as a pigmented shellac, to block the extractives, and then follow with two top coats.
- **Moisture content** is critical in determining the service life of paint. Allow wood to acclimate to site conditions (see "Conditioning Siding," **page 318**). Wood above 20% moisture content should never be painted, as the paint will most likely peel.
- **Surface texture.** Paint lasts longer on smooth, edge-grained surfaces than on smooth, flat-grained ones. However, paint will last longest on rough-sawn or rough-sanded wood, whether the wood is edge-grained or flat-grained. Sand smooth siding with 60-grit paper before painting.

PAINT VS. PRIMER

A primer contains a higher ratio of binder to pigment than a topcoat. This forms a protective surface film, seals in extractives, and keeps the wood from unevenly absorbing the topcoat.

For more information: "Making Paint Stick to Wood Siding and Trim," JLC 9/99

The topcoat, on the other hand, contains more pigment but less binder. This is necessary to shield the wood from the sun's ultraviolet rays.

Water and ultraviolet light are the biggest enemies of exterior wood, but a good paint job will protect against both. Binder seals out water, while the pigments form a shield against ultraviolet light. No single coating has enough binder and pigment to do both. However, the combination of primer and topcoat forms the most durable finish.

- **Repainting.** Where the existing paint is chalky, oily, or poorly adhered, use an oil-based or alkyd primer. Ideally, such areas should be sanded to bare wood before priming; but when this is not practical, an oil-based primer will provide the most secure base for the finish coat.
- **Priming tannin-heavy woods.** Oil-based primers also perform better in blocking the tannin stains characteristic of cedar and redwood.
- **Back-priming.** Paints do not penetrate the wood deeply, but form a film on the surface. This film can blister or peel if the wood becomes wet. Without back-priming, the back of the siding absorbs more moisture than the front, leading to cupping, warping, or even paint failure as the moisture looks for a way to escape.

OIL-BASED VS. LATEX PAINTS

The most durable house paints are the all-acrylic latex paints. Although oil-based paints usually provide the best protection from liquid water and water vapor, they are not necessarily the most durable because they become brittle over time. Latex paints, particularly the acrylic paints, remain flexible with age. Even though latex paints allow more water vapor to pass through, they hold up better by swelling and shrinking with the wood.

Figure 4-100. Suitability and Expected Life of Exterior Wood Finishes

Selecting Exterior Finishes

Type of Exterior Wood Surface	Water-Repellent Preservatives and Oil		Semi-Transparent Stain		Paint and Solid-Color Stain		
	Suitability	Expected Life (years)	Suitability	Expected Life (years)	Suitability	Expected Life (years)	
						Paint	Solid-color Stain
SIDING							
Cedar & redwood							
Smooth (vertical grain)	High	1-2	Moderate	2-3	High	4-6	3-5
Roughsawn	High	2-3	High	5-8	Moderate	5-7	4-6
Pine, fir, spruce							
Smooth (flat-grained)	High	1-2	Low	2-3	Moderate	3-5	2-4
Rough (flat-grained)	High	2-3	High	4-7	Moderate	4-6	3-5
Plywood (Douglas fir & Southern Pine)							
Sanded	Very Low	1-2	Low	2-4	Moderate	2-4	2-3
Textured (roughsawn)	Low	2-3	High	4-6	Moderate	4-6	3-5
Medium-density overlay	–	–	–	–	Excellent	6-8	5-7
Hardboard, medium density							
Smooth or Textured	–	–	–	–	High	4-6	3-5
MILLWORK (often pine)							
Windows, shutters, doors, exterior trim	High	–	Moderate	2-3	High	3-6	2-4
DECKING							
New (smooth)	High	1-2	Moderate	2-3	Very Low	2-3	1-2
Weathered (rough)	High	2-3	High	3-4	Very Low	2-3	1-2

Note: These data were compiled from the observations of many researchers. Expected life predictions are for one and two coats of each finish at an average location in the continental United States. Expected life will vary in extreme climates or exposure, such as desert, seashore, and deep woods.

GLOSSY VS. FLAT PAINT

Paints are available in different degrees of gloss, including flat, matte, semi-gloss, and gloss. Generally, high-gloss paints contain more paint resin and less pigment, and perform better and last longer than the low-gloss or flat paints. Flat paints tend to pick up dirt and absorb water more readily than the high-gloss paints. Because of this, mildew growth is often greater on the flat paints.

SOLID-COLOR STAINS

Solid-color stains are essentially thin paints, not true stains. Solid-color stains have a higher concentration of pigment than semitransparent penetrating stains but a somewhat lower concentration of pigment than standard paints. So solid-color stains cover the wood's natural color while retaining the wood's surface texture.

Solid-color stains will not last as long as paints (**Figure 4-100**), but often they are the finish of choice on textured or rough-sawn siding products. They also can be applied over existing paints and solid-color stains, and normally leave a flat finish appearance. Like paints, solid-color stains protect wood against UV degradation.

CLEAR EXTERIOR FINISHES

For more information: "Latex Enamel Problems and Solutions," JLC 1/97

A penetrating water-repellent preservative contains a preservative (a fungicide), a small amount of wax (or similar water repellent), a resin or drying oil, and a solvent (such as turpentine, mineral spirits, or paraffinic oil). Some may be lightly pigmented, and waterborne formulations are also available.

Unpigmented, or clear, exterior finishes provide minimal protection for wood. They can reduce warping and checking, prevent water staining at the edges and ends of wood siding, and help control mildew growth. Wood treated with preservative is easily refinished and usually requires minimal surface prep.

Paintable water-repellent preservative (such as DAP Woodlife II®,) is a good treatment for bare wood before priming and painting. This pretreatment keeps rain or dew from penetrating the wood, especially at joints and on end-grain, thus decreasing the shrinking and swelling of the wood. As a result, less stress is placed on the paint film and its service life is extended.

Penetrating Oils

Most penetrating oil-based and alkyd-based finishes available contain linseed or tung oil. These oils must have a mildewcide; otherwise, they serve as a food source for mildew. Oils also perform better if a water repellent, such as wax, is included in the formulation. All these oil systems will protect wood, but their average lifetime may be only one to three years.

SEMI-TRANSPARENT STAINS

Semi-transparent penetrating stains are pigmented water-repellent preservatives with a high resin content. Their life expectancy may vary from three to six years, depending on the texture of the wood surface and the quantity of stain applied.

Solvent-borne stains (oil- or alkyd-based) penetrate the wood surface to a degree, are porous, and do not form a surface film like paint. They will not trap moisture that may encourage decay, and they will not blister or peel even if moisture penetrates the wood. Better-quality solvent-borne penetrating stains contain a fungicide (preservative or mildewcide), a stabilizer to protect against ultraviolet radiation (an absorber), and a water repellent.

Waterborne stains (latex-based) do not penetrate the wood surface like their oil-based counterparts. Newer latex formulations are being developed that may provide some penetrating characteristics.

LANDSCAPE PROTECTION

Protecting Landscapes

To protect any plants during a job, saturate the ground and soak all the leaves with water. Then cover the plants with woven poly tarps to shed the bleach solution. Do not use plastic sheeting; the plants must breathe.

After spraying, rinse everything (including the windows and woodwork) thoroughly with water. The siding should dry for two or more days before applying the clear finish.

When using a cleaning solution containing bleach, cover any brass, copper, or aluminum fixtures to avoid corrosion. Also cover any stained or painted trim.

SURFACE PREP

Water-based finishes require more thorough surface preparation, cleaning, and careful application than oil-based finishes. Remove any loose paint, and clean any dirty or oily surfaces.

Cleaning Siding Surfaces

Dirty siding will require a good cleaning (a detergent wash followed by power washing and thorough drying) to remove chalk and dirt. For most surfaces, use Spic-and-Span® powdered cleaner. On a very oily surface, use painter's naptha.

Filling Nail Holes

Filling Nail Holes

Use hot-dipped galvanized finish nails for all trim work that will be filled. The material used to fill nail holes depends on the time available and the type of exterior finish:

Linseed oil-based filler, such as DAP Painter's Putty® (Dap Inc.; 800/543-3840, www.dap.com), works well under an oil-based primer. Add "whiting" (a thickening powder available through paint suppliers) to this somewhat gooey putty to make it more workable. However, it's important that the putty be allowed to dry for a few days before primer is applied.

Exterior spackle. If an oil or latex primer will be applied the same day, use an exterior spackle, such as UGL 222 Spackling Paste (United Gilsonite Laboratories; 800/272-3235, www.ugl.com). Exterior spackle shrinks as it dries, but the hole can be "overloaded" to account for this shrinkage. Sand any proud material flush after it dries.

Epoxy. Where a latex primer will be applied immediately after the holes are filled, use a quick-hardening, two-part filler, such as Bondo®.

APPLYING EXTERIOR FINISHES

Different finishes require different amounts of finish (**Figure 4-101**).

Spraying vs. Brushing

While spray guns deliver paint quickly, it is nearly impossible to spray the paint uniformly on uneven surfaces. The best solution for almost any siding is to use a sprayer to get the paint on the wall, and then brush it out evenly and on all surfaces and work it into joints.

Whether brushing, rolling, or spraying, keep a wet edge to prevent lap marks. This may mean painting smaller areas at a time.

Applying Stain and Clear Finishes

For the longest-lasting results, install the siding with the rough side out. Then apply two coats of a lightly-pigmented, semitransparent, oil-based stain that contains a water repellent and a preservative or mildewcide. Apply the first coat and let it soak into the wood 20 to 60 minutes and then apply the second coat. If you allow the first coat to dry, the second coat cannot penetrate into the wood. About an hour after applying the second coat, use a cloth, sponge, or dry brush to remove any excess stain. Otherwise, the stain that does not penetrate into the wood will form an unsightly film and glossy spots.

PRESERVING A NATURAL WOOD FINISH

Wood siding turns gray because of two factors: the degradation by sun and water of the outermost layer of wood cells, which turn gray as their natural oils dry out, and tiny mildew spores that grow on the wood's surface.

Clear finishes with UV inhibitors will slow the weathering

Figure 4-101. Application of Exterior Wood Finishes

	Initial Application		Maintenance Application	
Finish	**Prep**	**Amount**	**Prep**	**Timing**
Water-Repellent Preservative	Apply to bare wood only	1 coat prior to painting; avoid wax build-up on surface	Brush to remove surface dirt	1-3 yrs.
Oils (with or without preservative)	Apply to bare wood only	2 coats	Pressure wash	1-3 yrs. or when preferred
Semi-Transparent Stain	Apply to bare wood only	2 coats; use caution to avoid lap marks	Pressure wash	3-6 yrs. or when preferred
Clear Varnish	Apply over 1 coat semitransparent stain to extend service life	3 coats	Clean, sand and stain bleached areas; apply 2 more coats	2 yrs. or when finish breaks down
Paint and Solid-Color Stain	Apply over 1 coat water-repellent preservative and 1 coat primer	2 top coats	Clean and apply top coat, or scrape to remove and repeat initial application	7-10 yrs. for paint; 3-7 yrs. for solid-color stain

For more information: "Keeping the Natural Look of Wood Siding," JLC 4/92 process, but will need to be reapplied every two years or so. Eventually, the color will change. After that, the only option is to approximate the color of new wood by putting on a cedar-tone stain.

Restoring Natural Color

If wood siding is weathered and discolored, the new look can be restored temporarily by cleaning off the dirt and mildew with a solution of one-third cup liquid household detergent (be sure it is ammonia-free), one quart liquid household bleach (containing 4% to 5% sodium hypochlorite), and three quarts warm water. Follow this up with a water rinse, and then use an oxalic acid bleach solution made with about a half pound of oxalic acid per gallon of water. Be sure to rinse with water again. This oxalic acid bleach solution will draw out the tannins in the wood and revive the orange tone of the cedar. Again, be sure to cover plants to protect them from the bleach (see "Landscape Protection," page 377).

FINISHES FOR SHINGLES AND SHAKES

Cedar shingles are naturally resistant to insect damage and decay, so they can be left unfinished to weather naturally. However the "natural" look may be less uniform and darker than desired, depending on the wood and its exposure to sun and moisture. Other finishing options include paints, stains, bleaching oils, and various clear preservatives.

Left unfinished, cedar shingles will weather to colors ranging from silver gray to dark brown and even black where there is excessive moisture and limited sunshine. Redwood left unfinished tends to turn black. To ensure a uniform weathered appearance, apply a bleaching oil. A lightly pigmented stain will produce a similar result, offering protection while allowing the natural characteristics of the wood to show through.

Applying Finish

Other options include heavy-bodied stains or traditional paints, which should be applied only to thoroughly dry material. The texture of a shingle provides good adhesion for paints and stains.

PAINTING STUCCO

When painting stucco:

- Newly applied stucco should be allowed about 30 days to cure before painting, depending on the weather conditions.
- Existing stucco surfaces should be power-washed and allowed to thoroughly dry before painting. Do not use a water blaster, as the pressure can do severe damage to the stucco.
- Stucco should be painted only with acrylic-based paints with adequate vapor transmission characteristics, as these prevent new moisture-related problems within the house.

PAINTING FIBER-CEMENT

Fiber-cement siding does not expand and contract with humidity changes like wood siding does, so it holds paint very well. If the siding is already painted and the paint is in good condition, clean the surface, and then apply at least two coats of high-quality acrylic latex paint.

Priming may not be necessary if the old paint is in good condition (free from peeling, cracking, flaking, etc.). However, if the old paint is glossy, or there are bare spots, it may be best to prime first. Match the primer to the paint of the same brand.

PAINTING ASBESTOS SIDING

Old asbestos-cement shingles behave like fiber-cement. In general, they take and hold paint well. If the shingles are in good shape, follow the guidelines for fiber-cement.

If some of the siding is deteriorated, showing loose fibers or broken edges, special precautions may be required. Aggressive scraping or sanding of loose paint is not advised since this could release asbestos fibers.

A
Anchors
  foundation 48-51, 85, 116
  framing 112, 113, 116, 117, 144, 145, 155
  retrofit 48-51
  seismic 48, 165-170
  tile roof 236-240
Adhesives 122, 266
Angle of repose 12
Asphalt
  driveways 23, 24
  felt . . . . 296-299 (*see also* Underlayment)
  roll roofing . . . . 263, 264, 268, 269 (*see also* BUR)
  roof shingles 210, 212-221, 274-276

B
Backfill . . . . 4, 8, 72, 73, 85 (*see also* Soils)
Balloon-frame 174
Beams 98, 103, 105, 106, 136-138, 145, 150, 153, 179-183
Block, concrete 56-65, 66
  mortar 57, 58
  setting 60-65
  grouting 60, 61
  reinforcing 63-65
Blocking 148, 169
Board feet 97
Board siding 289-291, 318-325
Bracing
  foundations 54, 72
  roofs 144-147
  trusses 131, 132, 159-164
  walls 135, 137
Brick veneer 362-370
  estimating 295
BUR (built-up roofing) 262, 268-269

C
Cantilevers 120, 121, 128, 173
Caulk . . . . see Sealants
Clay tile . . . . *see* Tile, roofing
Collar ties 144-146
Compacting soil 9, 10
Concrete 24, 26-37, 38, 52, 74-80, 92, 181
  admixtures 30
  curing 33, 34
  finishing 32
  placing 31, 44
  poured walls 52-55, 92
  ready-mix 26
  water-cement ratio 26, 27
  weather conditions 35-37
Concrete block . . . . *see* Block
Concrete tile . . . . *see* Tile, roofing
Control joints 55, 65, 79-81
Corner boards 312, 313, 327, 338, 344
Cornices 314-316, 334, 335
Crown, exterior 315, 316

D
Dampproofing 66, 67
Design values 101
Doors
  flashing 301-304
  headers 98, 136-138, 189-194
  trim 310, 311, 335-337
Dormers 150-152, 278-280
Drainage
  driveway 18-20, 22
  foundation 68-71, 88
  soil 8
  surface 70, 71

INDEX

Drainage plane 296, 301, 302
  for brick veneer 364, 365
  for shingles and shakes. . 326
  for stucco 346, 347
  for wood, vinyl siding . . 321, 322, 325, 333, 344
Driveways 18-25
  asphalt. 23, 24
  concrete. 24, 25
  gravel 19-22
  temporary 16

E

EIFS 357, 358
EPDM, roofing 262-268
Eaves. 148, 154, 155, 157
Engineered wood 104-106
  beams and headers. 106
  floors. 112-113, 123-128
  rafters 153-155
  studs 105
Excavating. 11-14
Expansive clay 3-7, 84

F

Fascia 148, 149, 155, 314-316, 335-337
Fiber-cement siding 341-343, 379
Finish trim. 310-317
Finishes, exterior 378-380
Fire stopping 132, 139
Flashing
  roofs 244, 254-258, 270-282
  walls 276-278, 299-306, 310, 311, 315-317, 323, 337, 342, 359, 365, 367, 368
Flitch plates 184
Footings 38-42, 59
  for piers. 82
  repair. 42
  stepped, jump 40, 41
Foundations
  concrete block 56-65
  hillside. 53, 54
  pier-and-grade beam. . . . 82-85
  poured concrete 52-55
  see also Slabs
Frost-protected foundations 86-88

G

Gable end 174, 315, 317
Galvanized
  nails 109, 110, 307-309
  roofing 246-258, 271-273
Gutters 70, 71, 283, 284

H

Hardboard 293, 344, 345
Headers. 98, 136-138
  span tables. 189-194
Hip roofs. 100, 140-143, 156-158, 209, 228, 229, 241, 258
Housewrap . . . . *see* Sheathing wrap

I

Ice dams 288
Insulation
  cathedral ceiling. 175-178
  floor 86-88, 171, 173
  foundations 86-89
  walls 171-173

J

Joist hangers 112, 113
Joists
  ceiling 344-346
  estimating 97, 98
  floor 116-122
  I-joist. 105, 123-128, 154, 155
  span tables. 185-188

L

Ledger. 51, 111, 119, 131, 298, 299, 304, 366
Low-slope roofs. 209, 214, 218, 227, 244, 246, 262-269
Lumber 101-106
  engineered 104-106

M

Metal
  flashing 244, 271, 272, 275-282
  lath 348-351
  roofing 246-258, 271, 272
Mortar
  for brick veneer 362-364
  for concrete block 57-59
  for stucco 350-355
  for tile roofs. 236

N

Nails
  framing 107-111, 149, 165, 167
  roofing 210, 214, 215, 227, 228, 236-240, 245, 252, 259-261
  siding. 293, 307-309, 323, 324, 328-331, 337-339, 342, 343

P

Paint, exterior 253, 254, 263, 373-376
Pier foundations. 82-85
Plates
  raised-rafter 148
  sill 116
  wall. 98, 133, 134, 171
Pressure-treated
  wood foundations 89, 90

R

Radon 95, 96
Rafters 140-155
  hip and valley 100, 140-143
  raised. 148
  span tables. 195-208
Rainscreen. . . . . *see* Drainage plane
Rebar 43-47, 52, 63
  in foundations walls. . . . 45-47, 52, 53-63
  in slabs 79, 80
  splicing 44
Repointing brick 369-370
Reroofing 218-221, 231, 245, 258
Retaining walls 91-95
Ridge beam 150, 153, 170
Rim joist 98, 117, 124, 125
Roadway. . . . . *see* Driveways

S

Sealants. 81, 95, 254-258, 266, 267, 371-372
Sediment trap 17
Shear wall 103, 104, 165-170
Sheathing
  floor 122
  foam 135, 324, 325
  roofing 149, 162, 225-227, 235
  wall. 103, 104, 165-167

INDEX

Sheathing wrap 296-299
estimating 289
under brick veneer 364, 367
under stucco 346, 347
under wood 321, 326, 344
vinyl siding 312, 322, 326, 333
Shingles and shakes
roofing 210, 211, 222-231
siding. 291-293, 326-331, 379
Shoring 14, 15
Shrinkage 102, 103
Siding . . . . *see* Board siding; Brick veneer; Fiber-cement siding; Hardboard; Stucco; Vinyl siding
Silt barriers 16, 17
Sitework 11-16
erosion control. 11, 16, 17
excavating 11-14
Slabs 74-81
Slate roofing 259-262
Soffits 155, 285, 286, 314-316
Soils 3-9
bearing capacity of. 5
boring 9
compacting 9, 10
Sonotubes 2, 82
Stain, exterior 375-377
Stair openings 119, 127, 133-135
Steel, structural 179-183
Stucco 346-361, 379
estimating 294
Studs 98, 105, 133-135
Subbase and subgrade 18-22, 75-78, 96
Subfloor 122

T

Tile, roofing. 210, 232-245
Topsoil 5, 11, 17
Trenching 13, 14, 20
Trusses
floor 129-132
roof. 156-164
Tudor trim. 315, 317

U

Underlayment, roofing 213, 214, 224-227, 235-237, 242, 243, 270, 274-279
Unified Soil Classification
System 4

V

Valleys. 100, 140-142, 156-158, 209, 255, 274-276
Vapor barriers 78, 171-178
Ventilation
roof. 228, 229, 258, 285-288
siding. 296, 301, 302, 314-316
(*see also* Drainage plane)
Vinyl siding 332-340
cleaning 339, 340
trim for 333-336, 338

W

Waterproofing foundations 66-67, 90
Wet-site roadway 21
Windows
flashing 301-304
headers 98, 136-138, 189-194
trim. 310, 311, 335-337

ABDJ3